# Recognition Robotics

# Recognition Robotics

Editor

**José María Martínez-Otzeta**

Basel • Beijing • Wuhan • Barcelona • Belgrade • Novi Sad • Cluj • Manchester

*Editor*
José María Martínez-Otzeta
Computer Science and
Artificial Intelligence
University of the Basque
Country/Euskal Herriko
Unibertsitatea
Donostia—San Sebastián,
Spain

*Editorial Office*
MDPI
St. Alban-Anlage 66
4052 Basel, Switzerland

This is a reprint of articles from the Special Issue published online in the open access journal *Sensors* (ISSN 1424-8220) (available at: https://www.mdpi.com/journal/sensors/special_issues/Recognition_Robotics).

For citation purposes, cite each article independently as indicated on the article page online and as indicated below:

Lastname, A.A.; Lastname, B.B. Article Title. *Journal Name* **Year**, *Volume Number*, Page Range.

ISBN 978-3-0365-9570-2 (Hbk)
ISBN 978-3-0365-9571-9 (PDF)
doi.org/10.3390/books978-3-0365-9571-9

© 2023 by the authors. Articles in this book are Open Access and distributed under the Creative Commons Attribution (CC BY) license. The book as a whole is distributed by MDPI under the terms and conditions of the Creative Commons Attribution-NonCommercial-NoDerivs (CC BY-NC-ND) license.

# Contents

**José María Martínez-Otzeta**
Editorial for the Special Issue Recognition Robotics
Reprinted from: *Sensors* **2023**, *23*, 8515, doi:10.3390/s23208515 . . . . . . . . . . . . . . . . . . . . . 1

**Beril Yalçinkaya, Micael S. Couceiro, Salviano Soares and António Valente**
Human-Aware Collaborative Robots in the Wild: Coping with Uncertainty in Activity Recognition
Reprinted from: *Sensors* **2023**, *23*, 3388, doi:10.3390/s23073388 . . . . . . . . . . . . . . . . . . . . . 5

**Sara Casao, Pablo Azagra, Ana C. Murillo and Eduardo Montijano**
A Self-Adaptive Gallery Construction Method for Open-World Person Re-Identification
Reprinted from: *Sensors* **2023**, *23*, 2662, doi:10.3390/s23052662 . . . . . . . . . . . . . . . . . . . . . 35

**Gabriela Błażejowska, Łukasz Gruba, Bipin Indurkhya and Artur Gunia**
A Study on the Role of Affective Feedback in Robot-Assisted Learning
Reprinted from: *Sensors* **2023**, *23*, 1181, doi:10.3390/s23031181 . . . . . . . . . . . . . . . . . . . . . 53

**Muhammad Adnan Syed, Yongsheng Ou, Tao Li and Guolai Jiang**
Lightweight Multimodal Domain Generic Person Reidentification Metric for Person-Following Robots
Reprinted from: *Sensors* **2023**, *23*, 813, doi:10.3390/s23020813 . . . . . . . . . . . . . . . . . . . . . 71

**Luca Marchionna, Giulio Pugliese, Mauro Martini, Simone Angarano, Francesco Salvetti and Marcello Chiaberge**
Deep Instance Segmentation and Visual Servoing to Play Jenga with a Cost-Effective Robotic System
Reprinted from: *Sensors* **2023**, *23*, 752, doi:10.3390/s23020752 . . . . . . . . . . . . . . . . . . . . . 97

**John Lewis, Pedro U. Lima and Meysam Basiri**
Collaborative 3D Scene Reconstruction in Large Outdoor Environments Using a Fleet of Mobile Ground Robots
Reprinted from: *Sensors* **2023**, *23*, 375, doi:10.3390/s23010375 . . . . . . . . . . . . . . . . . . . . . 117

**Barak Pinkovich, Boaz Matalon, Ehud Rivlin and Hector Rotstein**
Finding a Landing Site in an Urban Area: A Multi-Resolution Probabilistic Approach
Reprinted from: *Sensors* **2022**, *22*, 9807, doi:10.3390/s22249807 . . . . . . . . . . . . . . . . . . . . . 135

**Praneel Chand and Sunil Lal**
Vision-Based Detection and Classification of Used Electronic Parts
Reprinted from: *Sensors* **2022**, *22*, 9079, doi:10.3390/s22239079 . . . . . . . . . . . . . . . . . . . . . 149

**Yeji Kim and Jeongho Cho**
AIDM-Strat: Augmented Illegal Dumping Monitoring Strategy through Deep Neural Network-Based Spatial Separation Attention of Garbage
Reprinted from: *Sensors* **2022**, *22*, 8819, doi:10.3390/s22228819 . . . . . . . . . . . . . . . . . . . . . 167

**Łukasz Rykała, Andrzej Typiak, Rafał Typiak and Magdalena Rykała**
Application of Smoothing Spline in Determining the Unmanned Ground Vehicles Route Based on Ultra-Wideband Distance Measurements
Reprinted from: *Sensors* **2022**, *22*, 8334, doi:10.3390/s22218334 . . . . . . . . . . . . . . . . . . . . . 183

**Jaekwang Lee, Kangmin Lim and Jeongho Cho**
Improved Monitoring of Wildlife Invasion through Data Augmentation by Extract–Append of a Segmented Entity
Reprinted from: *Sensors* **2022**, *22*, 7383, doi:10.3390/s22197383 . . . . . . . . . . . . . . . . . . . . 205

**Sara Marques-Villarroya, Jose Carlos Castillo, Juan Jose Gamboa-Montero, Javier Sevilla-Salcedo and Miguel Angel Salichs**
A Bio-Inspired Endogenous Attention-Based Architecture for a Social Robot
Reprinted from: *Sensors* **2022**, *22*, 5248, doi:10.3390/s22145248 . . . . . . . . . . . . . . . . . . . . 217

**David Cantero, Iker Esnaola-Gonzalez, Jose Miguel-Alonso and Ekaitz Jauregi**
Benchmarking Object Detection Deep Learning Models in Embedded Devices
Reprinted from: *Sensors* **2022**, *22*, 4205, doi:10.3390/s22114205 . . . . . . . . . . . . . . . . . . . . 243

**Dominykas Strazdas, Jan Hintz, Aly Khalifa, Ahmed Abdelrahman, Thorsten Hempel and Ayoub Al-Hamadi**
Robot System Assistant (RoSA): Towards Intuitive Multi-Modal and Multi-Device Human-Robot Interaction
Reprinted from: *Sensors* **2022**, *22*, 923, doi:10.3390/s22030923 . . . . . . . . . . . . . . . . . . . . 269

**José María Martínez-Otzeta, Itsaso Rodríguez-Moreno, Iñigo Mendialdua and Basilio Sierra**
RANSAC for Robotic Applications: A Survey
Reprinted from: *Sensors* **2023**, *23*, 327, doi:10.3390/s23010327 . . . . . . . . . . . . . . . . . . . . 293

*Editorial*

# Editorial for the Special Issue Recognition Robotics

José María Martínez-Otzeta

Department of Computer Science and Artificial Intelligence, University of the Basque Country, 20018 Donostia-San Sebastián, Spain; josemaria.martinezo@ehu.eus

**Citation:** Martínez-Otzeta, J.M. Editorial for the Special Issue Recognition Robotics. *Sensors* **2023**, *23*, 8515. https://doi.org/10.3390/s23208515

Received: 22 September 2023
Accepted: 7 October 2023
Published: 17 October 2023

**Copyright:** © 2023 by the author. Licensee MDPI, Basel, Switzerland. This article is an open access article distributed under the terms and conditions of the Creative Commons Attribution (CC BY) license (https://creativecommons.org/licenses/by/4.0/).

Perception of the environment is an essential skill for robotic applications that interact with their surroundings. Alongside perception often comes the ability to recognize objects, people, or dynamic situations. This skill is of paramount importance in many use cases, from industrial to social robotics. Robots that can accurately perceive and understand their environment are critical for tasks like manufacturing, delivery, healthcare, and assisting humans in homes or public spaces. Object recognition enables robots to identify items, tools, and obstacles in their vicinity. This allows industrial robots to select the right parts or manipulators, logistics robots to handle packages, and autonomous vehicles to avoid collisions. Activity recognition allows robots to interpret human motions and behavior. This facilitates safe and intuitive collaboration in shared workspaces. It also permits service robots to determine user intents and respond appropriately. Person recognition provides robots the means to identify individuals. This capability supports applications like personalized assistance, healthcare monitoring, and security surveillance. Altogether, these skills comprise the fundamental building blocks for robots to operate adaptively in the real world.

This Special Issue "Recognition Robotics" of *Sensors* seeks to explore new research proposals on this increasingly important topic. The fifteen accepted papers in this issue cover human–robot collaboration [1], person re-identification [2,3], human–robot interactions [4,5], visual servoing [6], cooperative mapping [7], semantic segmentation [8,9], object classification [10], multi-object tracking [11], robot path planning [12], embedded deep learning [13], activity recognition [14], and robust model fitting [15]. These works present novel techniques using tools such as fuzzy logic, deep learning, computer vision, ultrasonic sensing, spline optimization, and more to advance robot capabilities in real-world conditions. The research aims to overcome challenges in uncertainty, limited data, computational constraints, and complexity across various application domains. In summary, this Special Issue provides a sampling of the latest innovations and progress in enabling robots that can effectively perceive, learn, plan, manipulate, and collaborate in unstructured environments through advances in recognition capabilities.

In [1], Yalçinkaya et al. introduce a Fuzzy State-Long Short-Term Memory (FS-LSTM) approach for human–robot collaboration in dynamic fields like agriculture and construction. These tasks are time-consuming and risky for humans, making robotic assistance valuable. The method handles the ambiguity in human behavior by fuzzifying sensory data and employing a combined activity recognition system using state machines and LSTM. Experimental validation showed that FS-LSTM outperforms traditional LSTM in accuracy and computational efficiency.

In [2], Casao et al. introduce an unsupervised method for person re-identification, capable of automatically adding new identities to an adaptive gallery in open-world settings. The system compares current models to new unlabeled data and uses information theory to keep compact representative models. Experimental results, including comparisons to other unsupervised and semi-supervised methods, validate the effectiveness of their approach.

The authors of [3] propose a lightweight deep metric learning technique for reliable person re-identification, aimed at robot tracking. This method addresses challenges like clothing and pose changes by employing a novel attention mechanism. This focuses on specific body parts, retains global context, and enables cross-representations for robust identification. The experimental results show up to 80.73% and 64.44% top-rank accuracy, outperforming existing methods. The authors suggest that integrating this metric improves tracking reliability in dynamic environments.

In [4], Błażejowska et al. explore the impact of emotional feedback from the Miro-E robot on high school students during a programming education session. The robot monitored students' emotions via facial expression analysis and provided affective feedback like verbal praise and tail wagging. Compared to a control group with neutral robot responses, the emotional feedback positively impacted engagement, particularly for students with little prior programming experience. However, it also slightly reduced the robot's likeability, hinting at an uncanny valley effect. Due to a small sample size, the study focused on qualitative insights.

In [5], Marques-Villarroya et al. introduce a robotic perception architecture that employs bio-inspired endogenous attention to improve human–robot interactions. The architecture uses multisensory inputs and ranks stimuli based on their relevance to the robot's tasks, particularly emphasizing human presence and actions. By doing so, it optimizes the robot's focus and behavior, leading to more efficient interactions. Implemented on the Robot Operating System (ROS), the architecture demonstrates strong real-time performance and extensibility. The authors argue that this bio-inspired approach enhances the robot's responsiveness while reducing complexity.

In [6], Marchionna et al. demonstrate how a low-cost, six-axis robotic arm, e.Do, can play Jenga using instance segmentation and visual servoing. The system employs an affordable RGB-D camera and force sensor. A customized deep learning model is trained to identify each Jenga block, enabling precise visual tracking during manipulation. The force sensor helps decide if a block can be safely removed. Testing shows up to 14 consecutive successful block extractions before the tower collapses. The authors note that Jenga serves as a complex benchmark, driving advancements in multi-step reasoning, integrated sensory perception, and high-precision control.

The authors of [7] propose a decentralized framework for collaborative 3D mapping using mobile robots with LiDAR sensors in large-scale outdoor settings such as agriculture and disaster response. The real-time method allows robots to share and merge locally scanned submaps into a global map, even with limited communication bandwidth. A conditional peer-to-peer strategy is used for sharing map data over different distances. Experiments in a real-world solar power plant confirm the approach's efficiency and reliability for multi-robot mapping of extensive outdoor areas.

In [8], Pinkovich et al. address the challenge of autonomously selecting safe landing sites for delivery drones in dense urban areas. Their multi-resolution technique captures visual data at varying altitudes, enabling both wide exploration and high-resolution sensing. A semantic segmentation deep neural network processes this data, updating probability distributions for each ground patch's landing suitability. When a location's confidence exceeds a threshold, it is selected as viable. The authors find the method effectively balances the trade-off between exploration and resolution in constrained urban environments.

Lee et al. introduce in [9] an "Extract-Append" data augmentation technique to boost the accuracy of models detecting wild animals in agricultural fields. The method uses semantic segmentation to isolate animal shapes from sample images and combines them with new backgrounds to enrich the training dataset. Testing shows at least a 2.2% improvement in mean Average Precision over traditional methods, and the technique enables ongoing flexible data augmentation.

The authors of [10] present vision-based methods for automated recycling of used electronic components such as capacitors and voltage regulators. Using a custom object detection algorithm, they identify key areas in cluttered workspaces and compare three classification techniques: SNNs, SVMs, and CNNs. After hyperparameter tuning, CNNs prove to be the most accurate with a 98.1% success rate, making them the preferred method for reliable automated recycling.

Paper [11] proposes the use of deep neural networks to detect illegal garbage dumping in urban areas. They combine OpenPose for human pose estimation, YOLO for garbage bag classification, and DeepSORT for object tracking. The system measures the distance between a person's wrist and the garbage bag to determine illegal dumping. Experimental results show their method offers higher accuracy and lower false alarms compared to other approaches, making it effective for automated monitoring against unlawful waste disposal.

In the research presented by Rykała et al. [12], a path-planning method is developed for an unmanned ground vehicle (UGV) to follow a human guide using ultra-wideband (UWB) technology. They use smoothing splines to reconstruct the guide's path from periodic distance measurements. The approach is computationally efficient and can handle missing data, making it suitable for real-time applications.

The authors of [13] provide a comprehensive evaluation of how well state-of-the-art deep learning object detection models perform on embedded electronics. They assess multiple architectures and quantization techniques to make the models more efficient for embedded and robotics applications. The paper outlines the entire process from model conversion to deployment and performance measurement on embedded devices. It offers guidelines for choosing the right hardware and optimization strategies, and discusses the various factors that influence performance in real-time robotics systems.

In [14], Strazdas et al. introduce RoSA, a framework that facilitates human–robot interactions using speech and gestures. Running on ROS, the system incorporates speech recognition, face identification, and pose estimation. A user study revealed that RoSA's usability was on par with a human-controlled setup, suggesting it offers a natural interaction experience. The authors highlight the value of multi-sensory integration for more human-like and flexible robot interactions.

The authors of [15] review the RANSAC algorithm's applications in robotics, focusing on shape detection and feature matching. They explore various enhancements to RANSAC that improve its speed, accuracy, and robustness. The survey also discusses trade-offs between computational cost and performance, highlights recent robotics applications, and provides a list of open-source RANSAC libraries. The survey offers robotics researchers and developers an extensive reference on the state of the art in RANSAC techniques.

In summary, the fifteen papers in this Special Issue on "Recognition Robotics" demonstrate the tremendous progress being made in enabling robots to effectively perceive, understand, plan, and interact in the real world. However, significant challenges remain before these innovative techniques can be reliably deployed in unconstrained environments. Testing novel algorithms in controlled simulations or lab settings with simplified assumptions can be deceptively promising, because applying recognition capabilities on physical robotic platforms in complex dynamic scenarios reveals many subtleties. Interactive testing is critical to expose limitations around uncertainty, variability, and computational constraints. Moving innovations out of the lab or controlled scenarios requires addressing edge cases and graceful failure modes, and therefore there is still substantial effort needed to robustly handle the diversity and unpredictability of the real world. Nevertheless, the field continues steadily on an exciting path towards enabling robot assistants and coworkers that can perceive, learn, reason, and collaborate at a human level. These capabilities will lead to transformative applications, and the works presented in this Special Issue provide an inspiring snapshot of the road ahead.

**Conflicts of Interest:** The author declares no conflict of interest.

# References

1. Yalçinkaya, B.; Couceiro, M.S.; Soares, S.P.; Valente, A. Human-Aware Collaborative Robots in the Wild: Coping with Uncertainty in Activity Recognition. *Sensors* **2023**, *23*, 3388. [CrossRef] [PubMed]
2. Casao, S.; Azagra, P.; Murillo, A.C.; Montijano, E. A Self-Adaptive Gallery Construction Method for Open-World Person Re-Identification. *Sensors* **2023**, *23*, 2662. [CrossRef] [PubMed]
3. Syed, M.A.; Ou, Y.; Li, T.; Jiang, G. Lightweight Multimodal Domain Generic Person Reidentification Metric for Person-Following Robots. *Sensors* **2023**, *23*, 813. [CrossRef] [PubMed]

4. Błażejowska, G.; Gruba, Ł.; Indurkhya, B.; Gunia, A. A Study on the Role of Affective Feedback in Robot-Assisted Learning. *Sensors* **2023**, *23*, 1181. [CrossRef] [PubMed]
5. Marques-Villarroya, S.; Castillo, J.C.; Gamboa-Montero, J.J.; Sevilla-Salcedo, J.; Salichs, M.A. A Bio-Inspired Endogenous Attention-Based Architecture for a Social Robot. *Sensors* **2022**, *22*, 5248. [CrossRef] [PubMed]
6. Marchionna, L.; Pugliese, G.; Martini, M.; Angarano, S.; Salvetti, F.; Chiaberge, M. Deep Instance Segmentation and Visual Servoing to Play Jenga with a Cost-Effective Robotic System. *Sensors* **2023**, *23*, 752. [CrossRef] [PubMed]
7. Lewis, J.; Lima, P.U.; Basiri, M. Collaborative 3D Scene Reconstruction in Large Outdoor Environments Using a Fleet of Mobile Ground Robots. *Sensors* **2023**, *23*, 375. [CrossRef] [PubMed]
8. Pinkovich, B.; Matalon, B.; Rivlin, E.; Rotstein, H. Finding a Landing Site in an Urban Area: A Multi-Resolution Probabilistic Approach. *Sensors* **2022**, *22*, 9807. [CrossRef] [PubMed]
9. Lee, J.; Lim, K.; Cho, J. Improved Monitoring of Wildlife Invasion through Data Augmentation by Extract—Append of a Segmented Entity. *Sensors* **2022**, *22*, 7383. [CrossRef]
10. Chand, P.; Lal, S. Vision-Based Detection and Classification of Used Electronic Parts. *Sensors* **2022**, *22*, 9079. [CrossRef]
11. Kim, Y.; Cho, J. AIDM-Strat: Augmented Illegal Dumping Monitoring Strategy through Deep Neural Network-Based Spatial Separation Attention of Garbage. *Sensors* **2022**, *22*, 8819. [CrossRef] [PubMed]
12. Rykała, Ł.; Typiak, A.; Typiak, R.; Rykała, M. Application of Smoothing Spline in Determining the Unmanned Ground Vehicles Route Based on Ultra-Wideband Distance Measurements. *Sensors* **2022**, *22*, 8334. [CrossRef]
13. Cantero, D.; Esnaola-Gonzalez, I.; Miguel-Alonso, J.; Jauregi, E. Benchmarking Object Detection Deep Learning Models in Embedded Devices. *Sensors* **2022**, *22*, 4205. [CrossRef]
14. Strazdas, D.; Hintz, J.; Khalifa, A.; Abdelrahman, A.A.; Hempel, T.; Al-Hamadi, A. Robot System Assistant (RoSA): Towards Intuitive Multi-Modal and Multi-Device Human-Robot Interaction. *Sensors* **2022**, *22*, 923. [CrossRef]
15. Martínez-Otzeta, J.M.; Rodríguez-Moreno, I.; Mendialdua, I.; Sierra, B. RANSAC for Robotic Applications: A Survey. *Sensors* **2023**, *23*, 327. [CrossRef] [PubMed]

**Disclaimer/Publisher's Note:** The statements, opinions and data contained in all publications are solely those of the individual author(s) and contributor(s) and not of MDPI and/or the editor(s). MDPI and/or the editor(s) disclaim responsibility for any injury to people or property resulting from any ideas, methods, instructions or products referred to in the content.

*Article*

# Human-Aware Collaborative Robots in the Wild: Coping with Uncertainty in Activity Recognition

Beril Yalçinkaya [1,2,*], Micael S. Couceiro [1], Salviano Pinto Soares [2,3,4] and Antonio Valente [2,5]

1. Ingeniarius, Ltd., R. Nossa Sra. Conceição 146, 4445-147 Alfena, Portugal; micael@ingeniarius.pt
2. Engineering Department, School of Sciences and Technology, University of Trás-os-Montes and Alto Douro (UTAD), Quinta de Prados, 5000-801 Vila Real, Portugal; salblues@utad.pt (S.P.S.); avalente@utad.pt (A.V.)
3. Institute of Electronics and Informatics Engineering of Aveiro (IEETA), University of Aveiro, 3810-193 Aveiro, Portugal
4. Intelligent Systems Associate Laboratory (LASI), University of Aveiro, 3810-193 Aveiro, Portugal
5. INESC TEC, Campus da Faculdade de Engenharia da Universidade do Porto, Rua Dr. Roberto Frias, 4200-464 Porto, Portugal
* Correspondence: beril@ingeniarius.pt

**Abstract:** This study presents a novel approach to cope with the human behaviour uncertainty during Human-Robot Collaboration (HRC) in dynamic and unstructured environments, such as agriculture, forestry, and construction. These challenging tasks, which often require excessive time, labour and are hazardous for humans, provide ample room for improvement through collaboration with robots. However, the integration of humans in-the-loop raises open challenges due to the uncertainty that comes with the ambiguous nature of human behaviour. Such uncertainty makes it difficult to represent high-level human behaviour based on low-level sensory input data. The proposed Fuzzy State-Long Short-Term Memory (FS-LSTM) approach addresses this challenge by fuzzifying ambiguous sensory data and developing a combined activity recognition and sequence modelling system using state machines and the LSTM deep learning method. The evaluation process compares the traditional LSTM approach with raw sensory data inputs, a Fuzzy-LSTM approach with fuzzified inputs, and the proposed FS-LSTM approach. The results show that the use of fuzzified inputs significantly improves accuracy compared to traditional LSTM, and, while the fuzzy state machine approach provides similar results than the fuzzy one, it offers the added benefits of ensuring feasible transitions between activities with improved computational efficiency.

**Keywords:** human activity recognition and modelling; deep learning; human-robot collaboration; fuzzy logic; finite state machine; long short—term memory

## 1. Introduction

### 1.1. Importance of Human Activity Recognition for Human-Robot Collaboration

For years, robots and humans have been separated in different workspaces, whether it be industrial or field applications. The reason for this separation is primarily for safety. Even though robots have been designed for specific tasks, in most cases, they are not aware of the environment and surrounding dynamic agents. As a result, these robots are often placed in cages or in a completely separate environment from human operators [1]. This separation has resulted in issues, such as low adaptability in different environments, costly setup, and limited flexibility, which do not align with the ideals of Industry 4.0, which demands fast production and efficiency.

To address these demands, Human-Robot Collaboration (HRC) has become a major trend in robotics in recent years. The goal is to improve efficiency and productivity by combining the benefits of humans' critical thinking and empathy, with robots' physical robustness in demanding and often dangerous conditions [2]. The idea is for humans and robots to work together towards a common goal. Research has demonstrated that

interaction and collaboration between humans and robots are crucial factors in achieving ergonomic systems and enhancing the quality and efficiency of the production process [3].

Collaborative robots, also known as "co-bots", have become increasingly prevalent in industrial settings in recent years. However, they have also been utilized in a variety of other domains. For example, in the healthcare field, researchers are developing robotic walkers [4], wheelchairs [5], and elderly care robots [6]. Collaborative robots have the potential to assist humans in heavy and dangerous tasks as well, such as construction and search-and-rescue [7]. They can also be used in a range of industries and even in smart home applications. These robots can come in various forms, such as manipulators [8] and fully humanoid robots [2].

However, incorporating humans into the process presents many challenges, primarily due to the unpredictable nature of human behaviour. This can lead to difficulties with robots adaptability and robustness in changing and uncertain situations and environments. In HRC systems, robots are expected to understand human activities and intentions and, at times, even predict future human behaviour in order to efficiently achieve the shared goal. This can be a difficult task due to the inherent uncertainty of human behaviour.

Significant research has been dedicated to understanding human behaviour patterns through Human Activity Recognition (HAR), which involves analysing various sensor data to identify and detect simple and complex human activities. HAR has been applied not only to domains related to human daily life, such as healthcare, smart home applications, and elderly assistance [9], but also in robotics solutions where HRC is foreseen, being critical for the robot to have awareness of human actions. Traditional machine learning methods, such as Bayesian networks [10], random forest [11] and support vector machines [12] have been used to understand human behaviours. In addition to understanding human behaviour, some researchers have focused on predicting the most likely sequence of human actions. Probabilistic methods, such as Hidden Markov Models (HMM) [13] have been proposed to understand and predict human activities. Finite State Machines (FSM) have also been used as a tool to model dynamic changes over time and, when combined with fuzzy logic, to even handle uncertainty from sensor data through the use of linguistic variables [14]. Recently, deep learning has emerged as a new trend, as it has the ability to learn and identify complex patterns among large datasets. The major difference between deep learning and the previously described approaches is that it offers multiple hidden layers that are capable of feature extraction and transformation, thus significantly reducing the workload of human designers and developers. As a result, deep learning has been used in various other domains as well, such as image classification [15], speech recognition [16] and so on, and several deep learning algorithms, such as convolutional neural networks (CNNs) [17] and recurrent neural networks (RNNs) [18], have been key to improve the accuracy and robustness of HAR systems.

While these methods have shown promise, dealing with human uncertainty remains a challenge. One of the main difficulties is the high variability of human behaviour across different contexts, as well as the noise in the sensor data, which makes it difficult to generalize from training data. This uncertainty problem has a negative impact on trust and safety, which are critical measurements for any HRC system [19]: if the robot is unable to understand or anticipate human intention, this may lead it to make wrong decisions and even cause accidents and injuries, which will affect both acceptability and trustworthiness. Several authors point out to a panoply of solutions to eliminate uncertainty by extracting more information and rapid processing. However, there is no clear plan established for a constrained computing system, as robots and other facilitators of HRC (e.g., wearable technologies) often have. Given the complexity and addressed challenges associated with uncertainty in human behaviours, further research is still required to fully understand and address this problem.

*1.2. Research Question and Objectives*

This paper proposes a HAR framework capable of coping with uncertainty in human behaviours, resulting in positive improvements in trust and safety for HRC tasks. To this end, this paper presents three key incremental developments:

i. Enhance Long Short-Term Memory (LSTM) networks by incorporating fuzzy logic to model human uncertainty (Fuzzy-LSTM), building upon the work of [20]: the goal is to improve the performance of LSTM networks by incorporating fuzzy logic to model human uncertainty. In this method, features are extracted from sensor data, which may be uncertain due to the ambiguity of human behaviour or noise in the sensors. These features are then fuzzified using Tilt and Motion linguistic variables. This fuzzification step allows the model to handle uncertain data, making it more robust. The fuzzified features are then used as input to the LSTM network during training. The goal of this approach is to improve the accuracy of the LSTM network in handling uncertain sensor data.

ii. Further extend Fuzzy-LSTM representing the sequence of activities through finite-state machines (FSM), thus leading to the Fuzzy State LSTM (FS-LSTM): the goal of this method is to enhance the predictability of human activity sequences by combining the strengths of FSM and fuzzy logic in an LSTM-based model. In this approach, an LSTM network is trained for each state within the FSM. The output of the LSTM network is then used to determine the possible transitions between states.

iii. Estimate human uncertainty by aggregating predicted scores of the LSTM into a crisp output through defuzzification: this proposed method aims to estimate the uncertainty of the LSTM classifier's predictions by converting the classification scores into a crisp value through defuzzification. The classification scores are first converted into a fuzzy set to represent the degree of uncertainty in the predictions. Then, the fuzzy set is transformed into a crisp value to indicate the certainty of the classifier's predictions. This process allows for quantifying the uncertainty of the predictions, which is not only used within the FS-LSTM method to accept or reject transitions between states, but can also be useful in our future work in HRC, where certainty is important.

In addition to these three main contributions, a benchmark is presented to further investigate the impact of the proposed architecture, which compares the traditional LSTM and the incrementally developed novel architecture.

*1.3. Organization of the Article*

This article is structured as follows: In Section 2, a comprehensive review of relevant literature is provided. Section 3 outlines the use case and data collection through the developed simulator and a preliminary experimental study on generating synthetic data. The proposed method, including feature fuzzification and FSM learning with LSTM, is described in Section 4. At last, the results from the experimental studies are discussed in Section 5, followed by a description of future work and conclusions in Section 6.

**2. Literature Review**

HAR has gained significant attention for its ability to detect and identify human activities from sensor data [21]. The importance of HAR lies in its ability to handle the uncertainty that arises from the variability in human behaviour and ambiguity in activities. In robotics, understanding human behaviour and adapting accordingly is crucial for natural and safe interaction and collaboration with humans [22]. This literature review will examine the uncertainty problem in HAR and the methodologies used to address it, as well as the challenges and open research questions in the field.

Human uncertainty poses a significant challenge in HRC from various perspectives. One major aspect is the complex sequential decision-making required in dynamic environments during collaborative tasks, as discussed by Osman in her study [23] on complex dynamic control tasks. These tasks often require multiple decisions that have to accommodate many elements of the system to achieve a desired goal, which implies that there is a high degree of uncertainty introduced by humans regarding how they will behave in these changing conditions and environments. This makes it a difficult task for robots to predict and adapt. In addition to the dynamic nature of the environment and the complexity of collaborative tasks, the variability of human physical and cognitive abilities also contributes to their uncertainty. Human factors, such as fatigue, learning ability, and attentiveness can significantly impact a worker's efficiency and accuracy and even cause errors or safety issues in HRC systems [24]. This is aligned with the work of Vuckovic et al. [25], that highlighted the importance of human subjectiveness in creating uncertainty in human behaviours. According to the authors, individuals judge a stimulus and adapt their decisions accordingly to their judgments. This implies that human subjectivity has an important role in introducing uncertainty in human behaviours, as it leads them to perceive and react to situations differently based on their own experiences. Another important aspect in which uncertainty plays a role is in building trust between humans and robots in collaborative tasks. Trust is a crucial element in HRC, as it allows humans to rely on robots to safely perform tasks together. According to Law and Scheutz [26], understanding human needs and intentions, and effectively responding to them, is key to building trust.

Other researchers have emphasized the significant impact of human uncertainty on proactive planning for HRC. According to Kwon et al. [27], proactive planning involves a robot's ability to adapt to a dynamic environment by handling uncertainty. The authors note that the nature of the dynamic environment is not only affected by the robot's actions but also by human activities, which have complex temporal relationships. The uncertainty in these activities must be considered during planning as they are not easily predictable due to the robot's limited observation of the environment and the humans. Therefore, understanding and addressing uncertainty in collaborative tasks is essential for efficient planning in HRC.

Based on the literature reviewed, it is well understood that uncertainty introduced by humans poses a significant challenge in Human-Robot Collaboration (HRC). Therefore, a significant amount of research has been conducted in this field with the aim of mitigating the negative effects of uncertainty. These solutions mostly focus on the efficient and effective inference of human behaviour as a means of addressing uncertainty within the HRC context. One approach is the use of multimodal systems that combine different sensor types, such as video cameras, wearables, and even ambient sensors, such as infrared motion detectors. Video cameras are popular for HRC tasks, but they raise privacy concerns [28]. On the other hand, wearable sensors, such as inertial measurement units (IMUs), are widely used to cope with privacy and security concerns, but they also come with many challenges, such as limited representativeness of similar activities. Despite these challenges, wearable sensors are the most commonly used set of sensors in human activity monitoring. In [29], the authors proposed to use wearable sensors, such as accelerometers and gyroscopes worn at different positions on the human body, to capture activity data that are sampled at regular intervals to be used in HAR. Another study has been designing appropriate methodologies, such as utilizing data from individual accelerometers at the waist, which can identify basic daily activities, such as running, walking and lying down [30]. These works reported acceptable accuracy results for basic daily activities. However, they could not show good accuracy for more complex activities, such as transitions, e.g., standing up or sitting down. As said before, a way of improving such results would generally imply using a larger combination of sensors, although attaching many sensors to the human body is unfeasible and inconvenient for people's daily activities.

HAR is often treated as a pattern recognition problem, and many works have initially adopted machine learning techniques to recognize activities. Support Vector Machine (SVM) [31] and Hidden Markov Model (HMM) [32] classifiers are among the most commonly used methods for activity recognition. For example, Azim et al. [33] used an SVM classifier with trajectory features for activity classification and achieved an overall accuracy of 94.90% for the KTH online database and 95.36% for the Weizmann dataset (http://www.wisdom.weizmann.ac.il/~vision/SpaceTimeActions.html (accessed on 2 February 2023)). Kellokumpu et al. [34] used HMM and affine invariant descriptors, achieving an overall accuracy of 83.00%. While these works rely on offline data, Yamato et al. [35] used real-time sequential images and mesh features along with HMM, achieving a 90% accuracy. However, these traditional machine learning methods often rely on carefully designed and heuristic feature extraction methods, such as time-frequency transformation, statistical approaches, and symbolic representation. They lack a universal or systematic approach for effectively distinguishing human activities, and they are prone to overfitting and may perform poorly on unseen data [36].

To overcome these drawbacks, ensemble classifiers have been proposed, which involve training multiple models and combining their predictions to make a final decision. The aim of ensemble classifiers is to improve the performance of the model by combining the strengths of multiple models and mitigating their weaknesses [37]. Random forest is a popular ensemble classifier that is computationally efficient and commonly used in various domains, such as text and image classification. Random forest works by training multiple decision trees and combining their predictions through a voting procedure. This method is effective in addressing overfitting issues and has been shown to enhance accuracy by combining the outcome of each different classifier [38].

Both traditional machine learning and ensemble classifiers methods for feature extraction in HAR heavily rely on human experience and domain knowledge. However, these may not be effective for more general environments and may result in a lower chance of building an efficient recognition system. Additionally, the features learned by these methods are shallow, such as statistical information, and can only be used for low-level activity identification, such as walking or running, making it hard to detect high-level or context-aware activities, such as cooking. In contrast, in real-life scenarios, activity data comes in a stream and requires robust online learning from static data, which is a limitation of many of these traditional methods [39]. Deep learning methods, on the other hand, have been successful in learning complex activities due to their ability to learn features directly from the raw data hierarchically by performing nonlinear transformations. The layer-by-layer structure of deep models allows learning from simple to abstract features. Advances in computer resources have made it possible to use deep models to learn features from complex data from single or multimodal sensory systems. It is worth highlighting that deep neural networks can be detached and flexibly composed into a unified network, allowing for the integration of various deep learning techniques, such as deep transfer learning, deep active learning, and deep attention mechanism. This enables the integration of various effective solutions that can improve the performance of the recognition system [36].

Popular deep learning techniques include deep neural networks (DNN), convolutional neural networks (CNN), recurrent neural networks (RNN), and long short-term memory (LSTM) networks [28]. DNN are a type of Artificial Neural Network (ANN) that are characterized by a larger number of hidden layers. In contrast to traditional ANN, which often have only a few hidden layers, DNN can learn from large datasets more effectively. Hammerla et al. [40] adopted a five-hidden-layer DNN to perform automatic feature learning and classification. Vepakomma et al. [41] fed extracted hand-engineered features obtained from the sensors into a DNN model. CNNs are a type of neural network that exploit three key concepts: sparse interactions, parameter sharing, and equivariant representations. CNN have presented successful results in HAR application by utilizing local dependency, which refers to the nearby signals in a time-series that are most likely correlated. CNN also have shown the ability to handle variations in pace or frequency [39]. Several studies, such as [42,43] have employed one-dimensional (1D) on the individual univariate time-series signals for temporal feature extraction. Conventional 1D CNN have a fixed kernel size, which limits their ability to discover signal fluctuations over different temporal ranges. To address this, Lee et al. [17] combined multiple CNN structures of different kernel sizes to obtain the temporal features from different time scales. Nevertheless, this approach would demand more computational resources as well. Various deep learning methods have been applied to temporal information including RNN. While traditional RNN cells suffer from vanishing gradient problems, LSTM, as a specific type of RNN, overcomes this issue. A sliding window is generally used to divide the raw data into individual pieces, which are then used to feed LSTM. In a typical LSTM-based temporal feature extraction, it is essential to carefully tune the hyper-parameters, such as the length and moving step of the sliding window. Some researchers adopted The Bidirectional LSTM (Bi-LSTM) structure for extracting temporal dynamics from both forward and backward directions in HAR [44]. On the other hand, Guan and Plötz have combined multiple LSTM networks in an ensemble approach and obtained superior results [45].

Another trend in HAR is combining different deep learning approaches by developing hybrid models to exploit their different aspects. For instance, Ordóñez and Roggen have combined CNN and LSTM for both local and global temporal feature extraction [46]. The idea is to exploit CNN's ability to capture the spatial relationship, while LSTM can extract the temporal relationship. According to the reported results, CNN combined with LSTM outperforms CNN combined with dense layers. Differently, in [47], the authors presented a hybrid model for HAR which first identifies the abstract activity by using random forest to classify it as static and moving. For static activities the authors have used SVM, while for moving activities they have adopted 1D CNN. Even though the overall accuracy of the system was 97.71%, their system was evaluated over a dataset and has not been tested in real environments and/or in runtime.

Despite these models having shown significant accuracy in HAR, the uncertainty of the activities remains a challenge due to several reasons, such as noise in sensors and human factors. Several studies adopted different methodologies to investigate the degree of certainty, or uncertainty, of a given performed activity. One of the methods adopted was a dynamic Bayesian mixture model (DBMM), which is a type of ensemble probabilistic model that combines the likelihood of multiple classifiers into a single form by attaching different weights to each classifier. DBMM uses an uncertainty measure, such as the posterior probability, as a confidence level, which is updated during the online classification [48]. Therefore, the classifier with the highest confidence level is the outcome of the classification process. In [49], the authors presented an architecture that recognises seven different actions performed by athletes using a single-channel electromyography (EMG) combined with positional data obtained by benchmarking ANN, LSTM and DBMM. According to the results, ANN and LSTM models were not the most reliable choice to identify these actions due to the low number of trials in the dataset. On the other hand, DBMM led to better results, with 96.47% accuracy and 80.54% F1-score. Similarly, in [50], human daily activities were recognized by using DBMM. The authors proposed a set of spatio-temporal features, including geometrical, energy-based and domain frequency features to represent the different daily activities which were then fed into DBMM. The overall classification performance for DBMM and LSTM, in terms of precision and recall, was 86.63% and 85.01%, respectively.

Other studies have explored fuzzy-based architectures in HAR, which allows for the incorporation of uncertainty in the decision-making process. While traditional probabilistic models represent the likelihood of an event using crisp values, fuzzy-based models use fuzzy membership values to represent the degree of partial truth by providing semantic expressiveness through the use of linguistic variables to handle uncertain data. Karthigasri and Sornam [20] fuzzified the input features to be used in a fuzzy FSM (FFSM), which is a methodology used to model dynamic sequences of events. The reported results of the approach outperformed decision trees, K-nearest neighbors, SVM, Gaussian naïve Bayes and quadratic discriminant analysis. Mohmed et al. [14] proposed a HAR architecture using data obtained from low-level sensory devices by enhancing FFSM with deep learning methods, namely LSTM and CNN. While both models have shown high scores of accuracy, the CNN-FFSM model showed more robust and reliable performance when applied to a larger dataset, while LSTM-FFSM outperformed CNN-FFSM for simple scenarios with a short period of a dataset. Despite the paper presenting promising results for

HAR, the methodology is presented in a high-level manner, lacking relevant technical and scientific details, which makes it impossible for the reader to understand and fully asses its reproducibility.

In conclusion, the literature reviewed in this study highlights the importance for robots to understand human activities and cope with uncertainty in HRC applications. To this end, a variety of studies have been conducted in this field to understand human behaviour by exploring HAR architectures. However, it is clear that there is still a need for further research in this area in order to not only measure human uncertainty during collaborative tasks with robots in runtime, as well as to use such knowledge to adapt accordingly.

## 3. Use Case and Data Collection

### 3.1. Use Case: The FEROX Project

FEROX https://ferox.fbk.eu/ (accessed on 10 February 2023) is a project that aims to support workers collecting wild berries and mushrooms in wild and remote areas of Nordic countries by using robotic technologies. One of the key aspects of the project is its focus on HRC by deploying unmanned aerial vehicles (UAV) to monitor and assist groups of workers during field operations. This improves workers safety in remote environments, where access to help or assistance may be limited. The expected end results will be an increased worker trust in collaborating with robots, leading to larger number of berries harvested, higher quality berries for consumers, more efficient picking times, new level of worker safety in remote environments, and reduced worker exhaustion levels. Figure 1 depicts a view of the work field of the FEROX Project.

To achieve its aim, the FEROX project is exploring the use of wearable technology to infer the needs and states of the workers. One possible solution is to use a wearable device with integrated IMU (i.e., accelerometer, gyroscope and magnetometer) that can enable the identification of different activities, such as walking, running, sitting, collecting, and loading berries. Additionally, data from a global navigation satellite system (GNSS) (e.g., from the worker smartphone) can be used to infer activities performed over distance, such as driving a vehicle. It is foreseen that the combination of these two commonly adopted cheap devices would allow for more accurate and real-time monitoring of the workers' activities and needs, enabling the project to better support and assist them.

**Figure 1.** A view on the work field of the FEROX Project.

Figure 2 illustrates the conceptual overview of the architecture aimed to be implemented in the FEROX Project. As stated above, human workers are equipped with wearables and other technologies, which feed the herein proposed FS-LSTM architecture to assess their behaviors and the associated uncertainty for a high-level decision-making system. The system may integrate human physiological and kinematic data to identify human activities, such as (1) human locomotive activities, including idle and walking; (2) human work-related activities, such as berry picking; (3) potential detection of human injury, combining physiological data, such as heart rate, in the future; and (4) a multi-UAV system that provides assistance to the human workers based on the output of the high-level decision-making system, which takes into account the human state defined by FS-LSTM. The next phase of the study will focus on developing the high-level decision-making system to explore the areas, track human location, and assist with loading the collected berries to the collection point (see Section 6).

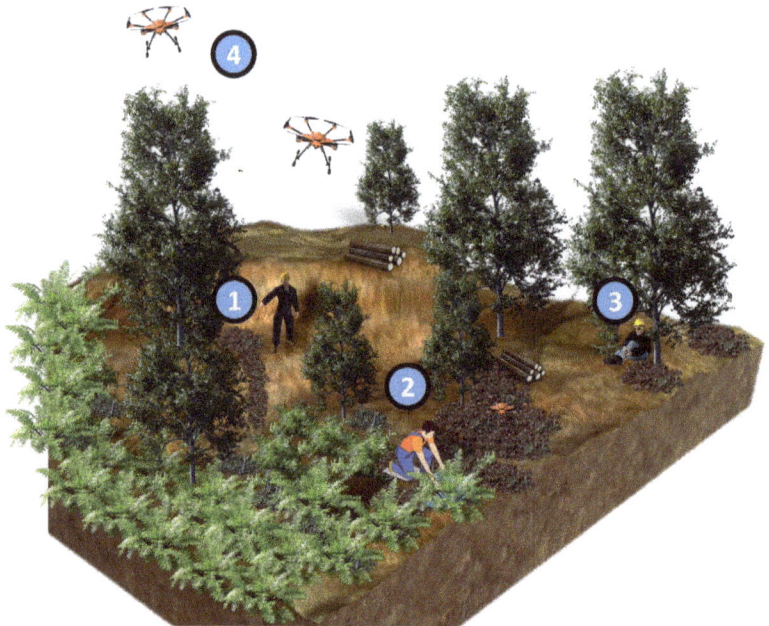

**Figure 2.** An overview of the use case scenario.

*3.2. Data Collection: FEROX Simulator and Synthetic Data*

In recent years, the performance of HAR systems has seen significant advancement due to the use of deep learning techniques. However, the acquisition and labelling of large datasets for training and evaluating these methods can be time-consuming and costly. To address these limitations, one solution is to use synthetic datasets that do not require manual labelling or expensive hardware for data capturing [51]. This approach has several advantages, such as producing labelled data without human input, being beneficial in fields where data acquisition is costly, such as field robotics [52].

As a preliminary study, we present a simulator that generates automatically labelled synthetic data by simulating a human character with a chest-worn virtual IMU and smartphone GNSS sensors. The first goal of this study is to develop a simulation environment that can produce synthetic human motion data to feed a HAR system capable of recognizing different locomotive actions. The focus of this research is on developing a system that can be trained using only synthetic labelled data, and then tested and evaluated with real data to justify its reliability for further studies.

3.2.1. FEROX Simulator Development and Virtual Sensor Modelling

We have developed the simulator using the Unity (https://unity.com/ (accessed on 10 February 2023)) game engine with the ultimate goal of creating a game-like environment for HRC. To achieve this, we initially focused on setting up the forestry scenario by using the Unity Terrain High-Definition Render Pipeline (https://assetstore.unity.com/packages/3d/environments/unity-terrain-hdrp-demo-scene-213198 (accessed on 10 February 2023)) and the avatar using the Mixamo library (https://www.mixamo.com/ (accessed on 10 February 2023)), contemplating simple actions, such as idle, walking, running, sitting, falling down and getting up as shown in Figure 3. We also implemented work-related actions, such as collecting and loading berries, driving a vehicle, etc. To generate the animations, we used a keyframe-based method that models connected virtual human body joints in a sequence of frames.

To establish communication between the different agents in the simulation and the developed framework, we integrated the ROS TCP Connector (https://github.com/Unity-Technologies/ROS-TCP-Connector (accessed on 10 February 2023)) to set up a TCP connection between Unity and the widely popular Robot Operating System (ROS) framework [53]. This allows us to generate C# classes to serialize and deserialize ROS messages, specifically the synthetic data obtained by the virtual IMU and GNSS sensors. The GNSS coordinate data is published at a rate of 5 Hz to the ROS network as a sensor_msgs/NavSatFix standard message. The synthetic IMU data is published at 50 Hz to the ROS network as a sensor_msgs/Imu standard message that stores the data over. Additionally, we also publish the current activity (label) being performed at 50 Hz to the ROS network under the message type `std_msgs/String`. All these three types of messages include a timestamp, which ensures that the activity labels can be synchronized with a given GNSS and IMU data stream.

**Figure 3.** The avatar performs locomotive actions.

In this study, we implemented a virtual model of the RION AH200C IMU sensor (http://en.rion-tech.net/products_detail/productId=158.html, (accessed on 10 February 2023)) which integrates an accelerometer, a gyroscope and a magnetometer, thus combining them and providing readings of linear acceleration, angular velocity and orientation. It is possible to place a virtual sensor in any desired position, as long as it is attached to a human joint. In our case, the virtual IMU sensor was placed on the chest of the avatar, as it is shown with a blue mark in Figure 3. The linear acceleration was calculated by taking into account the discrete derivative of the velocity with respect to the time as shown in Equation (1):

$$a(t_n) = K\left(\frac{v(t_n) - v(t_{n-1})}{T_s}\right) \quad (1)$$

where $a$, $K$, $v$, and $T$ stands for the linear acceleration, gain factor, velocity and time of the cycle, respectively. More particularly, we captured the position of the virtual IMU attached to the chest joint of the avatar and calculated its second discrete derivative every 20 ms on its own frame. In order to do that, we made use of the parent-child concept of Unity, which relies on a hierarchical structure between transform frames (position and rotation), as the pose of the child changes accordingly to the pose of the parent. The position of the child is applied from the current position of the avatar's chest. The rotation of the parent is adopted from the current rotation of the avatar's chest, while the position of the parent is adopted from the avatar's chest position in the previous frame as illustrated in the Figure 4 in which the transparent human figure represents the virtual IMU position of the previous cycle. This concept allows us to obtain the position of the virtual IMU in its local frame.

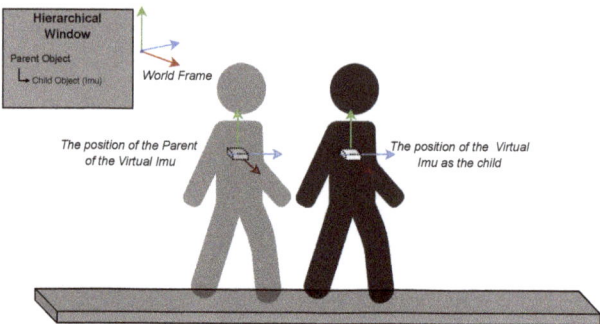

**Figure 4.** The parent-child relationship is adopted to obtain the position data in virtual IMU's local frame.

It is noteworthy that, due to the successive discrete derivative, the linear acceleration is greatly affected by noise, which is not observable in the data retrieved from the real IMU sensor. Therefore, we have applied a linear interpolation followed by smoothing the data using an exponential smoothing algorithm commonly employed in time-series data to remove high-frequency noises, as in Equation (2), where $x_t$ is the data sequence, $s_t$ is the output of the exponential smoothing algorithm, $t$ is time and $\alpha$ is the smoothing factor:

$$
\begin{aligned}
s_t &= x_t, & t &= 0 \\
s_t &= \alpha x_t + (1-\alpha)s_{t-1}, & t &> 0 \text{ and } 0 < \alpha \leq 1
\end{aligned}
\quad (2)
$$

Additionally, and because real-world accelerometers are generally affected by gravitational acceleration, we have calculated the gravity vector in the local frame of the sensor by making use of the Unity physics engine.

In order to have the orientation information, we extracted the virtual sensor's rotation in quaternions. Quaternions provide a convenient mathematical notation for representing the orientation of objects in space, being represented with complex numbers in the following form as shown in Equation (3), where $q_x$, $q_y$, $q_z$ are the vector units and $q_w$ is the scalar unit. Then, similarly to acceleration, we applied smoothing algorithm to smooth the quaternion data.

$$q = q_x + q_y + q_z + q_w \quad (3)$$

At last, to obtain the angular velocity, we made use of the orientation described above, converting quaternions to Euler angles, and applying the related discrete derivative at every 20 ms, as represented in Equation (4), where $\omega$, $\theta$, and $t$ represent angular velocity, rotation angle in radians, and time, respectively.

$$\omega = \frac{\Delta \theta}{\Delta t} \quad (4)$$

In line with the approach taken to model the virtual IMU, a similar methodology was adopted to model GNSS data. Specifically, a smartphone positioning system was utilized as a reference for modelling GNSS data. To simulate the GNSS data, the avatar's position in space was leveraged and converted into longitudinal and latitudinal coordinates using the GpsConverter package (https://github.com/MichaelTaylor3D/UnityGPSConverter (accessed on 10 February 2023)). It should be noted that the smartphone GNSS data had already undergone filtering, thus negating the need for additional data smoothing techniques [54].

While the data generated closely matches its real counterpart, real sensors are often affected by noise. Preliminary tests using the data generated from the aforementioned approach led to the overfitting of the models. The real IMU and GNSS sensors have noise characteristics due to some calibration errors or environmental noise that affects the sensor readings. Therefore, in order to make the synthetic data more realistic, a Gaussian Noise was injected on both sensors, more specifically affecting the longitude, latitude, linear acceleration and angular velocity variables. Noise was not added to quaternion data as the data provided by the real IMU sensor already comes from Extended Kalman Filter, which leads to a noiseless signal [55]. To add variability to the virtual IMU sensor data, the velocity and sequence of movements in the virtual avatar were adjusted in runtime. Different velocity levels can result in different patterns in the sensor data, while different sequences of activities can affect the overall variability of the data.

3.2.2. Data Preparation

To justify the reliability of the synthetic data, we conducted a preliminary study by using MatLab to deploy the sequence classifier for training and testing, benefiting from both Deep Learning Toolbox (https://www.mathworks.com/products/deep-learning.html (accessed on 10 February 2023)) (for sequence data classification) and ROS Toolbox (https://www.mathworks.com/products/ros.html (accessed on 10 February 2023)) (for seamless communication with the ROS master).

We have started by building our own dataset, containing the synthetic data obtained by the virtual IMU and GNSS sensors, as well as the real data obtained by the real RION AH200C IMU sensor and GNSS data of a smartphone. At this stage, the dataset included data from only four activities (Sit, Fall Down, Get Up and Idle) with automatic labelling being performed for the synthetic data as described in the previous section, and manual labeling being performed for the real data, which would be required to not only assess the feasibility of the virtual IMU and GNSS models, but also to validate and evaluate the classifier.

We have implemented a method for synchronizing the timestamps of IMU and GNSS sensors. For each timestamp of the label data, the closest GNSS and IMU timestamps are found. The GNSS data is associated with the IMU and label data over a short time of 10 timestamps. Then the GNSS route involving longitude ($\phi$) and latitude ($\lambda$) has been converted to Cartesian $x$ and $y$ coordinates.

Taking into account the use of a single IMU and GNSS, be it virtual or real, we have considered a feature vector $s(t)$ that includes linear acceleration ($a_x$, $a_y$, $a_z$), angular velocity ($\omega_x$, $\omega_y$, $\omega_z$), quaternion ($q_x$, $q_y$, $q_z$, $q_w$), $x$ and $y$ being represented as follows:

$$s(t) = \begin{bmatrix} a_x & a_y & a_z & \omega_x & \omega_y & \omega_z & q_x & q_y & q_z & q_w & x & y \end{bmatrix} \quad (5)$$

To tackle this classification problem, we adopted a Long Short-Term Memory (LSTM) network, which is known to be state-of-the-art supervised method for sequence data classification. As previously stated, LSTM is an improved type of recursive neural network and, instead of having a single neural network layer, it has four interacting layers, namely, cell state layer, input gate layer, forget gate layer and output gate layer. This enables it with the ability to "remember" information for a certain period, enabling learning-term dependencies [49]. Further detailed information on LSTM structure can be found in Section 4.2.

### 3.2.3. Synthetic Data Validation

We conducted a preliminary experiment to investigate if the synthetic data is as adequate as the real data. For real-world experiments, we have used a smartphone and the RION AH200C IMU sensor in a chest-worn sensor setup. For both real-world and virtual experiments, we have recorded 192 activities as 48 samples of each action, for up to 3 s, at 5 Hz and 50 Hz, respectively for GNSS and IMU data. We also created the categorical array that holds the labels corresponding to these actions. The LSTM network was trained with the synthetic data and subsequently tested with the real data. The adaptive moment estimation optimizer was adopted, with a maximum epoch of 200.

The results of these experiments are presented in Figure 5 with the confusion matrix depicting the accuracy of the experiment including the performance of each activity. A result of 84.9% indicates that although the model performs with acceptable accuracy, several Sit and Get Up actions were incorrectly classified as Idle. While the initial findings indicate that the model trained using synthetic data can accurately classify the four specified activities when presented with real-world data, the upcoming sections will delve deeper into the evaluation of HAR using synthetic data across a wider range of activities. Within the context of the FEROX project, and to propose a more encompassing architecture, more complex activities will be included, such as forestry-work-related ones. Therefore, due to the simulator feasibility for generating data, the next sections encompass data collection from 13 different activities and, likewise, a novel approach for HAR under uncertainty.

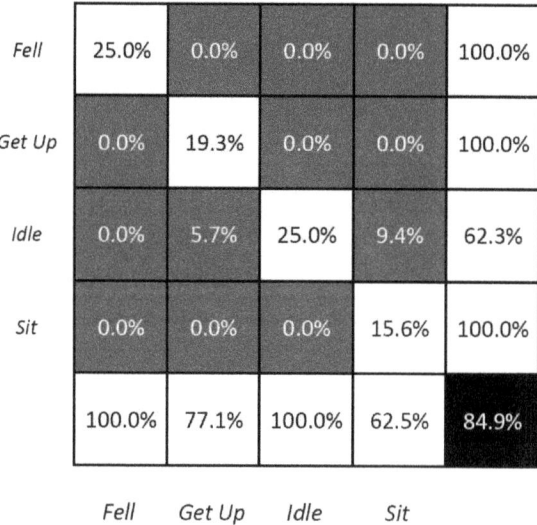

**Figure 5.** The confusion matrix of the LSTM network.

## 4. Fuzzy State Long-Short Term Memory (FS-LSTM)

The proposed FS-LSTM framework for HAR under uncertainty is presented in Figure 6. The framework is comprised of five blocks, labeled A, B, C, D, and E. Block A is responsible for collecting human-related data through multimodal sensors, hereby assessed using a chest-worn IMU and a GNSS smartphone positioning system. Block B processes the IMU and GNSS data, including linear acceleration, angular velocity, orientation from the IMU sensor, as well as longitude and latitude from the GNSS, which are published at 50 Hz and 5 Hz, respectively and as previously described in Section 3. This data is then transformed into linguistic labels for Motion and Tilt through a fuzzification process in Block C, as further described in Section 4.1. These fuzzified Motion and Tilt features serve as inputs for both Block D and E. In Block D, the fuzzified feature set is used as input in LSTM state machine learning, where multiple networks are trained for each state to be executed during runtime, including a recovery state called Lost. This process is further detailed in Section 4.2. In Block E, uncertainty is managed through defuzzification in a closed-loop. The fuzzified inputs are used in the classification network, which was established in the previous iteration, and the generated classification score is first fuzzified and then defuzzified into a crisp value to determine whether to progress to the next state or remain in the current one. The details of this process will be further explained in Section 4.3.

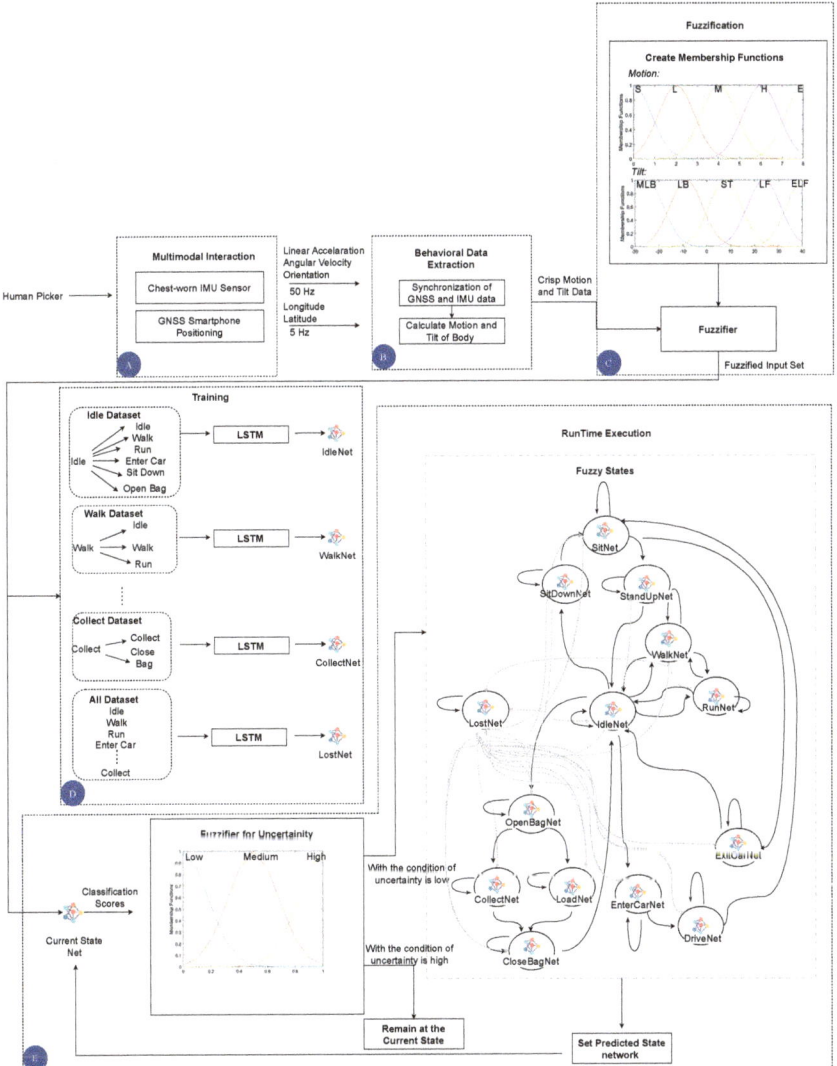

**Figure 6.** The diagram of the proposed architecture.

*4.1. Fuzzification of Features*

This section presents the design of the feature fuzzification process for converting IMU and GNSS data into linguistic variables that will serve as inputs in the proposed FS-LSTM model. This stage is marked as C block in Figure 6.

Fuzzy logic allows computer systems to mimic the human-like thinking and make decisions under uncertain and imprecise information. For instance, the subjective and ambitious statement "the food is good" is enough for a person to decide how much to tip. This way of handling uncertainty is important when the reliable exact information is not available.

The numeric data collected by the FEROX simulator, as described in Section 3, includes linear acceleration ($a_x$, $a_y$, $a_z$), angular velocity ($\omega_x$, $\omega_y$, $\omega_z$), and quaternion ($q_x$, $q_y$, $q_z$, $q_w$), as well as the $x$ and $y$ coordinates. The process of fuzzification involves converting these crisp numerical data from sensors (either synthetic or real) into linguistic variables for use in the proposed FS-LSTM model. In particular, we have selected Motion and Tilt variables for fuzzification. We obtained velocity information as crisp data inputs for Motion and y-axis Euler angle (or pitch) for Tilt. To calculate the velocity, we benefit from the synchronized GNSS and IMU data (previously addressed in Section 3) to first generate a smooth 2D Cartesian position, $x$ and $y$. The initial velocity is calculated as the derivative of the position:

$$V_{x_0}(t) = \frac{x(t) - x(t-1)}{\delta t},$$
$$V_{y_0}(t) = \frac{y(t) - y(t-1)}{\delta t} \quad (6)$$

At each time step, we then calculate the velocity along the $x$ and $y$ axis as:

$$V_x(t) = V_{x_0}(t) + a_x(t) * t,$$
$$V_y(t) = V_{y_0}(t) + a_y(t) * t \quad (7)$$

where $t$ is the elapsed time and $a_x$ and $a_y$ are the linear acceleration measurements along $x$ and $y$ axis, respectively, obtained after applying a rotation to the linear acceleration measurements provided by the IMU, so as to align the body frame with the world frame. Finally, $V$ is calculated as the magnitude of the velocity vector:

$$V(t) = \sqrt{V_x(t)^2 + V_y(t)^2} \quad (8)$$

To quantify Tilt, the y-axis Euler angle $\theta$ (pitch), which represents the tilt of the chest forward or backward was calculated. The extracted quaternion values were converted to Euler angles in degrees following the principle presented in Equation (9):

$$\phi = atan2(2(q_w q_x + q_y q_z), 1 - 2(q_x^2 + q_y^2))$$
$$\theta = arcsin(2(q_w q_y - q_z q_x)) \quad (9)$$
$$\psi = atan2(2(q_w q_z + q_x q_y), 1 - 2(q_y^2 + q_z^2))$$

The calculated $V$ and $\theta$ are inputs utilized in the fuzzification process. In particular, $V$ is mapped into five linguistic labels for Motion, while pitch ($\theta$) is translated into five linguistic labels for Tilt. The numerical data for both Motion and Tilt were mapped into five linguistic variables by Gaussian membership function using MatLab's Fuzzy Logic Toolbox as follows:

$$S_U(t) = \begin{cases} Motion \rightarrow \{S, L, M, H, E\} \\ Tilt \rightarrow \{MLB, LB, ST, LF, ELF\} \end{cases} \quad (10)$$

where $S$, $L$, $M$, $H$ and $E$ represent the linguistic variables Stopped, Low, Medium, High, and Extreme for Motion, respectively, while $MLB$, $LB$, $ST$, $LF$ and $ELF$ represent Medium Lean Back, Lean Back, Straight, Lean Front, and Extremely Lean Front, for Tilt, respectively. $S_U(t)$ represents the feature vector which then is used as input for the FS-LSTM method.

In Figure 7, the five linguistic variables for Motion and Tilt are displayed, alongside an activity sequence in which the human is first idle, then walks, then becomes idle again, and finally opens the bag and collects some berries.

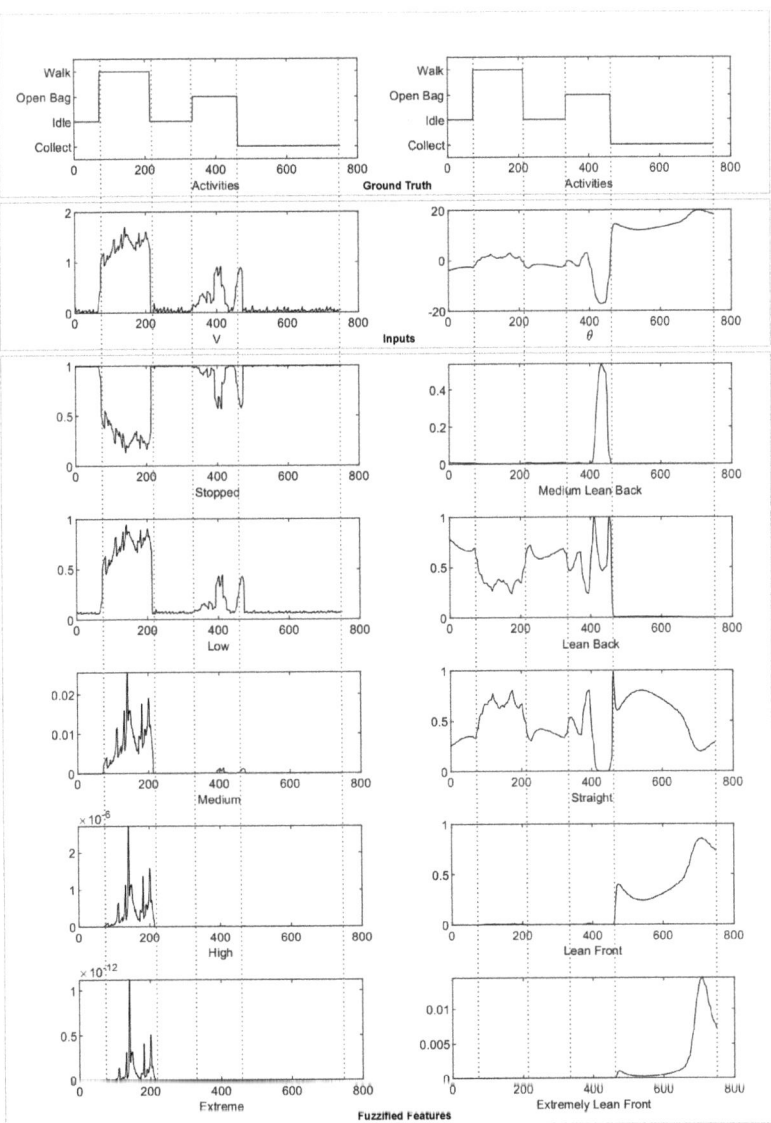

**Figure 7.** Motion and Tilt plotting along an activity sequence.

The activity sequence in this short period is represented as: Idle, Walk, Idle, Open Bag, and Collect, which are shown in the figure in the ground truth section. Additionally, the figure shows the precise numerical values of the Velocity ($V$) and pitch ($\theta$) as the inputs for Motion and Tilt, respectively. It can be seen from the figure that the chosen linguistic variables effectively depict the activities. For instance, one can easily recognize when the human is walking by noticing that the Motion value falls between Low and Medium. The Open Bag and Collect activities also showcase distinct characteristics, such as leaning back while opening the bag and leaning forward while collecting berries. Thus, these activities and their fuzzified representations are clearly illustrated and may be used as features to the HAR architecture.

## 4.2. State Machine Learning

This section presents the modelling of human activity by exploring the transitions between different activity states. These states may involve specific activity sequences or individual activities. As an example, the use case discussed in Section 3 is depicted through a state diagram in Block E of Figure 6. This particular use case models the sequential activities of a human picker, including locomotive movements (such as Idle, Walk, Run), berry collection and loading, and vehicle driving. The flow is represented by 14 states (with the recovery state Lost) and includes states that occur in a sequential manner, such as sitting and driving a car, and transitions that occur between these states, such as Sit Down to Sit, Sit to Stand Up or Enter Car to Drive. There are also states that lead to multiple possibilities, such as the Idle state, which can transition to activities such as Open Bag, Walk, Run, Enter Car, or Sit Down.

The state transitions in a given process are modelled individually through the use of LSTM networks. As previously mentioned, LSTM is capable of extracting hidden patterns from long-term sequential data by handling gradient exploding or vanishing gradients problems [56]. In more detail, the internal structure of an LSTM network consists of multiple gates, including input gate $i_t$, forget gate $f_t$, and output gate $o_t$, that control the flow of information towards the final output. The input gate updates the information, while the forget gate processes information from both the input gate $i_t$ and the previous state $C_{t-1}$, selectively removing information from the current state $C_t$ when necessary. The output gate forwards the final output to the next LSTM unit and retains the output value for subsequent sequence predictions. The recurrent unit, on the other hand, estimates the state of the previous cell $C_{t-1}$ and the current input $x_t$ using a $tanh$ activation function. The value of $h_t$ can then be calculated as the scalar product of the output gate $o_t$ and the $tanh$ of the $C_t$. The ultimate output is obtained by passing $h_t$ to a softmax classifier [57].

Each LSTM network within the proposed HAR architecture is designed to receive sequential data as input (fuzzy features addressed in the previous section) and generate output predictions for only the feasible transition states. The objective of this approach is to ensure that there are no impossible transitions between states. For example, in a use case where the activities include Sit Down and Stand Up, or Collect and Load, it is possible that similar characteristics may result in an incorrect transition from one activity to the other. However, these transitions should not occur according to the expert-designed state flow, hereby represented as an FSM. The aim of this modelling approach is to guarantee that the predicted next state will be one of the possible states, providing important information for decision-making in an HRC system. Moreover, the number of possible classes significantly affects the size of the model, and a larger number of possible classes leads to a more complex model structure. Such a complex model demands additional computational resources, which would result in a longer runtime execution [58]. For instance, a Sit Down LSTM network with a reduced number of possible classes, such as Sit Down (remains in the same state) and Sit (moves to only the next possible state), is more efficient when compared to an LSTM network trained with all 13 possible activities, particularly when many of the outputs are unlikely to be the actual state.

The implementation of the LSTM layer in MatLab was carried out using the Deep Learning Toolbox. For the purpose of predicting the subsequent activity sequence, a sequence-to-sequence classification approach was employed. Each LSTM network corresponding to a specific state was trained using the same sequential input data. The input vector $S_U(t)$ comprises ten features, including the fuzzified values for Motion and Tilt, as described in Section 4.1 and depicted in Equation (11):

$$S_U(t) = [\ S\quad L\quad M\quad H\quad E\quad MLB\quad LB\quad ST\quad LF\quad ELF\ ] \qquad (11)$$

where,

$$\begin{aligned} S &= \text{Stopped} & MLB &= \text{Medium Lean Back} \\ L &= \text{Low} & LB &= \text{Lean Back} \\ M &= \text{Medium} & ST &= \text{Straight} \\ H &= \text{High} & LF &= \text{Lean Front} \\ E &= \text{Extreme} & ELF &= \text{Extreme Lean Front} \end{aligned} \quad (12)$$

It is noteworthy, however, that each state-based LSTM network output sets differ, with each output set corresponding to a specific state, including the potential transitions, and including the possibility of remaining in the same state, as it is shown in Figure 8. Any non-feasible state names are labelled as Lost. The Lost network, which uses the same sequential input data and an output dataset of all states, serves as a recovery mechanism, being only activated when such class is output by the previous LSTM network.

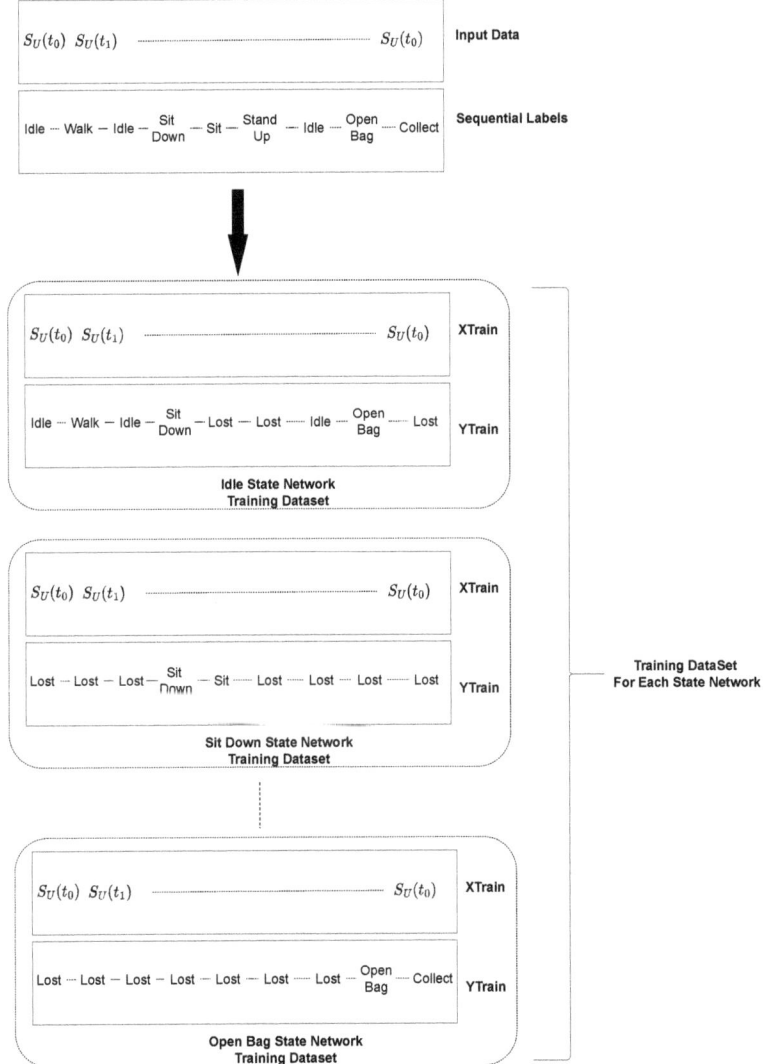

**Figure 8.** LSTM training datasets.

The data generated in this study have been made publicly available (https://gitlab.ingeniarius.pt/ingeniarius_public/ferox/hrc-ferox.git (accessed on 22 February 2023)). The dataset includes the following features, as shown in Equation (13): $t_s$ is the timestamp, $s(t)$ is the feature vector containing the raw sensor data shown in Equation (5), $s_U(t)$ is the feature vector containing the fuzzified sensor data shown in Equation (11), and $G$ is the ground truth of activities presented as a string array where each element represents the activity label at a specific timestamp.

$$[\ t_s\quad s(t)\quad S_U(t)\quad G\ ] \tag{13}$$

$$G = [\text{"Idle"}, \text{"Idle"}, \text{"Sit"}, \ldots, \text{"Close Bag"}] \tag{14}$$

*4.3. Coping with the Uncertainty through Defuzzification*

The trained network models described in the preceding section are employed in a closed-loop architecture to predict the subsequent state, as depicted in Block E of Figure 6. The fuzzified outputs generated by Block C serve as inputs to the trained network model associated with the current state, which is referred to as the Lost network in the first iteration and is updated in subsequent iterations in accordance with the flow of the architecture.

The proposed architecture makes use of network model classification scores as posterior probabilities based on the fuzzified input set. These probabilities are calculated based on Bayes' Theorem:

$$\hat{P}(B\mid A) = \frac{P(A\mid B)P(B)}{\sum_{j=1}^{R} P(A\mid j)P(j)} \tag{15}$$

where $\hat{P}(B\mid A)$ is the posterior probability that an observation $A$ of given class $B$, $P(A\mid B)$ is the conditional probability of $A$ given class $B$, $P(B)$ is the prior probability for class $B$ and $R$ is the number of classes in the response variable [59]. The classification scores are represented as an $z$-by-$R$ matrix, where $z$ is the number of observations in the data and $R$ is the number of unique classes. The matrix indicates the probability of each observation belonging to a specific class, with the predicted class being determined by the class with the highest score.

FS-LSTM falls on the assumption that state network models may struggle to make confident predictions if the highest score is low or if the scores are similar across classes, leading to a certain level of uncertainty. Yet, it uses such level of uncertainty to still generate a prediction based on the class with the highest score. It is therefore important to consider the level of confidence in the prediction and interpret the scores before making any decisions based on the model's output. The uncertainty in the scores is evaluated using the fuzzy logic system.

Hence, similar as carried out for Motion and Tilt fuzzy variables, classification scores are used for the fuzzification to produce fuzzy linguistic labels for Uncertainty as Low, Medium, and High. The triangular fuzzifier is used to determine the degree of membership for each value. Unlike Motion and Tilt fuzzy variables, however, inference is followed by adopting a set of rules designed for each state, which are then later utilized to assess the uncertainty in the classification scores generated by the current state network. An example of fuzzy rules is presented for the Close Bag state. The uncertainty is assessed in the classification scores generated for the possible transition outputs of the Idle, Close Bag, and Lost recovery state by implementing these rules:

1. IF Idle is Low and Close Bag is Low and Lost is Low THEN Uncertainty is High
2. IF Idle is Medium and Close Bag is Medium and Lost is Medium THEN Uncertainty is High
3. IF Idle is High and Close Bag is High and Lost is High THEN Uncertainty is High
4. IF Idle is High and Close Bag is Low and Lost is Low THEN Uncertainty is Low
5. IF Idle is Low and Close Bag is High and Lost is Low THEN Uncertainty is Low
6. IF Idle is Low and Close Bag is Low and Lost is High THEN Uncertainty is Low

7. IF Idle is High and Close Bag is not Low and Lost is not Low THEN Uncertainty is High
8. IF Idle is not Low and Close Bag is High and Lost is not Low THEN Uncertainty is High
9. IF Idle is not Low and Close Bag is not Low and Lost is High THEN Uncertainty is High

In the fuzzy inference, the rules are applied to the fuzzified inputs to calculate the degree of fulfilment for each rule through aggregation. The following step is defuzzification, which transforms the fuzzy outputs into crisp outputs by using the fuzzy sets and their corresponding membership degrees. The result of the aggregation is converted into a crisp output value through the centroid method. The output of this system expresses the uncertainty as a crisp value between 0 and 1. This crisp value coming from the fuzzy logic system is then used to assess the confidence level of each network model before determining the next state. This is performed by comparing it to an experimentally-defined threshold. If the crisp value of uncertainty generated through defuzzification is lower than the threshold, this implies the model is confident in the prediction and the classification is carried out based on the highest score, which is identified as the next state. This state could be a different state or remain unchanged. The network model that corresponds to the predicted state is then selected for use in the next iteration. If the level of uncertainty exceeds the established threshold, the system remains in the same state. In this case, the next iteration performs the classification utilizing the current network model. This iterative process continues in accordance with the closed-loop architecture until the model reaches a sufficient level of confidence in its prediction.

Figure 9 shows a small section of the timeline of activity recognition handling the uncertainty. Between 62 and 65 samples, the model experiences low confidence in its predictions as the uncertainty is above the threshold and remains in the same state for subsequent selections. At sample 184, the model erroneously classifies the data as Walk while the ground truth remains Run. While it may seem like a premature classification, it could be a coincidence since transitions from Run to Walk are possible. However, at sample 185, it is observed that the Walk state network model lacks certainty in the classification, causing it to remain in the Walk state. The uncertainty level drops below the threshold once the ground truth and the predicted classes align at sample 186.

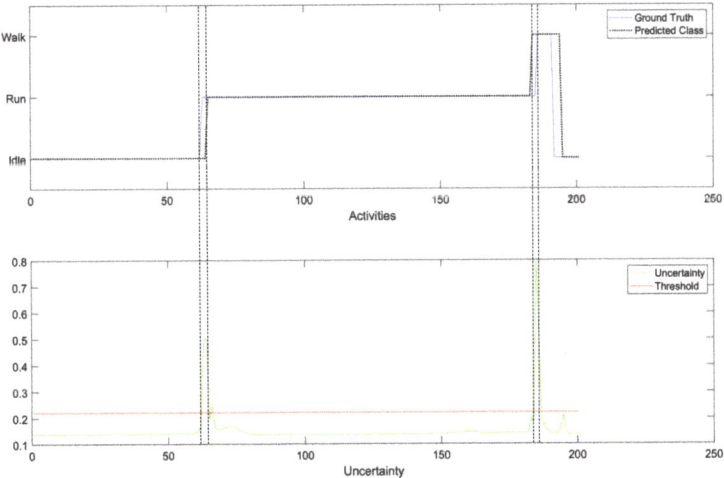

**Figure 9.** Accessing uncertainty in the activity recognition.

## 5. Results and Discussion

### 5.1. Benchmark of Activity Recognition Performance

In this section, we present the results of our experimental study designed to evaluate the effectiveness of the proposed approach. We used a training dataset consisting of 13 different activities (Idle, Walk, Run, Sit Down, Sit, Stand Up, Exit Car, Drive, Enter Car, Open Bag, Collect, Load, Close Bag) as depicted in Figure 6 block E. The training dataset was collected over a period of 17.85 min and includes 539 activities, with a total of 53,553 samples of virtual IMU data and 5355 samples of virtual GNSS data. The testing dataset consists of 168 activities and comprises 21,188 samples of virtual IMU data and 2118 samples of virtual GNSS data, spanning a total duration of 7.06 min. In this section, the number of hidden layers was set to 64 for all LSTM networks considered.

We have benchmarked three different methodologies:

(a) Traditional LSTM: a LSTM model that was trained using raw sensor input, similar to what has been presented in Section 3, though with all 13 states instead of four.

(b) Fuzzy LSTM: a LSTM model that was trained with the fuzzified features described in Section 4.1 and outputs all 13 states.

(c) FS-LSTM: the multiple LSTM models that were trained for each state, using fuzzified features, and each model only outputs the feasible states, as per presented in Figure 6 and described in the previous Section 4.

For the Traditional LSTM model, the input feature vector $s(t)$ is the one previously presented in Equation (5), consisting of linear acceleration values ($a_x$, $a_y$, $a_z$), angular velocity values ($\omega_x$, $\omega_y$, $\omega_z$), quaternion values ($q_x$, $q_y$, $q_z$, $q_w$), and the $x$ and $y$ Cartesian coordinates. The input vector for both the Fuzzy LSTM and FS-LSTM models, $s_U(t)$, includes the five linguistic labels previously described in Section 4.1 for each of the extracted Motion and Tilt variables (Equation (11)).

The results of the three methodologies are presented in Figure 10 as confusion matrices, which depict the outcomes of (i) Traditional LSTM; (ii) Fuzzy-LSTM; and (iii) FS-LSTM. The rows represent the target classes, while the columns represent the output classes. Superior classification accuracy results are identified in bold. The results indicate that utilizing solely raw sensor data leads to a significantly low accuracy of 23.2%. Conversely, by utilizing fuzzified inputs, the Fuzzy-LSTM approach markedly enhances the accuracy to 93.2%. This demonstrates how a fuzzy logic system handles data ambiguity and achieves correct classification, where the Traditional LSTM method frequently falls short. The proposed FS-LSTM methodology achieved an accuracy of 90.9%. This result was obtained by treating samples classified as Lost as unchanged. This approach ensures that the system waits until it recovers from the Lost state, which runs the same model as the Fuzzy-LSTM approach (with 13 state outputs), before transmitting the predicted output to a higher-level decision-making system to ensure the correctness of the transmitted output.

Although the overall accuracy does not show a significant difference from the Fuzzy-LSTM, being even slightly inferior in terms of accuracy, the FS-LSTM prevents transitions that should not occur from happening. This might result in slightly superior performance on Sit Down-Sit-Stand Up transitional states when compared to Fuzzy-LSTM. This not only avoids passing incorrect information to the higher-level management system but also improves the probability of predicting the next state. Figure 11 illustrates this prevention of wrong transitions more clearly for one of the many sequences generated by the aforementioned models. In the figure, a sequence of activities is given with their ground truth in the blue dotted line and predicted outputs with the straight black line obtained through Traditional LSTM (top), Fuzzy-LSTM (middle), and FS-LSTM (bottom). Once again, it is shown that Traditional LSTM is unable to classify activities correctly, alternating between Walk and Idle in this sequence. For the Fuzzy-LSTM, around sample 150, the predicted class is Enter Car, while the ground truth class is Idle. In contrast, in the same sample in FS-LSTM, the classification of the current network model (Idle state network) is either correct, or the uncertainty is high, and the system remains unchanged in Idle. This prevents the system from making infeasible transitions to Enter Car, which would break the sequence of predicted activities. Fuzzy-LSTM, however, makes infeasible transitions multiple times, such as in sample 665, where it transitions from Idle to Close Bag, which should never happen. FS-LSTM handles such infeasible transitions through uncertainty defuzzification and the state machine approach generally well. As it is described earlier, FS-LSTM handles misclassifications via recovery of the Lost state. In the confusion matrix Figure 10, the Lost classified outputs are not shown as they were treated as unchanged states. However, in practice, when the network is unable to classify a sample in any of the possible transitions, the decision-making moves to the Lost state. For instance, in the sample around 956 marked with a red circle, the system is in the Lost state. This occurred because the Idle network predicted Enter Car instead of Sit Down, which was feasible but wrong. The Enter Car state network model was unable to classify the samples in any feasible transitions, and the system went to the Lost state before recovering to the Sit Down state.

In comparison, while Fuzzy-LSTM mostly performs poorly in the presented time window, it still achieves overall accurate classification, as demonstrated by the corresponding confusion matrix. However, it is important to note that Fuzzy-LSTM should be regarded as a sequence of activity flow prediction rather than individual sample prediction, underscoring the superiority of FS-LSTM. Furthermore, FS-LSTM is expected to achieve the same level of accuracy, of even higher, under constrained resources, since the multiple LSTM networks it encompasses are expected to not require the same number of hidden layers given the reduced number of outputs foreseen by each. This is further explored in the next section.

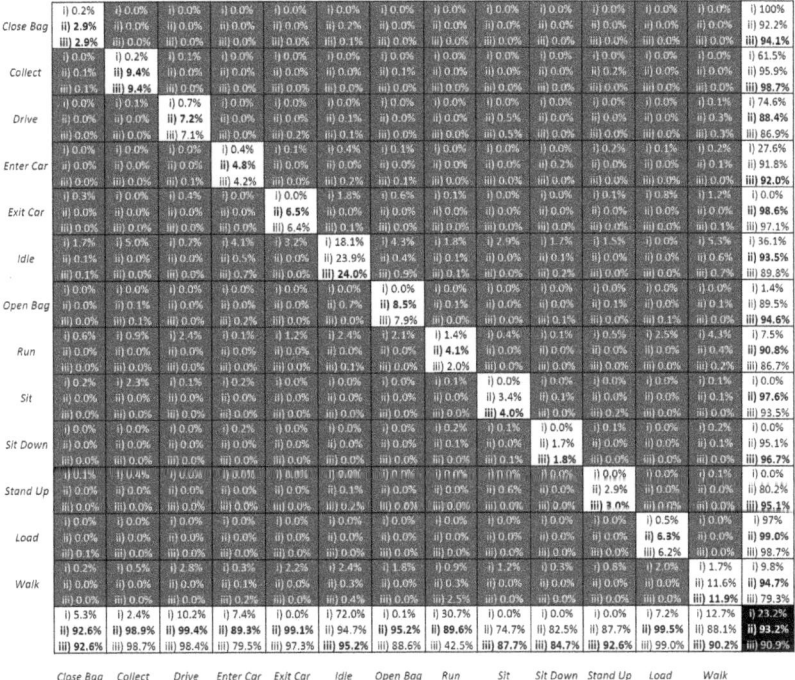

**Figure 10.** The confusion matrix of Traditional LSTM, Fuzzy-LSTM and FS-LSTM. Superior classification accuracy results are identified in bold.

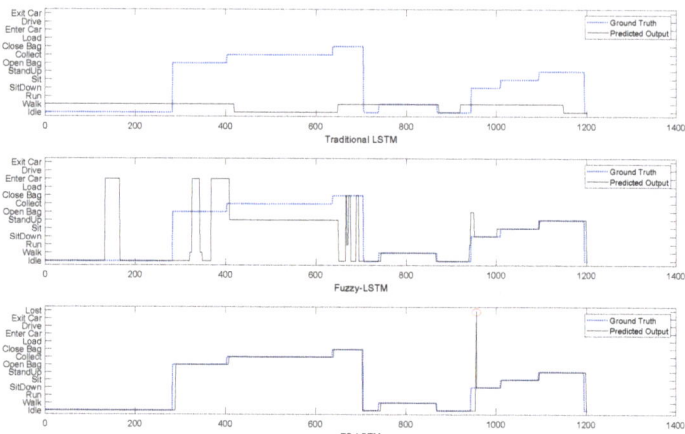

**Figure 11.** The benchmark of three predicted output sequences via Traditional LSTM, Fuzzy-LSTM and FS-LSTM. The Lost state is marked with the red circle.

## 5.2. Benchmark of Efficiency

In this section, we evaluate the classification performance of the proposed FS-LSTM compared to Fuzzy-LSTM from the perspective of GPU resource efficiency. In the previous section, both models were trained with 64 hidden layers, and while Fuzzy-LSTM classified 13 states, FS-LSTM only used this complex network in the Lost state. To better understand the impact of this difference on computer resources, we monitored GPU utilization and power consumption during the classification runtime process for a period of 25 minutes. Our experiments were conducted using an NVIDIA GeForce GTX 1050. The benchmark chart for both models is presented in Figures 12 and 13.

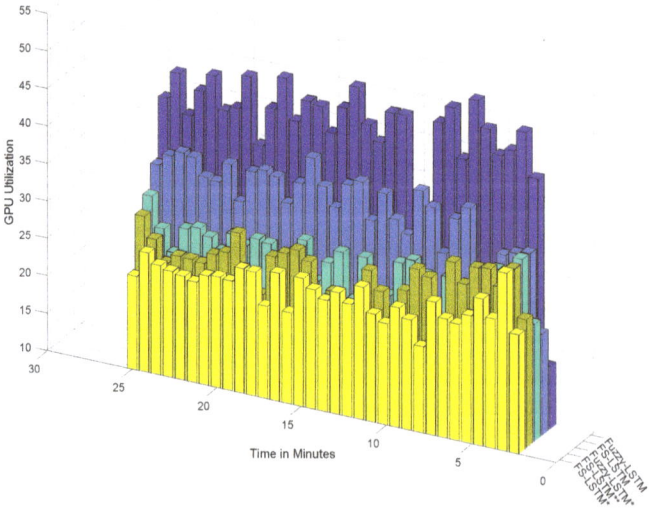

**Figure 12.** The benchmark of GPU utilization during testing. Fuzzy-LSTM and FS-LSTM networks trained with 32 layers (denoted with *), and a hybrid version of FS-LSTM (denoted with **).

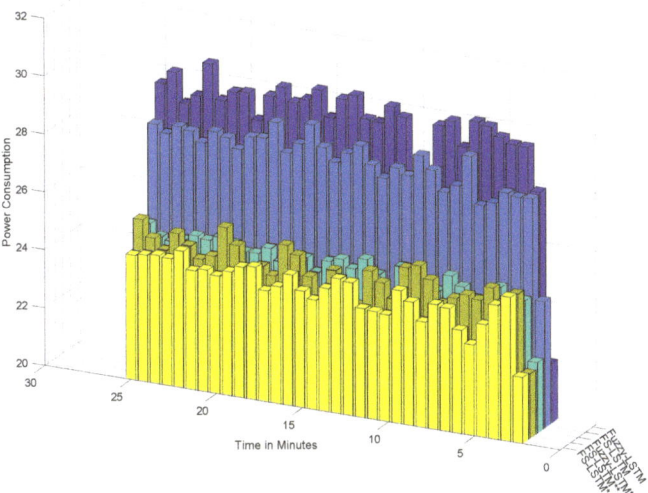

**Figure 13.** The benchmark of power consumption during testing. Fuzzy-LSTM and FS-LSTM networks trained with 32 layers (denoted with *), and a hybrid version of FS-LSTM (denoted with **).

As shown in the last two rows of columns from Figure 12, despite both methods performing with high accuracy, Fuzzy-LSTM (represented by the dark blue bars) relies more heavily on GPU resources than the FS-LSTM approach (represented by the light blue bars). This is a critical consideration in any long-term outdoor application where computer resources are constrained. While in the last two rows of columns from Figure 13 the power consumption did not show any significant difference between these methods over a short-term test, high GPU load is a critical consideration in any long-term outdoor application where computer resources are constrained. A high GPU load indicates that the GPU is being heavily utilized to complete the classification task, which can cause the GPU to generate more heat and consume more power, ultimately affecting the overall performance and energy efficiency of the system over an extended period.

A question may arise as to whether reducing the number of hidden layers in both models can reduce its complexity, then leading to better efficiency, while still maintaining the desired accuracy. To answer this, we extended the benchmark study to five different combinations: both Fuzzy-LSTM and FS-LSTM networks trained with 32 layers (denoted with *), and a hybrid version of FS-LSTM (denoted with **) trained with 32 layers, except for the Lost network, which was trained with 64 layers. As shown in Figures 12 and 13, reducing the number of layers significantly decreases GPU utilization and power consumption. Table 1 further compares the mean and standard deviations of GPU utilization and power consumption, demonstrating that a smaller model size with fewer output and hidden layers leads to significantly higher efficiency. However, reducing the number of hidden layers sacrifices accuracy, as shown in the confusion matrix in Figure 14.

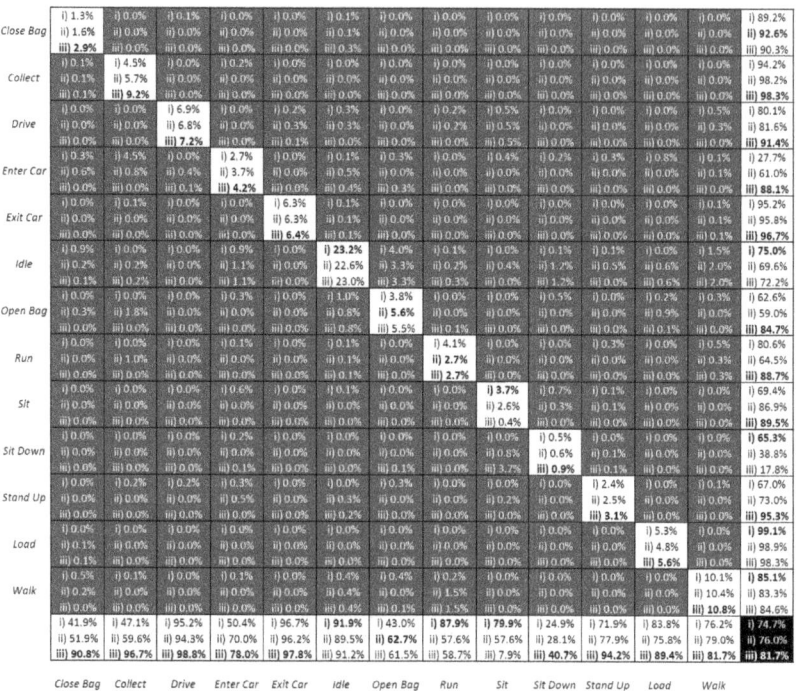

**Figure 14.** The confusion matrix of Fuzzy-LSTM* (i), FS-LSTM* (ii) and FS-LSTM** (iii). Superior classification accuracy results are identified in bold.

These conclusions are not new, though what can be seen is that Fuzzy-LSTM* (i) and FS-LSTM* (ii), show a decrease in accuracy of 74.7% and 76.0%, with FS-LSTM slightly dethroning Fuzzy-LSTM under a lower number of resources (32 hidden layers instead of 64 hidden layers), while still requiring less GPU utilization. Furthermore, while the FS-LSTM 64 hidden layer Lost network (with 32 hidden layers for all other states), FS-LSTM** (iii), presents a similar GPU utilization and power consumption than the Fuzzy-LSTM with a 32 hidden layer (Fuzzy-LSTM*), its accuracy rises to 81.7% (versus the 74.7% of Fuzzy-LSTM*). While this is not an outstanding result, it can be a useful compromise between model size, performance, and energy efficiency in certain applications. One such application is covered by the FEROX Project, where the system is expected to run on smaller portable devices, such as smartphones and wearables. In such scenarios, an efficient utilization of computing resources becomes crucial, and the FS-LSTM model may offer a viable solution without significantly reducing the human activity recognition accuracy.

**Table 1.** The mean ± standard deviation of GPU utilization and power consumption. Fuzzy-LSTM and FS-LSTM networks trained with 32 layers (denoted with *), and a hybrid version of FS-LSTM (denoted with **).

| Mean ± SD | Fuzzy-LSTM | FS-LSTM | Fuzzy-LSTM * | FS-LSTM * | FS-LSTM ** |
|---|---|---|---|---|---|
| GPU Utilization | 44.55 ± 9.85 | 35.21 ± 9.57 | 28.05 ± 9.56 | 25.58 ± 9.46 | 27.54 ± 10.16 |
| Power Consumption | 29.25 ± 2.04 | 28.14 ± 1.91 | 24.66 ± 0.89 | 24.25 ± 0.89 | 24.70 ± 1.34 |

## 5.3. Discussion

As described in Section 1.2, this paper presented three key incremental developments, which have been successfully achieved and summarized as it follows:

i   Enhance Long Short-Term Memory (LSTM) networks by incorporating fuzzy logic to model human uncertainty (Fuzzy-LSTM): the objective of this research was to enhance the accuracy of LSTM networks by incorporating fuzzy logic to model human uncertainty. Even though the preliminary results, as shown in Figure 5, depicted an accuracy of 84.9% for recognizing four activities, such accuracy dropped to 23.3% when trying to recognize 13 different activities. By utilizing fuzzified Motion and Tilt features, the Fuzzy-LSTM model was capable to effectively handle uncertain data. The initial results of the study showed that Fuzzy-LSTM improved the accuracy of activity recognition by a significant margin, achieving 93.2% accuracy compared to the traditional LSTM model using raw sensor data.

ii  Extend the Fuzzy-LSTM approach by incorporating finite-state machines (FSM) to model activity sequences, resulting in the Fuzzy State LSTM (FS-LSTM) model: the primary objective of this research was to improve the predictability of human activity sequences by identifying possible transitions between states. While Fuzzy-LSTM achieved 93.2% accuracy under unconstrained GPU resources, approximately 3% more than FS-LSTM, it often resulted in infeasible transitions. However, when using constrained resources such as embedded systems or limited GPU resources, the FS-LSTM showed superior performance compared to Fuzzy-LSTM. As shown in Figure 14, FS-LSTM has a trade-off between accuracy and computational resources, but it offers significant benefits for long-term real-time outdoor applications.

iii Develop a defuzzification-based method to estimate human uncertainty by aggregating predicted scores of the LSTM model: this proposed approach aimed to estimate the uncertainty associated with the LSTM classifier's predictions through defuzzification. By waiting until the prediction became certain, the system could achieve an accuracy of 90.0%. This development is crucial to prevent the system from making wrong transitions between states before the model becomes certain, thereby improving the overall performance of the system. Although this study did not show the direct impact of this issue, it could significantly affect the high-level decision-making process for robots, where the system needs to consider the current human state. Any infeasible transitions could compromise the performance of the system, causing trust and safety issues.

This study presents a few drawbacks and limitations that should be taken into consideration. Firstly, the results were obtained using only synthetic data. Although the preliminary results (Figure 5) show that models trained with such data are transferable to real-world data, this may still not accurately reflect real-world scenarios or domains without further modification and customization. Therefore, caution should be exercised when applying the proposed approach to other contexts. Furthermore, while synthetic data provides advantages such as low-cost data generation and easy labelling, it may not fully capture the complexity and variability of real-world data. Thus, a further study is needed to ensure the generation of data with the same characteristics as real data when using a different type of sensor. Additionally, in a real-world setup, the sensor placement and other aspects of the experimental setup may need to be adjusted to account for different environmental conditions and potential interferences.

## 6. Conclusions and Future Work

This study presents an architecture for human activity recognition and modelling intended for use in human-robot collaboration in field applications. The approach employs multiple LSTM networks, each trained to recognize feasible states within an FSM architecture. The FSM architecture is further enhanced with fuzzy logic to determine the uncertainty level of the classification made by the LSTM, thereby preventing unfeasible activity transitions in high-level decision-making systems. The proposed approach is compared to a traditional LSTM model trained on raw sensory data and a Fuzzy-LSTM model that used fuzzified sensory data as inputs to train a single LSTM network.

The proposed approach achieves high accuracy, with a rate of 90.9%, while efficiently utilizing computer resources. The system's performance is evaluated using synthetic data generated from a berry collection use case developed in a simulator. Future work will involve assessing the system's performance using real-world data within the FEROX project, as well as optimizing the developed work to operate on a small platform such as a smartphone.

As a continuation of this work, the next step will involve developing a high-level decision-making system that utilizes the human state predicted by the proposed FS-LSTM, as well as its associated uncertainty, to make informed decisions for each agent in a multi-robot and multi-human system. The decision-making system will be developed based on explicit and implicit relationships between agents, building upon the presented study's understanding of human activity.

**Author Contributions:** Conceptualization, B.Y. and M.S.C.; methodology, B.Y. and M.S.C.; software, B.Y.; validation, B.Y.; investigation, B.Y. and M.S.C.; resources, S.P.S. and A.V.; writing—original draft preparation, B.Y.; writing—review and editing, M.S.C., S.P.S. and A.V.; visualization, B.Y.; supervision, M.S.C., S.P.S. and A.V.; project administration, M.S.C.; funding acquisition, S.P.S. and A.V. All authors have read and agreed to the published version of the manuscript.

**Funding:** This work has been partly funded by the EU FEROX project (https://ferox.fbk.eu/ (accessed on 10 February 2023)). FEROX has received funding from the European Union's Horizon Europe Framework Programme under Grant Agreement No. 101070440. Views and opinions expressed are however those of the authors only and the European Commission is not responsible for any use that may be made of the information it contains. This work was also partially funded by FCT—Fundação para a Ciência e a Tecnologia (FCT) I.P., through national funds, within the scope of the UIDB/00127/2020 project (IEETA/UA, http://www.ieeta.pt/ (accessed on 25 February 2023)). This work was also supported by Ingeniarius Ltd with provided materials and co-funded by UTAD.

**Data Availability Statement:** The data generated in this study are openly available here: https://gitlab.ingeniarius.pt/ingeniarius_public/ferox/hrc-ferox.git (accessed on 22 February 2023)).

**Conflicts of Interest:** The authors declare no conflict of interest.

## Abbreviations

The following abbreviations are used in this manuscript:

| | |
|---|---|
| HRC | Human Robot Collaboration |
| HAR | Human Activity Recognition |
| FS-LSTM | Fuzzy State-Long Short-Term Memory |
| Bi-LSTM | Bidirectional Long Short-Term Memory |
| SVM | Support Vector Machines |
| HMM | Hidden Markov Models |
| FSM | Finite State Machines |
| RNN | Recurrent Neural Networks |
| CNN | Convolutional Neural Networks |
| DNN | Deep Neural Networks |
| ANN | Artificial Neural Networks |
| DBMM | Dynamic Bayesian Mixture Model |
| IMU | Inertial Measurement Units |
| ROS | Robot Operating System |
| GPU | Graphics Processing Unit |

## References

1. Villani, V.; Pini, F.; Leali, F.; Secchi, C. Survey on human–robot collaboration in industrial settings: Safety, intuitive interfaces and applications. *Mechatronics* **2018**, *55*, 248–266. [CrossRef]
2. Ajoudani, A.; Zanchettin, A.M.; Ivaldi, S.; Albu-Schäffer, A.; Kosuge, K.; Khatib, O. Progress and prospects of the human–robot collaboration. *Auton. Robot.* **2018**, *42*, 957–975. [CrossRef]
3. Chueshev, A.; Melekhova, O.; Meshcheryakov, R. Cloud Robotic Platform on Basis of Fog Computing Approach. In Interactive Collaborative Robotics, Proceedings of the Interactive Collaborative Robotics, Leipzig, Germany, 18–22 September 2018; Ronzhin, A., Rigoll, G., Meshcheryakov, R., Eds.; Springer International Publishing: Cham, Switzerland, 2018; pp. 34–43.
4. Rodriguez-Losada, D.; Matia, F.; Jimenez, A.; Galan, R.; Lacey, G. Implementing Map Based Navigation in Guido, the Robotic SmartWalker. In Proceedings of the 2005 IEEE International Conference on Robotics and Automation, Barcelona, Spain, 18–22 April 2005; pp. 3390–3395. [CrossRef]
5. Jia, P.; Hu, H. Head gesture based control of an intelligent wheelchair. In Proceedings of the 11th Annual Conference of the Chinese Automation and Computing Society in the UK [CACSUK05], Sheffield, UK, 10 September 2005; pp. 85–90.
6. Montemerlo, M.; Pineau, J.; Roy, N.; Thrun, S.; Verma, V. Experiences with a mobile robotic guide for the elderly. *AAAI/IAAI* **2002**, *2002*, 587–592.
7. Bauer, A.; Wollherr, D.; Buss, M. Human–robot collaboration: A survey. *Int. J. Humanoid Robot.* **2008**, *5*, 47–66. [CrossRef]
8. Albu-Schäffer, A.; Haddadin, S.; Ott, C.; Stemmer, A.; Wimböck, T.; Hirzinger, G. The DLR lightweight robot: Design and control concepts for robots in human environments. *Ind. Robot. Int. J.* **2007**, *34*, 376–385. [CrossRef]
9. Nweke, H.F.; Teh, Y.W.; Mujtaba, G.; Al-garadi, M.A. Data fusion and multiple classifier systems for human activity detection and health monitoring: Review and open research directions. *Inf. Fusion* **2019**, *46*, 147–170. [CrossRef]
10. Xiao, Q.; Song, R. Action recognition based on hierarchical dynamic Bayesian network. *Multimed. Tools Appl.* **2018**, *77*, 6955–6968. [CrossRef]
11. Hu, C.; Chen, Y.; Hu, L.; Peng, X. A novel random forests based class incremental learning method for activity recognition. *Pattern Recognit.* **2018**, *78*, 277–290. [CrossRef]
12. Abidine, B.; Fergani, L.; Fergani, B.; Oussalah, M. The joint use of sequence features combination and modified weighted SVM for improving daily activity recognition. *Pattern Anal. Appl.* **2018**, *21*, 119–138. [CrossRef]
13. Ronao, C.A.; Cho, S.B. Human activity recognition using smartphone sensors with two-stage continuous hidden Markov models. In Proceedings of the 2014 10th International Conference on Natural Computation (ICNC), Xiamen, China, 19–21 August 2014; IEEE: New York, NY, USA, 2014; pp. 681–686.
14. Mohmed, G.; Lotfi, A.; Pourabdollah, A. Enhanced fuzzy finite state machine for human activity modelling and recognition. *J. Ambient. Intell. Humaniz. Comput.* **2020**, *11*, 6077–6091. [CrossRef]
15. Tan, T.H.; Gochoo, M.; Huang, S.C.; Liu, Y.H.; Liu, S.H.; Huang, Y.F. Multi-resident activity recognition in a smart home using RGB activity image and DCNN. *IEEE Sens. J.* **2018**, *18*, 9718–9727. [CrossRef]
16. Young, T.; Hazarika, D.; Poria, S.; Cambria, E. Recent trends in deep learning based natural language processing. *IEEE Comput. Intell. Mag.* **2018**, *13*, 55–75. [CrossRef]
17. Lee, S.M.; Yoon, S.M.; Cho, H. Human activity recognition from accelerometer data using Convolutional Neural Network. In Proceedings of the 2017 IEEE International Conference on Big Data and Smart Computing (Bigcomp), Jeju, Republic of Korea, 13–16 February 2017; IEEE: New York, NY, USA, 2017; pp. 131–134.
18. Inoue, M.; Inoue, S.; Nishida, T. Deep recurrent neural network for mobile human activity recognition with high throughput. *Artif. Life Robot.* **2018**, *23*, 173–185. [CrossRef]
19. Devitt, S. Trustworthiness of autonomous systems. *Foundations of Trusted Autonomy (Studies in Systems, Decision and Control, Volume 117)*; Springer: Berlin/Heidelberg, Germany, 2018; pp. 161–184.
20. Karthigasri, R.; Sornam, M. Evolutionary Model and Fuzzy Finite State Machine for Human Activity Recognition. Available online: http://www.ijcnes.com/documents/%20V8-I1-P7.pdf (accessed on 1 March 2023).
21. Kong, Y.; Fu, Y. Human action recognition and prediction: A survey. *Int. J. Comput. Vis.* **2022**, *130*, 1366–1401. [CrossRef]
22. Kostavelis, I.; Vasileiadis, M.; Skartados, E.; Kargakos, A.; Giakoumis, D.; Bouganis, C.S.; Tzovaras, D. Understanding of human behavior with a robotic agent through daily activity analysis. *Int. J. Soc. Robot.* **2019**, *11*, 437–462. [CrossRef]
23. Osman, M. Controlling uncertainty: A review of human behavior in complex dynamic environments. *Psychol. Bull.* **2010**, *136*, 65. [CrossRef]
24. Golan, M.; Cohen, Y.; Singer, G. A framework for operator–workstation interaction in Industry 4.0. *Int. J. Prod. Res.* **2020**, *58*, 2421–2432. [CrossRef]
25. Vuckovic, A.; Kwantes, P.J.; Neal, A. Adaptive decision making in a dynamic environment: A test of a sequential sampling model of relative judgment. *J. Exp. Psychol. Appl.* **2013**, *19*, 266. [CrossRef]
26. Law, T.; Scheutz, M. *Trust: Recent Concepts and Evaluations in Human-Robot Interaction*; Academic Press: New York, NY, USA, 2021; pp. 27–57. [CrossRef]
27. Kwon, W.Y.; Suh, I.H. Planning of proactive behaviors for human–robot cooperative tasks under uncertainty. *Knowl.-Based Syst.* **2014**, *72*, 81–95. [CrossRef]
28. Ramasamy Ramamurthy, S.; Roy, N. Recent trends in machine learning for human activity recognition—A survey. *Wiley Interdiscip. Rev. Data Min. Knowl. Discov.* **2018**, *8*, e1254. [CrossRef]

29. Dua, N.; Singh, S.N.; Semwal, V.B. Multi-input CNN-GRU based human activity recognition using wearable sensors. *Computing* **2021**, *103*, 1461–1478. [CrossRef]
30. Narayanan, M.R.; Scalzi, M.E.; Redmond, S.J.; Lord, S.R.; Celler, B.G.; Lovell, N.H. A wearable triaxial accelerometry system for longitudinal assessment of falls risk. In Proceedings of the 2008 30th Annual International Conference of the IEEE Engineering in Medicine and Biology Society, Vancouver, BC, Canada, 20–25 August 2008; pp. 2840–2843. [CrossRef]
31. Anguita, D.; Ghio, A.; Oneto, L.; Parra, X.; Reyes-Ortiz, J.L. Human activity recognition on smartphones using a multiclass hardware-friendly support vector machine. In *Proceedings of the International Workshop on Ambient Assisted Living*; Springer: Berlin/Heidelberg, Germany, 2012; pp. 216–223.
32. Kolekar, M.H.; Dash, D.P. Hidden markov model based human activity recognition using shape and optical flow based features. In Proceedings of the 2016 IEEE Region 10 Conference (TENCON), Singapore, 22–25 November 2016; IEEE: New York, NY, USA, 2016; pp. 393–397.
33. Abdul-Azim, H.A.; Hemayed, E.E. Human action recognition using trajectory-based representation. *Egypt. Inform. J.* **2015**, *16*, 187–198. [CrossRef]
34. Kellokumpu, V.; Pietikäinen, M.; Heikkilä, J. Human activity recognition using sequences of postures. In Proceedings of the MVA, Tsukuba Science City, Japan, 16–18 May 2005; pp. 570–573.
35. Yamato, J.; Ohya, J.; Ishii, K. Recognizing human action in time-sequential images using hidden Markov model. In Proceedings of the Proceedings 1992 IEEE Computer Society Conference on Computer Vision and Pattern Recognition, Champaign, IL, USA, 15–18 June 1992; pp. 379–385. [CrossRef]
36. Chen, K.; Zhang, D.; Yao, L.; Guo, B.; Yu, Z.; Liu, Y. Deep learning for sensor-based human activity recognition: Overview, challenges, and opportunities. *Acm Comput. Surv.* **2021**, *54*, 77. [CrossRef]
37. Parmar, A.; Katariya, R.; Patel, V. A review on random forest: An ensemble classifier. In *International Conference on Intelligent Data Communication Technologies and Internet of Things*; Springer: Berlin/Heidelberg, Germany, 2018; pp. 758–763.
38. Song, Q.; Liu, X.; Yang, L. The random forest classifier applied in droplet fingerprint recognition. In Proceedings of the 2015 12th International Conference on Fuzzy Systems and Knowledge Discovery (FSKD), Zhangjiajie, China, 15–17 August 2015; IEEE: New York, NY, USA, 2015; pp. 722–726.
39. Wan, S.; Qi, L.; Xu, X.; Tong, C.; Gu, Z. Deep learning models for real-time human activity recognition with smartphones. *Mob. Netw. Appl.* **2020**, *25*, 743–755. [CrossRef]
40. Hammerla, N.Y.; Halloran, S.; Plötz, T. Deep, convolutional, and recurrent models for human activity recognition using wearables. *arXiv* **2016**, arXiv:1604.08880.
41. Vepakomma, P.; De, D.; Das, S.K.; Bhansali, S. A-Wristocracy: Deep learning on wrist-worn sensing for recognition of user complex activities. In Proceedings of the 2015 IEEE 12th International conference on wearable and implantable body sensor networks (BSN), Cambridge, UK, 9–12 June 2015; IEEE: New York, NY, USA, 2015; pp. 1–6.
42. Bai, L.; Yao, L.; Wang, X.; Kanhere, S.P.S.; Xiao, Y. Prototype similarity learning for activity recognition. In Proceedings of the Pacific-Asia Conference on Knowledge Discovery and Data Mining, Zhangjiajie, China, 15–17 August 2015; Springer: Berlin/Heidelberg, Germany, 2020; pp. 649–661.
43. Duffner, S.; Berlemont, S.; Lefebvre, G.; Garcia, C. 3D gesture classification with convolutional neural networks. In Proceedings of the 2014 IEEE International Conference on Acoustics, Speech and Signal Processing (ICASSP), Barcelona, Spain, 4–8 May 2020; IEEE: New York, NY, USA, 2014; pp. 5432–5436.
44. Ishimaru, S.; Hoshika, K.; Kunze, K.; Kise, K.; Dengel, A. Towards reading trackers in the wild: Detecting reading activities by EOG glasses and deep neural networks. In *UbiComp '17: Proceedings of the 2017 ACM International Joint Conference on Pervasive and Ubiquitous Computing and Proceedings of the 2017 ACM International Symposium on Wearable Computers*; Association for Computing Machinery: New York, NY, USA, 2017; pp. 704–711.
45. Guan, Y.; Plötz, T. Ensembles of deep lstm learners for activity recognition using wearables. *Proc. ACM Interact. Mob. Wearable Ubiquitous Technol.* **2017**, *1*, 11. [CrossRef]
46. Ordóñez, F.J.; Roggen, D. Deep convolutional and lstm recurrent neural networks for multimodal wearable activity recognition. *Sensors* **2016**, *16*, 115. [CrossRef]
47. Hossain Shuvo, M.M.; Ahmed, N.; Nouduri, K.; Palaniappan, K. A Hybrid Approach for Human Activity Recognition with Support Vector Machine and 1D Convolutional Neural Network. In Proceedings of the 2020 IEEE Applied Imagery Pattern Recognition Workshop (AIPR), Washington, DC, USA, 13–15 October 2020; pp. 1–5. [CrossRef]
48. Faria, D.R.; Premebida, C.; Nunes, U. A probabilistic approach for human everyday activities recognition using body motion from RGB-D images. In Proceedings of the 23rd IEEE International Symposium on Robot and Human Interactive Communication, Edinburgh, UK, 25–29 August 2014; IEEE: New York, NY, USA, 2014; pp. 732–737.
49. Nunes Rodrigues, A.C.; Santos Pereira, A.; Sousa Mendes, R.M.; Araújo, A.G.; Santos Couceiro, M.; Figueiredo, A.J. Using artificial intelligence for pattern recognition in a sports context. *Sensors* **2020**, *20*, 3040. [CrossRef]
50. Vital, J.P.; Faria, D.R.; Dias, G.; Couceiro, M.S.; Coutinho, F.; Ferreira, N.M. Combining discriminative spatiotemporal features for daily life activity recognition using wearable motion sensing suit. *Pattern Anal. Appl.* **2017**, *20*, 1179–1194. [CrossRef]
51. Martinez-Gonzalez, P.; Oprea, S.; Garcia-Garcia, A.; Jover-Alvarez, A.; Orts-Escolano, S.; Garcia-Rodriguez, J. Unrealrox: An extremely photorealistic virtual reality environment for robotics simulations and synthetic data generation. *Virtual Real.* **2020**, *24*, 271–288. [CrossRef]

52. Puig, X.; Ra, K.; Boben, M.; Li, J.; Wang, T.; Fidler, S.; Torralba, A. VirtualHome: Simulating Household Activities via Programs. In Proceedings of the 2018 IEEE/CVF Conference on Computer Vision and Pattern Recognition, Salt Lake City, UT, USA, 18–22 June 2018.
53. Quigley, M.; Gerkey, B.; Conley, K.; Faust, J.; Foote, T.; Leibs, J.; Berger, E.; Wheeler, R.; Ng, A.Y. ROS: An open-source Robot Operating System. In Proceedings of the ICRA Workshop on Open Source Software, Kobe, Japan, 12–17 May 2009; Volume 3, p. 5.
54. Zangenehnejad, F.; Gao, Y. GNSS smartphones positioning: Advances, challenges, opportunities, and future perspectives. *Satell. Navig.* **2021**, *2*, 24. [CrossRef] [PubMed]
55. Kim, A.; Golnaraghi, M. A quaternion-based orientation estimation algorithm using an inertial measurement unit. In Proceedings of the PLANS 2004. Position Location and Navigation Symposium (IEEE Cat. No.04CH37556), Monterey, CA, USA, 26–29 April 2004; pp. 268–272. [CrossRef]
56. Haq, I.U.; Ullah, A.; Khan, S.U.; Khan, N.; Lee, M.Y.; Rho, S.; Baik, S.W. Sequential learning-based energy consumption prediction model for residential and commercial sectors. *Mathematics* **2021**, *9*, 605. [CrossRef]
57. Khan, I.U.; Afzal, S.; Lee, J.W. Human activity recognition via hybrid deep learning based model. *Sensors* **2022**, *22*, 323. [CrossRef] [PubMed]
58. Han, S.; Kang, J.; Mao, H.; Hu, Y.; Li, X.; Li, Y.; Xie, D.; Luo, H.; Yao, S.; Wang, Y.; et al. Ese: Efficient speech recognition engine with sparse lstm on fpga. In Proceedings of the 2017 ACM/SIGDA International Symposium on Field-Programmable Gate Arrays, Washington, DC, USA, 14–18 August 2017; pp. 75–84.
59. Berrar, D. Bayes' theorem and naive Bayes classifier. *Encycl. Bioinform. Comput. Biol. ABC Bioinform.* **2018**, *403*, 412.

**Disclaimer/Publisher's Note:** The statements, opinions and data contained in all publications are solely those of the individual author(s) and contributor(s) and not of MDPI and/or the editor(s). MDPI and/or the editor(s) disclaim responsibility for any injury to people or property resulting from any ideas, methods, instructions or products referred to in the content.

*Article*

# A Self-Adaptive Gallery Construction Method for Open-World Person Re-Identification

**Sara Casao *, Pablo Azagra, Ana C. Murillo and Eduardo Montijano**

Department of Computer Science and Systems Engineering, Universidad de Zaragoza, 50018 Zaragoza, Spain
* Correspondence: scasao@unizar.es

**Abstract:** Person re-identification, or simply re-id, is the task of identifying again a person who has been seen in the past by a perception system. Multiple robotic applications, such as tracking or navigate-and-seek, use re-identification systems to perform their tasks. To solve the re-id problem, a common practice consists in using a *gallery* with relevant information about the people already observed. The construction of this gallery is a costly process, typically performed offline and only once because of the problems associated with labeling and storing new data as they arrive in the system. The resulting galleries from this process are static and do not acquire new knowledge from the scene, which is a limitation of the current re-id systems to work for open-world applications. Different from previous work, we overcome this limitation by presenting an unsupervised approach to automatically identify new people and incrementally build a gallery for open-world re-id that adapts prior knowledge with new information on a continuous basis. Our approach performs a comparison between the current person models and new unlabeled data to dynamically expand the gallery with new identities. We process the incoming information to maintain a small representative model of each person by exploiting concepts of information theory. The uncertainty and diversity of the new samples are analyzed to define which ones should be incorporated into the gallery. Experimental evaluation in challenging benchmarks includes an ablation study of the proposed framework, the assessment of different data selection algorithms that demonstrate the benefits of our approach, and a comparative analysis of the obtained results with other unsupervised and semi-supervised re-id methods.

**Keywords:** person recognition; open-world recognition; incremental clustering

## 1. Introduction

Person re-identification, or simply re-id, addresses the problem of matching people across non-overlapping views in a multi-camera system [1,2]. Solutions to this problem benefit many robotic applications where people are involved, such as tracking [3,4], navigation [5] or searching [6,7]. An extensive number of studies have focused on obtaining the best feature representation in supervised close-world scenarios (e.g., [8–11]) where the problem is narrowed to seek a query person from an existing pool of labeled people images, generally called *gallery*. While they obtain high performance in commonly used benchmarks, from the viewpoint of practical re-id systems, people identity annotation to obtain sufficient ground truth data could be extremely inefficient [12]. Hence, there is a tendency in the research community to address other alternatives and still open problems in re-identification, such as unsupervised [13–15], domain adaptation [16–18] or open-set in open-world [19–21]. The vast majority of these works use a static and preset gallery in their development that restrains the dynamic nature of the open-world, where raw data from camera systems collect new people, detection errors, or junk data. In order to solve problems related to open-world recognition, the system needs to deal with unknown classes but also be able to incrementally self-adapt by acquiring new knowledge [22,23]. Therefore, an open-world re-identification system should automatically evolve its gallery, be able to identify new identities and update known people's data. To the best of our knowledge, existing approaches in person re-identification have not yet considered this fundamental problem of building a self-adaptive gallery. Thus, the lack of methods that address this problem motivates our research to propose a re-identification framework that focuses on the applicability of re-id approaches in open-world settings without any human assistance.

This work presents a novel framework for person re-identification focusing on a self-adaptive gallery that evolves over time in an unsupervised fashion. The presented framework is able to dynamically expand to identify new individuals and build their appearance models with representative information. Figure 1 gives an overview of the differences between a labeled and static gallery traditionally used and our proposed adaptive gallery. Unlike the static gallery, we start with an empty gallery and update its structure as new samples arrive (unlabeled person images) to acquire new knowledge. The samples that provide the most representative appearance description of each person are selected to be included in the gallery. This selection is fully unsupervised and assembled using concepts of active learning techniques. Specifically, we analyze the uncertainty and diversity of each sample to evaluate its informativeness, keeping only those that present a good balance between low uncertainty and high diversity (less likely to be failures but not redundant with the rest). The main contributions of this work are: (1) A novel approach to build a self-adaptive gallery for person re-identification in open-world scenarios. The appearance model of each person is kept small and representative by selecting those samples that are most representative using information theory concepts. (2) A thorough evaluation of the posed problem. We include a metric based on the standard precision and recall to evaluate the quality of the gallery structure. This metric provides an intuition of the final quality of the gallery structure when the problem is complex and identifying the total number of classes is highly challenging.

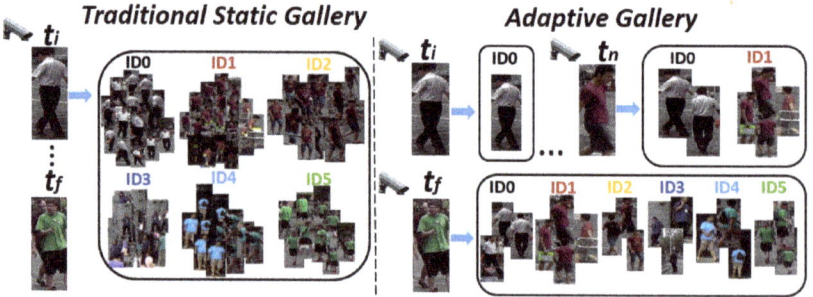

**Figure 1.** Simplified comparison between a large static gallery, traditionally used, and our small self-adaptive gallery. Both have a set of images representing each identity (ID0, ID1, . . . ), i.e., each person. The traditional gallery is the same for every person query that arrives at different times ($t_i$, $t_f$). However, because the adaptive gallery is being built and updated as new data arrives, we can appreciate a more comprehensive gallery for later times ($t_i < t_n < t_f$).

The experiments section provides a detailed analysis of the main parameters defined in the method, along with a comparison of different data selection algorithms commonly used in incremental settings. A comparison with other unsupervised and semi-supervised re-id methods is also discussed.

The rest of the paper is organized as follows. Section 2 details the related work. Section 3 describes the problem addressed, along with the main stages of the proposed framework. Section 4 presents a complete evaluation of the presented method on two challenging benchmarks. The first subsection analyzes the influence of the key parameter defined in the algorithm. Then, a comparison of different data selection methods demonstrates the benefits of our approach, and a discussion compares the proposed method with traditional approaches to re-identification. Finally, Section 5 concludes the work.

## 2. Related Work

The problem of person re-identification has been widely studied through time, as shown in [24]. Early works defined the problem as tracking [25], then moved to image-based classification [26] and video-based classification [27]. With the success of deep learning, works have shifted from hand-crafted descriptors [28] to deep learning methods [29]. The next step in person re-identification research was the shift from close-world (complete known classes and correctly annotated data) to open-world (multiple modalities, limited and noisy annotations, an undefined number of people, etc.) and has raised interesting new research challenges [22] relating the problem to other fields.

### 2.1. Unsupervised and Semi-Supervised Re-Id Methods

Several works attempt to tackle the re-id problem by building the re-id models in an unsupervised or semi-supervised manner. For example, Panda et al. [30] present a method to add a new camera to a multi-camera re-id system using unsupervised transfer learning from the knowledge obtained on the other cameras. Unsupervised algorithms typically focus on modeling the spatiotemporal information to match the people images between them [14,31], generate new data from unlabeled samples [32,33], or reduce the error in hard pseudo-labels using softer adaptable pseudo-labels [15]. Semi-supervised methods leverage the available annotated information by gradually refining the descriptors with the unlabeled data most similar to the labeled one [34] or by generating virtual samples based on the annotated data [35]. Different from these, we propose a method that focuses on creating a gallery that incrementally adds new unsupervised data, and we do not retrain the feature descriptors.

### 2.2. Incremental Person Re-Id

Incremental person re-identification has been approached from two main perspectives. First, the incremental adaptation of the learned model as new data arrives at the system [36]. This perspective trains the model in the same domain as the queries that will be analyzed later and uses a human in the loop to label the most representative data for the model adaptation through active learning techniques. Second, instead of adapting the feature representation, the goal is to perform a re-ranking in the gallery as new queries are matched with the labeled images [37]. Both perspectives use a static large gallery that ensures a match for the query person.

### 2.3. Gallery Construction

The construction of the gallery is based on the principle that instances of the same class are close in the feature space. This problem is often solved using clustering algorithms [31], which have been studied thoroughly in the literature [38,39] and applied in many fields. Close to our approach, DeCann et al. [40] present a work that updates the reference database (gallery) if the new data is not similar to any user by adding new users. However, they focus on different multi-modal information (face and finger) and an unlimited amount of data stored. To deal with the gallery construction problem in incremental scenarios, the available system resources should be taken into account since storing all the information received in a limitless fashion is not feasible. Therefore, the imposition of a bounded memory is commonly applied in many of these approaches [41,42]. Some works address the dynamical expansion of the classes aided by the labeling of the novel samples [43,44], while others also consider receiving new instances of already known classes, facing the challenges related to the update of existing class models [45,46]. They perform the update of each class model using a scoring system and controlling the size limit of each class by merging the most similar elements. This scenario is the most similar to our approach, but different from these existing works, our approach updates the model by analyzing not only the diversity of the samples but also the global uncertainty of the gallery. The result sought by combining both properties, obtaining a more varied model, is similar to that of prior work [47], which selects data with different levels of uncertainty from a set of labeled images. Different from all these methods, our approach deals with incremental and unlabeled information in an open-world scenario.

## 3. Method

This section describes in detail the addressed problem, the method overview, and the main stages of the proposed system.

### 3.1. Problem Description

We define the gallery as a set of classes, $\mathcal{C} = \{\mathcal{C}_1, \ldots, \mathcal{C}_N\}$, where each class, $\mathcal{C}_i \in \mathcal{C}$, represents one person. Each class is represented by a set of at most $m$ features $\mathcal{C}_i = \{f_i^1, \ldots, f_i^m\}$ with $f_i^j$ the $j$th feature of the class, respectively. The features are extracted from sample images, named samples for simplicity, and comprise an appearance descriptor, obtained from a generic re-id neural network, $x_i^j$, and the skeleton joints visible in the sample, $s_i^j$, $f_i^j = (x_i^j, s_i^j)$. Specifically in this work, we select the re-identification Osnet model [9] to extract the appearance descriptors, and the OpenPose network [48] to obtain the skeleton joints.

The problem is to devise a method able to incrementally create the gallery from an empty initialization as new samples arrive in the system, considering an unknown (possibly unlimited) number of classes, $N$.

### 3.2. Method Overview

The overall idea of the proposed method is represented in Figure 2. First, whenever a new sample is acquired, the associated feature, $f_q$, is obtained. Then, the method performs a classification by computing the class probability distribution of the new sample through a similarity evaluation. Based on the confidence of the classification, the system decides whether to conduct a dynamic expansion or not. Samples with high confidence enter the gallery, while samples with low confidence are sent to the unknown data manager for further analysis. The set of unknown data is periodically clustered to generate new potential classes that are compared with the existing ones to identify and initialize new classes. Finally, since there is a limit in the memory budget of $m$ features per class, the gallery optimization handles the efficient use of memory resources by deciding the relevant data to keep.

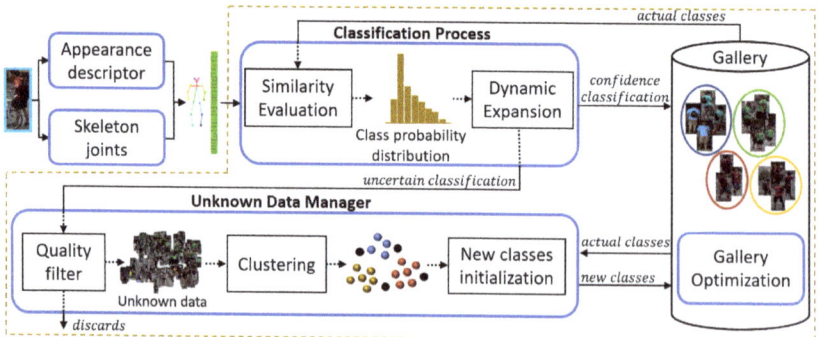

**Figure 2.** Self-adaptive gallery construction method overview. The person bounding box undergoes a pre-processing where the sample features are obtained with existing deep neural network encoders. Then, the proposed method analyzes the features obtained to decide which ones are used to adapt and evolve the gallery with the new information.

## 3.3. Classification Process

### 3.3.1. Initialization Stage

In the initial phase of the gallery construction, the low number of classes initialized does not allow to work properly with probability distributions in the general regime. Therefore, the proposed system runs a short initialization stage. In order to perform this initialization, following the incremental setup, a set of candidate-classes, $\mathcal{B} = \{\mathcal{B}_1, \ldots, \mathcal{B}_k\}$, is defined, where the first candidate-class is created with the arrival of the first sample $\mathcal{B}_1 = \{f_1^1\}$. Then, the similarity of incoming samples is evaluated by computing the cosine similarity between $x_q$ and those appearance descriptors already included in $\mathcal{B}$. If the maximum cosine similarity is greater than a threshold, $\varepsilon$, the sample is included in the corresponding candidate-class set; otherwise, a new candidate-class is initialized. As soon as a candidate-class reaches a minimum size of $l$, it becomes a person-class, i.e., a real class, belonging to the gallery $\mathcal{C} = \{\mathcal{C}_1\}$. Once the gallery reaches a minimum number of person-classes, $Q$, the proposed decision-making based on the class probabilistic distribution of the samples is run as detailed next.

### 3.3.2. General Regime

Once the gallery is initialized, the system **evaluates the similarity** of each new sample with the current gallery to obtain a probability distribution over the set of existing classes. This is accomplished using the softmax operator

$$p(x_q \in \mathcal{C}_i) \equiv p_i(x_q) = \frac{\exp(\bar{x}_i^\top x_q / v)}{\sum_{j=1}^N \exp(\bar{x}_j^\top x_q / v)}, \quad (1)$$

where $v$ is a temperature parameter that controls the softness of probability distribution over classes [31], $x_q$ is the normalized appearance descriptor of the new sample, and $\bar{x}_i$ is the weighted centroid of $\mathcal{C}_i$. Working with normalized vectors, the product of both descriptors, $\bar{x}_i^\top x_q$, is equivalent to the cosine similarity between them. In this work, the weighted centroid $\bar{x}_i$ is defined as

$$\bar{x}_i = \frac{\sum_{j=1}^m r_i^j x_i^j}{\sum_{j=1}^m r_i^j}, \quad (2)$$

$r_i^j = s_i^j / s_T$ being the ratio of joints visible in the person image bounding box with $s_i^j$ the number of detected joints and $s_T$ the total number of joints in a complete skeleton. Weighting the samples according to the number of joints favors the selection of samples with more body parts shown.

In a similar fashion to existing techniques for incremental learning [23,49], a threshold is used to control the dynamic expansion of the classes identified in the current gallery. More concretely, a simple and intuitive condition is used to measure the classification confidence of $x_q$ through its class probability distribution,

$$\frac{\max_i p_i(x_q)}{\max_{j \neq i} p_j(x_q)} \geq \tau, \quad (3)$$

where $\tau$ is the expansion threshold. Samples whose probability distribution does not comply with the condition (3) are considered doubtful and go into the pool of unknown data. Conversely, if the confidence of the classification obtained with (1) is higher or equal than $\tau$, the pseudo-label assigned to the sample corresponds to the class with maximum probability, $i^* = \arg\max_i p_i(x_q)$, and will be considered to be part of its representation model, $\mathcal{C}_{i^*}$.

## 3.4. Unknown Data Manager

Samples that do not satisfy the classification confidence criteria (3) are defined as unknown. The role of the Unknown Data Manager is to identify new identities as well as to recover samples that could not be previously classified with enough certainty. To avoid the initialization of new classes with sets of poorly-explained features, i.e., images showing only one arm or one leg, all the unknown samples first undergo a quality filter to ensure that the appearance descriptors represent at least half of a person, formally $r \geq 0.5$, $r$ being the ratio of joints.

The identification of new classes is tackled through the periodic **clustering** of the unknown data. In open-world scenarios, the number of classes is unbounded, making the use of clustering methods such as K-Means unfeasible. Thus, to partition the set of unknown data, we use a DBSCAN algorithm [50] based on sample density and can deal with noisy information. The resulting clusters that reach the minimum size of $l$ are compared with the current classes in the gallery to check whether they belong to an existing class or represent a new one. Following the analysis performed in [31] on criteria methods to decide which pair of clusters to merge, the minimum distance criterion is used to verify if a potential new class, $\mathcal{C}_w$, shares identity with any of the existing in the gallery. The minimum distance criterion takes the shortest distance between samples from the new cluster, $\mathcal{C}_w$, and all elements of the gallery, $\mathcal{C}$,

$$D(\mathcal{C}_w, \mathcal{C}) = \min_{\mathcal{C}_i \in \mathcal{C}} \left( \min_{x_j \in \mathcal{C}_i, x \in \mathcal{C}_w} \left(1 - x^\top x_j\right) \right). \tag{4}$$

Since the computational cost of this process is considerably high, we compute an approximation limiting the number of existing classes that are compared with $\mathcal{C}_w$ from the set $N$ to a subset of $k$. To select which classes are analyzed, for each $x \in \mathcal{C}_w$, we compute the $k$-Nearest centroids of the gallery and then select the $k$ most frequent classes among all of them. Using only these classes in the first minimum of (4), the computational cost remains constant with the size of the gallery.

Finally, if the approximated minimum distance is higher than $\alpha$, the cluster $\mathcal{C}_w$ is **initialized in the gallery as a new class**. Otherwise, the new cluster and the class with the closest sample represent the same identity and are merged, complying with the memory budget by means of the gallery optimization process.

## 3.5. Gallery Optimization

Our approach performs an intelligent decision-making process with the goal of storing representative features of each existing class and making efficient use of memory resources. In order to address this goal, we use two metrics that describe the relationship of each appearance descriptor with those in the same class and with all the rest.

The **first metric is the intra-class diversity** of the samples. For a descriptor, $x$, that belongs to class $\mathcal{C}_i$, we define its diversity through the minimum cosine distance among all the other descriptors that belong to the same class,

$$D_i(x) = \min_{x_j \in \mathcal{C}_i \setminus x} \left(1 - x^T x_j\right). \tag{5}$$

The diversity of the whole class is then defined as the minimum diversity among all of its features,

$$D(\mathcal{C}_i) = \min_{x_j, x_k \in \mathcal{C}_i, x_j \neq x_k} \left(1 - x_j^T x_k\right). \tag{6}$$

This metric is useful to identify redundant information, i.e., similar samples within a class. Leveraging this information, when a new sample is classified and assigned to an existing class of the gallery, $C_i$, it is only added to the representation model of the class if its diversity is greater than the current diversity of the class,

$$D_i(x_q) \geq D(C_i). \tag{7}$$

The **second metric is the uncertainty** of the sample with respect to the whole gallery, which is measured through Shannon's entropy by

$$H(x) = -\sum_{i=1}^{N} p_i(x) \log(p_i(x)), \tag{8}$$

where $N$ is the number of classes at the moment in the gallery, and $p_i(x)$ is the probability described in (1). High entropy values stand for appearance descriptors that can be easily confused with those of other classes. In contrast, a feature with low entropy indicates high confidence in belonging to a certain class. Therefore, this metric provides an intuition of the relative distance between the feature and the rest of the classes of the gallery (inter-class).

The dependency on all the classes in (8), together with the constant evolution of the class centroids required for (1), makes the computation of this metric very heavy. For efficient computation, we keep a matrix for each class, $\mathbf{R}_i$, with the cosine similarity between its samples, $x_i^j$, and all the weighted centroids of the gallery,

$$\mathbf{R}_i = \begin{bmatrix} \bar{x}_1^\top x_i^1 & \bar{x}_2^\top x_i^1 & \cdots & \bar{x}_N^\top x_i^1 \\ \bar{x}_1^\top x_i^2 & \bar{x}_2^\top x_i^2 & \cdots & \bar{x}_N^\top x_i^2 \\ \vdots & \vdots & \ddots & \vdots \\ \bar{x}_1^\top x_i^m & \bar{x}_2^\top x_i^m & \cdots & \bar{x}_N^\top x_i^m \end{bmatrix}, \tag{9}$$

as well as a list of the classes that have changed since the last gallery optimization of $C_i$ was performed. This list is used to update only the columns associated with classes with changes, noting that the other distances have not changed and can be reused. Note that the $\mathbf{R}_i$ matrix is the changing element of (1) since $v$ is a constant value. Once we compute the update of the probability distribution of the samples belonging to $C_i$, obtaining entropy with (8) is straightforward.

When the memory budget of a class is exceeded, because of a merge caused by the Unknown Data Manager or the insertion of a new sample, an optimization process using both metrics is run to decide which sample to drop. In particular, the sample to drop is

$$x^* = \arg\max_{x \subseteq C_i} \left( \gamma \frac{H(x)}{\log(1/N)} - (1-\gamma) D_i(x) \right), \tag{10}$$

where $\gamma \in [0,1]$ is a parameter to weigh the relevance of the uncertainty and the diversity terms. The logarithm, $\log(1/N)$, normalizes the entropy to a value between zero and one, equivalent to the diversity. The proposed optimization function seeks a balance between how much a given feature mixes the different classes (entropy) and how distinctive it is with respect to the rest of the features of the same class (diversity).

Figure 3 shows a simplified example with two clusters, $C_1$ and $C_2$, where $C_1$ has exceeded its size constraint $m = 3$, and two examples of the final appearance models obtained with the proposed process. Note the balance between uncertainty and diversity even though the two identities look very similar.

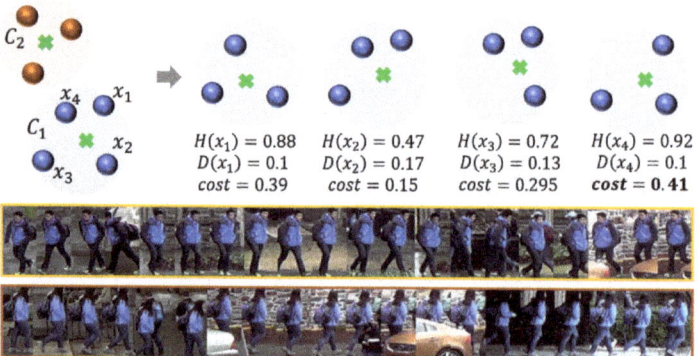

**Figure 3.** Gallery optimization. Upper area: example simplified where $C_1$ exceeds the memory budget and the gallery optimization selects the feature with the maximum cost to be dropped, $x_4$. Lower area: visual sample of two appearance models from similar identities that are correctly separated in the *DukeMTMC-VideoReID* dataset. The yellow edge corresponds to identity 86 and the orange edge to identity 194, both ground truth identities.

## 4. Experiments

This section analyzes the influence of the main parameters defined in the system, the algorithm selected to model the person's appearance and compares the performance of the proposed framework with other unsupervised and semi-supervised re-id approaches.

*4.1. Experimental Setup*

The evaluation is performed with two public benchmarks, MARS [51] and DukeMTMC-VideoReID [34]. In both of them, we use the official test set, which is split into the *query set* and the *gallery set*.

Two experiments are performed in this section. First, the analysis of the *gallery construction* process assesses the key aspects of our approach. The second experiment, *query re-identification*, runs a conventional evaluation for re-id methods in order to compare the proposed framework with other unsupervised and semi-supervised approaches. For both experiments, the settings for our approach configuration are: similarity threshold in the initialization stage $\varepsilon = 0.9$, temperature parameter in the softness operator $v = 0.1$, the $k$-Nearest centroids with $k = 3$ used by the Unknown Data Manager, distance threshold to initialize a new cluster $\alpha = 0.1$, gallery size to run the probabilistic decision making $Q = 20$, the re-identification network used in cross-domain is an OsNet model [9] trained with the MSMT17 Benchmark [52], and the OpenPose network [48] is used to obtain the skeleton joints. The setup for both experiments is detailed next.

4.1.1. Gallery Construction

The *gallery set* from both datasets is used to evaluate the self-adaptive gallery construction process. As in traditional incremental settings, the tracklets are randomly shuffled, and then, the images from each tracklet are provided one by one to simulate an incremental input to the self-adaptive gallery.

In order to evaluate the global performance of the proposed approach, we consider the following three metrics based on the classic precision, recall, and F1 score:

- **Gallery Structure**: The perfect gallery structure has one (and only one) class per ground truth identity (GT-ID). This GT-ID is set for each class with the mode of all the sample identities present at the class initialization. In order to evaluate the quality of the final gallery structure, we compute the **precision (P)**, **recall (R)**, and **F1 score** metrics as

$$P = \frac{TP}{TP+FP}, \quad R = \frac{TP}{TP+FN} \quad \text{and} \quad F1 = \frac{2 \cdot (P \cdot R)}{P+R}, \tag{11}$$

where we define the false negatives ($FN$) as those GT-ID not associated with any class, i.e., identities not found, the true positives ($TP$) as all GT-IDs associated with at least one class, i.e., identities found, and the false positives ($FP$) as the additional classes with the same GT-ID associated, i.e, two classes associated to the same GT-ID count as one $FP$ and one $TP$.

- **Class Precision**: This metric assesses the precision of the samples that enter the gallery over time. The true positives ($TP$) are the samples whose identity matches the GT-ID of the class they have been assigned, and the false positives ($FP$) are the samples that do not.
- **Sample Classification F1**: This metric evaluates the pseudo-label assigned to every sample that arrives to the system. Considering that the gallery structure often has redundancy due to the unsupervised nature of the system, we deem a limited number of redundant classes for each identity. In particular, for a given GT-ID, we only consider the $K$ classes with the highest number of samples associated with them, discarding the rest. The true positives ($TP$) are the samples that match the GT-ID with the assigned class. The false positives ($FP$) are the samples with mismatching GT-IDs, and the false negatives ($FN$) are samples classified as unknown or assigned to the discarded classes.

4.1.2. Query Re-Identification

In order to compare the proposed framework with other unsupervised and semi-supervised approaches, we use the *query set* to evaluate the gallery obtained at the end of the *gallery construction* process. Thus, the *query set* is matched with the limited size gallery created in the previous experiment, which remains static during this evaluation. The conventional evaluation for re-identification [9] is performed including the **Rank-1** and **Rank-5** metrics.

*4.2. Gallery Construction: Parameter Evaluation*

We first study the effect of the three key parameters for the gallery construction process: (1) the weight used in Equation (10) to balance the influence of the uncertainty and the diversity, $\gamma$, (2) the expansion threshold, $\tau$, in Equation (3), and (3) the minimum size required to initialize a class, $l$, used during the initialization stage and the clustering process, along with the memory budget per identity, $m$, defined in Section 3.1. In this evaluation, we use $K = 4$ for the *sample classification F1*. The goal of this analysis is to choose the parameters that yield balanced galleries based on the defined metrics.

The results of the analysis are shown in Figure 4. The influence of each parameter at the end of the process is analyzed in Figure 4a–c, where it can be seen that the trend of the quality *gallery structure F1* is inverse to the tendency of the *class precision* and the *sample classification F1*.

Figure 4a shows the effect of **weighting the uncertainty and diversity** with $\gamma$, fixing all the other parameters to $\tau = 2$, $l = 20$ and $m = 50$. The increase in $\gamma$ favors the selection of samples with low entropy but less diverse ones in the appearance models. The balance between uncertainty and diversity in the gallery is attained at $\gamma = 0.6$.

The **expansion threshold**, $\tau$, is analyzed in Figure 4b. We keep $l = 20$, $m = 50$, and from the former analysis, $\gamma$ is set to 0.6. When this parameter increases, more samples are sent to the Unknown Data Manager, resulting in the initialization of more classes. The trade-off between the metrics analyzed is accomplished at $\tau = 2$.

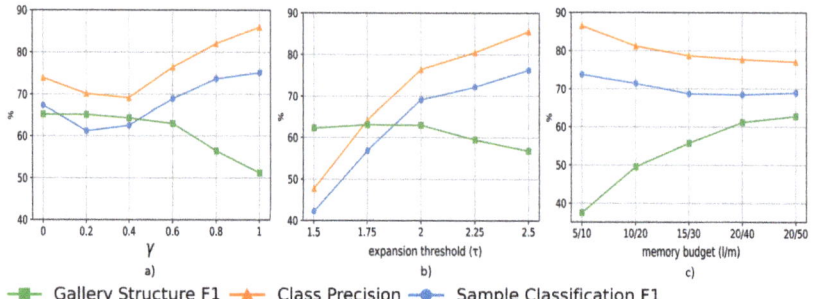

**Figure 4.** Parameter evaluation in the gallery construction process using the MARS dataset: (**a**) effect of the weight assigned to uncertainty and diversity ($\gamma$); (**b**) influence of the expansion threshold ($\tau$); (**c**) effect of the minimum size to create a class and the memory budget ($l/m$).

Finally, the influence of the minimum size to create a class, $l$, and the **memory budget** per identity, $m$, is evaluated in Figure 4c. The rest of the parameters are set to $\gamma = 0.6$, $\tau = 2$. The increase in the *gallery structure* F1 is caused by the reduction in the initialization, leading to fewer redundant classes. This implies greater confidence in the classification of the samples as $m$ increases. Therefore, the selected memory budget configuration is the one that generates the highest *gallery structure* F1, $l = 20$ and $m = 50$, the influence being not highly significant in the other metrics analyzed.

Figure 5 shows the evolution over time of the metrics with the final parameters set, $\gamma = 0.6$, $\tau = 2$, $l = 20$, and $m = 50$. Since it is an evaluation over time, in this particular case we consider $K = \infty$ for the *sample classification* F1. All the metrics settle after processing 20% of the samples. Then, it can be fairly assumed that the method's behavior is stable beyond that stage.

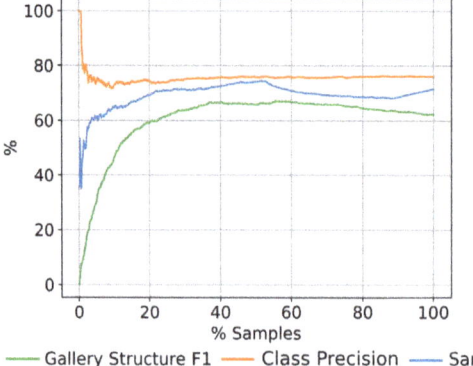

**Figure 5.** Evolution over time of the metrics with the final parameters set in the gallery construction process using the MARS dataset.

*4.3. Gallery Construction: Data Selection Method Comparison*

Following the analysis from the previous section, this experiment sets $\gamma = 0.6$, $\tau = 2$, $l = 20$, and $m = 50$. We study different gallery optimization processes that decide which sample to remove from the appearance model when the memory budget is exceeded. The compared techniques are algorithms used in incremental clustering works that have to deal with memory budget requirements. They are evaluated at the end of the gallery construction process. The first method is uniform sampling (Uniform) which saves a new feature for every $U = 5$ instance. When the size limit is exceeded, the oldest data is dropped to save a newer one. Another typical process is random decision-making (Random) which removes a random index when the memory reaches its budget. Regarding more sophisticated methods, we compare the two closest approaches in the literature, the method proposed in [45], called Incremental Object Model (IOM), and the ExStream method [46]. In both cases, we use the implementation provided by the authors to evaluate the effect of the data dropped in the gallery in our overall method. Moreover, due to the influence on the final results of the data arrival order in incremental setups, three different iterations are run (i.e., three different random data arrival orders). To make a fair comparison, all five methods use the same features extracted from OsNet [9].

First, a comprehensive analysis of the final quality of the *gallery structure* is performed. The number of classes created per GT-ID and the *gallery structure* metrics are shown in Figure 6a,b, respectively. The results in Figure 6a indicate that the ExStream and the Uniform algorithms create a high number of redundant classes in the gallery. This means that the appearance models resulting from these methods are significantly less representative, leading to more uncertain classifications. Thus, they send a high number of samples to the unknown pool and create new classes for already existing identities. The proposed optimization process (Ours) creates only one class for the same number of GT-IDs as IOM while identifying more people in the scene, which is represented by a smaller number of GT-IDs with 0 classes created. Then, derived from this analysis and verified in the *F*1 results on Figure 6b, the methods which provide a gallery structure of better quality are IOM, Random, and Ours, being Ours the one that identifies the most people in the scene among them, as measured with the *gallery structure recall*.

Second, Figure 6c shows the analysis of varying $K$ in the *sample classification F1*, and Figure 6d shows the *class precision* results. As expected, the *sample classification F1* improves in all algorithms with the increment of $K$. Comparing the methods that generate a gallery with a suitable structure, i.e., IOM, Random, and Ours, the results shown in Figure 6c,d demonstrate that the proposed gallery optimization process (Ours) outperforms IOM and Random in both metrics. Our approach is able to create more reliable people models without losing diversity, thus enhancing the classification of the samples. The ExStream and Uniform methods obtain high values in these metrics because of the large number of redundant classes, limiting in practice the actual ability to re-identify known people.

As a summary of the experiment, our algorithm is the one that maintains the best balance between having a good *gallery structure* and providing good classification metrics of the individual samples with it. The rest of the methods either generate galleries with worse quality structure, i.e., ExStream and Uniform, or obtain worse *class precision* and *sample classification F1* results, i.e., IOM and Random.

*4.4. Gallery Construction: Final Results*

A detailed evaluation on MARS and DukeMTMC-VideoReID is provided using the same hyperparameter values from the previous section for both benchmarks.

Table 1 shows the final results of the complete self-adaptive gallery construction approach on both datasets. In the *gallery structure analysis*, the table includes the number of GT-IDs, classes created and the gallery structure *F*1, the precision, and the recall. The larger number of people in DukeMTMC-VideoReID makes it more challenging to identify most of them, causing lower recall metrics than in the MARS dataset, i.e., the 80.06% of the people have been correctly identified in DukeMTMC-VideoReID against the 89.43% in MARS (*gallery structure recall*). In terms of *class precision*, note that the proposed framework obtains similar and consistent results for both datasets, 76.69% in MARS and 80.1% in DukeMTMC-VideoReID. Thus, the method creates robust appearance models, being able to correctly distinguish the people in the scene, which in turn helps in the *sample classification* obtaining precision results of 72%.

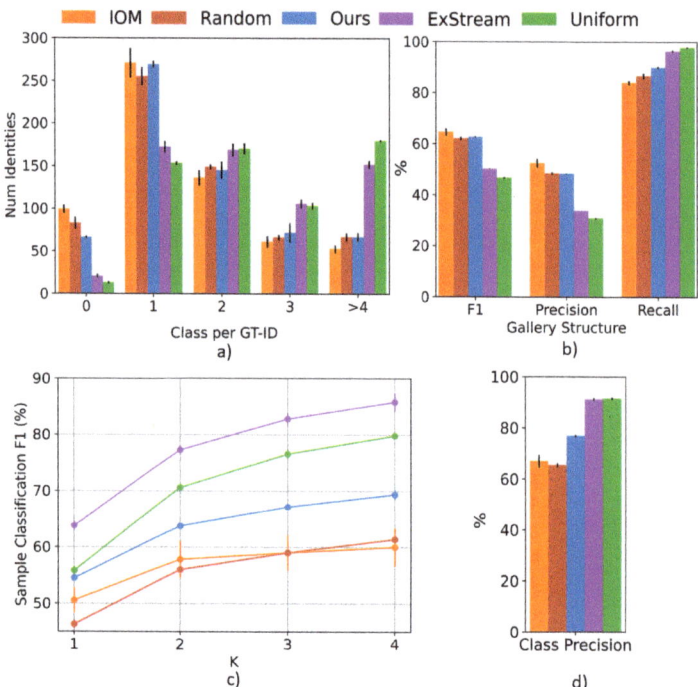

**Figure 6.** Data selection method comparison with the MARS dataset. We analyze (**a**) the number of classes created (*x*-axis) per GT-ID in the dataset (*y*-axis) showing the number of GT-ID with more than one class associated or those GT-ID that have not been correctly found, i.e., 0 classes associated; (**b**) *gallery structure* metrics: F1, precision, and recall; (**c**) *sample classification* F1 analyzing the influence of varying $K$; (**d**) *class precision*.

**Table 1.** Detailed results of the proposed framework on the MARS and DukeMTMC-VideoReID datasets. The results show the mean and the standard deviation of the three iterations performed, mean (±std).

| Metrics | Dataset | |
|---|---|---|
| | MARS | DukeMTMC-VideoReID |
| *Gallery Structure* | | |
| Total IDs (GT) | 620 | 1110 |
| Classes Created | 1147.6 (±2.5) | 1337.33 (±16) |
| F1 | 62.67 (±0.19) | 72.62 (±0.26) |
| Precision | 48.24 (±0.12) | 66.45 (±0.51) |
| Recall | 89.43 (±0.4) | 80.06 (±0.47) |
| *Class Precision* | 76.9 (±0.36) | 80.1 (±0.60) |
| *Sample Classification* | | |
| F1 | 69.4 (±0.86) | 62.6 (±0.87) |
| Precision | 72.23 (±0.12) | 72.43 (±1.12) |
| Recall | 66.8 (±1.69) | 55.21 (±0.69) |

Finally, Figure 7 includes samples of the gallery for one identity per dataset at three different times during their construction, showing in each row the person model at different times. The left identity includes an example of corruption that the gallery can suffer remarked by a discontinuous red line.

In both cases, the third row shows how our resulting gallery presents high variability of samples, resulting in a representative model for each identity. More detailed qualitative results of the proposed self-adaptive gallery can be seen in the Supplementary Material, where the identification of new classes and the evolution of the people's appearance models are shown.

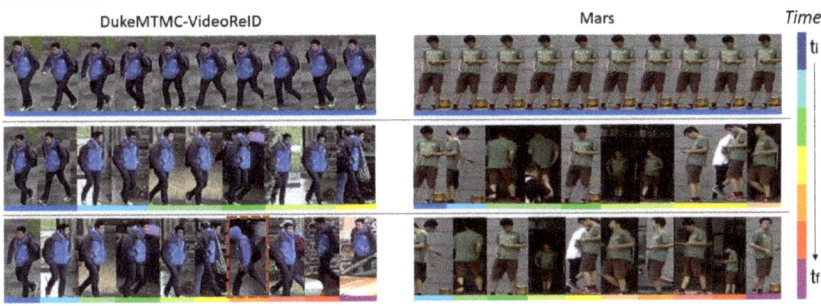

**Figure 7.** Visualization of the evolution of appearance models in the gallery. Each row corresponds to gallery samples at a certain time. The different colors represent the time stamp of the samples included in the gallery (best viewed in color).

### 4.5. Query Re-Identification

This final experiment performs the traditional evaluation of person re-id, i.e., obtains the expectation that the true match is found within the first $R$ ranks [53]. However, instead of matching the *query set* with a completely labeled gallery, the *query set* is matched with the resulting gallery from the *gallery construction* process. In this experiment, the gallery remains static. The proposed method obtains its results in an incremental unsupervised cross-domain setting (IUCD). Table 2 shows the results of this experiment, including the setting in which the different methods operate. Our offline baseline is the Full-gallery method, which has the whole gallery available and manually labeled using the same descriptors as our approach. This method is our upper bound result in the cross-domain setting. Moreover, due to the unsupervised component of our approach, we present the results of unsupervised and semi-supervised systems that perform offline training in the same domain as the *query set*. The unsupervised methods that included BUC [31], softened sim [15] and GLC+ [32] do not use any labeled data in the whole process (None). Concerning the semi-supervised approaches, they use one tracklet labeled per identity (OneEx). Note that we are the only algorithm working on the incremental unsupervised cross-domain (IUCD) setting, while the rest perform the entire process offline. Thus, although Table 2 is not a fair or direct comparison for our approach, we believe that it is interesting to see how close the proposed approach results are with respect to existing methods, despite the much more challenging and realistic scenario of our approach. The resulting values for our approach are the average and the standard deviation for the three random iterations performed previously, i.e., mean ($\pm$std). Besides, since the proposed gallery deals with memory requirements, the percentage of the gallery size used with respect to the total (GS) is shown. In this case, the standard deviation is not included, but we remark that it is lower than 0.01 in all cases.

**Table 2.** Comparison with re-id approaches in DukeMTMC-VideoReID and MARS *query set*.

| Method | Setting | DukeMTMC-VideoReID | | | MARS | | |
|---|---|---|---|---|---|---|---|
| | | GS (%) | Rank-1 | Rank-5 | GS (%) | Rank-1 | Rank-5 |
| Full-gallery | Cross-Domain | 100 | 63.2 | 72 | 100 | 66.4 | 73.3 |
| EUG [34] | OneEx | 100 | 72.7 | 84.1 | 100 | 62.67 | 74.94 |
| SCLU [54] | OneEx | 100 | 72.7 | 85 | 100 | 63.74 | 78.44 |
| BUC [31] | Unsp. (None) | 100 | 76.2 | 88.3 | 100 | 57.9 | 72.3 |
| Softened Sim [15] | Unsp. (None) | 100 | 76.4 | 88.7 | 100 | 62.7 | 77.2 |
| GLC+ [32] | Unsp. (None) | 100 | 80.9 | 91.5 | 100 | 66.5 | 78.7 |
| Ours | IUCD | 18.4 | 59.5 (±1.2) | 69.1 (±1.04) | 8.1 | 60.1 (±0.78) | 69.8 (±0.38) |

The DukeMTMC-VideoREID results show the impact of the different goals sought. In our case, the correct identification of the 1110 people that compose the gallery is a really challenging task, where some of the queries analyzed in this evaluation do not have corresponding models in the gallery. In contrast, the methods that focus on improving the feature representation obtain better results than in the MARS dataset due to the lack of distractors in the gallery. Regarding the MARS dataset, which is closer to an open-world scenario, the results with our approach are close to the unsupervised or semi-supervised approaches using two orders of magnitude less in the amount of data stored in the gallery. Finally, considering the difference between the Full-gallery baseline and our approach, we see how the proposed approach achieves comparable performance despite a much smaller (one or two orders of magnitude less) and unsupervised built gallery.

## 5. Conclusions

This work has presented a novel framework to address the problem of person re-id in open-world able to detect new identities and update the model about existing identities in the system. To deploy and evaluate intelligent systems in open-world settings, it is essential to be able to bridge certain gaps, such as lack of supervised data or lack of computational resources. In particular, the proposed approach shows how to build a self-adaptive gallery for person re-identification in a fully unsupervised fashion, while managing limited memory resources. Low supervision and resource requirements are key to robotics applications in the real world, so our self-adaptive gallery can boost robotic tasks that involve people in real-world applications, such as information gathering or searching. The main limitations of the presented work are those inherent to the re-identification systems, concerning long-term person re-id when there is a change of clothing or strong appearance changes in the people being monitored. In this situation, our system is likely to start a new class under the assumption that a new identity has appeared on the scene. Future steps to improve this aspect may include re-identification models focused on long-term robustness. In the short-term person re-identification problem, our framework can identify more than 80% of the people presented in the challenging scenarios evaluated by comparing the new unlabeled data and the existing classes in the gallery. The existing classes in the gallery are modeled with an optimization process that selects the most representative information to represent each class, balancing the uncertainty (inter-class) and the diversity (intra-class) of the samples. The experiments carried out demonstrate that the proposed optimization process returns a *class precision* of about 80% while encouraging the variability inside the classes, generating well-balanced and more structured galleries than those of the similar existing methods analyzed. The high *class precision* maintained over time aids the continuous person re-id by obtaining an *F1 sample classification* of 62.6% and 69.4% in the Mars and Duke datasets, respectively. Compared to existing re-id algorithms, our method obtains similar results to the fully labeled galleries storing one or two orders of magnitude less data.

**Supplementary Materials:** The following supporting information can be downloaded at: https://unizares-my.sharepoint.com/:v:/g/personal/scasao_unizar_es/EZWQrO_Qk4ZNjaSdFT63AQIBEHD672uPJKKTLVi6OhcdXQ?e=maXZHo, Video: Supplementary.mp4.

**Author Contributions:** Conceptualization, S.C., P.A., A.C.M. and E.M.; data curation, S.C.; formal analysis, S.C., P.A., A.C.M. and E.M.; funding acquisition, A.C.M. and E.M.; investigation, S.C., A.C.M. and E.M.; methodology, S.C., P.A., A.C.M. and E.M.; project administration, A.C.M. and E.M.; resources, S.C.; software, S.C.; supervision, P.A., A.C.M. and E.M.; validation, S.C., P.A., A.C.M. and E.M.; visualization, S.C.; Writing—original draft, S.C.; writing—review and editing, S.C., P.A., A.C.M. and E.M. All authors have read and agreed to the published version of the manuscript.

**Funding:** Funded by FEDER/Ministerio de Ciencia, Innovación y Universidades – Agencia Estatal de Investigación project PID2021-125514NB-I00, DGA T45 20R/FSE, and the Office of Naval Research Global project ONRG-NICOP-N62909-19-1-2027.

**Institutional Review Board Statement:** Not applicable.

**Informed Consent Statement:** Not applicable.

**Data Availability Statement:** The MARS people re-identification dataset is available at: http://zheng-lab.cecs.anu.edu.au/Project/project_mars.html (accessed on 18 January 2023). The DukeMTMC-VideoReID people re-identification dataset is available at: https://github.com/Yu-Wu/DukeMTMC-VideoReID (accessed on 18 January 2023).

**Acknowledgments:** This work has been supported by FEDER/Ministerio de Ciencia, Innovación y Universidades – Agencia Estatal de Investigación project PID2021-125514NB-I00, DGA T45 20R/FSE, and the Office of Naval Research Global project ONRG-NICOP-N62909-19-1-2027.

**Conflicts of Interest:** The authors declare no conflict of interest. The funders had no role in the design of the study; in the collection, analyses, or interpretation of data; in the writing of the manuscript; or in the decision to publish the results.

## Abbreviations

The following abbreviations are used in this manuscript:

| | |
|---|---|
| DBSCAN | Density-Based Spatial Clustering of Applications with Noise |
| MARS | Motion Analysis and Re-identification Set |
| GT-ID | Ground Truth Identity |
| IOM | Incremental Object Model |
| IUCD | Incremental Unsupervised Cross-Domain |
| BUC | Bottom-Up Clustering |
| EUG | Exploit Unknown Gradually |
| SCLU | one-Shot person re-identification with Collaboration between the Labeled and Unlabeled |

## References

1. Wu, D.; Zheng, S.J.; Zhang, X.P.; Yuan, C.A.; Cheng, F.; Zhao, Y.; Lin, Y.J.; Zhao, Z.Q.; Jiang, Y.L.; Huang, D.S. Deep learning-based methods for person re-identification: A comprehensive review. *Neurocomputing* **2019**, *337*, 354–371. [CrossRef]
2. Ye, M.; Shen, J.; Lin, G.; Xiang, T.; Shao, L.; Hoi, S.C. Deep learning for person re-identification: A survey and outlook. *IEEE Trans. Pattern Anal. Mach. Intell.* **2021**, *44*, 2872–2893. [CrossRef]
3. De Langis, K.; Sattar, J. Realtime multi-diver tracking and re-identification for underwater human-robot collaboration. In Proceedings of the International Conference on Robotics and Automation, Paris, France, 31 May–31 August 2020; pp. 11140–11146.
4. Chang, F.M.; Lian, F.L.; Chou, C.C. Integration of modified inverse observation model and multiple hypothesis tracking for detecting and tracking humans. *IEEE Trans. Autom. Sci. Eng.* **2015**, *13*, 160–170. [CrossRef]

5. Truong, X.T.; Ngo, T.D. Toward socially aware robot navigation in dynamic and crowded environments: A proactive social motion model. *IEEE Trans. Autom. Sci. Eng.* **2017**, *14*, 1743–1760. [CrossRef]
6. Shree, V.; Chao, W.L.; Campbell, M. Interactive Natural Language-Based Person Search. *IEEE Robot. Autom. Lett.* **2020**, *5*, 1851–1858. [CrossRef]
7. Mohamed, S.C.; Rajaratnam, S.; Hong, S.T.; Nejat, G. Person finding: An autonomous robot search method for finding multiple dynamic users in human-centered environments. *IEEE Trans. Autom. Sci. Eng.* **2019**, *17*, 433–449. [CrossRef]
8. Wen, Z.; Sun, M.; Li, Y.; Ying, S.; Peng, Y. Asymmetric Local Metric Learning with PSD Constraint for Person Re-identification. In Proceedings of the International Conference on Robotics and Automation, Montreal, QC, Canada, 20–24 May 2019; pp. 4862–4868.
9. Zhou, K.; Yang, Y.; Cavallaro, A.; Xiang, T. Learning generalisable omni-scale representations for person re-identification. *Trans. Pattern Anal. Mach. Intell.* **2021**, *44*, 5056–5069. [CrossRef]
10. Luo, H.; Jiang, W.; Zhang, X.; Fan, X.; Qian, J.; Zhang, C. Alignedreid++: Dynamically matching local information for person re-identification. *Pattern Recognit.* **2019**, *94*, 53–61. [CrossRef]
11. Hou, R.; Ma, B.; Chang, H.; Gu, X.; Shan, S.; Chen, X. Feature completion for occluded person re-identification. *IEEE Trans. Pattern Anal. Mach. Intell.* **2021**, *44*, 4894–4912. [CrossRef]
12. Leng, Q.; Ye, M.; Tian, Q. A survey of open-world person re-identification. *IEEE Trans. Circuits Syst. Video Technol.* **2019**, *30*, 1092–1108. [CrossRef]
13. Wang, D.; Zhang, S. Unsupervised person re-identification via multi-label classification. In Proceedings of the Conference on Computer Vision and Pattern Recognition, Seattle, WA, USA, 14–19 June 2020; pp. 10981–10990.
14. Sridhar Raj S, M.V.P.; Balakrishnan, R. Spatio-Temporal association rule based deep annotation-free clustering (STAR-DAC) for unsupervised person re-identification. *Pattern Recognit.* **2022**, *122*, 108287.
15. Lin, Y.; Xie, L.; Wu, Y.; Yan, C.; Tian, Q. Unsupervised person re-identification via softened similarity learning. In Proceedings of the Conference on Computer Vision and Pattern Recognition, Seattle, WA, USA, 14–19 June 2020; pp. 3390–3399.
16. Feng, H.; Chen, M.; Hu, J.; Shen, D.; Liu, H.; Cai, D. Complementary pseudo labels for unsupervised domain adaptation on person re-identification. *IEEE Trans. Image Process.* **2021**, *30*, 2898–2907. [CrossRef] [PubMed]
17. Wang, G.; Lai, J.H.; Liang, W.; Wang, G. Smoothing adversarial domain attack and p-memory reconsolidation for cross-domain person re-identification. In Proceedings of the Conference on Computer Vision and Pattern Recognition, Seattle, WA, USA, 14–19 June 2020; pp. 10568–10577.
18. Zheng, Y.; Tang, S.; Teng, G.; Ge, Y.; Liu, K.; Qin, J.; Qi, D.; Chen, D. Online pseudo label generation by hierarchical cluster dynamics for adaptive person re-identification. In Proceedings of the International Conference on Computer Vision, Montreal, BC, Canada, 11–17 October 2021; pp. 8371–8381.
19. Huang, Y.; Zha, Z.J.; Fu, X.; Hong, R.; Li, L. Real-world person re-identification via degradation invariance learning. In Proceedings of the Conference on Computer Vision and Pattern Recognition, Seattle, WA, USA, 14–19 June 2020; pp. 14084–14094.
20. Zhao, Y.; Li, Y.; Wang, S. Open-world person re-identification with deep hash feature embedding. *IEEE Signal Process. Lett.* **2019**, *26*, 1758–1762. [CrossRef]
21. Martini, M.; Paolanti, M.; Frontoni, E. Open-world person re-identification with rgbd camera in top-view configuration for retail applications. *IEEE Access* **2020**, *8*, 67756–67765. [CrossRef]
22. Bendale, A.; Boult, T. Towards open world recognition. In Proceedings of the Conference on Computer Vision and Pattern Recognition, Boston, MA, USA, 7–12 June 2015; pp. 1893–1902.
23. Fontanel, D.; Cermelli, F.; Mancini, M.; Bulo, S.R.; Ricci, E.; Caputo, B. Boosting deep open world recognition by clustering. *IEEE Robot. Autom. Lett.* **2020**, *5*, 5985–5992. [CrossRef]
24. Zheng, L.; Yang, Y.; Hauptmann, A.G. Person re-identification: Past, present and future. *arXiv* **2016**, arXiv:1610.02984.
25. Zajdel, W.; Zivkovic, Z.; Krose, B.J. Keeping track of humans: Have I seen this person before? In Proceedings of the International Conference on Robotics and Automation, Barcelona, Spain, 18–22 April 2005; pp. 2081–2086.
26. Gheissari, N.; Sebastian, T.B.; Hartley, R. Person re-identification using spatio-temporal appearance. In Proceedings of the Conference on Computer Vision and Pattern Recognition, New York, NY, USA, 17–22 June 2006; Volume 2; pp. 1528–1535.
27. Bazzani, L.; Cristani, M.; Perina, A.; Farenzena, M.; Murino, V. Multiple-shot person re-identification by hpe signature. In Proceedings of the International Conference on Pattern Recognition, Istanbul, Turkey, 23–26 August 2010; pp. 1413–1416.
28. Zhao, R.; Ouyang, W.; Wang, X. Unsupervised salience learning for person re-identification. In Proceedings of the Conference on Computer Vision and Pattern Recognition, Portland, OR, USA, 23–28 June 2013; pp. 3586–3593.
29. Yi, D.; Lei, Z.; Liao, S.; Li, S.Z. Deep metric learning for person re-identification. In Proceedings of the International Conference on Pattern Recognition, Stockholm, Sweden, 24–28 August 2014; pp. 34–39.
30. Panda, R.; Bhuiyan, A.; Murino, V.; Roy-Chowdhury, A.K. Adaptation of person re-identification models for on-boarding new camera(s). *Pattern Recognit.* **2019**, *96*, 106991. [CrossRef]
31. Lin, Y.; Dong, X.; Zheng, L.; Yan, Y.; Yang, Y. A bottom-up clustering approach to unsupervised person re-identification. In Proceedings of the AAAI Conference on Artificial Intelligence, Honolulu, HI, USA, 27–1 February 2019; Volume 33; pp. 8738–8745.
32. Chen, H.; Wang, Y.; Lagadec, B.; Dantcheva, A.; Bremond, F. Learning Invariance from Generated Variance for Unsupervised Person Re-identification. *arXiv* **2022**, arXiv:2301.00725v1.

33. Zhang, X.; Li, D.; Wang, Z.; Wang, J.; Ding, E.; Shi, J.Q.; Zhang, Z.; Wang, J. Implicit sample extension for unsupervised person re-identification. In Proceedings of the Proceedings of the Conference on Computer Vision and Pattern Recognition, New Orleans, LO, USA, 19–24 June 2022; pp. 7369–7378.
34. Wu, Y.; Lin, Y.; Dong, X.; Yan, Y.; Ouyang, W.; Yang, Y. Exploit the unknown gradually: One-shot video-based person re-identification by stepwise learning. In Proceedings of the Conference on Computer Vision and Pattern Recognition, Salt Lake City, UT, USA, 18–22 June 2018; pp. 5177–5186.
35. Han, H.; Zhou, M.; Shang, X.; Cao, W.; Abusorrah, A. KISS+ for rapid and accurate pedestrian re-identification. *IEEE Trans. Intell. Transp. Syst.* **2020**, *22*, 394–403. [CrossRef]
36. Martinel, N.; Das, A.; Micheloni, C.; Roy-Chowdhury, A.K. Temporal model adaptation for person re-identification. In Proceedings of the European Conference on Computer Vision, Amsterdam, The Netherlands, 11–14 October 2016; pp. 858–877.
37. Wang, Z.; Jiang, J.; Yu, Y.; Satoh, S. Incremental re-identification by cross-direction and cross-ranking adaption. *IEEE Trans. Multimed.* **2019**, *21*, 2376–2386. [CrossRef]
38. Xu, R.; Wunsch, D. Survey of clustering algorithms. *IEEE Trans. Neural Netw.* **2005**, *16*, 645–678. [CrossRef] [PubMed]
39. Xu, D.; Tian, Y. A comprehensive survey of clustering algorithms. *Ann. Data Sci.* **2015**, *2*, 165–193. [CrossRef]
40. DeCann, B.; Ross, A. Modelling errors in a biometric re-identification system. *IET Biom.* **2015**, *4*, 209–219. [CrossRef]
41. Rebuffi, S.A.; Kolesnikov, A.; Sperl, G.; Lampert, C.H. icarl: Incremental classifier and representation learning. In Proceedings of the Conference on Computer Vision and Pattern Recognition, Honolulu, HI, USA, 21–26 July 2017; pp. 2001–2010.
42. Belouadah, E.; Popescu, A. IL2M: Class incremental learning with dual memory. In Proceedings of the International Conference on Computer Vision, Seoul, South Korea, 27 October–2 November 2019; pp. 583–592.
43. Mancini, M.; Karaoguz, H.; Ricci, E.; Jensfelt, P.; Caputo, B. Knowledge is never enough: Towards web aided deep open world recognition. In Proceedings of the International Conference on Robotics and Automation, Montreal, QC, Canada, 20–24 May 2019; pp. 9537–9543.
44. Valipour, S.; Perez, C.; Jagersand, M. Incremental learning for robot perception through HRI. In Proceedings of the International Conference on Intelligent Robots and Systems, Vancouver, BC, Canada, 24–28 September 2017; pp. 2772–2777.
45. Azagra, P.; Civera, J.; Murillo, A.C. Incremental Learning of Object Models From Natural Human–Robot Interactions. *IEEE Trans. Autom. Sci. Eng.* **2020**, *17*, 1883–1900. [CrossRef]
46. Hayes, T.L.; Cahill, N.D.; Kanan, C. Memory efficient experience replay for streaming learning. In Proceedings of the International Conference on Robotics and Automation, Montreal, QC, Canada, 20–24 May 2019; pp. 9769–9776.
47. Bang, J.; Kim, H.; Yoo, Y.; Ha, J.W.; Choi, J. Rainbow Memory: Continual Learning with a Memory of Diverse Samples. In Proceedings of the Conference on Computer Vision and Pattern Recognition, Nashville, TN, USA, 19–25 June 2021; pp. 8218–8227.
48. Cao, Z.; Hidalgo Martinez, G.; Simon, T.; Wei, S.; Sheikh, Y.A. OpenPose: Realtime Multi-Person 2D Pose Estimation using Part Affinity Fields. *arXiv* **2019**, arXiv:1812.08008v2.
49. Rao, D.; Visin, F.; Rusu, A.; Pascanu, R.; Teh, Y.W.; Hadsell, R. Continual Unsupervised Representation Learning. *Adv. Neural Inf. Process. Syst.* **2019**, *32*, 7647–7657.
50. Ester, M.; Kriegel, H.P.; Sander, J.; Xu, X. A density-based algorithm for discovering clusters in large spatial databases with noise. In Proceedings of the KDD, Portland OR, USA, 2–4 August 1996; Volume 96; pp. 226–231.
51. Zheng, L.; Bie, Z.; Sun, Y.; Wang, J.; Su, C.; Wang, S.; Tian, Q. Mars: A video benchmark for large-scale person re-identification. In Proceedings of the European Conference on Computer Vision, Amsterdam, The Netherlands, 11–14 October 2016; pp. 868–884.
52. Wei, L.; Zhang, S.; Gao, W.; Tian, Q. Person transfer gan to bridge domain gap for person re-identification. In Proceedings of the Conference on Computer Vision and Pattern Recognition, Salt Lake City, UT, USA, 18–22 June 2018; pp. 79–88.
53. Hirzer, M.; Beleznai, C.; Roth, P.M.; Bischof, H. Person re-identification by descriptive and discriminative classification. In Proceedings of the Scandinavian Conference on Image Analysis, Ystad, Sweden, 23–27 May 2011; pp. 91–102.
54. Yin, J.; Li, B.; Wan, F.; Zhu, Y. A new data selection strategy for one-shot video-based person re-identification. In Proceedings of the International Conference on Image Processing, Taipei, Taiwan, 22–25 September 2019; pp. 1227–1231.

**Disclaimer/Publisher's Note:** The statements, opinions and data contained in all publications are solely those of the individual author(s) and contributor(s) and not of MDPI and/or the editor(s). MDPI and/or the editor(s) disclaim responsibility for any injury to people or property resulting from any ideas, methods, instructions or products referred to in the content.

Article

# A Study on the Role of Affective Feedback in Robot-Assisted Learning

Gabriela Błażejowska [1], Łukasz Gruba [2], Bipin Indurkhya [3,*] and Artur Gunia [3]

1 Nextbank Software, 30-085 Krakow, Poland
2 Kitopi, 30-383 Krakow, Poland
3 Cognitive Science Department, Institute of Philosophy, Jagiellonian University, 31-007 Krakow, Poland
* Correspondence: bipin.indurkhya@uj.edu.pl

**Abstract:** In recent years, there have been many approaches to using robots to teach computer programming. In intelligent tutoring systems and computer-aided learning, there is also some research to show that affective feedback to the student increases learning efficiency. However, a few studies on the role of incorporating an emotional personality in the robot in robot-assisted learning have found different results. To explore this issue further, we conducted a pilot study to investigate the effect of positive verbal encouragement and non-verbal emotive behaviour of the Miro-E robot during a robot-assisted programming session. The participants were tasked to program the robot's behaviour. In the experimental group, the robot monitored the participants' emotional state via their facial expressions, and provided affective feedback to the participants after completing each task. In the control group, the robot responded in a neutral way. The participants filled out a questionnaire before and after the programming session. The results show a positive reaction of the participants to the robot and the exercise. Though the number of participants was small, as the experiment was conducted during the pandemic, a qualitative analysis of the data was carried out. We found that the greatest affective outcome of the session was for students who had little experience or interest in programming before. We also found that the affective expressions of the robot had a negative impact on its likeability, revealing vestiges of the uncanny valley effect.

**Keywords:** human–robot interaction; programming education; social robots; Miro-E; emotion recognition; affective computing

## 1. Introduction

There is a long history of using robots to teach computer programming to children and college students [1–5]. A robot is a tangible, physical device that can be programmed to make different movements, and display different behaviours. This makes robots more interesting to novice programmers compared to writing "Hello World" on a display.

In using a robot as a vehicle to teach programming, one critical factor is what kind of personality should be given to the robot to make it more effective. Previous research on intelligent tutoring systems has demonstrated that an affective interface yields better learning outcomes [6]. However, for a robot tutor, the results are mixed. Some studies have found that endowing a robot with an emotional personality is effective [7,8], but others were not able to find any significant effect [9,10].

To explore this issue further, we conducted a pilot study where a dog-like robot (Miro-E) was used to teach programming to children (11–15 yrs) under two different conditions: a neutral-personality condition and an emotional-feedback condition. In the emotional-feedback condition, the robot sensed the emotional state of the students through their facial expressions, and gave encouragement through verbal and non-verbal modalities. The verbal feedback took the form of praising the student, and non-verbal feedback included wagging the tail, moving the head, and wiggling the ears. Throughout the experiment, we monitored the emotional state of the students through their facial expressions.

This study was conducted during the pandemic, so the number of participants was small. However, we carried out a qualitative analysis of our observations, which is reported here.

This paper is organised as follows. In Section 2 we review the related research. Section 3 presents our experimental design, followed by the details of the experiment in Section 4. The results and discussion are presented in Sections 5 and 6, respectively, followed by the conclusions in Section 7.

## 2. Related Work

### 2.1. Robots in Education

Over the years, there have been many attempts at using robots for educational purposes. For example, in an older study [7], a robot was used to teach an artificial language to primary-school students. The robot was designed to offer two levels of social behaviours—neutral and supportive. Participants who studied with the supportive robot achieved significantly higher results and reported higher motivation levels.

In an earlier survey [3], it was observed that 74% of the reviewed studies found support for robots as an effective teaching tool. A later survey [5] reported that the introduction of robotics in the curriculum increases children's interest in engineering, and allows children to engage in interactive and engaging learning experiences.

Sharma et al. [11], while studying how collaboration and engagement affect children's attitudes towards programming, asked the children to manipulate digital robots (avatars) as a priming activity before starting programming exercises. Van den Berghe et al. [12] directly compared how robots as opposed to avatars affected children's cooperation while learning programming, and found robots to be more effective than avatars.

According to a meta-analysis of studies on the efficacy of social robots in education [13], robot tutors are not at the same level as human tutors: students show lower learning outcomes when directly comparing studying with a robot versus a human tutor. However, there are some benefits of robots over humans in education. It is more economically viable to provide devices to each student than one-on-one tutoring with a human teacher. Technology also allows the curriculum to be customised to the learning pace of each student.

Robots have an advantage over screen-based educational applications, because they increase cognitive learning gains [14] and elicit more social engagement from students [15], compared to screen educational content. The use of robots also appears to be effective for interactive courses where technology is the subject of the course. In this case, the robots engage students in critical and computational thinking, problem solving, and collaboration [16–19]. Moreover, the use of robots is motivating for both the student and the teacher designing the course [20,21].

In particular, when it comes to teaching programming, physical devices have an advantage that the student can see the effect of executing an algorithm. Devices such as micro:bit [22] and robots [19,23] have been found to be effective.

As robots do not yet have the capability to be general all-round teachers and perform better than humans, many studies choose to compare two different robot behaviours with each other, instead of measuring one robot behaviour against a human tutor. For example, one could compare a socially supportive behaviour that engages in a social dialogue with a neutral behaviour that focuses on a plain knowledge transfer [7]. Or, one could compare the tutor condition—a robot that is focused on guiding a learner in solving increasingly complex problems in a scaffolding fashion—with the peer-like behaviour to support engagement [7,8]. This is the approach adapted in our study.

### 2.2. The Role of Affective Feedback

Affective feedback is known to have a major impact when a human is the teacher or the trainer [24,25]. In computer-based tutoring systems, affective feedback is also found to be effective [6].

In human–robot interaction, however, incorporating an emotional personality into the robot has yielded mixed results. For instance, Saerbeck et al. [7] found that incorporating a life-like social personality in a virtual actor increases the learning efficiency of students. Zaga et al. [8] compared the effect of two different social personalities of a robot—a peer and a tutor—and found the peer personality to be more effective. However, Konijn & Hoorn [10] used the humanoid robot NAO to teach primary-school children the multiplication table. Their study compared a robot using neutral language (providing feedback with only variations of 'correct' and 'incorrect') to a more social and encouraging robot. Techniques used to create social interaction included addressing the participant by name, having the robot follow their gaze, and using encouraging gestures and language (such as 'well done', 'fantastic'). The results showed no significant difference between the two groups when comparing across all participants. Students with below-average test scores performed worse with social robot tutoring than with a neutral robot. Similarly, another study [9] did not find benefits of social behaviours of the robot for a lesson on prime numbers.

One reason for this effect might be that social behaviours from a robot can distract from the lesson and increase cognitive load. It could also be that students are surprised or unsettled by robots showing such behaviours. Studies showing a lack of benefit from social behaviours in a robot [10,26] have compared the cognitive outcomes, measuring differences in test scores. In our study, we chose to focus on qualitative feedback from the participants to assess the effect of affective feedback on learning.

*2.3. Emotion Recognition*

Knowing the emotional state of a student is important from the point of view of teaching. Having the ability to recognise if a student is bored, frustrated, excited, or in any other emotional state is a valuable skill for every teacher. For example, if a student is bored, it could be an indicator that they have lost focus or that they may already be familiar with a particular topic and are ready to move on [27]. Another example could be when a student is frustrated, which most probably means that they are experiencing some difficulties with the learning material.

There are several aspects of emotions and many available techniques for measuring them [28]. One key issue is the dimensions of emotions, and the literature [29] provides the following list: (1) arousal—whether an emotion turns on, activates an action, or inhibits it; (2) value—whether an emotion has positive or negative value for a person; (3) intensity—whether the strength with which the emotion is perceived is low or high; (4) duration—time duration of a given emotion; (5) frequency of occurrence—how often does a given emotion occur; (6) time dimension—whether the emotion is retrospective (e.g., relief), real (e.g., pleasure), or prospective (e.g., hope). Another factor is the set of basic emotions in terms of which all other emotions can be expressed. The Plutchik model [30] provides one such set of basic emotions.

Tools for measuring emotions can be divided into three groups: (1) psychological, mainly subjective, and retrospective reporting of one's own emotional states (e.g., via verbal reports, questionnaires); (2) physiological objective tests that measure physiological responses using sensors (e.g., electrocardiogram (ECG), electroencephalogram (EEG), galvanic skin response (GSR)); and (3) behavioural objective measures based on bodily manifestations (e.g., facial expression, voice prosody, body posture) [31].

A commonly used emotion recognition technique is based on the set of emotions proposed by Paul Ekman [32,33], and it has been used successfully in educational tests [34,35]. It is a discrete model with six basic emotions: anger, disgust, fear, happiness, sadness, and surprise. Later on, the list was expanded to include emotions of contempt, guilt, embarrassment, relief, or satisfaction. However, the original model is often used, especially for automated emotion recognition [36], as it is based on a relatively small set of well-defined and easily distinguishable states.

The Ekman model can be used in automated techniques for detecting emotions, which in practice consist of detecting emotions from changes in a facial micro-expression (a facial expression that only lasts for a short moment). Detecting emotion from a micro-expression is not without its drawbacks, such as the ability to assess only basic emotions or the fact that it does not always work for all respondents. However, thanks to information technology (IT) solutions, it is a quick and relatively simple method used to evaluate emotions in changing conditions. Though emotions can be recognised from facial images using automated techniques, some sort of an image and pattern recognition algorithm has to be involved. Creating any image recognition algorithm manually can be difficult and error prone. In recent years, the most popular approach to this is to use machine learning [37–39].

Dimensional models present a different approach to classifying emotions. As opposed to discrete models, where emotions are defined as distinct states, in dimensional models all emotional states are described by two or more dimensions. Thus, emotional states form a spectrum rather than separate groups. The dimensions used to describe emotions are usually based on intensity and whether the emotion is positive or negative. One of the dimensional models was proposed by James Russell [40]. It is known as the circumplex model of emotions [40]. It is a two-dimensional model where the dimensions are valence (whether the emotion is positive or negative) and arousal (the level of activation, e.g., calm vs. excited). Placing valence on the horizontal axis and arousal on the vertical axis, all emotions are placed in the circular space defined by these two dimensions [41]. This is the model used in our study.

## 3. Study Design

The main objective of this research was to study how the emotional response of the robot affects the learning process and the emotional attitude of the student. More specifically, our aim was to address whether the process of learning can be more effective when assisted by an AI that can engage emotionally with the student by managing the difficulty of the task based on the emotional feedback and by providing encouraging verbal feedback. Consequently, the hypothesis of our study was that participants who receive encouragement and emotional support from an empathetic robot will be more engaged in the lessons, will be less frustrated by failures, and will have a higher interest in continuing their development in the field of computer science.

To validate this hypothesis, we conducted a study where children (11–15 yrs) were asked to complete the task of programming a robot while interacting with it. A Miro-E robot was used in the study; this is described below.

### 3.1. The Miro-E Robot

Miro-E is a small robot developed by Consequential Robotics (Figure 1), and has animal-like features. It is designed to look like a hybrid of different pet animals. It has an articulated head with ears and eyes, haptic sensors in the head and the body that react to touch, and a microphone to detect sound. The eyes, ears, tail, and head can be moved to express affective states. The robot's behaviour can be programmed using a block interface based on the Blockly library. By combining sensor readings with conditional logic, one can create programs so that the robot reacts when touched or when it hears a clap. The programming interface also allows for running separate scripts on the robot in parallel: this feature is used so that the robot can run programs written by the participants during the study while running reaction scripts to provide affective feedback at the same time.

The reaction scripts were written in Python (URL: https://www.python.org/ accessed on 4 January 2023), using the Rospy library (URL: http://wiki.ros.org/rospy accessed on 4 January 2023) and the Miro Interface modules to control the hardware. The script files were uploaded to the robot's memory, and were executed on demand by running them with a Python interpreter from the command line via secure shell protocol (SSH).

**Figure 1.** Miro-E robot.

The study participants used MiroCode (URL: https://www.miro-e.com/mirocode accessed on 4 January 2023) (Figure 2) to program the robot. MiroCode is a visual interface that uses a block representation of the robot's actions. A program is created by chaining together blocks that describe sequential actions of the robot. This web application was created by Consequential Robotics specifically for teaching programming using Google's open-source library Blockly (URL: https://developers.google.com/blockly accessed on 4 January 2023).

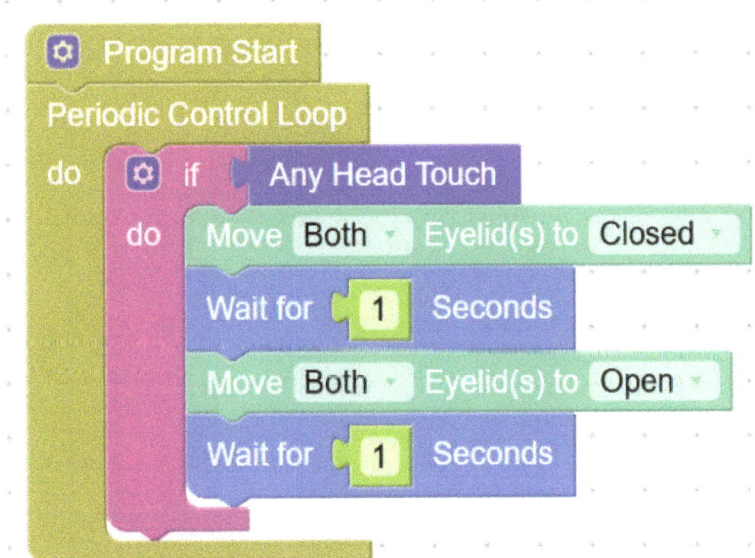

**Figure 2.** An example program written in MiroCode's visual interface.

Robot capabilities are divided into separate modules grouped by themes (motion, time sequence, sensors, etc.). The programming interface is friendly to beginners as it requires little knowledge of syntax, and each block explains the action taken by the robot. The blocks can be executed in a sequence or in a loop. The cloud version of MiroCode, MiroCloud, was used in this study.

## 3.2. Emotion Recognition Module

The emotion recognition module implemented for this study was divided into two modules running in parallel. The first module was a standalone client-side application that is responsible for collecting facial images from the laptop camera and generating their valence and arousal values. The second module was a server-side application responsible for storage, managing the collected data, and computing end results. The architecture is shown in Figure 3.

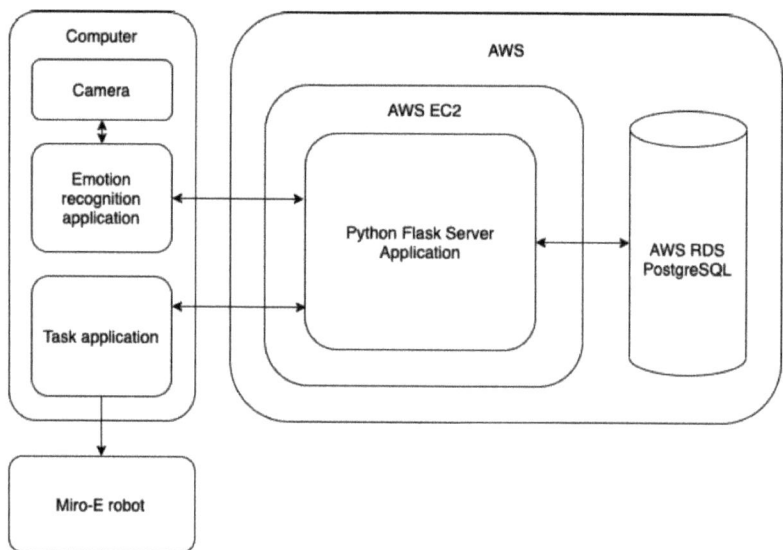

**Figure 3.** Architecture diagram. Please refer to the text for an explanation of the acronyms.

The client-side module ran on the open-source software library Keras (Keras library accessed on 15 December 2022; URL: https://keras.io/about/) with trained models for valence and arousal. The video was captured from the laptop camera and the frames were continuously fed into the model, which returned the valence and arousal values for each frame. These values were sent to the server-side application via the http protocol. It prepared the records and saved them in the database for further calculations. The server-side application ran on Amazon Web Services (AWS) (cloud computing with AWS accessed on 15 December 2022; URL: https://aws.amazon.com/what-is-aws/). A single Amazon Elastic Compute Cloud (EC2) (Amazon EC2 accessed on 15 December 2022; URL: https://aws.amazon.com/ec2/) instance served as the host for a Python application implemented in the Flask (Flask accessed on 15 December 2022; URL: https://palletsprojects.com/p/flask/) framework. It exposed a set of REST services (What is REST, accessed on 15 December 2022; URL: https://restfulapi.net/) which were being called by the client-side application (emotion recognition application). An instance of the AWS Relational Database Service (RDS) (Amazon Relational Database Service (RDS) accessed on 15 December 2022; URL: https://aws.amazon.com/rds/) was hosted in the cloud, and was used to store all the data collected through the experiment in the open-source database system PostgreSQL (URL: https://www.postgresql.org/ accessed on 15 December 2022).

The server-side application also returned the results for a given time-frame. Timestamps were taken when a student started and ended solving a given task. These were attached to the request sent to the server-side application, which retrieved all the data during this time period to compute the end result according to the algorithm explained in Section 3.2.2.

### 3.2.1. Training the Emotion Recognition Model

For recognising emotions from facial images, a machine learning model was trained using AffectNet, which is currently the largest facial expression data set, with each image annotated with a categorical label, and its valence and arousal values based on the circumplex model [42].

The AffectNet data set contains more than one million images, of which 440,000 were annotated manually, and the rest were annotated automatically. We observed that these images are distributed unevenly across the valence and arousal spectrum: most of the images covered a small range of valence and arousal values, and there were few images with very high or very low valence and arousal values. Such an uneven distribution of data is not ideal for training.

To address this problem, we divided the entire range of values (from $-1.0$ to $1.0$) into small intervals (in steps of 0.01, so $-1.00$ to $-0.99$, $-0.99$ to $-0.98$, and so on) and considered how many images fell in each interval. With trial and error, we found that by taking at most 400 images from each interval (when an interval had less than 400 images, we took all of them), we could create a more uniform distribution across the entire spectrum, and still have a large enough dataset to train the model. This procedure was performed once across the valence spectrum and once across the arousal spectrum.

To address the problem that the images were of different resolutions, we scaled all the images down to $200 \times 200$ pixels. As the Xception network is designed for images of size $299 \times 299$ pixels, the first input layer had to be readjusted to work with different image sizes. As a separate data set had to be used for training the valence model and the arousal model, all the operations mentioned before had to be repeated twice, once for each model. This resulted in two final data sets that could be used during training, one for valence and the other for arousal.

A random split was performed to divide the data sets into training and validation categories: 80% were assigned to the training data set and the remaining 20% to the validation data set.

The parameter values used for the network were as follows. The network used was an Xception pre-trained on the ImageNet data set (ImageNet accessed on 15 December 2022; URL: https://www.image-net.org/). The batch size was set to 32 images. The loss function used during training was mean absolute error (MAE) [43]. The Paperspace (About Paperspace Gradient, accessed on 15 December 2022; URL: https://docs.paperspace.com/gradient/) platform was used to provide more computing capacity. Training occurred on a single Free GPU + instance (Instance Types available in the Free Tier, accessed on 15 December 2022; URL: https://docs.paperspace.com/gradient/more/instance-types/free-instances#instancetypes-available-in-the-free-tier) equipped with 8 CPUs, 30 GB RAM, and a Quadro M4000 GPU. The value loss achieved after training was 0.244 for the valence model and 0.258 for the arousal model. Considering that the values for both parameters have a range from $-1.0$ to $1.0$, this translates into a 12.2% error rate for the valence model and 12.9% for the arousal model.

### 3.2.2. Emotion Computing Algorithm

An algorithm was created to aggregate valence and arousal values over time into one of the three categories, positive, negative, or neutral. These aggregated values were used by the experiment control system to control the task flow.

Only measurements taken while solving a particular task were considered. We assumed that the participants would have a neutral facial expression most of the time. A high-pass filter was used to filter out measurements of low significance. Euclidean distance was used for filtering as follows:

$$\sqrt{valence^2 \times arousal^2} \geq 0.3 \qquad (1)$$

Different threshold values were tried, and in the end, 0.3 was chosen as a good compromise between filtering out noise and not filtering too much. This allowed most of the unimportant measurements to be filtered out.

Weights were used to give more significance to the measurements taken at the end of the task, as they would be related to the participant finding the final solution. A hyperbolic tangent function was used to calculate the weights as follows:

$$weight(t) = tanh(rel\_time(t) \times \pi) \qquad (2)$$

where *rel_time* is a function to compute the relative time of the measurement compared to the entire duration of the task execution. The relative time was calculated using the formula:

$$rel\_time(t) = \frac{t - t_{min}}{t_{max} - t_{min}} \qquad (3)$$

where $t_{min}$ and $t_{max}$ are the start and end times of task execution.

After computing the weights for every measurement taken for a particular task, the final result was computed as follows.

$$result = \frac{\sum_{i=1}^{m\_size} m_i \times weight(t_i)}{\sum_{j=1}^{m\_size} weight(t_j)} \qquad (4)$$

where *m_size* is the size of the measurement set, $m_i$ is a particular measurement value, and $t_i$ is the measurement time.

This formula was applied to the valence and arousal measurements separately, and the calculated values were used to determine the final outcome depending on where the results fell in the circumplex model. A visualisation of this is shown in Figure 4.

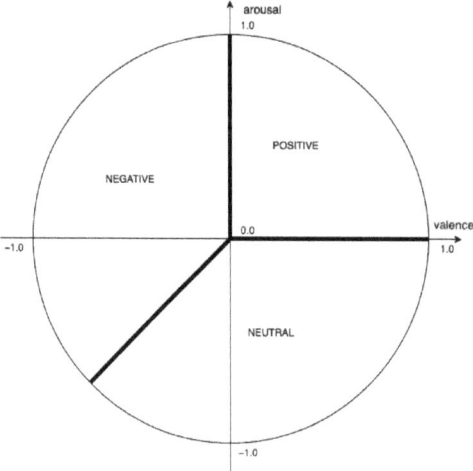

**Figure 4.** Visualisation of the valence-arousal plane for the final result calculation.

*3.3. Experimental Set-Up*

The main activity in this study was for the participants to solve programming tasks to control the Miro-E robot in a MiroCode environment. While they were engaged in this task, their emotional states were analysed from their facial images (taken by the laptop camera). Based on these emotional states, the robot provided appropriate affective feedback in both verbal and non-verbal modalities. The feedback could be praising the participant for completing a task successfully, or congratulating them on finishing a tricky task. Non-verbal feedback included wagging the tail, moving the head, and wiggling the ears. The robot then presented the next task to the participant.

The system also decided whether the participant should skip some tasks. If the participant completed the current task in a short time (less than the preset threshold), and the emotion recognition module found that the participant had a positive reaction, then the system would skip over the next task.

For the control group, the Miro-E robot was only a vehicle for the programming tasks (it only performed the actions programmed by the participant): it took no actions of its own. The robot announced the next task in a neutral manner: 'Start Task 3'.

All participants were also asked to fill out a pre-test questionnaire to assess their previous programming experience, and a post-test questionnaire to gather information about their experience during the study. Data collected from the questionnaires were used to determine the impact of affective feedback provided to participants in the experimental group.

## 4. Experiment

### 4.1. Participants

The experiment was conducted in the period June–July 2021. Because of the COVID-19 pandemic, it was difficult to find participants, but we managed to recruit nine participants (3F, 6M; 11–15 yrs) to take part in the activity of learning to program a robot. All participants had sufficient English knowledge to allow them to use the robot's interface. English proficiency was determined from the participant's self-declaration—the call for participation mentioned that English would be required. Moreover, all the participants had attended several years of primary school with mandatory English lessons. Four participants were placed in the control group and five in the experimental group.

### 4.2. Coding Tasks

The participants were given the task of writing programs to control the Miro-E Robot. A set of ten programming tasks were prepared, which were expected to take about 40 min to complete. The tasks were progressively more difficult, with later tasks building on earlier tasks, and with each section of the worksheet introducing new concepts or block modules.

The first task in each section was to open an example program that had been saved on the laptop, read the code, and explain what it does to the researcher. Then, the participant was asked to run the program on the robot and observe if their predictions were correct. This familiarised the participant with what the blocks in the program did, and provided context for the next task.

The second task in each section was a coding exercise. The participant was given a desired behaviour of the robot (for example, to make Miro-E walk in a square), and was asked to write a program to make Miro-E behave in that way. The participants were asked to do this independently: the researcher helped (if needed) only with language difficulties.

The final section, *Functions*, did not have a given outcome of the task, but introduced the concept of functions as reusable blocks of code that are defined once, but then they can be called from different places in the program. The example code contained three function slots that participants could complete as they wanted using the knowledge gained from the previous sections. The objective of this task was for the participants to understand how a function code is triggered from the main program.

### 4.3. Procedure

Informed consent was obtained from the legal guardian of each participant after explaining to them the following aspects of the study:

- The video of facial expressions from the laptop camera is used for emotion-recognition.
- No identifying information of the participants is disclosed in the study.
- Data from the participants are used anonymously and in aggregate.

Before starting with the programming tasks, each participant was asked to fill out a pre-test questionnaire containing the following questions (items 3 and 4 used a five-point Likert scale):

1. Participant identification number.
2. Grade in school.
3. General interest in programming (1: 'no interest'; 5: 'great interest').
4. Previous coding experience (1: 'no experience'; 5: 'much experience').
5. Familiar programming environments.

Then, the participant was asked to sit at a large table with a laptop that showed the MiroCode interface, with the Miro-E robot on an adjacent table. Throughout the study, a researcher sat next to the participant to explain the experiment, help with language problems, and provide input to the system when starting or finishing a task.

The experiment was started with the researcher showing the participant the worksheet with tasks, and explaining the structure and the goal of each section. The participant was then given a tour of the MiroCode interface: where to find blocks, how to run the code on the robot, and how to open example programs. Participants were encouraged to explore solutions even when they were not sure about their correctness. Finally, the researcher told the participant to feel free to ask any questions about the language or meaning of certain blocks during the experiment.

The participant then started the first task. When the participant started a task, the researcher entered this into the system. The participant then completed the task, usually running several versions of their code on the Miro-E robot before succeeding. In tasks that involved moving the robot, the researcher positioned the robot next to the participant, or in a location where the robot could complete the movement without encountering an obstacle.

When the participant completed a task, the researcher entered the task-end response into the system using a mobile device.

This procedure was repeated until all the tasks were completed. The procedure was the same for both the experimental and the control groups, with the only difference being that (as explained above in Section 3.3) for the experimental group, Miro-E provided affective feedback, and the progression of tasks depended on the participant's affective state while completing the tasks.

After the participant finished all the tasks, they were asked to complete the following post-test questionnaire (items (1)–(4) used a five-point Likert scale):

1. Rate Miro-E's behaviour (1: 'unpleasant/rude'; 5: 'pleasant/nice').
2. Rate your enjoyment of the session (1: 'didn't like it at all'; 5: 'liked it a lot').
3. Rate your general interest in programming (1: 'no interest'; 5: 'great interest').
4. Would you be interested in another session with Miro? (1: 'no interest'; 5: 'great interest').
5. Did any task make you feel frustrated?
6. If yes, which task(s)?
7. Did any task make you feel accomplished?
8. If yes, which task(s)?

## 5. Results

As this study was conducted during the COVID-19 pandemic, the groups were relatively small: four participants were in the control group and five were in the experimental group. Nonetheless, we analysed the affective outcomes: the feelings of the participants towards the Miro-E robot, towards the experiment, and towards programming in general. The results of the post-test questionnaire are summarised in Table 1.

All the participants finished all the given tasks—no one stopped the experiment before it ended. For one participant in the control group, all the data could not be recorded due to a technical difficulty, so this participant was excluded from the analysis. None of the participants reported feeling frustrated by any of the tasks.

Table 1. Average responses to the survey.

| Question | Total | Control | Experimental |
|---|---|---|---|
| Miro-E likeability | 4.78 | 5 | 4.6 |
| Session enjoyment | 4.44 | 4.25 | 4.6 |
| Programming experience | 2.22 | 1.75 | 2.6 |
| Interest in programming (pre-test) | 3.77 | 3.75 | 3.8 |
| Interest in programming (post-test) | 4.11 | 4 | 4.2 |

Six participants (4 from the experimental group, 2 from the control group) reported that a task made them feel accomplished or happy. One participant from the experimental group reported that task one, moving the robot, affected her or him in this way. Five participants (3 from the experimental group, 2 from the control group) pointed to task 10 as making them feel accomplished.

The aggregate (over the participants) of the data collected by the emotion recognition module is shown as heat maps in Figures 5 and 6. These heat maps can be interpreted qualitatively by comparing a segment of the obtained valence/arousal predictions to the ground truth values. Therefore, the heat map illustrates, in the 2-D valence and arousal space, the histograms of the ground truth labels of the test set and the corresponding predictions of the trained model. We can see that the heat points are mostly in the middle because the measurements were mostly neutral or shifted towards negative valence and arousal.

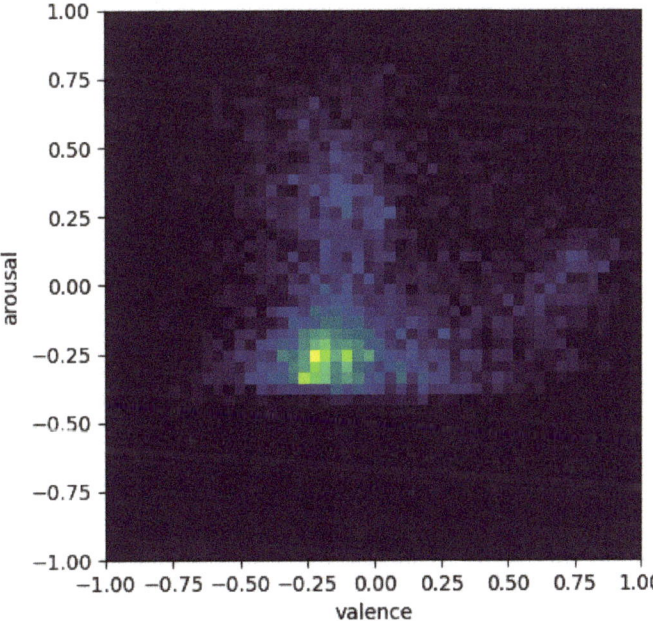

Figure 5. Heatmap showing the valence and arousal for the control group participants.

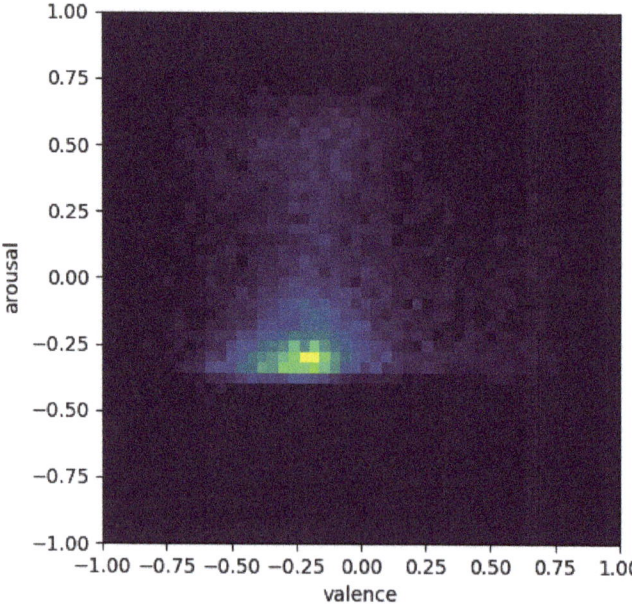

**Figure 6.** Heatmap showing the valence and arousal for the experimental group participants.

## 6. Discussion

The results of the survey show that all the participants rated Miro-E's behaviour as friendly: on a five-point Likert scale with a range from "unpleasant, unfriendly" to "nice, friendly", the average response was 4.8. In the control group, where the Miro-E robot exhibited only neutral behaviour and language, all the participants responded with a 5. The average in the experimental group was 4.6 (with two participants rating Miro-E's friendliness at a 4). This suggests that the robot's attempts at friendliness had a negative impact on its likeability. One participant in particular seemed visibly surprised, and moved away from the robot when starting the script for completing a task. This could be due to the uncanny valley effect [44].

These responses show that the Miro-E robot is perceived as friendly by itself, even when no additional behaviour to support this is programmed. This is by the design of the manufacturer, as the robot is aimed at younger children, and looks like a pet animal. Moreover, the tasks the participants were performing made the robot appear more friendly. The participants themselves were in charge of programming the robot and used its emotive features in their programs—making the robot wag its tail and wink in response to being touched. This kind of social behaviour did not trigger the same surprise reaction, as it was expected and programmed by the participant.

Another question asked in the survey related to the participants' enjoyment of the programming session and whether they would like to take part in another session with the Miro-E robot. The average for the control group was 4.25, and for the experimental group was 4.6. Thus, fewer participants from the experimental group expressed an interest in a future lesson with Miro-E. One reason for this could be that prior experience with programming was higher in the experimental group compared to the control group, and for participants having more prior experience with programming, the tasks seemed easy, so they were not so interested in another session with Miro-E.

The aggregated heat maps of valence and arousal for participants in the control and the experimental groups (Figures 5 and 6) show that the participants in the control group experienced higher overall valence and arousal values, while the experimental group's heat map is concentrated mostly in the neutral region.

This suggests that the control group experienced more positive emotions compared to the experimental group. However, this could also mean that emotion recognition based on computer vision and facial micro-expressions was not very effective. In future, we need to incorporate other measures that are indicative of attention and interest besides the emotional state of the user.

The results from the survey conflict with the results of the valence and arousal graphs. This may be due to the courtesy bias, as the participants were rating the study while the researchers were in the room. Moreover, it is hard to ascertain satisfaction with the robot-based learning session by just comparing the results of the pretest and post-test surveys, especially as the emotional state of the participants was changing during the session.

Nonetheless, these results can be explained as follows. The survey results show considerable interest and satisfaction with the robot-assisted learning. However, the emotional recognition based on facial expressions suggest that a robotic assistant does not trigger a strong emotional state. This can also be interpreted positively in that the robot assistant does not distract from the required task.

The affective outcome of the study was measured by asking the participants about their interest in programming before and after the session (on a five-point Likert scale) with the robot. This showed an overall increase of 0.34 for the entire group: the increase was 0.25 for the control group and 0.4 for the experimental group. It should be noted that most of the participants did not change their answer (from pretest to post-test), but the participants with a low pre-test interest in programming showed an increase in the post-test.

In response to the question about which task made them feel most accomplished, most participants chose the last task. This was also confirmed by a graph of valence and arousal for one of the participants, as shown in Figure 7. This could be due to the recency bias, or because the last task was a free-form task where the participants could implement a behaviour of their own choosing.

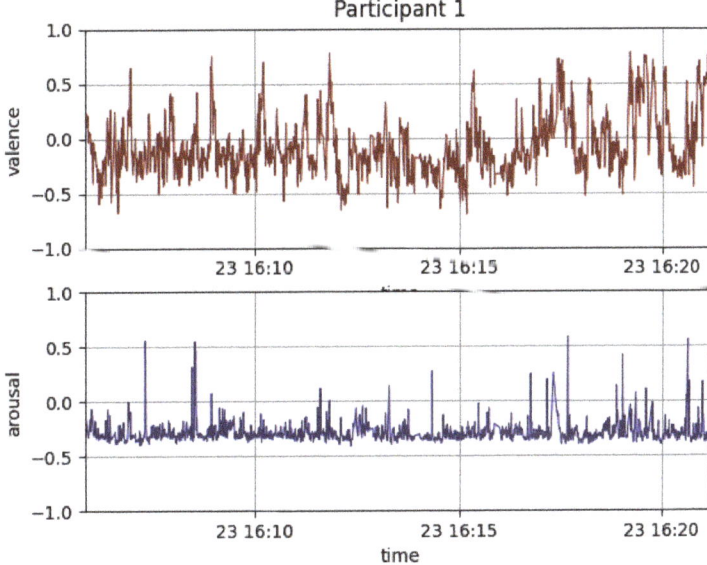

**Figure 7.** Graphs of valence and arousal over time.

Figure 7 shows that the valence increased towards the end of the study. Peaks in the arousal value were more frequent in the later part of the session. In the free-form task 10, the participants were most interested in using emotive features of the robot—wagging the tail, closing eyes, and moving ears. This suggests that the participants preferred to program social behaviours in a robot.

The emotion recognition module worked well when the facial expression clearly indicated a strong emotion like happy, angry, or sad. However, with micro-expressions, the changes in valence and arousal were very small and could be considered as noise. It was observed that the participants' faces were mostly neutral during the experiment, and their facial expressions barely changed regardless of whether they were doing well with the tasks or were facing difficulties.

## 7. Conclusions

The goal of this research was to study how affective feedback by an educational robot impacts learning outcomes; for example, in our particular case, does it lead to more interest in programming and computer science?

Our experimental data suggest that the Miro-E robot was perceived as friendly and likeable. However, we did not find a significant impact of the affective feedback of the robot on the participants—most of the differences can be attributed to other factors, such as programming experience or interest.

We did not find a strong positive or negative correlation of the robot's behaviour to the participants' responses. Previous studies in this area also report conflicting outcomes [7,9,45], suggesting that the link between social behaviour from robots and better outcomes for students is not so straightforward.

Our results show that the greatest affective outcome of the session was for students who had little experience or interest in programming before. This suggests that to maximise the impact of robot-assisted learning, it should be introduced early on. We also found that the affective expressions of the robot had a negative impact on its likeability.

### 7.1. Limitations

The sample size of nine participants was too small to draw statistical conclusions.

The feedback given by the robot was short and simple. This might have not been enough to generate observable effects, as most of the time the robot acted the same in both the experimental and the control groups.

As the study involved only one forty-minute session, the participants did not have much time to become confident with the robot's programming interface.

Having a more uniform level of programming experience among the participants would have allowed creating tasks that were not too easy or too difficult for any participant, thereby eliminating one source of variations in the responses.

Participants were assigned to the experimental and the control groups before filling out the pre-test survey. Two participants with the most programming experience were placed in the experimental group. Due to the small sample size, the groups ended up with an unbalanced skill level and this difference had a visible effect on the answers.

### 7.2. Future Research

A study involving more participants over several sessions would answer the questions posed in this study with higher confidence. Future research could explore whether students who are learning programming with a robot would benefit more from the robot exhibiting social behaviours on its own, or from programming the robot to behave in social ways. Comparing a friendly-looking robot like Miro-E with a less inherently friendly-looking robot could give insight into how much the appearance of the robot influences its effectiveness as an educational tool.

Another issue for the future research is to study the potential drawbacks of affective feedback. It has been shown that too much emotional engagement from the robot can be a disadvantage and can lead to an increased cognitive load on the participant and worse outcomes [46]. Ethicists also argue that too much emotionality can lead to false relationships [47], or that the emotionality of robots is simply false [48]. These considerations must be taken into account when designing an affective educational robot.

**Author Contributions:** Conceptualization, G.B., Ł.G., A.G. and B.I.; methodology, G.B., Ł.G., A.G. and B.I.; software, G.B. and Ł.G.; validation, A.G. and B.I.; formal analysis, G.B. and Ł.G.; investigation, G.B. and Ł.G.; resources, G.B., Ł.G., A.G. and B.I.; data curation, G.B. and Ł.G.; writing—original draft preparation, G.B., Ł.G., A.G. and B.I.; writing—review and editing, A.G. and B.I.; visualization, Ł.G., G.B. and A.G.; supervision, A.G. and B.I.; project administration, A.G. and B.I. All authors have read and agreed to the published version of the manuscript.

**Funding:** This research was supported by a grant from the Priority Research Area DigiWorld PSP: U1U/P06/NO/02.19 under the Strategic Programme Excellence Initiative at the Jagiellonian University.

**Institutional Review Board Statement:** Ethical review and approval were waived for this study because no personal data was being collected and no invasive tests were being carried out.

**Informed Consent Statement:** Informed consent was obtained from the parents of the participants before the study.

**Data Availability Statement:** The experimental data is available by contacting the corresponding author.

**Conflicts of Interest:** The authors declare no conflict of interest.

# References

1. Belmar, H. Review on the Teaching of Programming and Computational Thinking in the World. *Front. Comput. Sci.* **2022**, *4*, 128. [CrossRef]
2. Holo, O.; Kveim, E.; Lysne, M.; Taraldsen, L.; Haara, F. A review of research on teaching of computer programming in primary school mathematics: Moving towards sustainable classroom action. *Educ. Inq.* **2022**, 1–16. [CrossRef]
3. Major, L.; Kyriacou, T.; Brereton, O.P. Systematic literature review: Teaching novices programming using robots. *IET Softw.* **2012**, *6*, 502–513. [CrossRef]
4. Pachidis, T.; Vrochidou, E.; Kaburlasos, V.; Kostova, S.; Bonković, M.; Papić, V. Social robotics in education: State-of-the-art and directions. In Proceedings of the International Conference on Robotics in Alpe-Adria Danube Region, Patras, Greece, 6–8 June 2018; Springer: Berlin/Heidelberg, Germany, 2018; pp. 689–700.
5. Toh, L.P.E.; Causo, A.; Tzuo, P.W.; Chen, I.M.; Yeo, S.H. A review on the use of robots in education and young children. *J. Educ. Technol. Soc.* **2016**, *19*, 148–163.
6. Jiménez, S.; Juárez-Ramírez, R.; Castillo, V.H.; Ramírez-Noriega, A.; Márquez, B.Y.; Alanis, A. The Role of Personality in Motivation to use an Affective Feedback System. *Program. Comput. Softw.* **2021**, *47*, 793–802. [CrossRef]
7. Saerbeck, M.; Schut, T.; Bartneck, C.; Janse, M.D. Expressive robots in education: Varying the degree of social supportive behavior of a robotic tutor. In Proceedings of the SIGCHI Conference on Human Factors in Computing Systems, Atlanta, GA, USA, 10–15 April 2010; pp. 1613–1622.
8. Zaga, C.; Lohse, M.; Truong, K.P.; Evers, V. The effect of a robot's social character on children's task engagement: Peer versus tutor In Proceedings of the International Conference on Social Robotics, Paris, France, 26–30 October 2015; Springer: Berlin/Heidelberg, Germany, 2015; pp. 704–713.
9. Kennedy, J.; Baxter, P.; Belpaeme, T. The robot who tried too hard: Social behaviour of a robot tutor can negatively affect child learning. In Proceedings of the 2015 10th ACM/IEEE International Conference on Human-Robot Interaction (HRI), Portland, OR, USA, 2–5 March 2015; pp. 67–74.
10. Konijn, E.A.; Hoorn, J.F. Robot tutor and pupils' educational ability: Teaching the times tables. *Comput. Educ.* **2020**, *157*, 103970. [CrossRef]
11. Sharma, K.; Papavlasopoulou, S.; Giannakos, M. Coding games and robots to enhance computational thinking: How collaboration and engagement moderate children's attitudes? *Int. J. Child-Comput. Interact.* **2019**, *21*, 65–76. [CrossRef]
12. Van den Berghe, R.; Petersen, H.; Hellendoorn, A.; van Keulen, H. Programming a robot or an avatar: A study on learning outcomes, motivation, and cooperation. In Proceedings of the Companion of the 2020 ACM/IEEE International Conference on Human-Robot Interaction, Cambridge, UK, 23–26 March 2020; pp. 496–498.
13. Belpaeme, T.; Kennedy, J.; Ramachandran, A.; Scassellati, B.; Tanaka, F. Social robots for education: A review. *Sci. Robot.* **2018**, *3*, eaat5954. [CrossRef]

14. Leyzberg, D.; Spaulding, S.; Toneva, M.; Scassellati, B. The physical presence of a robot tutor increases cognitive learning gains. In Proceedings of the Annual Meeting of the Cognitive Science Society, Sapporo, Japan, 1–4 August 2012; Volume 34.
15. Kennedy, J.; Baxter, P.; Belpaeme, T. Comparing robot embodiments in a guided discovery learning interaction with children. *Int. J. Soc. Robot.* **2015**, *7*, 293–308. [CrossRef]
16. Durak, H.Y.; Yilmaz, F.G.K.; Yilmaz, R. Computational thinking, programming self-efficacy, problem solving and experiences in the programming process conducted with robotic activities. *Contemp. Educ. Technol.* **2019**, *10*, 173–197. [CrossRef]
17. Fanchamps, N.; Specht, M.; Hennissen, P.; Slangen, L. The Effect of Teacher Interventions and SRA Robot Programming on the Development of Computational Thinking. In Proceedings of the International Conference on Computational Thinking Education, Online, 19–20 August 2020; The Education University of Hong Kong: Hong Kong, China, 2020; pp. 69–72.
18. Lee, P.T.; Low, C.W. Implementing a computational thinking curriculum with robotic coding activities through non-formal learning. In Proceedings of the International Conference on Computational Thinking Education, Online, 19–20 August 2020 ; The Education University of Hong Kong: Hong Kong, China, 2020; pp. 150–151.
19. Noh, J.; Lee, J. Effects of robotics programming on the computational thinking and creativity of elementary school students. *Educ. Technol. Res. Dev.* **2020**, *68*, 463–484. [CrossRef]
20. Bernstein, D.; Mutch-Jones, K.; Cassidy, M.; Hamner, E. Teaching with robotics: Creating and implementing integrated units in middle school subjects. *J. Res. Technol. Educ.* **2020**, 161–176. [CrossRef]
21. Kucuk, S.; Sisman, B. Pre-Service Teachers' Experiences in Learning Robotics Design and Programming. *Inform. Educ.* **2018**, *17*, 301–320. [CrossRef]
22. Kalelioglu, F.; Sentance, S. Teaching with physical computing in school: The case of the micro: Bit. *Educ. Inf. Technol.* **2020**, *25*, 2577–2603. [CrossRef]
23. Kong, S.C.; Wang, Y.Q. Nurture interest-driven creators in programmable robotics education: An empirical investigation in primary school settings. *Res. Pract. Technol. Enhanc. Learn.* **2019**, *14*, 20. [CrossRef]
24. Trewick, N.; Neumann, D.L.; Hamilton, K. Effect of affective feedback and competitiveness on performance and the psychological experience of exercise within a virtual reality environment. *PLoS ONE* **2022**, *17*, e0268460. [CrossRef]
25. Yu, S.; Jiang, L. Doctoral students' engagement with journal reviewers' feedback on academic writing. *Stud. Contin. Educ.* **2022**, *44*, 87–104. [CrossRef]
26. Schouten, A.P.; Portegies, T.C.; Withuis, I.; Willemsen, L.M.; Mazerant-Dubois, K. Robomorphism: Examining the effects of telepresence robots on between-student cooperation. *Comput. Hum. Behav.* **2022**, *126*, 106980. [CrossRef]
27. Pekrun, R. Emotions and Learning. Educational Practices Series-24; UNESCO International Bureau of Education. 2014. Available online: https://eric.ed.gov/?id=ED560531 (accessed on 16 January 2023).
28. Jiménez, S.; Juárez-Ramírez, R.; Castillo, V.H.; Armenta, J.J.T. *Affective Feedback in Intelligent Tutoring Systems: A Practical Approach*; Springer: Berlin/Heidelberg, Germany, 2018.
29. Hascher, T. Learning and emotion: Perspectives for theory and research. *Eur. Educ. Res. J.* **2010**, *9*, 13–28. [CrossRef]
30. Plutchik, R. The nature of emotions: Human emotions have deep evolutionary roots, a fact that may explain their complexity and provide tools for clinical practice. *Am. Sci.* **2001**, *89*, 344–350. [CrossRef]
31. Feidakis, M.; Daradoumis, T.; Caballé, S. Emotion measurement in intelligent tutoring systems: What, when and how to measure. In Proceedings of the 2011 Third International Conference on Intelligent Networking and Collaborative Systems, Fukuoka, Japan, 30 November–2 December 2011; pp. 807–812.
32. Ekman, P. Facial expressions of emotion: New findings, new questions. *Psychol. Sci.* **1992**, *3*, 34–38.
33. Ekman, P. Emotions revealed. *BMJ* **2004**, *328*, 0405184. [CrossRef]
34. Arroyo, I.; Cooper, D.G.; Burleson, W.; Woolf, B.P.; Muldner, K.; Christopherson, R. Emotion sensors go to school. In Proceedings of the Artificial Intelligence in Education, Brighton, UK, 6–10 July 2009; Ios Press: Amsterdam, The Netherlands, 2009; pp. 17–24.
35. O'Sullivan, M.; Ekman, P.; Geher, G. Facial expression recognition and emotional intelligence. In *Face Recognition: New Research*; Leeland, K.B., Ed.; Nova Science Publishers: Hauppauge, NY, USA, 2008; pp. 25–40.
36. Oh, S.; Lee, J.Y.; Kim, D.K. The design of CNN architectures for optimal six basic emotion classification using multiple physiological signals. *Sensors* **2020**, *20*, 866. [CrossRef] [PubMed]
37. Jain, D.K.; Shamsolmoali, P.; Sehdev, P. Extended deep neural network for facial emotion recognition. *Pattern Recognit. Lett.* **2019**, *120*, 69–74. [CrossRef]
38. Liliana, D.Y. Emotion recognition from facial expression using deep convolutional neural network. *J. Phys. Conf. Ser.* **2019**, *1193*, 012004. [CrossRef]
39. Ozdemir, M.A.; Elagoz, B.; Alaybeyoglu, A.; Sadighzadeh, R.; Akan, A. Real time emotion recognition from facial expressions using CNN architecture. In Proceedings of the 2019 IEEE Medical Technologies Congress (tiptekno), Izmir, Turkey, 3–5 October 2019; pp. 1–4.
40. Russell, J.A. A circumplex model of affect. *J. Personal. Soc. Psychol.* **1980**, *39*, 1161. [CrossRef]
41. Scherer, K.R. What are emotions? And how can they be measured? *Soc. Sci. Inf.* **2005**, *44*, 695–729. [CrossRef]
42. Mollahosseini, A.; Hasani, B.; Mahoor, M.H. Affectnet: A database for facial expression, valence, and arousal computing in the wild. *IEEE Trans. Affect. Comput.* **2017**, *10*, 18–31. [CrossRef]
43. Willmott, C.J.; Matsuura, K. Advantages of the mean absolute error (MAE) over the root mean square error (RMSE) in assessing average model performance. *Clim. Res.* **2005**, *30*, 79–82. [CrossRef]

44. Mori, M.; MacDorman, K.F.; Kageki, N. The uncanny valley [from the field]. *IEEE Robot. Autom. Mag.* **2012**, *19*, 98–100. [CrossRef]
45. Leite, I.; Castellano, G.; Pereira, A.; Martinho, C.; Paiva, A. Empathic robots for long-term interaction. *Int. J. Soc. Robot.* **2014**, *6*, 329–341. [CrossRef]
46. Lanteigne, C. Social Robots and Empathy: The Harmful Effects of Always Getting What We Want. 2019. Available online: https://montrealethics.ai/social-robots-and-empathy-the-harmful-effects-of-always-getting-what-we-want/ (accessed on 16 January 2023).
47. Sharkey, A.; Sharkey, N. We need to talk about deception in social robotics! *Ethics Inf. Technol.* **2021**, *23*, 309–316. [CrossRef]
48. Elder, A. False friends and false coinage: A tool for navigating the ethics of sociable robots. *ACM SIGCAS Comput. Soc.* **2016**, *45*, 248–254. [CrossRef]

**Disclaimer/Publisher's Note:** The statements, opinions and data contained in all publications are solely those of the individual author(s) and contributor(s) and not of MDPI and/or the editor(s). MDPI and/or the editor(s) disclaim responsibility for any injury to people or property resulting from any ideas, methods, instructions or products referred to in the content.

*Article*

# Lightweight Multimodal Domain Generic Person Reidentification Metric for Person-Following Robots

Muhammad Adnan Syed [1,2], Yongsheng Ou [1,3,*], Tao Li [2] and Guolai Jiang [1]

1. Shenzhen Institutes of Advanced Technology, Chinese Academy of Sciences, Shenzhen 518055, China
2. Konka R&D Department, Konka Group Co., Ltd., Shenzhen 518053, China
3. Guangdong Provincial Key Laboratory of Robotics and Intelligent System, Shenzhen 518055, China
* Correspondence: ys.ou@siat.ac.cn

**Abstract:** Recently, person-following robots have been increasingly used in many real-world applications, and they require robust and accurate person identification for tracking. Recent works proposed to use re-identification metrics for identification of the target person; however, these metrics suffer due to poor generalization, and due to impostors in nonlinear multi-modal world. This work learns a domain generic person re-identification to resolve real-world challenges and to identify the target person undergoing appearance changes when moving across different indoor and outdoor environments or domains. Our generic metric takes advantage of novel attention mechanism to learn deep cross-representations to address pose, viewpoint, and illumination variations, as well as jointly tackling impostors and style variations the target person randomly undergoes in various indoor and outdoor domains; thus, our generic metric attains higher recognition accuracy of target person identification in complex multi-modal open-set world, and attains 80.73% and 64.44% *Rank*-1 identification in multi-modal close-set PRID and VIPeR domains, respectively.

**Keywords:** person re-identification; impostor resisting metric; multi-modal re-identification metric; lightweight domain generic metric; part-wise attention learning

## 1. Introduction

With the advent of deep learning, human–robot interaction (HCI) is increasing rapidly in many applications. A robot following a person is one such application [1,2], where the person-following robots assist humans in elderly assistance and healthcare, work as service robots in industrial use, and also serve as autonomous carts in shopping malls.

Clearly, for all the applications above, it is required to track the person, and for tracking the person, the fundamental step is to first accurately identify the target person *P*1 shown in Figure 1, and then robustly track the target *P*1 in real-world. However, the dynamic real world is highly nonlinear and multimodal where the appearance of target person *P*1 shown in Figure 1 drastically varies from indoor home environment or domain to outdoor domain, such as walking across *Road*1, *Road*2, and *Road*3 in Figure 1, as well as across different outdoor domains, such as shopping mall or airport due to continually varying styles, illumination, poses, and viewpoints. In Figure 1, it can also be noted that the real world is also crowded where the target person *P*1 is also either occluded by other persons, say occluded by distractor *D*1 at time *t*2 on walking across *Road*1, or is also occluded by impostor *I*1 at time *t*4. Thus, in the real world in Figure 1, due to occlusion and nearby impostors, tracking the target person *P*1 is a very difficult task, hence, state-of-the-art trackers [3–6] could lose the tracking of target person *P*1, and either wrongly start following distractor *D*1 at time *t*4, or wrongly start following an impostor person *I*1 at time *t*5, as shown in red rectangles in Figure 1. The real target person *P*1 is walking on *Road*1 at time *t*4, and on *Road*4 at time *t*5, respectively. Therefore, robustly tracking the target person *P*1 in the nonlinear open world is still an unsolved problem, and it requires a robust target person identification in real time to reliably follow the person *P*1 in the real world.

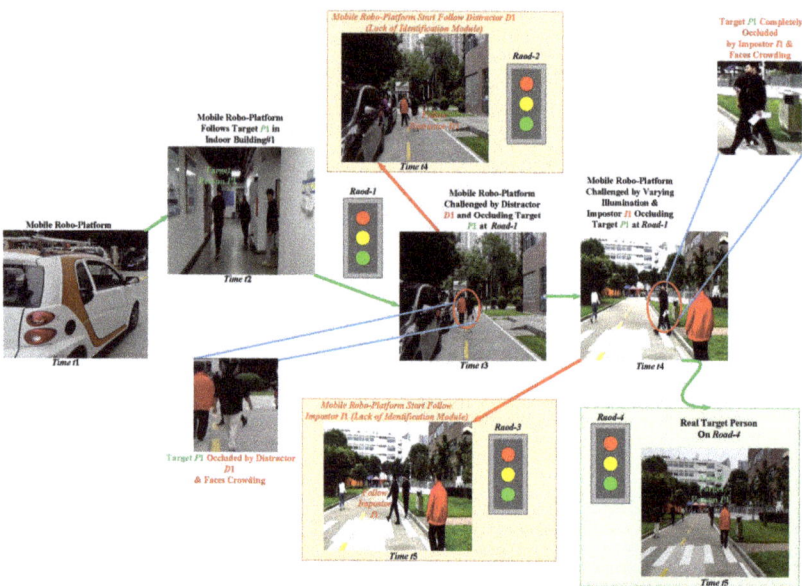

**Figure 1.** Person-follow robot scenario in outdoor world. The Mobile Roboplatform is Distracted in the Real Multi-Modal Open-Set World During Tracking Target P1.

In past, several works have addressed target person tracking and following. These past works use Laser Range Finder (LRF) [7,8], stereo camera [9], and RGB-D sensors [10] to track and follow the person. These trackers [7–10] have successfully addressed the person following problem, but these methods still get distracted in nonlinear and noisy outdoor environments due to occlusion, whereas these methods also lack the ability to reliably discriminate the target against similar-looking distractors in the outdoor world. Recently, deep visual trackers [11] have also learned to overcome the problem of reliably tracking the person in noisy environments; however, these deep visual trackers have still lacking to specifically address the target person identification problem during tracking and following. Visual trackers [11] are thereby still prone to impostors, and appearance changes the target person $P1$ undergoing in different nonlinear indoor and outdoor environments or domains.

Therefore, to overcome these shortcomings in visual trackers [11], recently, reidentification metrics have been learned and integrated with visual trackers [12] to follow the target person [1,2]. These reidentification metrics are learned by matching color-histograms and gait features [1,13], as well as extracting deep CNN features to learn deep similarity metrics [2,3,14–16]. However, the reidentification metrics in the present works [2,3,14–16] are all learned assuming the naïve world, i.e., it is assumed that the outside world is close-set, unimodal (it is assumed the robot only uses RGB sensor), and the target person $P1$ appearance remains unchanged across moving different domains. Due to these naïve assumptions, not only do the generalization capabilities of the learned reidentification metrics suffer largely in tracking the target in the outside world, but these metrics also get distracted due to the impostors observed across complex scenes and across nonlinear domains during tracking target $P1$, as shown in Figure 1. In Figure 1, at times $t3$ and $t4$ when the actual target $P1$ is completely occluded by distractor $D1$ and impostor $I1$ or if target $P1$ is completely moved out from the perception of mobile roboplatform in Figure 1, then the mobile roboplatform wrongly identifies a distractor $D1$ or impostor $I1$ as target and starts tracking the distractor $D1$ or impostor $I3$ at time $t4$ and $t5$, respectively.

Consequently, it is clear that to robustly track the target person $P1$ across different nonlinear indoor and outdoor environments, improvement of the trackers and integration of robust yet generic reidentification metrics in tracking are needed. The learned robust and generic reidentification metric, thus, identifies the target person in each consecutive image frame and thereby improves the tracking of the target while largely preventing the tracker to wrongly follow the distractors or impostors. Therefore, in our work, we have learned a lightweight domain generic reidentification metric, referred as $M_G$, for following the target in outside world. Furthermore, a lightweight novel part-attention learning method is also proposed to accurately identify the target $P1$ across different nonlinear domains, as well as to further improve the tracker to reliably track the target $P1$. The purpose of the lightweight metric is to run the metric on the compact robotic platform, while domain generic metric is learned to tackle (i) the appearance and style changes of target $P1$ across different domains, (ii) to tackle impostors of $P1$ in the scenes, and (iii) to reliably recapture the target person $P1$ again using the novel attention features, if target $P1$ is lost due to occlusion or completely moved out from the robot perception during tracking. The generalization capability of the learned reidentification metric in our work is further improved in a way that our domain generic metric $M_G$ is learned under realistic open-set scenarios, i.e., it is assumed that the appearance of the target person $P1$ varies when $P1$ moves across different indoor and outdoor environments. Then, the novel proposed attention module extracts the attention features from each single body part of the target person $P1$ to learn the deep cross-representations among the different images of the target person $P1$ undergoing appearance changes due to varying styles and modals. Therefore, the learned cross-representations are used to jointly address the pose and occlusion and also used to reject the large number of impostors during identifying the target in outside world, consequently largely improving the tracking of the target in outdoor world. In last, our contributions are:

- A robust vision-based target reidentification metric is proposed for target tracking. Compared to previous reidentification metrics proposed to target tracking, our reidentification metric is cross-modal and can address the style changes across large number of varying environments.
- Our reidentification learns part attention features, and unlike past works, the attention features are more stable to style changes and more robust against impostors. This is because the attention maps in our work are learned locally for each individual part, while during attention learning, it also uses global contextual and semantic information of the individual part. The global contextual provide useful relations among parts, while semantic information provides structural cues.
- Furthermore, the proposed reidentification uses the cross-representation module to jointly address pose and viewpoint changes and learns discriminative cross-view representations to tackle a large number of impostors in the open-set world.
- Finally, the learned metric is learned for the purpose of target tracking; therefore, it is designed with a lightweight backbone, while it is generic to help tracking the target in different nonlinear environments.

## 2. Related Work

The aim of our work is to learn a reidentification metric to integrate with tracking. Therefore, in this section, we review the recent works learning the reidentification metric for target tracking. Furthermore, in this section, we also review the recent works learning the reidentification metric for target tracking on mobile robots, and in the last, we also review the state-of-the-art related work in person reidentification that learns robust person identification metrics. In the beginning, we first cover the present works learning the reidentification metric for the visual tracking purpose.

### 2.1. Reidentification Metrics in Visual Tracking

Here, we cover the related work that learns reidentification metric for visual tracking purpose. However, the most of the learned models are used for offline tracking purpose. In [17], Neeti et.al. learned a LSTM-based CNN tracker using person reidentification module. They have learned spatiotemporal features of the person for reidentification;

however, during training, it is assumed that the real world is close-set and unimodal, and hence, the model performance could be challenged when deployed in the multimodal open-set world.

In another work in [18], tracking with person reidentification is learned, where the learned tracker tracks the target person in traffic scenes. Similarly, in [19], the authors follow the tracking-by-detection method, where the CNN-based feature matching is used to identify the target person in consecutive frames, and thus, the tracking of the target is done. The frameworks in [18,19] are simple, but these works require fine-tuning the reidentification metric every time for for every unseen domain, and thus have low generalization. Other works including [20,21] solve tracking in the multicamera network for smart city applications. [Edge Video] focuses to learn a lightweight reidentification metric to implement target tracking on edge devices, while Ref. [21] solves the problem of retracking the target after occlusion. Ref. [21] uses reidentification metric to reidentify the target when the target is recaptured again after the occlusion. Even [20,21] have good performance, however, ignored to solve reidentification problem in multimodal open-set world.

Furthermore, Ref. [22] also addressed the problem of target tracking in large scale scenario and proposed to learn an unsupervised reidentification metric for this purpose. The authors believed for the large-scale scenario that unsupervised reidentification requires no label of identities for training, and thus perform better than supervised metrics. Although the above works [17–22] used the learned reidentification metric for visual tracking, the methods are not intended for tracking the target on mobile robots. In order to track the target on mobile robots, lightweight and efficient reidentifications are needed; therefore, in the next subsection, we cover the recent works that specifically learn the reidentification metrics for tracking the target in real time and on mobile robots.

### 2.2. Reidentification Metrics for Visual Tracking on Mobile Robots

Here, we cover the recent state-of-the-art works learning reidentification metric for person identification and tracking for mobile robots. In [1] height, gait, and appearance features are used to learn an online person classifier to identify the target person to follow, while, Ref. [23] uses human pose estimation to detect the person indoors and then identify the target person using an appearance-features-based reidentification metric to follow. In addition, both [24] and [25] also use the appearance-features-based reidentification metric to detect and track the target person in the indoor environment. Although the method from [1,23–25] tracks well indoors, it lacks the ability to handle nonlinear style and appearance variations that the target person undergoes in outdoor world, and it is largely distracted due to impostors in the outdoor environment.

On the other hand, Ref. [15] also uses convolutional channel features to first identify the target person and then follow the identified target person using the mobile robot. Both [15] and [26] first use laser range finder to track the person position, then Ref. [15] learns the convolutional-channel-features-based classifier to verify the target to follow, whereas Ref. [26] uses monocular camera to perform appearance matching. In another work [2], an online person classifier is also learned to track the target person, but in the robot coordinate space. The authors believed tracking the person in robot coordinate space is more accurate than tracking in the real-world space. The methods [1,2,15] follow the person in both indoor and outdoor environments; however, the learned person classifiers in their works are not generic, while their works also fail to address resisting the impostors in the outdoor world.

Furthermore, some works have used depth sensing to track the position of target. Both [27] and [28] use kinect depth sensor to track the person position. The robot tracks the person, however, in the crowded environment; due to distractors, the accuracy of tracking is largely challenged in the outdoor setting due to the sensitivity of the Kinect sensor.

In other works, authors, e.g., those of [14], propose to track the target person using the Kalman filter. Once the bounding box of target person predicted, the state of Kalman filter is updated by identifying the target bounding box, and the target is then followed. In [3], another problem of tracking the target in uniform crowd environment is solved. Their method depends on accurate face identification because the target and other persons all have similar appearances in the scene. Even the method performs well, but the method has still not addressed generalization problem across all outdoor environments, as well as Ref. [14] can be distracted due to impostors, whereas Ref. [3] can be distracted in crowded environment due to poor depth sensing.

In addition, there are a few works in reidentification-based person tracking on mobile robots aim to track the person for social and virtual game environments. In [29], both depth and laser range finder sensors are used to track the person position, however, only in the indoor environment, whereas Ref. [30] track and follow the person for the virtual game environment. Even their methods can track the target, but their methods are optimal for close-set scenarios, whereas the real-world scenario is multimodal and open-set. Now, covering the recent works in reidentification-based person tracking on mobile robots, we now further explore the state-of-the-art metrics learned for real-world person reidentification and are covered in detail in the subsection below.

*2.3. State-of-the-Art Reidentification Metrics*

2.3.1. Deep Metrics

Deep metric learning for Person ReID has been extensively studied in past works, e.g., [31,32]. These metrics, though, aim to address pose, viewpoint, occlusion, and misalignment of parts to attain high similarity; however, they underperform against unseen domains.

2.3.2. Domain Adaptation Metrics

To improve poor generalization, the authors of a few works proposed unsupervised domain adaptation (UDA) [33,34]. UDA adapts the learned ReID metric from labeled source domains to unlabeled target domain, but it is still time-consuming due to collecting data and fine-tuning the metric for each new unseen target domain.

2.3.3. Domain Generalizable Metrics

Therefore, recently, domain generalizable metrics have gained a great amount of attention in ReID [35–38]. Domain generic metrics are learned once and then directly applied for identification on previously unseen domains; however, existing metrics still ignore that the real world is multimodal and open-set, where the same person is seen in several different styles and modals. Therefore, it is desired that the generic metric in the real world (i) matches different Probe and Gallery images of the same person in different modals and styles and (ii) jointly addresses pose, viewpoint, and displacement of parts across views, while (iii) is also robust against impostors in open-world, and (iv) is lightweight, thus it can run on devices in the real world.

## 3. Methodology

In this work, our aim is to learn person reidentification metric to identify the target person and to integrate the learned reidentification metric with pretrained state of the art visual tracker to track the target; therefore, in this section, we will cover the details of learning the cross-modal domain generic open-set person reidentification metric and will describe its complete framework as shown below in Figure 2. In Figure 2, first realistic training data from open-set multi-modal world is generated, as shown in Figure 2a. In Figure 2a, a large number of images of $N$ different person identities, say person $P1$, $P2$, and $P_N$, as shown in Figure 2a, are taken from $D_T$ different source/camera domains.

Then, in the next step, a large number of images of each person identity are generated in different random poses. The purpose of generating images in different random poses for each different identity is to augment the training images of each person identity in different poses to learn its pose invariant features for identification. Next, in Figure 2a, taking all the original images of $N$ persons and their generated images in random poses, we now randomly transfer different images of different persons into different random styles. This is done because in the real world, when tracking the person in the outdoor world, it could move across several different environments, and in each different environment, it could undergo nonlinear style variations; thus, in order to identify the target person in different environments and in different styles, it is needed to learn a style robust person reidentification metric, and therefore, the images of a person are generated in multiple styles to train the reidentification metric in such a way that it could match all the images of the same person in all the different random styles. Finally, in the real world, it is also needed to obtain a cross-modal reidentification metric because it is possible that different mobile robot platforms use different imaging sensors, such as RGB and IR cameras. Therefore, in Figure 2a, to learn a cross-modal reidentification metric, our work takes different original images of $N$ persons and takes their generated images in random poses and in random styles to randomly transform these different training images into different modals, say RGB, Grayscale, and Sketch modals. In our work, RGB images are transformed into Grayscale and Sketch modals due to the reason that in the real world, it is not necessary to always use an RGB sensor on a mobile robot; already, a large number of works have used IR modality [12]. However, a large number of public reidentification datasets have no IR images; therefore, we opted to transform RGB images into grayscale images. Sketch images are also generated to further improve the feature extraction power of the learned metric as well as to help in improving the cross-modality matching. After getting the realistic training data, a lightweight deep CNN backbone is then used to extract the features from the training images, as shown in Figure 2b. This deep CNN feature extraction backbone is designed using efficient residual module as shown in Figure 2b and is described in detail in Section 3.2.1. Furthermore, the feature extraction backbone also uses a novel part-attention module, as shown in Figure 2b, to extract the subtle features of the different individual parts. The details of novel part attention module are covered in Section 3.2.2. Next, using the learned attention features of different individual parts, our work learns cross-representations using the cross-representation learning module as shown in Figure 2b. The purpose of cross-representation learning is to minimize both the style and modality differences across cross-view features, which then feed to fully connected layers to learn the complex feature relationships to finally predict the similarity between the pair of images, as shown in Figure 2c.

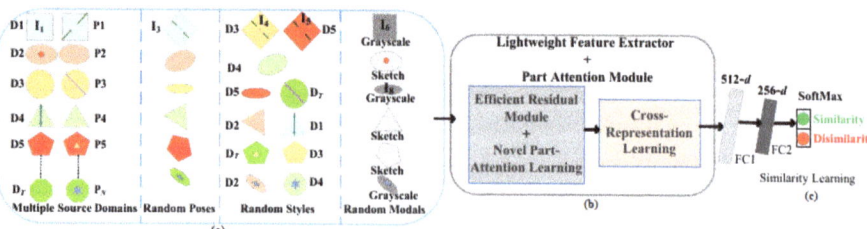

**Figure 2.** Framework of domain generic metric $M_G$. (**a**) Real Open-world Training Data. (**b**) Lightweight Features Extractor with Novel Attention and Cross-Representation. (**c**) Similarity Learning.

Now, below in Section 3.1, we will cover the details of generating training data in the complex nonlinear multimodal open-set world.

*3.1. Multimodal Open-World Training Data*

Here, we describe the details of generating training data in nonlinear multimodal open-set world. The real world is nonlinear, multimodal, and open-set where a person can be seen across several different domains in different styles and can also experiences pose, viewpoint, and parts displacement. Therefore, to train a robust metric for the real world, it is first needed to obtain realistic world data for training. For this purpose, first, $N$ number of different persons are randomly chosen from $D_T$ source domains, as shown in Figure 2a. In order to generate realistic open-world training data, in the first step, as shown in Figure 2a, images in random poses are generated for each person. Taking the image $I_1$ of person $P1$ seen in domain $D1$, as shown in Figure 2a, a new image instance $I_3$ for $P1$ in random pose $\theta_p$ is generated as:

$$I_3 = \theta(I_1, \theta_p), \tag{1}$$

here $\theta$ is pose generation model [39]. Now, Equation (??) is used to generate images in random poses for all the $N$ IDs in training data (in our work, images are generated in 8 random poses), and a few generated images for $ID1$ are shown below in Figure 3. Getting images for all the $N$ persons in random poses, our work now generates images in varying styles for all $N$ IDs. In Figure 2a, two instances $I_4$ and $I_5$ for ID $P1$ are generated in the styles of domain $D3$ and $D5$, respectively, as:

$$[I_4, I_5] = G\left(\left[w_{G_{D_3}}, w_{G_{D_5}}\right], I_3\right), \tag{2}$$

here $w_{G_{D_3}}$ and $w_{G_{D_5}}$ are the parameters of translation model $G$ [33] for domains $D3$ and $D5$, respectively, and $I_3$ is the input image. In Equation (??), the purpose of generating images for $P1$ in varying styles [33,34] is to exploit the diverse and varying styles images of $P1$ to learn its style generic representation; thus, the learned generic metric could distinguish $P1$ seen in any random style in the open world. Hence, our work generates images of $P1$ (i) in varying styles in different disjoint views of the same domain $D1$ [34], (ii) in varying styles across random Re-ID domains [33], e.g., in styles of domains $D3$ and $D5$ as shown in Figure 2a, and in varying styles of random detection and recognition datasets, such as Imagenet [40]. Images in varying styles and poses are obtained for $N$ persons; however, the real world is actually multimodal. Therefore, Grayscale image $I_6$ and Sketch image $I_8$ in Figure 2a for IDs $P1$ and $P4$ are respectively generated as:

$$\begin{aligned} I_6 &= \varphi(I_1), \\ I_8 &= \phi(I_7), \end{aligned} \tag{3}$$

here, function $\varphi$ and $\phi$ convert RGB images $I_1$ and $I_7$ of $P1$ and $P4$ into Grayscale and Sketch modals, respectively. Now, in the next Section 3.2, we now describe the details of learning person features using the proposed novel part attention module.

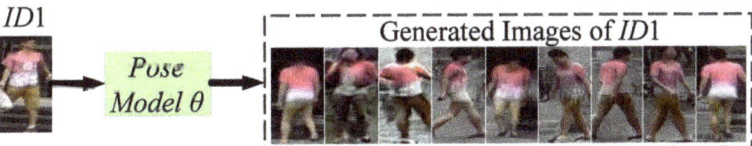

**Figure 3.** A Few Generated images of ID1 During training. Images are Generated using Pose Model $\theta$ [39].

*3.2. Novel Part-Attention Feature Learning*

Now, using the generated cross-modal open-world training data, here, in this section, we cover the details of features extraction and novel part attention module. In the first, as shown in Figure 2b, person features are extracted with efficient lightweight CNN backbone and then, as shown in Figure 2c, the similarity between extracted pair of part attention features are learned. However, before covering the learning of features extraction, in our methodology, we first cover the details and design process of lightweight CNN backbone shown in Figure 2b below in Section 3.2.1.

### 3.2.1. Lightweight Backbone Design

Unlike [35–38], efficient residual module [40] is designed in Figure 4 to build a lightweight CNN backbone for learning cross-representations as shown in Figure 2b and learning the similarity between pair of learned features as shown in Figure 2c. Lightweight backbone as shown in Figure 2b is a Siamese network with 10-layers (details of layers are listed in Table 1), but for simplicity, only one stream is shown in Figure 2b. Each Convolution layer in Figure 2b is then realized with efficient residual module, where all convolutions are implemented as mixed depthwise separable convolutions [41] following Wider ResNet [42] strategy, i.e., the number of filters in each successive convolution layer are increased 2× times than previous convolution layer, thereby improving the features representation power with minimal computational cost. In addition, channel shuffle and channel split [43] are also used in efficient residual module in Figure 4 to enable information mixing across different filters and layers, thus further increasing diversity in features. After every convolution layer in Figure 4, Batch Normalization (BN) and ELU activation function are used for faster network convergence. Realizing the lightweight CNN backbone in Figure 2b, then, deep cross-representations are learned for each individual part of a person complimented with novel part-attention learning mechanism, as shown in Figure 2b.

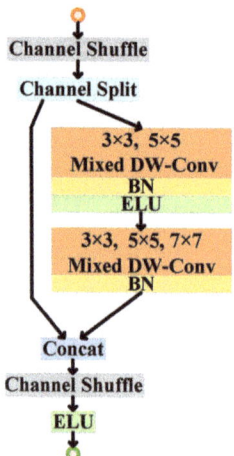

**Figure 4.** Our Efficient Residual Module Design.

**Table 1.** Detail of each layer, here, DW: Depthwise and MDW: Mixed-Depthwise. All different convolutions in MDW are 1-layer.

| Layer Name | Output Size | 10-Layer |
|---|---|---|
| conv1 | 112 × 112 | 7 × 7 (DW-Conv.), 64, stride 2 |
| conv2_x | 56 × 56 | 3 × 3, Max Pool, stride 2 <br> 3 × 3 (DW-Conv.), 128 |
| conv3_x | 28 × 28 | $\begin{bmatrix} 3 \times 3, 5 \times 5 (\text{MDW} - \text{Conv.}), 256 \\ 3 \times 3, 5 \times 5, 7 \times 7 (\text{MDW} - \text{Conv.}), 256 \end{bmatrix} \times 1$ |
| conv4_x | 14 × 14 | $\begin{bmatrix} 3 \times 3, 5 \times 5 (\text{MDW} - \text{Conv.}), 512 \\ 3 \times 3, 5 \times 5, 7 \times 7 (\text{MDW} - \text{Conv.}), 512 \end{bmatrix} \times 1$ |
| conv5_x | 7 × 7 | $\begin{bmatrix} 3 \times 3, 5 \times 5 (\text{MDW} - \text{Conv.}), 1024 \\ 3 \times 3, 5 \times 5, 7 \times 7 (\text{MDW} - \text{Conv.}), 1024 \end{bmatrix} \times 1$ |
| | 1 × 1 | Global Average Pool, 512-d FC1, 256-d FC2, Softmax |
| FLOPs | | $1.1 \times 10^9$ |

### 3.2.2. Novel Part Attention Cross-Representations

Attention learning has been proved promising in Re-ID; however, methods from past works [35,44,45] learn attention globally for the whole body as shown in Figure 5a, and thereby, certain valuable and unique features from different individual parts are loss. Therefore, we argue to learn attention features for each individual part, as shown in Figure 5b, to prevent from loss, as well as, highlight the unique cues of different parts.

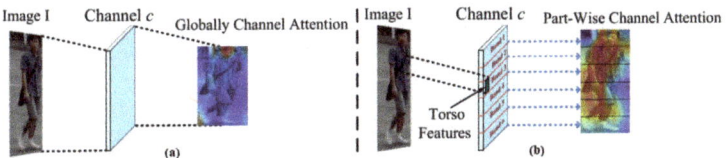

**Figure 5.** (a) Global channel attention [44,45]. (b) Our part-wise channel attention.

Novel Channel Attention for Individual Parts: Taking input features $F \in \mathbb{R}^{C \times H \times W}$, where $C$ is the number of filters, while, $H \times W$ are feature spatial dimensions, channel attention for each individual part is then learned by horizontally dividing all the $C$ filters into six spatial bands, as shown in Figure 6a, because each spatial band carries features of different part, as shown in Figure 5b, therefore, learning the attention for different corresponding bands, consequently, learns the attention weights for different corresponding parts. Hence, each single channel is partitioned into six horizontal bands, then the channel attention weight for each single band from each single channel is learned, but before the attention weight is learned it is first needed to capture the spatial, structural, and semantic relations of each band with its $C$ neighboring bands, as shown in Figure 6b. Thus, similar to [44] relationships $r$ and $r'$ between bands $x_1$ and $y_1$, and between bands $x_1$ and $z_1$ are learned as:

$$r_{(x_1,y_1)} = \vartheta_c(f_{x_1}) \cdot \mu_c(f_{y_1}), \tag{4}$$

here, $f_{x_1}$ and $f_{y_1}$ are features of $x_1$ and $y_1$ with dimensions $\mathbb{R}^{1 \times \frac{H}{6} \times \frac{W}{6}}$, and the value $r_{(x_1,y_1)}$ computed as dot product between embedding functions $\vartheta_c$ and $\mu_c$ [44]. Embedding functions $\vartheta_c$ and $\mu_c$ are implemented by first flattening features $f_{x_1}$ and $f_{y_1}$, then, apply $1 \times 1$ convolution followed by BN and ELU activation. Now, using Equation (??), all the $C$ relations of band $x_1$ with all the $C$ neighboring bands are then obtained to form the relation vector $\mathbf{r}_1$ for $x_1$ as: $\mathbf{r}_1 = \left[r_{c_{(x_1,:)}}\right]_{c=1,\ldots,C}$. Relation vector $\mathbf{r}_1$ is then embed with features $f_{x_1}$ as:

$$f'_{x_1} = [pool_{Av}(v_c(f_{x_1})), pool_{Mx}(v_c(f_{x_1})), v_c(r_1)]. \tag{5}$$

here $pool_{Av}$ and $pool_{Mx}$ are global average and max pooling operations. Embedding function $v_c$ first flattens features $f_{x_1}$, then, both $v_c$ and $v_c$ are implemented as $1 \times 1$ Conv followed by BN and ELU activation. Now, features $f'_{x_1}$ are used to learn channel attention weight $a_{c_{x_1}}$, shown in Figure 6c, for band $x_1$ as:

$$a_{c_{x_1}} = Sigmoid(W_2 ELU(W_1 f'_{x_1})), \tag{6}$$

here, $W1$ and $W2$ are $1 \times 1$ Conv followed by BN. Now using Equations (4)–(6), first, all the six weights $a_{c_{x_1}}, a_{c_{x_2}}, a_{c_{x_3}}, a_{c_{x_4}}, a_{c_{x_5}}$, and $a_{c_{x_6}}$, shown in Figure 6c, for all the six bands in Channel-1 are computed, then, similarly, all the six weights for all the six bands in all the $C$ channels are computed. Now, computing all the six weights for all the $C$ channels, the six weights of each channel, e.g., Channel-1 weights shown in Figure 6c are taken and then each weight of each corresponding band is broadcasted similar to [45] over the spatial dimensions of each corresponding band to finally obtain channel attention $a'_c \in \mathbb{R}^{H \times W}$ for Channel-1. Following this, the channel attention $a'_c$ for all the $C$ channels are then obtained, and finally, the attention maps for all $C$ channels are concatenated together to form matrix $A_c$ as: $A_c = [a'_{c'}]_{c'=1....C}$. Channel Attention features are now computed as:

$$F_{a_c} = F \otimes A_c, \tag{7}$$

here $\otimes$ denotes elementwise multiplication [45] between weights $A_c$ and features $F$.

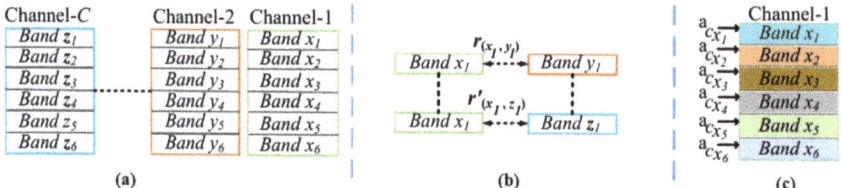

**Figure 6.** Novel channel attention learning for individual parts. (**a**) Dividing $C$ channels into six bands; (**b**) Learning relations among bands; (**c**) Channel attention weight $a'_c$.

Novel Spatial Part Attention Features: Unlike past [35,44,45], our work learns spatial attention for pixel $(i,j)$ for every 8 filters, i.e., weight $a_{s_k}$ for every 8 filters as shown in Figure 7. This is done to improve spatial attention while preventing the loss of vital patterns that are largely diminish when spatial attention for pixel $(i,j)$ is learned globally over all the $C$ filters, e.g., $a_s(i,j) = -1$ in Figure 7. Now, every time taking 8 filters, e.g., filters $c = 1$ to $c = 8$ in Figure 7, spatial attention $a_{s_k}$ for pixel $(i,j)$ is learned by first learning the relations of pixel $(i,j)$ with all the $(H \times W)-1$ pixels in the corresponding 8 filters as:

$$r_{k,l} = \vartheta_s(f_k) \cdot \mu_s(f_l), \tag{8}$$

here $f_k$ and $f_l$ are 8-dimension feature vectors of pixels $(i,j)$ and $(i',j')$ as shown in Figure 7, while, $r_{k,l}$ is the leaned relation, and embedding functions $\vartheta_s$ and $\mu_s$ are implemented as $1 \times 1$ spatial convolution followed by BN and ELU activation. Now, Equation (8) is used to learn all the $H \times W$ relations of vector $f_k$ of pixel $(i,j)$ to form the relation vector $\mathbf{r}_k$ as: $\mathbf{r}_k = [r_{r'}]_{r'=1,...,H \times W}$, then, vector $\mathbf{r}_k$ embed with vector $f_k$ to form feature $f'_k$ as:

$$f'_k = [pool_{Av}(\nu_s(f_k)), pool_{Mx}(\nu_s(f_k)), v_s(\mathbf{r}_k)]. \tag{9}$$

here, in Equation (9), embedding functions $\nu_s$ and $v_s$ are learned as $1 \times 1$ Conv followed by BN and ELU activation. Now, the attention $a_{s_k}$ for pixel position $(i,j)$ is learned as:

$$a_{s_k} = Sigmoid(W_2 ELU(W_1 f'_k))), \tag{10}$$

here W1 and W2 are $1 \times 1$ Conv followed by BN. Now, Equations (8)–(10) are first used to learn the spatial weights $a_{s_k}$ for pixel $(i,j)$ for every 8 filters, as shown in Figure 7, and then, similarly, Equations (8)–(10) are also used to learn the spatial weights $a_{s_k}$ for all the $H \times W$ pixels in every 8 filters. Getting the spatial weights $a_{s_k}$ for all the $H \times W$ pixels in all the $C$ filters, first, the learned corresponding $H \times W$ weights for every corresponding 8 filters are broadcasted [45] over the spatial dimensions $H \times W$, and then the attention maps $a_{s_g} \in \mathbb{R}^{8 \times H \times W}$ (here $g = 1$ to $C/8$) for every corresponding 8 filters are obtained. These attentions maps are then concatenated together to form spatial attention weights matrix $A_s$ for all $C$ filters as: $A_s \in \mathbb{R}^{C \times H \times W}$. Then, finally, spatial attention features $F_{a_s} \in \mathbb{R}^{C \times H \times W}$ are obtained as:

$$F_{a_s} = F_{a_c} \otimes A_s. \tag{11}$$

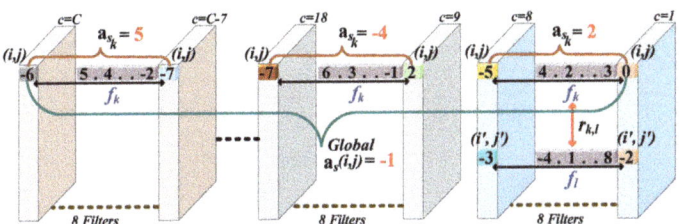

**Figure 7.** Novel Spatial Attention Learning for Individual Parts.

*3.3. Multi-Modal Open-Set Generic Metric*

Using the attention features of each single part, cross-representation module [46] shown in Figure 2b now learns the cross-representations for pair of features $f_{q_1}$ and $f_{q_2}$ as:

$$g(q_1, q_2) = CRM(f_{q_1}, f_{q_2}). \tag{12}$$

here, cross-representation module CRM in Equation (12) not only learns the complex nonlinear relationships between features $f_{q_1}$ and $f_{q_2}$ to minimize the existing domain, style, and modality gaps in different environments between positive pair $(q_1, q_2)$, but, at the same time also addresses the pose, viewpoint, and spatial miss-alignment across views. For the given quadruplet $q_1, q_2, q_3, q_4$, all the cross-representations $g(q1, q2)$, $g(q1, q3)$, and $g(q2, q4)$ are learned using Equation (12), and the representations are then sent to fully-connected layers FC1 and FC2 shown in Figure 2c with 2-dimension softmax classifier to learn similarity between $(q_1, q_2)$. The learned similarity value is then used to compute the quadruplet loss [47] $L_{quad}$ as:

$$L_{quad} = \sum_{n=1}^{N} \left[ g(q_1, q_2)^2 - g(q_1, q_3)^2 + \alpha_1 \right] \\ + \sum_{n=1}^{N} \left[ g(q_1, q_2)^2 - g(q_2, q_4)^2 + \alpha_2 \right], \tag{13}$$

here, N are total number of quadruplets, $\alpha_1 = 1$ and $\alpha_2 = 0.3$ are the margin values used, while $q_3$ and $q_4$ are the impostors of $q_1$ and $q_2$, and randomly seen in any domain in any style and modality, respectively.

## 4. Experiments and Analysis

*4.1. Datasets and Data Augmentation*

For training metric $M_G$ Market1501, DukeMTMC-reID, CUHK03, and CUHK02 are used as training source domains, while $M_G$ is comprehensively evaluated using domains VIPeR, PRID, GRID, and i-LIDS. For cross-modal evaluation, SYSU-MM01 dataset used the following settings in [48]. In addition, random cropping, horizontal flipping, random rotation, color jittering, random contrast, brightness, and label smoothing regularization [49] are used in training to prevent $M_G$ from overfitting.

*4.2. Implementation Details*

Lightweight CNN backbone in Figure 2b is trained from scratch with randomly initialized weights [50] for 600 epochs on single 16G NVIDIA RTX 2080 Ti GPU. The during-training image resolution is 224 × 224, with the Adam optimizer used with initial learning rate $8 \times 10^{-5}$, mini-batch size 64, and weight decay $5 \times 10^{-4}$. All of the code was written in Pytorch.

*4.3. Evaluation Metrics and Protocols*

Unlike [35–38], $M_G$ is comprehensively evaluated in challenging multimodal, multi-style, close-set, and open-set scenarios. Cumulative Matching Characteristics (CMC) at

*Rank*-1 and mean average precision (mAP) are the metrics used for all close-set experiments, while for all open-set experiments, true target rate (TTR) is measured against false target rate (FTR) [51] as a performance metric. All results obtained after averaging over 10 trials.

4.3.1. Naïve Close-Set Scenario

Unlike [35–38], in this scenario, $M_G$ is evaluated in two different and difficult settings, in setting#1 and setting#2. In both settings, Probe/Gallery images splits are: VIPeR: 316/316; PRID: 100/649; GRID: 125/900; i-LIDS: 60/60, respectively. Here, VIPeR: 316/316 means there are 316 person identities in test Query view, and similarly, there 316 same person identities in the test Gallery view. In setting#1, matches of Queries observed in a corresponding domain, e.g., observed in CUHK-03 domain, are found from the Gallery view of the corresponding CUHK-03 domain only. Below in Figure 8, the testing scenario of setting#1 is shown.

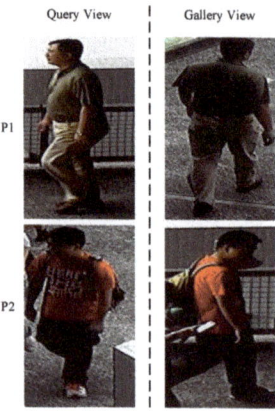

**Figure 8.** In Setting#1 it can be seen the matches of all Queries exist in the Gallery View, while the match for each person Seen in VIPeR domain is found from the Gallery of VIPeR domain domain.

On contrary, as shown in Figure 9, the matches of Queries in setting#2 are found from a joint Gallery containing Gallery images from all the test domains, i.e., VIPeR, PRID, GRID, and i-LIDS. Thus, in setting#2, $M_G$ is tested in more realistic and challenging scenario to find the matches of given Query by resisting large number of impostors from different outdoor environments. Identification results of $M_G$ in Setting#1 are summarized in Table 2, where $M_G$ attains 80.73%, 64.44%, and 88.99% Rank-1 identification on PRID, VIPeR, and i-LIDS, respectively.

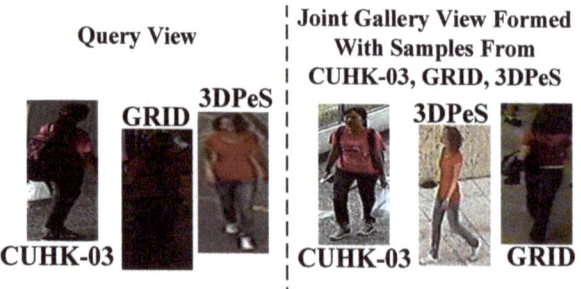

**Figure 9.** In Setting#2 the match for given Query image, the match for Query from GRID domain is found from the joint Gallery, where the Gallery contains images from all test domains, i.e., GRID, CUHK-03, and 3DPes.

Table 2. Results on Naïve Close-Set Setting#1.

| Methods | Target:PRID | | Target:GRID | | Target:VIPeR | | Target:i-LIDS | |
|---|---|---|---|---|---|---|---|---|
| | Rank-1 | mAP | Rank-1 | mAP | Rank-1 | mAP | Rank-1 | mAP |
| CBAM [45] | 47.52 | 53.89 | 33.07 | 39.17 | 45.80 | 51.87 | 75.08 | 81.0 |
| HLGAT [52] | 48.74 | 56.84 | 35.22 | 43.94 | 48.96 | 58.03 | 79.67 | 82.64 |
| RAGA [44] | 50.69 | 60.47 | 37.79 | 45.51 | 49.03 | 58.33 | 83.91 | 85.2 |
| SRN [35] | 52.1 | 66.5 | 40.2 | 47.7 | 52.9 | 61.3 | 84.1 | 89.9 |
| D.Norm [36] | 60.4 | - | 41.1 | - | 53.9 | - | 74.8 | - |
| MoE [38] | 57.7 | 67.3 | 46.8 | 54.2 | 56.6 | 64.6 | 85.0 | 90.2 |
| Meta [37] | 74.2 | 81.0 | 48.4 | 57.9 | 59.9 | 68.6 | 81.3 | 87.0 |
| Our $M_G$ | 80.73 | 86.91 | 55.07 | 61.23 | 64.44 | 75.1 | 88.99 | 91.0 |

Clearly, metric $M_G$ attains higher recognition than [35–38,44,52]; it is because $M_G$ learns cross-representations among different indoor and outdoor images of a person complimented with part-attention learning, where learned part-attention pays large focus on unique features from each different part, thus, $M_G$ jointly minimizes pose, viewpoint, and spatial displacement of parts, as well as jointly addresses style, modality, and domain gaps to resist large number of impostors in outdoor world.

Furthermore, in setting#2, though, setting#2 is difficult than setting#1, but $M_G$ still surmounts the challenges in setting#2, and in Table 3, $M_G$ retrieves 77.89%, 52.64%, 61.92%, and 86.23% matches of the Queries at *Rank*-1 from the joint Gallery for test domains PRID, GRID, VIPeR, and i-LIDS, respectively. These results clearly reveal $M_G$ can inherently tackle pose, viewpoint, style, and modality transforms across both nonlinear indoor and outdoor environments, while, complemented with the part-attention mechanism, $M_G$ also learns cross-representations in a way to jointly address occlusion and misalignment of parts; therefore, in Figure 10d, $M_G$ declines the large number of impostors and improves the identification accuracy from *Rank* = 5 in Figure 10a to *Rank* = 1.

Table 3. Results on Naïve Close-Set Setting#2.

| Methods | Target:PRID | | Target:GRID | | Target:VIPeR | | Target:i-LIDS | |
|---|---|---|---|---|---|---|---|---|
| | Rank-1 | mAP | Rank-1 | mAP | Rank-1 | mAP | Rank-1 | mAP |
| CBAM [45] | 39.42 | 51.33 | 25.21 | 37.99 | 32.61 | 44.04 | 69.24 | 73.0 |
| HLGAT [52] | 43.95 | 50.07 | 28.77 | 39.5 | 37.66 | 49.72 | 70.77 | 76.95 |
| RAGA [44] | 44.49 | 50.1 | 30.92 | 40.32 | 38.55 | 51.79 | 72.14 | 78.63 |
| SRN [35] | 47.85 | 51.72 | 34.87 | 43.67 | 44.99 | 52.05 | 72.1 | 78.41 |
| D.Norm [36] | 53.70 | 60.1 | 37.22 | 44.33 | 49.54 | 54.41 | 71.04 | 80.03 |
| MoE [38] | 51.51 | 63.91 | 39.35 | 47.69 | 48.74 | 57.0 | 79.0 | 83.1 |
| Meta [37] | 70.82 | 78.07 | 42.61 | 50.95 | 51.74 | 59.97 | 79.68 | 84.78 |
| Our $M_G$ | 77.89 | 84.78 | 52.64 | 58.85 | 61.92 | 73.08 | 86.23 | 90.0 |

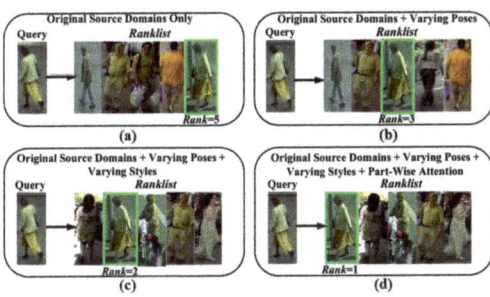

Figure 10. Retrieval results obtained in close-set setting#2 using different training data. (a) Training data only contains original image samples with no data augmentation; (b) Training data contains original image samples, and image samples are generated in different random poses too; (c) Training data contains original image samples, and image samples are generated in different random poses and in different random styles too; (d) Training data contains original image samples, and image samples are also generated in different random poses and styles, as well as, part-wise attention is learned to improve features power.

### 4.3.2. Challenging Close-Set Scenario

The joint Gallery in setting#2 is very challenging; however, the target person being followed by robot when moving across different outdoor environments could undergo style, illumination, pose, and viewpoint changes in the real world. Therefore, to obtain a robust reidentification metric, $M_G$ is tested in the real-world environment where Probe-Gallery pairs could be seen in different modals [53] and in different styles [35]. The real-world testing scenario for Challenging Close-Set is shown below in Figure 11.

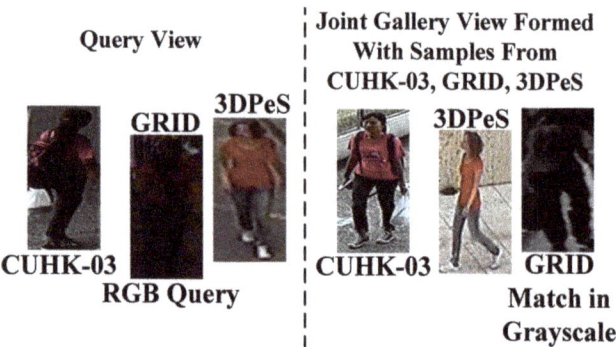

**Figure 11.** In the Challenging Close-Set Scenario, the Query image, the image of GRID domain could be seen in the RGB modalily (or different style), whereas its match in Gallery view has Grayscale modality.

Therefore, during testing in the Challenging Close-Set Scenario, images in Query and in joint Gallery views are randomly transformed into different modals and styles, then, $M_G$ finds the matches from joint multimodal multistyle Gallery, and the results are given in Table 4.

**Table 4.** Results on Challenging Close-Set Scenario.

| Methods | Target:PRID | | Target:GRID | | Target:VIPeR | | Target:i-LIDS | |
|---|---|---|---|---|---|---|---|---|
| | Rank-1 | mAP | Rank-1 | mAP | Rank-1 | mAP | Rank-1 | mAP |
| CBAM [45] | 30.98 | 40.88 | 21.8 | 28.4 | 25.74 | 32.37 | 58.37 | 62.66 |
| D.Norm [36] | 31.49 | 42.41 | 23.54 | 30.68 | 27.73 | 35.20 | 60.56 | 69.03 |
| RAGA [44] | 34.93 | 42.99 | 24.12 | 32.34 | 32.64 | 37.85 | 64.05 | 73.08 |
| HLGAT [52] | 36.79 | 43.35 | 24.31 | 33.14 | 35.25 | 46.39 | 65.43 | 76.33 |
| SRN [35] | 40.23 | 49.43 | 25.06 | 34.84 | 38.98 | 50.27 | 70.47 | 76.23 |
| MoE [38] | 45.12 | 53.33 | 28.49 | 37.19 | 44.46 | 57.96 | 75.64 | 80.27 |
| Meta [37] | 62.85 | 71.33 | 30.94 | 39.0 | 44.05 | 57.37 | 71.01 | 79.95 |
| Our $M_G$ | 75.04 | 82.19 | 49.38 | 57.83 | 60.17 | 71.93 | 84.04 | 88.13 |

In Table 4 $M_G$ attains 3.47% drop at *Rank*-1 accuracy than close-set setting#2. Therefore, the reasons for this drop are analyzed in retrieval results in Figure 12. In Figure 12a,b, it is evident that $M_G$ is robust against style changes and impostors and finds matches at *Rank* = 2; however, in Figure 12c $M_G$ lags in cross-modal matching due to impostors. Clearly, color images dominate intensity images, thus, to improve the multimodal recognition capability of metric $M_G$ in the outdoor environment, it is needed to optimally represent each person in RGB, Grayscale, and Sketch modals during training; hence, $M_G$ can resist a large number of multimodal impostors in outdoor environment.

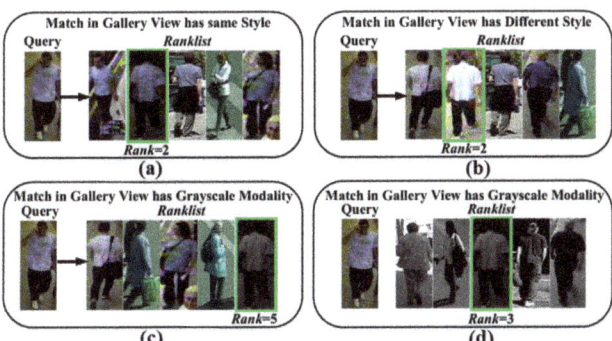

**Figure 12.** Retrieval results obtained in challenging close-set scenario. (**a**) Finding the match in Gallery view where the correct Gallery image has same style and modality with Query image; (**b**) Finding the match in Gallery view where the correct Gallery image has different style, but, same modality with Query image; (**c**) Finding the match in Gallery view where the correct Gallery image has different style, and different modality with Query image; (**d**) Finding the match in Gallery view where all the images in Gallery view has different style, and different modality with Query image.

Therefore, to find the optimal representation of persons in different modals, our work performed different experiments, and the results are shown in Figure 13. In Figure 13, for domains VIPeR and GRID, it is observed that $M_G$ declines a large number of impostors when a number of images for a large number of training persons have representation ratio 1:1:3 for Grayscale (G.Sc.) vs. Sketch (Sk.) vs. RGB modals. Therefore, $M_G$ is retrained with the representation ratio 1:1:3 to regain the performance, and in Figure 12d, $M_G$ successfully declines impostors to find match at $Rank = 3$.

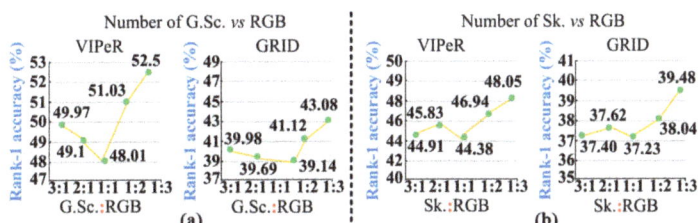

**Figure 13.** Rank-1 Result Comparison for Different Representation Ratios on VIPeR & GRID for (**a**) G.Sc. vs. RGB (**b**) Sk. vs. RGB.

4.3.3. Open-World Scenario

This is the scenario where the target person $P1$ in the real world moves out from the robot perception, and thus the robot losses the target person. While there is no target person in the robot perception, it is required that the robot following the person has robust reidentification capability to resist both impostors in the open world and at the same time has inherent discriminating ability to reidentify the real target person as soon as the target person is recaptured into the robot perception. Therefore, our work also evaluates metric $M_G$ in the realistic open world. In Figure 14, we have shown the high-level overview of this scenario.

However, unlike [51], for the open-world testing in our work, randomly, 48 person IDs from each testing domain, i.e., from VIPeR, PRID, GRID, and i-LIDS domains are chosen to form the realistic world joint open-set Gallery, then $M_G$ finds the matches for target Query images from the joint open-set Gallery, and the results are given in Table 5. In Table 5, $M_G$ in the open world optimizes the part-attention weights in a way to learn cross-representations to simultaneously decline a large number of impostors in the open world and also discriminates difficult nontarget Queries to attain 68.02%, 56.09%, and 76.57% *Rank*-1 identification at FTR 0.1% on PRID, VIPER, and i-LIDS, respectively.

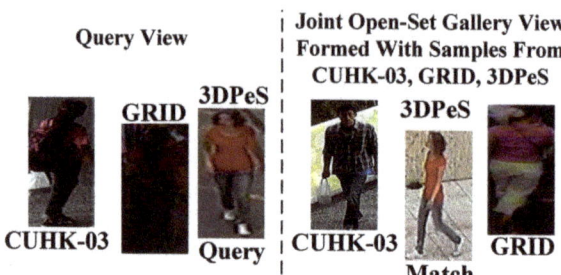

**Figure 14.** Open-World Scenario where it is not necessary that all the Queries have match in Gallery view. Thus, the metric $M_G$ needs to be robust against unknown persons seen in the open world while the target is moved out of perception.

**Table 5.** Results on Open-World Re-ID.

| Methods | Target:PRID | | Target:GRID | | Target:VIPeR | | Target:i-LIDS | |
|---|---|---|---|---|---|---|---|---|
| | Rank-1 | mAP | Rank-1 | mAP | Rank-1 | mAP | Rank-1 | mAP |
| CBAM [45] | 26.95 | 51.00 | 18.43 | 34.26 | 23.48 | 44.38 | 41.07 | 68.98 |
| APN [51] | 28.63 | 53.29 | 19.34 | 36.60 | 24.80 | 46.21 | 43.32 | 70.87 |
| RAGA [44] | 31.05 | 55.74 | 19.92 | 38.34 | 27.42 | 49.49 | 45.82 | 73.03 |
| HLGAT [52] | 32.98 | 57.06 | 20.27 | 40.97 | 28.14 | 51.64 | 47.12 | 75.55 |
| SRN [35] | 37.34 | 60.49 | 23.13 | 44.17 | 30.43 | 53.42 | 50.38 | 77.00 |
| MoE [38] | 43.01 | 65.09 | 26.36 | 47.15 | 33.01 | 56.10 | 52.18 | 80.67 |
| Meta [37] | 45.70 | 65.39 | 28.43 | 49.68 | 32.11 | 55.27 | 51.35 | 80.29 |
| Our $M_G$ | 68.02 | 80.86 | 41.3 | 54.4 | 56.09 | 78.1 | 76.57 | 86.42 |

Furthermore, in Figure 15, attention maps and the corresponding rise in *Rank*-1 identification accuracy are analyzed. In Figure 15b, it is revealed that $M_G$ exploits part-attention module and learns cross-representations that declines large number of impostors and nontarget Queries in the open world; thus, *Rank*-1 accuracy at FTR 0.1% rises to 41.3% from 35.87% in Figure 15a.

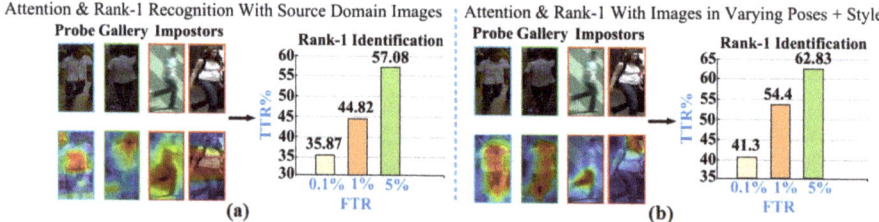

**Figure 15.** Rank-1 Result Comparison on VIPeR & GRID for (**a**) G.Sc. vs. RGB (**b**) Sk. vs. RGB.

### 4.3.4. Challenging Open-World Scenario

Though, metric $M_G$ is evaluated in the open world; however, open-set Gallery in real-world is far more challenging where Probe-Gallery pairs can be seen in different modals and styles. In Figure 16, this complex scenario is shown visually.

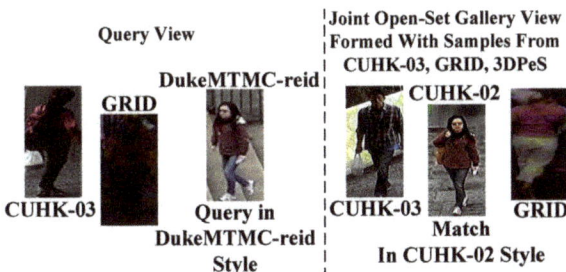

**Figure 16.** In the Challenging Open-World Scenario, the Query image can be seen in different domains, styles, or modalities, while the match is needed to be found from any other random domain, e.g., the CUHK-02 domain, while the matching image has a style or modality different from that of the Query image.

Therefore, images in Query and in joint Gallery are randomly transformed into different modals and styles during testing, then $M_G$ finds the matches from the joint multimodal multistyle open-set Gallery, and the results are summarized in Table 6.

**Table 6.** Results on Challenging Open-World Re-ID.

| Methods | Target:PRID | | Target:GRID | | Target:VIPeR | | Target:i-LIDS | |
|---|---|---|---|---|---|---|---|---|
| | Rank-1 | mAP | Rank-1 | mAP | Rank-1 | mAP | Rank-1 | mAP |
| CBAM [45] | 19.34 | 38.27 | 10.66 | 20.03 | 17.58 | 33.71 | 28.0 | 52.14 |
| APN [51] | 21.01 | 47.75 | 12.12 | 25.34 | 18.97 | 36.40 | 32.31 | 59.86 |
| RAGA [44] | 22.65 | 49.38 | 12.23 | 25.96 | 19.72 | 38.04 | 34.54 | 60.43 |
| HLGAT [52] | 23.54 | 51.03 | 14.18 | 28.81 | 21.17 | 40.47 | 36.84 | 64.46 |
| SRN [35] | 26.09 | 54.80 | 16.83 | 34.34 | 23.24 | 43.22 | 39.13 | 68.08 |
| MoE [38] | 29.37 | 59.0 | 19.04 | 38.71 | 25.79 | 48.35 | 42.86 | 71.53 |
| Meta [37] | 32.45 | 60.07 | 21.22 | 40.48 | 25.35 | 47.98 | 42.24 | 71.22 |
| Our $M_G$ | 65.11 | 80.34 | 37.51 | 51.77 | 52.08 | 76.9 | 76.04 | 86.0 |

$M_G$, in contrast to [35,37,38,44], is an open-set metric with inherent ability to match cross-modal and cross-style Probe-Gallery pairs in the nonlinear outdoor environment, while the $M_G$ also simultaneously rejects large number of impostors; therefore, in Table 6 $M_G$ attains 65.11%, 37.51%, 52.08%, and 76.04% *Rank*-1 identification at FTR 0.1% on PRID, GRID, VIPeR, and i-LIDS datasets, respectively.

Furthermore, in Figure 17, attention maps and in Figure 18 retrieval results are analyzed in the open-world. Our attention maps in Figure 17 are more focused on individual parts and do not discard unique valuable cues from different filters, therefore, our attention maps are more robust against interenvironment and intraenvironment style and modality transforms than SRN [35] and RAGA [44]. Consequently, $M_G$ successfully identifies cross-modal pair (Q1,G1), and cross-style pairs (Q1,G2) and (Q2,G2) at *Rank*=3, *Rank*=1, and *Rank*=1, respectively, in Figure 18; there exists large number of impostors in Gallery in scenario#1, scenario#2, and scenario#3, whereas [35,37,38] inherently lack matching cross-modal and cross-style images. In scenario#4 $M_G$, it also matches cross-style pair (Q2,G3) at *Rank*=2; even the G3 is seen in COCO domain style, where underlying nonlinear transforms and impostors in COCO domain affect the retrieval results.

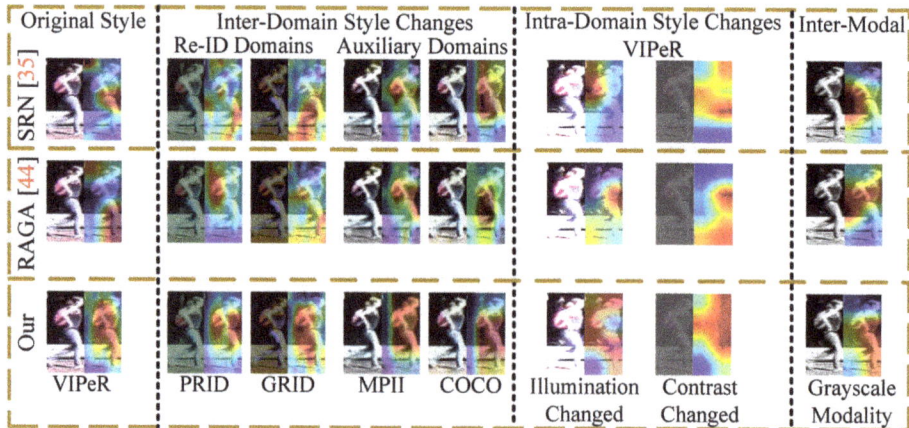

**Figure 17.** Comparison of Our Part Attention Maps with SRN [35] and RAGA [44].

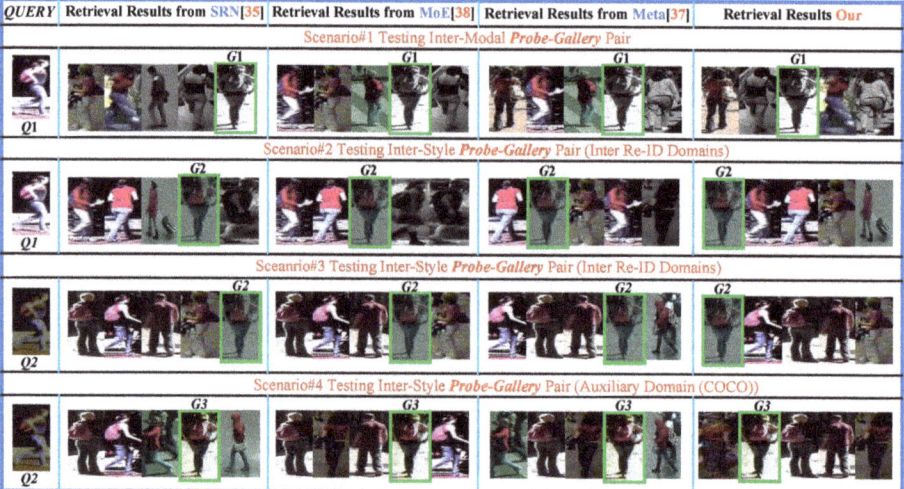

**Figure 18.** Retrieval results of Our $M_G$, SRN [35], Meta [37], and MoE [38] in real-world scenarios. Green Rectangles contain correct matches.

### 4.3.5. Cross-Modal Scenario

Our $M_G$ is also evaluated against RGB-Infrared matching on SYSU-MM01 dataset. $M_G$ compared to [48,54] is trained in real multi-modal open-world to tackle complex nonlinear transforms and can resist large number of impostors; thus, in Table 7, $M_G$ attains 64.93% and 72.58% *Rank*-1 identification on All-search and Indoor-Search, respectively.

**Table 7.** Cross-Modal Comparison on SYSU-MM01.

| Methods | SYSU-MM01 | | | | | |
|---|---|---|---|---|---|---|
| | All-Search | | | Indoor-Search | | |
| | R = 1 | R = 10 | mAP | R = 1 | R = 10 | mAP |
| Hi-CMD [53] | 34.94 | 77.58 | 35.94 | - | - | - |
| GECNet [48] | 53.37 | 89.86 | 51.83 | 60.60 | 94.29 | 62.89 |
| cm-SSFT [54] | 61.6 | 89.2 | 63.2 | 70.5 | 94.9 | 72.6 |
| Our $M_G$ | 64.93 | 92.31 | 66.04 | 72.58 | 94.89 | 72.98 |

### 4.3.6. Computational Complexity

To run $M_G$ in the real world, all convolutions are implemented as mixed depthwise separable convolutions [41]. The computation burden for depthwise separable convolution is $K \times K \times M \times H \times W + M \times N \times H \times W$, whereas the computation burden for standard convolutions are $K \times K \times M \times N \times H \times W$. If kernel size $K \times K$ is $3 \times 3$, and feature map dimensions $H \times W$ are $56 \times 56$, then, the cost of depthwise separable convolution is $3 \times 3 \times 256 \times 56 \times 56 + 256 \times 256 \times 56 \times 56 = 212{,}746{,}240$ multiplications, while the cost for standard convolutions are $3 \times 3 \times 56 \times 56 \times 256 \times 256 = 1{,}849{,}688{,}064$ multiplications. Clearly, depthwise separable convolution lowers the computation cost by 88%, and thus $M_G$ is realized on smart embedded camera $Hi3516DV300$.

### 4.3.7. Computation Time

Running time of metric $M_G$ for different image sizes for one forward pass on $Hi3516DV300$ are given in Table 8, where for image size $224 \times 224$ $M_G$ takes 29.4ms to process one single pass and obtain similarity.

**Table 8.** Computation Time for Different Image Sizes.

| Image Size | Time/Image (ms) | Device Platform |
| --- | --- | --- |
| $324 \times 324$ | 52 | $Hi3516DV300$ |
| $256 \times 256$ | 38.4 | $Hi3516DV300$ |
| $224 \times 224$ | 29.4 | $Hi3516DV300$ |

### 4.4. Reidentification-Based Tracking Experiments

To evaluate the learned generic metric $M_G$ in the person tracking applications, our work performed several experiments in outdoor to track target. In Figure 19, the complete framework is shown where the learned reidentification metric $M_G$ is integrated with pretrained CNN tracker [4].

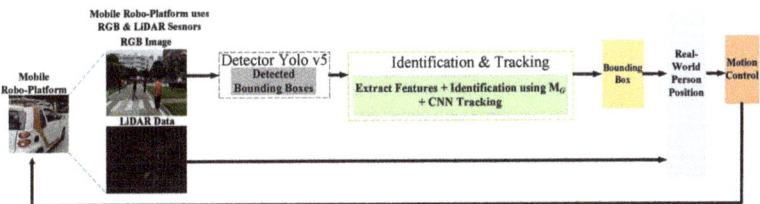

**Figure 19.** System Overview of Person-Following Robot in Our Work.

The input in Figure 19 is RGB image with LiDAR data, where the RGB image is first sent to Yolov5 detector [55] to obtain the bounding boxes. Reidentification module with the tracker in the next step in Figure 19 then takes the detected bounding boxes and identifies the target using the generic metric $M_G$ and sends the identified box to the pretrained tracker [4] as the input dynamic template of the person to be tracked. To perform the identification of target, Re-ID metric $M_G$ in our work uses prestored features of the target person. Finally, using the tracker prediction box and LiDAR data, the real-world position of the target is then updated. Motion control in last uses the updated position to generate the actuation signals for the robocar.

#### 4.4.1. Experiment Setup and Testing Scenarios

All the experiments in our work are conducted in outdoor, where the person is tracked in different environments, while undergoing illumination and background variations.

In Figure 20, it can be seen the person is tracked in three challenging and complex outdoor environments, which are referred to as Scenario#1 (Seq#1), Scenario#2 (Seq#2), and

Scenario#3 (Seq#3). In each of these scenarios, the person is tracked for 115 s, 115 s, and 120 s, respectively, while the person in three different scenarios experiences crowding and occlusion with distractors (distractors in our work are the person occluding the target but are the not the impostors of target) and impostors in the scenes, and at several occasions, the person also completely moved out of the perception of mobile roboplatform. The mobile roboplatform visual system consists of RGB camera and LiDAR sensors, both are mounted on top of roboplatform, as shown in Figure 21.

**Figure 20.** Person-following roboplatform in outdoor world. (**a**) Following the target in black clothing on Road-1; (**b**) Following the target in crowded and occluded scenario on Road-2; (**c**) Continue following the target on Road-3 in varying pose; (**d**) Continue following the target on Road-3 even even there are distractor and impostor.

**Figure 21.** Robocar Robotic Platform mounted with Senors in Outside World.

Mounted RGB sensor take images with resolution 1080P with 30 frames per second. Furthermore, the depth and RGB data both are processed with on board CPU i7 8700k on the roboplatform, where the memory size is 16G.

4.4.2. Evaluation Metrics and Comparison

To make fair evaluation and comparison with other state-of-the-art works, our work uses four standard metrics [2] for quantitative evaluation. These metrics both evaluate reidentification performance and tracking success and are defined as: Correctly Identified and Tracked (CT): meaning the person is correctly identified and tracked successfully, Correctly Loss (CL): meaning it is correctly identified the person is not in the scene (either moved out or completely occluded) and thus successfully loss, Wrongly Identified and Tracked (WT): meaning the identification metric wrongly identified the impostor or distractor, and the tracker wrongly tracks the distractor or impostor, and the last metric is Wrongly Loss (WL): meaning the identification metric assumes there is no target in the scene; however, there is target present in the scene, and the tracker wrongly loss tracking.

4.4.3. Results and Analysis

Our work compares the performance with three state-of-the-art trackers, which are STARK [4], DiMP [5], and ATOM [6]. Table 9 below summarizes the results for tracking in all the three scenarios and presents the tracking results for STARK [4], DiMP [5], and ATOM [6], as well as results of our model for the evaluation metrics CT, CL, WT, and WL with the total tracking time in seconds and the percentage of identified and tracked frames in the given Seq#1, Seq#2, and Seq#3.

Table 9. Person Reidentification Results for Tracking in the Outdoor World.

|  |  | Methods | | | | | | | |
|---|---|---|---|---|---|---|---|---|---|
|  |  | DiMP [5] | | ATOM [6] | | STARK [4] | | Ours | |
|  |  | Time (Sec) | Frames (%) | Time (Sec) | Frames (%) | Time (Sec) | Frames (%) | Time (Sec) | Frames (%) |
| Seq. #1 | CT | 52 | 45.2 | 57 | 49.57 | 62 | 53.91 | 80 | 69.6 |
|  | CL | 20 | 17.39 | 17 | 14.78 | 16 | 13.92 | 25 | 21.74 |
|  | WT | 13 | 11.29 | 15 | 13.05 | 13 | 11.31 | 6 | 5.21 |
|  | WL | 30 | 26.09 | 26 | 22.61 | 24 | 20.87 | 4 | 3.48 |
| Seq. #2 | CT | 31 | 26.96 | 44 | 38.26 | 56 | 48.69 | 73 | 63.48 |
|  | CL | 17 | 14.78 | 10 | 8.69 | 12 | 10.43 | 25 | 21.74 |
|  | WT | 22 | 19.13 | 19 | 16.52 | 15 | 13.04 | 8 | 6.96 |
|  | WL | 45 | 39.13 | 42 | 27.83 | 32 | 26.08 | 9 | 7.82 |
| Seq. #3 | CT | 28 | 24.35 | 38 | 33.04 | 40 | 34.78 | 69 | 60.02 |
|  | CL | 14 | 12.17 | 8 | 6.95 | 10 | 8.69 | 29 | 25.22 |
|  | WT | 26 | 22.61 | 23 | 20 | 20 | 17.39 | 12 | 10.43 |
|  | WL | 52 | 45.22 | 51 | 44.35 | 50 | 43.48 | 9 | 7.83 |

Results in Seq#1: In Table 9 in Seq#1, our work Correctly Tracked (CT) the target person in 69.6% of total frames, whereas original STARK [4] with no identification module Correctly Tracked (CT) target person in 53.91% of total frames. This shows that tracking when complemented with reidentification module can largely improve the target tracking in different nonlinear scenes.

Furthermore, in Figure 22, visual comparison is shown, where it can be seen that when there are no distractors in the frames frame#90 and frame#146, then the identification of target and its tracking is easier; however, when the target is occluded with object in frame#155 in Figure 22, then a few trackers including STARK [4], DiMP [5], and ATOM [6] start tracking distractors. The reason is obvious: STARK [4], DiMP [5], and ATOM [6] all lack the identification ability to distinguish between target and distractor. In addition, when the target is completely moved out of the perception, such as in frames frame#164, frame#165, and frame#222 in Figure 22, then these trackers still continue wrongly tracking either distractors in frame#164 and in frame#165 or impostors in frame#222 since there are no reidentification model to verify if the detected person is a real target or impostor. In last, in Seq#1, it is also evaluated that how the model performs under the scenarios when the person is completely occluded with impostor, such as in frames frame#767 and frame#779. In Figure 22, in frame#767 and frame#779, it can be seen that our model with reidentification metric can successfully identify the person and thereby successfully update the dynamic image template of target to improve tracking, and in last, using both detected bounding box and LiDAR data it can robustly track the person during occlusion. In contrast, in Figure 22, when the person is occluded in frame#767 and frame#779, then DiMP [5] and ATOM [6] wrongly start tracking the impostor in frames frame#767 and frame#779.

Figure 22. Person-Following Tests in Scenario#1 (seq#1), where the Target moves out of Robocar perception and faces crowding and distractions.

Results in Seq#2: In Table 9 in Seq#2, our work Correctly Tracked (CT) the target person in 63.48% of total frames, whereas original STARK [4] with no identification module Correctly Tracked (CT) target person in 48.69% of total frames.

The results are lower than in Seq#1 because Seq#2 is more challenging than Seq#1, where in Seq#2 person moves in a varying illumination environment, while the background noise also affect reidentification, as shown in Figure 23. Furthermore, in Figure 23, the person is also occluded by distractor and impostor in the scene. In Seq#2 in frame#151 and frame#186, our reidentification metric $M_G$ can identify the target successfully in the presence of impostor while other trackers DiMP [5] and ATOM [6] follow impostor person. Furthermore, in frame#191 in Figure 23, when both impostor and target are seen in the scene, the target is occluded by both impostor and distractor; then, in such a scenario, still our identification metric $M_G$ can address both occlusion and illumination variation to identify the target and thereby improve its tracking. Similarly, in other complex scenes, when target is completely moved out of perception in frame#211 in Figure 23 and when the target is occluded by both impostor and distractor in frame#606 in Figure 23, then, in both scenes, our reidentification metric $M_G$ helps tracker to not follow impostor, and the mobile roboplatform stops and wait for the target to reappear and identified.

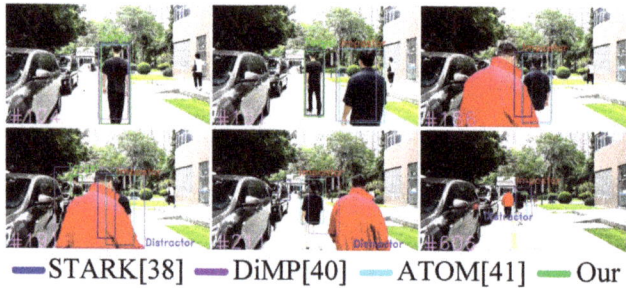

**Figure 23.** Person-Following Tests in Scenario#2 (seq#2), where the Target moves in varying illumination environment, and is occluded with impostors and distractors, while, Target also moves out of perception.

Results in Seq#3: In Table 9 in Seq#3, our work Correctly Tracked (CT) the target person in 60.02% of total frames, whereas original STARK [4] with no identification module Correctly Tracked (CT) target person in 34.78% of total frames. Seq#3 is far more challenging than Seq#2 and Seq#1, where in Seq#3 the person simultaneously undergoes pose and viewpoint changes, illumination and background variations, as well as, distracted and occluded by distractors and impostors, as shown in Figure 24. However, still, our reidentification metric $M_G$ successfully identified and tracked the person in 60.02% of total frames, compared to the state of art STARK [4], DiMP [5] and ATOM [6]. This is mainly due to the correct identification of the target, which helps the tracking. In Figure 24, in frame#174 in Seq#3, the target person undergoing varying posture, while impostors are in the seen; however, still the learned generic metric $M_G$ discriminates the target well. Furthermore, in frame#282 and frame#286 in Figure 24, when the target is fully occluded by distractor, while an impostor is nearby, still $M_G$ successfully discriminates the impostor, whereas DiMP [5] and ATOM [6] track the impostors. In addition, in frame#293 in Figure 24, when the target reappears after full occlusion, then both our model and STARK [4] track the target well; however, STARK [4] has a little higher localization error in tracking than our model.

Last, in Figure 24 in frames frame#354 and frame#362, the target again occluded by impostor, while, also undergo illumination variations. Though, the situation is challenging, but the metric $M_G$ is trained to address both illumination variations and style variations, and it is resistant against impostors; therefore, $M_G$ continues identifying the person in consecutive frames, i.e., in frame#354, frame#362, and in frame#373.

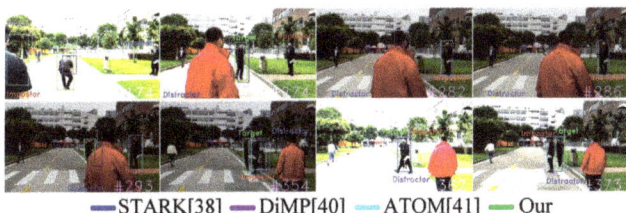

**Figure 24.** Person-Following Tests in Scenario#3 (seq#3), where the Target moves in varying illumination environment, while the Target is completely occluded with impostors and distractors, and also suffers due to changing backgrounds.

## 5. Conclusions & Future Directions

This work learns lightweight domain generic metric in the multimodal open world for person-following robots to address the practical world challenges face by person-following robots including nonlinear pose, viewpoint, style, and multimodal transforms, and a novel part-attention module is proposed to learn attention weighted cross-representations to address displacement and occlusion of parts. Thereby, the learned generic metric can resist large number of impostors and nontarget queries in the open world, while the learned metric is also lightweight and can run on robotic platform. Furthermore, future research will focus to improve the learned domain generic reidentification metric to solve multiscale and night reidentification problems for person-following robots.

**Author Contributions:** Conceptualization, M.A.S., Y.O., T.L., G.J.; methodology, M.A.S.; software, M.A.S.; validation, M.A.S., G.J.; writing—original draft preparation, M.A.S.; writing—review and editing, Y.O., T.L., G.J.; supervision, Y.O., T.L. All authors have read and agreed to the published version of the manuscript.

**Funding:** This research is funded by the National Key Research and Development Program of China (No. 2018AAA0103001), National Natural Science Foundation of China (No. U1813208, 62173319, 62063006), Guangdong Basic and Applied Basic Research Foundation (No. 2020B1515120054), Shenzhen Fundamental Research Program (No. JCYJ20200109115610172).

**Institutional Review Board Statement:** Not applicable.

**Informed Consent Statement:** Not applicable.

**Data Availability Statement:** Not applicable.

**Conflicts of Interest:** The authors declare no conflict of interest.

## References

1. Koide, K.; Miura, J. Identification of a specific person using color, height, and gait features for a person following robot. *Robot. Auton. Syst.* **2016**, *84*, 76–87. [CrossRef]
2. Koide, K.; Miura, J.; Menegatti, E. Monocular person tracking and identification with on-line deep feature selection for person following robots. *Robot. Auton. Syst.* **2020**, *124*, 103348. [CrossRef]
3. Ghimire, A.; Zhang, X.; Javed, S.; Dias, J.; Werghi, N. Robot Person Following in Uniform Crowd Environment. *arXiv* **2022**, arXiv:2205.10553.
4. Yan, B.; Peng, H.; Fu, J.; Wang, D.; Lu, H. Learning Spatio-Temporal Transformer for Visual Tracking. In Proceedings of the IEEE International Conference on Computer Vision (ICCV), Montreal, QC, Canada, 10–17 October 2021; pp. 10448–10457.
5. Bhat, G.; Danelljan, M.; Gool, L.V.; Timofte, R. Learning Discriminative Model Prediction for Tracking. In Proceedings of the IEEE International Conference on Computer Vision (ICCV), Seoul, Republic of Korea, 27 October–2 November 2019; pp. 6181–6190.
6. Danelljan, M.; Bhat, G.; Khan, F.S.; Felsberg, M. ATOM: Accurate Tracking by Overlap Maximization. In Proceedings of the IEEE Conference on Computer Vision and Pattern Recognition (CVPR), Long Beach, CA, USA, 15–20 June 2019; pp. 4655–4664.
7. Arras, K.O.; Mozos, O.M.; Burgard, W. Using Boosted Features for the Detection of People in 2D Range Data. In Proceedings of the IEEE International Conference on Robotics and Automation (ICRA), Rome, Italy, 10–14 April 2007; pp. 3402–3407.
8. Leigh, A.; Pineau, J.; Olmedo, N.; Zhang, H. Person tracking and following with 2D laser scanners. In Proceedings of the IEEE International Conference on Robotics and Automation (ICRA), Seattle, WA, USA, 26–30 May 2015; pp. 726–733.

9. Chen, B.X.; Sahdev, R.; Tsotsos, J.K. Person Following Robot Using Selected Online Ada-Boosting with Stereo Camera. In Proceedings of the Conference on Computer and Robot Vision (CRV), Edmonton, AB, Canada, 16–19 May 2017; pp. 48–55.
10. Munaro, M.; Menegatti, E. Fast RGB-D people tracking for service robots. *Auton. Robot.* **2014**, *37*, 227–242. [CrossRef]
11. Chen, Z.; Zhong, B.; Li, G.; Zhang, S.; Ji, R. Siamese Box Adaptive Network for Visual Tracking. In Proceedings of the IEEE Conference on Computer Vision and Pattern Recognition (CVPR), Seattle, WA, USA, 13–19 June 2020; pp. 6668–6677.
12. Flores, S.; Jost, J. Person Re-Identification on a Mobile Robot Using a Depth Camera. In Proceedings of the IEEE International Symposium on Industrial Electronics (ISIE), Anchorage, AK, USA, 1–3 June 2022; pp. 115–122.
13. Tsai, T.H.; Yao, C.H. A robust tracking algorithm for a human-following mobile robot. *IET Image Process.* **2020**, *15*, 786–796. [CrossRef]
14. Pang, L.; Cao, Z.; Yu, J.; Guan, P.; Chen, X.; Zhang, W. A Robust Visual Person-Following Approach for Mobile Robots in Disturbing Environments. *IEEE Syst. J.* **2020**, *14*, 2965–2968. [CrossRef]
15. Koide, K.; Miura, J. Convolutional Channel Features-Based Person Identification for Person Following Robots. In *International Conference on Intelligent Autonomous Systems*; Springer: Berlin/Heidelberg, Germany, 2018; Volume 867, pp. 186–198.
16. Liu, J.; Chen, X.; Wang, C.; Zhang, G.; Song, R. A person-following method based on monocular camera for quadruped robots. *Biomim. Intell. Robot.* **2022**, *2*, 100058. [CrossRef]
17. Narayan, N.; Sankaran, N.; Setlur, S.; Govindaraju, V. Re-identification for Online Person Tracking by Modeling Space-Time Continuum. In Proceedings of the IEEE Conference on Computer Vision and Pattern Recognition Workshops (CVPRW), Salt Lake City, UT, USA, 18–22 June 2018; pp. 1519–1527.
18. Nissimagoudar, P.C.; Iyer, N.C.; Gireesha, H.M.; Pillai, P.; Mallapur, S. Multi-pedestrian Tracking and Person Re-identification. In *International Conference on Soft Computing and Pattern Recognition*; Springer: Berlin/Heidelberg, Germany, 2022; pp. 178–185.
19. Babaee, M.; Li, Z.; Rigoll, G. A dual CNN–RNN for multiple people tracking. *Neurocomputing* **2019**, *368*, 69–83. [CrossRef]
20. Neff, C.; Mendieta, M.; Mohan, S.; Baharani, M.; Rogers, S.; Tabkhi, H. REVAMP2T: Real-Time Edge Video Analytics for Multicamera Privacy-Aware Pedestrian Tracking. *IEEE Internet Things J.* **2020**, *7*, 2591–2602. [CrossRef]
21. Fleuret, F.; Shitrit, H.B.; Fua, P. Re-identification for Improved People Tracking. In *Person Re-Identification*; Springer: London, UK, 2014; pp. 309–330.
22. Liu, Q.; Chen, D.; Chu, Q.; Yuan, L.; Liu, B.; Zhang, L.; Yu, N. Online multi-object tracking with unsupervised re-identification learning and occlusion estimation. *Neurocomputing* **2022**, *483*, 333–347. [CrossRef]
23. Welsh, J.B. Real-Time Pose Based Human Detection and Re-identification with a Single Camera for Robot Person Following. Ph.D. Thesis, University of Maryland, College Park, MD, USA, 2017.
24. Thakran, A.; Agarwal, A.; Mahajan, P.; Kumar, S. Vision-Based Human-Following Robot. In Advances in Data Computing, Communication and Security. Springer: Singapore, 2022; pp. 443–449.
25. Algabri, R.; Choi, M.T. Deep-Learning-Based Indoor Human Following of Mobile Robot Using Color Feature. *Sensors* **2020**, *20*, 2699. [CrossRef] [PubMed]
26. Chebotareva, E.; Hsia, K.H.; Yakovlev, K.; Magid, E. Laser Rangefinder and Monocular Camera Data Fusion for Human-Following Algorithm by PMB-2 Mobile Robot in Simulated Gazebo Environment. In Proceedings of the 15th International Conference on Electromechanics and Robotics "Zavalishin's Readings", Ufa, Russia, 15–18 April 2020; pp. 357–369.
27. Condés, I.; Cañas, J.M. Person Following Robot Behavior Using Deep Learning. In *Advances in Physical Agents*; Springer: Cham, Switzerland, 2019; pp. 147–161.
28. Anuradha, U.A.D.N.; Kumari, K.W.S.N.; Chathuranga, K.W.S. Human detection and following robot. *Int. J. Sci. Technol. Res.* **2020**, *9*, 6359–6363.
29. Montesdeoca, J.; Toibero, J.M.; Jordan, J.; Zell, A.; Carelli, R. Person-Following Controller with Socially Acceptable Robot Motion. *Robot. Auton. Syst.* **2022**, *153*, 104075. [CrossRef]
30. Gemerek, J.; Ferrari, S.; Wang, B.H.; Campbell, M.E. Video-guided Camera Control for Target Tracking and Following. *IFAC-PapersOnLine* **2019**, *51*, 176–183. [CrossRef]
31. Yan, S.; Zhang, Y.; Xie, M.; Zhang, D.; Yu, Z. Cross-domain person re-identification with pose-invariant feature decomposition and hypergraph structure alignment. *Neurocomputing* **2022**, *467*, 229–241. [CrossRef]
32. Li, Y.; He, J.; Zhang, T.; Liu, X.; Zhang, Y.; Wu, F. Diverse part discovery: Occluded person re-identification with part-aware transformer. In Proceedings of the IEEE Conference on Computer Vision and Pattern Recognition (CVPR), Nashville, TN, USA, 20–25 June 2021; pp. 2898–2907.
33. Deng, W.; Zheng, L.; Ye, Q.; Guoliang, D.W.; Zheng, L.; Ye, Q.; Kang, G.; Yang, Y.; Jiao, J. Image-Image Domain Adaptation with Preserved Self-Similarity and Domain-Dissimilarity for Person Re-identification. In Proceedings of the IEEE Conference on Computer Vision and Pattern Recognition (CVPR), Salt Lake City, UT, USA, 18–23 June 2018; pp. 994–1003.
34. Zhong, Z.; Zheng, L.; Zheng, Z.; Li, S.; Yang, Y. Camera Style Adaptation for Person Re-identification. In Proceedings of the IEEE Conference on Computer Vision and Pattern Recognition (CVPR), Salt Lake City, UT, USA, 18–23 June 2018; pp. 5157–5166.
35. Jin, X.; Lan, C.; Zeng, W.; Chen, Z.; Zhang, L. Style Normalization and Restitution for Generalizable Person Re-Identification. In Proceedings of the IEEE Conference on Computer Vision and Pattern Recognition (CVPR), Seattle, WA, USA, 13–19 June 2020; pp. 3140–3149.
36. Jia, J.; Ruan, Q.; Hospedales, T.M. Frustratingly Easy Person Re-Identification: Generalizing Person Re-ID in Practice. In Proceedings of the British Machine Vision Conference (BMVC), Cardiff, UK, 9–12 September 2019; pp. 1–14.

37. Choi, S.; Kim, T.; Jeong, M.; Park, H.; Kim, C. Meta Batch-Instance Normalization for Generalizable Person Re-Identification. In Proceedings of the IEEE Conference on Computer Vision and Pattern Recognition (CVPR), Nashville, TN, USA, 20–25 June 2021; pp. 3424–3434.
38. Dai, Y.; Li, X.; Liu, J.; Tong, Z.; Duan, L.Y. Generalizable Person Re-identification with Relevance-aware Mixture of Experts. In Proceedings of the IEEE Conference on Computer Vision and Pattern Recognition (CVPR), Nashville, TN, USA, 20–25 June 2021; pp. 16140–16149.
39. Zhu, Z.; Huang, T.; Shi, B.; Yu, M.; Wang, B.; Bai, X. Progressive Pose Attention Transfer for Person Image Generation. In Proceedings of the IEEE Conference on Computer Vision and Pattern Recognition (CVPR), Long Beach, CA, USA, 15–20 June 2019; pp. 2347–2356.
40. He, K.; Zhang, X.; Ren, S.; Sun, J. Deep Residual Learning for Image Recognition. In Proceedings of the IEEE Conference on Computer Vision and Pattern Recognition (CVPR), Las Vegas, NV, USA, 27–30 June 2016; pp. 770–778.
41. Tan, M.; Le, Q.V. MixConv: Mixed Depthwise Convolutional Kernels. In Proceedings of the British Machine Vision Conference (BMVC), Cardiff, UK, 9–12 September 2019; pp. 1–13.
42. Zagoruyko, S.; Komodakis, N. Wide Residual Networks. In Proceedings of the British Machine Vision Conference (BMVC), York, UK, 19–22 September 2016; pp. 1–12.
43. Ma, N.; Zhang, X.; Zheng, H.T.; Sun, J. ShuffleNet V2: Practical Guidelines for Efficient CNN Architecture Design. In Proceedings of the European Conference on Computer Vision (ECCV), Munich, Germany, 8–14 September 2018; pp. 122–138.
44. Zhang, Z.; Lan, C.; Zeng, W.; Jin, X.; Chen, Z. Relation-Aware Global Attention for Person Re-Identification. In Proceedings of the IEEE Conference on Computer Vision and Pattern Recognition (CVPR), Seattle, WA, USA, 13–19 June 2020; pp. 3183–3192.
45. Woo, S.; Park, J.; Lee, J.Y.; Kweon, I.S. CBAM: Convolutional Block Attention Module. In Proceedings of the European Conference on Computer Vision (ECCV), Munich, Germany, 8–14 September 2018; pp. 3–19.
46. Wang, F.; Zuo, W.; Lin, L.; Zhang, D.; Zhang, L. Joint Learning of Single-Image and Cross-Image Representations for Person Re-identification. In Proceedings of the IEEE Conference on Computer Vision and Pattern Recognition (CVPR), Las Vegas, NV, USA, 27–30 June 2016; pp. 1288–1296.
47. Chen, W.; Chen, X.; Zhang, J.; Huang, K. Beyond Triplet Loss: A Deep Quadruplet Network for Person Re-identification. In Proceedings of the IEEE Conference on Computer Vision and Pattern Recognition (CVPR), Honolulu, HI, USA, 21–26 July 2017; pp. 1320–1329.
48. Zhong, X.; Lu, T.; Huang, W.; Ye, M.; Jia, X.; Lin, C.W. Grayscale Enhancement Colorization Network for Visible-infrared Person Re-identification. IEEE Trans. Circuits Syst. Video Technol. 2022, 32, 1418–1430. [CrossRef]
49. Szegedy, C.; Vanhoucke, V.; Ioffe, S.; Shlens, J.; Wojna, Z. Rethinking the Inception Architecture for Computer Vision. In Proceedings of the IEEE Conference on Computer Vision and Pattern Recognition (CVPR), Las Vegas, NV, USA, 27–30 June 2016; pp. 2818–2826.
50. He, K.; Zhang, X.; Ren, S.; Sun, J. Delving Deep into Rectifiers: Surpassing Human-Level Performance on ImageNet Classification. In Proceedings of the IEEE International Conference on Computer Vision (ICCV), Santiago, Chile, 7–13 December 2015; pp. 1026–1034.
51. Li, X.; Wu, A.; Zheng, W.S. Adversarial Open-World Person Re-Identification. In Proceedings of the European Conference on Computer Vision (ECCV), Munich, Germany, 8–14 September 2018; pp. 287–303.
52. Zhang, Z.; Zhang, H.; Liu, S. Person Re-identification using Heterogeneous Local Graph Attention Networks. In Proceedings of the IEEE Conference on Computer Vision and Pattern Recognition (CVPR), Nashville, TN, USA, 20–25 June 2021; pp. 12131–12140.
53. Choi, S.; Lee, S.; Kim, Y.; Kim, T.; Kim, C. Hi-CMD: Hierarchical Cross-Modality Disentanglement for Visible-Infrared Person Re-Identification. In Proceedings of the IEEE Conference on Computer Vision and Pattern Recognition (CVPR), Seattle, WA, USA, 13–19 June 2020; pp. 10254–10263.
54. Lu, Y.; Wu, Y.; Liu, B.; Zhang, T.; Li, B.; Chu, Q.; Yu, N. Cross-Modality Person Re-Identification With Shared-Specific Feature Transfer. In Proceedings of the IEEE Conference on Computer Vision and Pattern Recognition (CVPR), Seattle, WA, USA, 13–19 June 2020; pp. 13376–13386.
55. Jocher, G. YOLOv5 by Ultralytics. Released date: 2020-5-29. Available online: https://github.com/ultralytics/yolov5 (accessed on 1 November 2022).

**Disclaimer/Publisher's Note:** The statements, opinions and data contained in all publications are solely those of the individual author(s) and contributor(s) and not of MDPI and/or the editor(s). MDPI and/or the editor(s) disclaim responsibility for any injury to people or property resulting from any ideas, methods, instructions or products referred to in the content.

Article

# Deep Instance Segmentation and Visual Servoing to Play Jenga with a Cost-Effective Robotic System

Luca Marchionna [1,†], Giulio Pugliese [1,†], Mauro Martini [1,2,*], Simone Angarano [1,2], Francesco Salvetti [1,2] and Marcello Chiaberge [1,2]

1 Department of Electronics and Telecommunications (DET), Politecnico di Torino, 10129 Torino, Italy
2 PIC4SeR Interdepartmental Centre for Service Robotics, 10129 Torino, Italy
* Correspondence: mauro.martini@polito.it; Tel.: +39-3936286325
† These authors contributed equally to this work.

**Abstract:** The game of Jenga is a benchmark used for developing innovative manipulation solutions for complex tasks. Indeed, it encourages the study of novel robotics methods to successfully extract blocks from a tower. A Jenga game involves many traits of complex industrial and surgical manipulation tasks, requiring a multi-step strategy, the combination of visual and tactile data, and the highly precise motion of a robotic arm to perform a single block extraction. In this work, we propose a novel, cost-effective architecture for playing Jenga with e.Do, a 6DOF anthropomorphic manipulator manufactured by Comau, a standard depth camera, and an inexpensive monodirectional force sensor. Our solution focuses on a visual-based control strategy to accurately align the end-effector with the desired block, enabling block extraction by pushing. To this aim, we trained an instance segmentation deep learning model on a synthetic custom dataset to segment each piece of the Jenga tower, allowing for visual tracking of the desired block's pose during the motion of the manipulator. We integrated the visual-based strategy with a 1D force sensor to detect whether the block could be safely removed by identifying a force threshold value. Our experimentation shows that our low-cost solution allows e.DO to precisely reach removable blocks and perform up to 14 consecutive extractions in a row.

**Keywords:** Jenga; robotic arm; deep instance segmentation; visual servoing; sensor fusion

## 1. Introduction

In recent years, visual-based control strategies have successfully spread in a wide variety of robotics contexts [1]. Nowadays, advances in computer vision for robotic perception are strictly tied to deep learning (DL). DL has been used in many robotics applications where objects must be detected [2] or segmented [3] to address manipulation tasks, demonstrating competitive advantages compared to classic image processing algorithms in terms of accuracy and robustness. For instance, relevant works in the precision agriculture field have been proposed in recent years to support autonomous navigation [4,5], harvesting [6], and spraying [7]. Intelligent DL-based behaviors are also desired for visual-based robotic surgery to detect and segment instruments [8,9], and in many industrial robotic tasks [10,11]. Nonetheless, to fill the gap between robot and human perception, multisensory approaches have recently been studied and evolved in novel robotic platforms, combining visual data with vocal interfaces [12,13], or tactile sensors [14–16].

The game of Jenga is a perfect example of a challenging benchmark for robotic perception and control. In recent years, researchers have tackled the game with disparate platforms and approaches, adopting sophisticated manipulators [17,18] and complex control systems [19,20]. The contribution to an effective robotic solution for Jenga goes beyond the fascinating dynamics of this popular game. Indeed, it can support the evolution of cutting-edge visual and multisensory control strategies for complex real-world tasks requiring human-level precision. The case of Jenga is not isolated in the historical advancement of artificial intelligence (AI), where games are often used as a common benchmark for newly proposed learning algorithms [21–23]. The complexity of a round of Jenga resides in two different factors: first, it requires a multi-step policy to select a feasible block in the tower, approach it, and finalize its extraction. Second, all of these steps are based on the combination of real-time visual and tactile data processing and the highly precise motion of the end-effector for a single block extraction. According to this, it can be surely compared to real-world industrial [11], surgical [24], and agricultural [25,26] manipulation tasks.

This work presents a cost-effective system to play Jenga with the educational robotic manipulator—e.DO by Comau and a custom pushing finger as an end-effector. Our proposed solution is based on the combination of visual and tactile perception to handle the human-level complexity and the high precision required by the task. In particular, compared to previous attempts to play Jenga with a manipulator, we adopted a single RGB-D camera and a basic 1D force sensor as complementary hardware to the robot arm, considerably reducing the cost and complexity of the solution. An illustrative sequence of frames depicting our robotic system in action is shown in Figure 1. Overall, our perception and control system is composed of the following:

- A DL-based instance segmentation model fine-tuned on a custom Jenga tower dataset realized in simulation, which effectively allows the system to segment and select single blocks;
- An eye-in-hand visual control strategy that carefully handles the block extraction;
- A 1D force sensor to correctly evaluate the removability of a specific block.

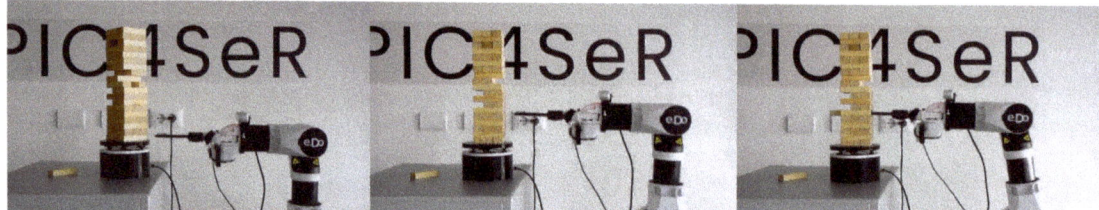

**Figure 1.** Our proposed solution in action: from left to right in the sequence, the e.DO robot selects a block to extract in the tower, adopting a visual-based control to approach it. Then, it starts pushing for the block extraction if it detects a low reaction force.

Our extensive experimentation validates all the sub-components of the proposed system. First, we study the force reaction on the 1D sensor and obtain a threshold-based decision policy to classify the extraction of the block as feasible or not. Then, we provide details on the training and testing of the instance segmentation model, comparing results obtained on simulated and real-world images. Moreover, we test our visual perception pipeline composed of segmentation and pose estimation of a group of blocks, measuring the tracking time during several runs. The adopted visual servoing control strategy is tested on the two major behavioral features of interest: precision, in terms of distance between the point of contact and the center of the block, and efficiency, estimated as the time required to align the end-effector to a block. Finally, we evaluate the overall performance of our solution by counting consecutive successful block extractions.

The paper is organized as follows. In Section 2, related works are presented in three subsections, discussing previous attempts to play Jenga with a manipulator and the state-of-the-art of deep instance segmentation, visual servoing, and a multisensory control strategy in robotics applications. Section 3 illustrates the overall strategy adopted to tackle the Jenga game and all the specific components of our solution. Finally, in Section 4, we present and discuss all the conducted experiments and the obtained results. Section 5 draws some conclusions and potential suggestions for future works.

## 2. Related Works

In this section, we first describe previous studies aimed at playing Jenga with a robot. Then we introduce the computer vision task of instance segmentation and its state-of-the-art and report similar works adopting visual servoing and multisensory control in robotic applications.

### 2.1. Playing Jenga with a Robot

Jenga is a common benchmark for robotic systems, allowing for a direct comparison of methods, experiments, and results. So far, few works have proposed a complete visual-based robotic system to play the game autonomously. At the same time, several studies focus on a partial aspect of the game with a specific solution.

Recently and most notably, Fazeli et al. [19] delved into the details of the manipulation and artificial intelligence capabilities needed to learn Jenga by sight and touch, achieving 20 consecutive block extractions; their system is made up of an end-effector that can push and pick blocks, a long-reach 6DOF industrial robot arm, an expensive 6-axis force sensor, and a fixed camera. The method is focused on learning from multisensory fusion and is compared to state-of-the-art learning paradigms with simulation and experiments, resulting in high performance and fast convergence. Their approach adheres to all the rules of Jenga with a tower in standard conditions and considers both block removal and placement on top of the tower. However, the hierarchical control strategy they adopt to carefully push the block during the extraction strongly relies on the use of an expensive 6-axis force sensor and a professional industrial arm, drastically increasing the overall cost of the solution.

Kroger et al. [17], who adopt similar expensive hardware, achieved 29 extractions with a 6DOF industrial robot arm, a 6-axis force sensor, a 6-axis acceleration sensor, a laser distance sensor, and two static CCD cameras. The authors develop a modular control system based on a generic number of sensors with a primitive manipulation programming interface to play a standard Jenga game, manually recoloring the blocks to help the vision system. Both reaction force and tower-perturbed movements seen from cameras provide extraction feedback when a block is randomly chosen to be extracted. Pose estimation from cameras is refined with a laser distance profiler before gripping the block, which is then placed on top with force feedback.

Wang et al. [27] propose a simpler system to leverage inexpensive vision and manipulation hardware to develop a strategic planner based only on visual feedback. They achieve up to five consecutive extractions. The limitations of using classical computer vision with two CMOS cameras and a 5-DOF pioneer short-reach robotic arm without force measurements led the authors to choose a quite different and simplified Jenga setup compared to the real one. Target blocks were partially pre-pulled and distanced one from the other, and the tower had half the levels.

A two-fingers, anthropomorphic, 7-DOF industrial robot arm is used in [18]. An eye-in-hand omnidirectional camera detects the tower configuration, and a block is chosen using a stability criterion. The block is grasped by the robot hand, which mounts a 6-axis force sensor on each fingertip. The Jenga setup is not standard, as the tower has only 10 layers. However, the system can detect and place blocks on the top of the tower, presenting a pretty high autonomy level.

A fine-grained kinematic analysis of the physics behind weight, friction interactions, and stability of the Jenga tower is studied in [28], both during and after the extraction. With a 6DOF manipulator, a custom gripper, and a 6-axis force sensor, they compare the real forces with the modeled ones and achieve 14 consecutive extractions before breaking the tower. No vision or pose estimation system is used, so a human operator must provide poses and manually rotate the tower.

Differently, Yoshikawa et al. [20] investigated a reinforcement learning approach, using a deep Q-Network and a 6DOF manipulator to correctly push a block without a priori knowledge of the kinematics and stability of the tower. The work is done in simulation on an ideal Jenga tower. Negative rewards are extrapolated from how humans play the game, such as pushing in the wrong direction and touching other blocks. The result is a policy for precise movement.

Similar to most related studies, our solution uses a force sensor to detect push failures and empirically check the removability of blocks. On the other hand, our perception system presents several novelties: a block identification approach based on an instance segmentation neural network, an eye-in-hand camera configuration, and a visual servoing control for the manipulator.

## 2.2. Deep Learning for Object Recognition

During the last few years, deep learning [29] has achieved state-of-the-art performance on various computer vision tasks. Different methods can be used to extract knowledge from visual data and give them a semantic interpretation. The literature refers to classification as the task of assigning a descriptive label to the whole image. Several approaches have been proposed to solve this problem, introducing architectural methodologies, such as convolutions [30], residual connections [31], feature [32], and space attention [33], or the more recent transformer-based architecture with self-attention [34,35].

Suppose a more fine-grained semantic description is needed. In that case, the object detection task has the objective of localizing instances that belong to target classes with the regression of bounding boxes. This detection allows the system to understand the scene hierarchically depicted in the image, assign multiple labels, and spatially identify the objects in the image reference frame. Popular methodologies for object detection [36–39] have focused on efficient and real-time execution to be used on continuous streams of images.

On the other hand, the semantic segmentation task aims at assigning a semantic label at the pixel level by predicting masks that identify the portion of the image belonging to a certain class. Classical approaches to this task are based on fully convolutional networks organized in an encoder–decoder fashion [40,41], which adopt successive downsampling and upsampling operations to predict labels at the pixel level. The main difference between semantic segmentation and object detection is that the former does not identify single instances but only regions that depict objects of the target classes.

Instance segmentation aims to localize single instances by predicting masks. This approach allows the most precise interpretation of the input image since it avoids coarse bounding box localization by identifying masks at a pixel level. Several methods have been proposed in the literature to solve this task. mask-RCNN [42] extends an object detection method called faster R-CNN [43] and is based on a two-stage approach that first proposes possible regions of interest (ROI) and predicts segmentation masks and classes in the second stage. Other approaches are based on one-stage architectures [44,45] and are inherently faster than two-stage methods. Other approaches solve a semantic segmentation task and then perform instance discrimination with boundary detection [46], clustering [47], or embedding learning [48]. Recently proposed YOLACT [49] and YOLACT++ [50] focus on a real-time approach to instance segmentation that extends an object detection approach with mask proposals. The combination of mask proposals and bounding boxes gives pixel-level instance localization. In this work, we adopted this approach due to its computational efficiency and ability to detect many near objects, typical of Jenga block segmentation.

## 2.3. Visual Servoing in Robotics

Industrial robotic tasks, such as assembly, welding, and painting represent standard scenarios where manipulators execute repeatable point-to-point motion by using off-line trajectories [51]. However, the variability and disturbances of different environments may affect the estimation of the target pose and lead to a degradation of task accuracy. Moreover, there are better strategies than this open-loop control technique for motion-based objects due to the target position and orientation variations.

Visual-based control strategies recently emerged as valid candidates for real-time trajectory computation and correction. In particular, visual servoing was introduced in 1979 [52] and refers to closed-loop systems where visual measurements are fed back into the controller to enhance task precision. Several works have proposed this approach for robotics applications in medical [53], agricultural [54], and aerospace contexts. The ability to move a robotic arm flexibly in high-precision surgery operations [55–57] confirms the potential of this technique in complex scenarios where small errors can compromise human health.

Visual servoing taxonomy distinguishes two approaches [58] according to the type of tracker used to generate visual features. Image-based visual servoing reconstructs the relative pose of the target in the manipulator reference frame using the camera field of view. This approach is widely used in agriculture applications [59,60], where occlusions of the camera can lead to poor visual feature extraction.

On the other hand, position-based visual servoing leverages a priori geometrical information on the object to derive the corresponding visual features. Recently, hybrid schemes have tried combining the two techniques and leveraging 2D and 3D visual features. In [61], an aerial vehicle with a robotic arm presents a hybrid visual servoing scheme to plug a bar into a fixed base. In this case, a marker detector provides the pose information of the object to be grasped, narrowing the possible field of use. Indeed, the robustness of tracking algorithms remains a central problem for visual-based control. Recently, in [62], the authors proposed a novel approach that uses augmented reality (AR) to generate 3D model-based tracking or 3D model-free tracking techniques to enhance the system's robustness.

## 3. Methodology

In this section, we frame the Jenga game and translate the rules of the game into a methodological set of requirements for the robotic system. First, a player's final goal is safely removing blocks from the tower. In the original game setting, a player has to place each extracted block at the top of the tower to validate its round and continue with a new block extraction. In our robotic experimental setting, we remove this rule and aim to extract as many blocks as possible from the tower without reallocation. This choice is mainly related to the limited workspace of the e.DO manipulator, designed for educational purposes, since it cannot reach the fallen blocks behind the tower. Moreover, our custom end-effector, similar to a human finger, cannot perform grasping and instead extracts blocks by pushing. As the first practical task, a Jenga player should be able to select one of the blocks of the tower to be extracted. To this end, each block of the tower is identified in our system using an instance segmentation deep neural network. Moreover, we define a heuristic block selection policy based on the idea that extracting multiple blocks from the same tower level is not recommended. Moreover, as better detailed in Section 3.2, blocks at different tower layers present diverse frictions and effects on the tower's configuration.

Therefore, our solution is based on the following assumptions:

- The identification and pose estimation of each block of the tower is the first step to selecting a suitable piece and approaching it;
- A removable block cannot be identified only by visual analysis: the integration of a tactile perception system is needed;
- A sufficiently precise alignment between the end-effector and the center of a target block allows the arm to extract it successfully by simply pushing.

A complete illustrative schematic of the system proposed to play Jenga is depicted in Figure 2.

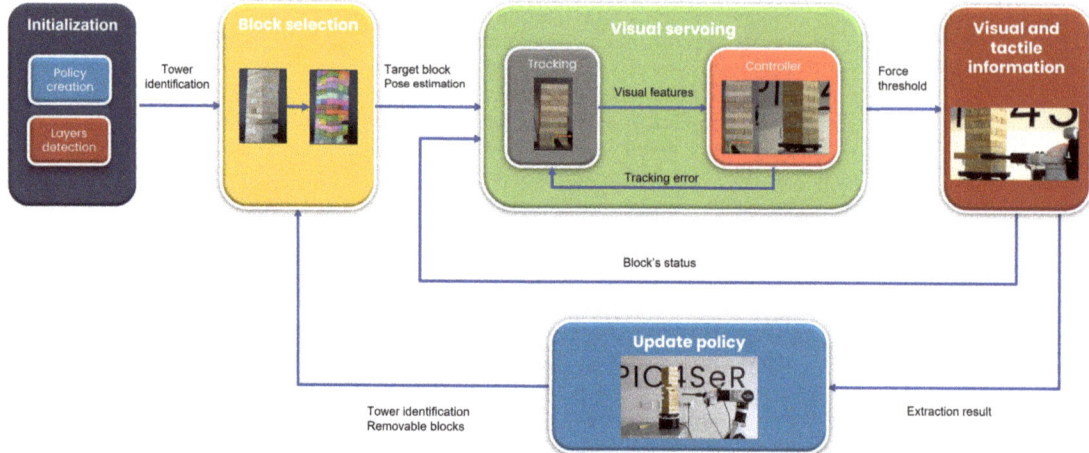

**Figure 2.** Block diagram of the proposed system architecture for Jenga: after tower identification, an RGB image of the tower is first used to extract the segmentation masks of each block with the deep instance segmentation neural network. A block selection policy chooses the block to extract while considering blocks that have already been tried. The selected block is precisely approached by the end-effector adopting a visual-based control strategy. Finally, the end-effector pushes the block, and an inexpensive 1D force sensor verifies its removability to stop or approve the extraction attempt.

### 3.1. Deep Instance Segmentation and Pose Estimation

The proposed methodology's first step relies on identifying the blocks present in the tower by visual analysis. We adopt a deep learning-based method to perform instance segmentation. The necessity of using a deep learning approach is caused by the fact that wooden blocks do not have easily-distinguishable features while having predictable positions in the camera frame instead. In this setup, the vision system can benefit from the ability of neural networks to generalize to different light conditions and points of view. Moreover, training the model on an exhaustive dataset can make it robust towards missing blocks and tower misplacement during the game. Among different approaches for image semantic analysis, instance segmentation aims at detecting all the objects of interest in an image at the pixel level. Given an input image of the tower, the model should output a set of possible Jenga block candidates, together with their respective pixel masks. Instance segmentation can therefore be seen as a combination of an object detection task with a mask prediction task. By translating the position of the blocks from the image reference frame to the robot reference frame, it is possible to achieve block localization.

Model Architecture

We select the YOLACT++ [50] architecture to implement the instance segmentation algorithm. This architecture has been chosen for its computational speed, capability to handle occlusion, and efficiency in detecting a high number of tightly packed same-class objects. The model is based on a ResNet-50 [31] + FPN [63] feature extraction backbone followed by two branches, one dealing with object detection and the other with mask prototype production. The detection branch outputs a set of anchor predictors $a$ as possible Jenga block candidates. Each anchor prediction consists of the class confidence $c$, four bounding box regressors, and $k$ mask coefficients. These mask coefficients are used to weigh the mask prototypes produced by the second branch. Both branches are based on convolutional layers applied to the features extracted by the common backbone.

The two branches are finally followed by a mask assembly block, which combines the masks with the coefficients predicted by each anchor to reach the final instances prediction. At inference time, anchor predictions are thresholded with a certain value $t_c$ on their confidence $c$ score to produce the actual output. Moreover, as in standard object detection algorithms [36,37], an NMS (non-maximum suppression) method is applied to remove redundant predictions.

*3.2. Block Selection Policy*

Our solution defines a heuristic block selection policy based on physical and empirical considerations. Visual information cannot provide sufficient information to determine the status of a block. The imperceptible tolerances of each block cause minuscule variations in pressure that prevent visual-based systems from understanding which blocks are truly removable. The blocks in the higher layers are easier to extract, i.e., they can be pushed out of the tower by applying a smaller force. However, pushing from a decentralized contact point contributes to the formation of torques on the block that causes asymmetry in the tower. This effect is amplified on upper layers due to lower friction forces and may affect the stability of either adjacent blocks or the tower itself. Instead, the friction force increases for blocks located at lower levels, making the extraction harder and risking the Jenga tower falling.

Such considerations imply the need for a policy to select the block to extract. However, the policy must also consider the physical dimensions of our 6DOF anthropomorphic manipulator. Indeed, during the extraction primitive, the robot has to be parallel to the block, which implies a loss of DOFs. These orientation constraints restrict the robot's workspace, so the manipulator can only reach a limited range of tower levels. In order to overcome these issues, the policy divides the entire tower into two subspaces according to the robot's workspace. In particular, the subspaces correspond to the upper and lower levels, each with a predetermined number of tower levels. By convention, the numeration of tower levels carries in ascending order, where one corresponds to the lowest level. In addition to this vertical division, the policy is also initialized with the block direction for each level. Indeed, two possible tower orientations have a relative rotation of 90 deg between the two. The pose estimation described in Section 3.3, applied by the convention on the top, provides the reference to infer the actual orientation of the tower and all its levels.

As human players usually do, we initially adopt a random policy that selects the Jenga pieces in one of the subspaces. Then, the manipulator approaches the chosen block and starts pushing. At this point, force sensor data are collected to evaluate the block's status according to the adopted force threshold, as explained in Section 3.5. After each trial, a memory list updates the information on extracted and tested blocks. More in detail, a memory buffer stores the information about the blocks: the status (present, tested, or extracted), the threshold force to apply, and the current number of extraction attempts. In addition, it also keeps track of additional layers as the game goes on. Only one piece per level can be extracted as a further safety measure.

The policy repeats the process until all levels contained in the subspace are tested. After that, the system changes the subspace to test additional levels until the tower collapses. More in detail, selecting the higher layers as the starting subspace leads to three subspaces in total, with extra layers being included in the final subspace. Alternatively, if the process begins by extracting blocks from the lower layers, there will be only two subspaces, with extra layers being included in the high subspace. Figure 3 provides a minimal representation of the policy strategy. Except for the pick-and-place operation, the game implementation does not neglect any official rule.

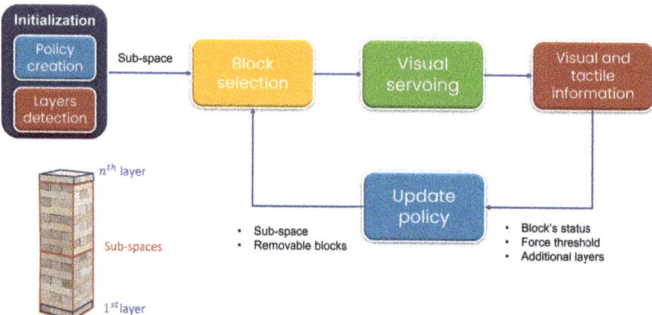

**Figure 3.** High-level input-output diagram describing the logic flow of the policy used to select the next block to extract from the Jenga tower. The layers of the tower are numbered from the 1st to the $n$th starting from the bottom, as shown in the schematic on the left. The policy starts to extract the blocks from one of the two subspaces of the tower, randomly chosen. It proceeds until all levels contained in the subspace are tested, to finally switch to the other subspace trying to avoid the tower to collapse.

*3.3. Block Pose Estimation*

Predicted block masks are the input of a post-processing unit implemented using OpenCV. It estimates the position and orientation of the desired block to be extracted with respect to the camera. This information is the first requirement to approach the block's face starting the manipulator's tracking and visual servo control. The pose estimation is performed with an intrinsic calibrated camera through the Perspective n Points (PnP) algorithm optimized for planar points [64]. PnP is applied to the four corners of the front face of the target block, separately identified from the others thanks to the predicted segmentation mask, and paired with the known dimensions of the block. To increase the robustness of the tracking algorithm and the overall precision of the visual control of the manipulator, corners of nearby blocks are also considered (if present). Adjacent blocks may have diverse mask shapes: smaller for occluded front-facing blocks and larger for side-facing ones. Hence, two different sets of points are considered to estimate their poses. Side-facing blocks can provide a significant advantage in the subsequent tracking phase, offering a more stable visual reference during the motion of the end-effector.

Therefore, the PnP algorithm provides an initial estimate of the 6DoF pose of the group of blocks (target and adjacent blocks) using segmentation mask corners to initialize the model-based tracker.

*3.4. Tracking and Visual Servoing*

The geometric dimensions of a standard piece of Jenga ($25 \times 15 \times 75$ mm) require precise movements to perform extraction successfully. Considering such dimensions and the width of the custom fingertip (i.e., 11 mm), it can be shown that the maximum position error from the center of the block must be smaller than 7 mm. Therefore, the maneuver of the end-effector requires accurate trajectory planning to approach a block precisely. Standard point-to-point planning and online control [65] can be considered valid methodologies. However, these control schemes adopt an open-loop control system which requires high precision on pose estimation and tiny mechanical tolerances to reach the desired point with a small error. On the other hand, visual servoing is a closed-loop control strategy that exploits visual measurements to correct the pose of the end-effector with respect to the target in real time. Continuous visual feedback is used to correct the end-effector trajectory and to align the relative pose of the camera with the target block. For this reason, visual servoing is a competitive and flexible strategy to accurately approach the desired object (in this case, the Jenga block to extract).

In this regard, a robust tracking algorithm is a fundamental and challenging component of visual servoing control, which is required to guarantee smooth trajectories and allow a faster convergence to the target. As described in the previous sections, the segmentation masks are used to estimate the position and orientation of the desired block. At this stage, the tracking system receives both the initial pose estimation and the segmentation masks and combines them with a 3D model of a Jenga block. Therefore, it detects the target to establish a continuous mapping of the 6DOF pose of the Jenga block in the camera field of view. Specifically, the adopted stereo model-based tracking method ViSP [66] combines several visual features, such as moving edges, key points, and depth information, to improve stability and robustness. The ViSP tracker requires the 3D model of a generic block to project its geometry into the image space and generate the visual features accordingly.

However, as mentioned in the previous Section 3.3, the tracking performance is only partially reliable if only the visible face of the desired block is used (Figure 4). For this reason, the tracking algorithm does not rely only on the visual information of the single block to extract, but it integrates adjacent blocks in a unique group model. As already discussed in the previous Section 3.3, this choice leads to more robust tracking thanks to the higher number of visual features extracted, especially from side-facing blocks.

**Figure 4.** The initial pose estimation of the target block obtained by exploiting the segmented mask is shown on the left. The group model composed of multiple blocks to increase the robustness of the tracker is shown on the right.

Moreover, the visual features of the group of blocks of interest are collected from different perspectives before the tracking is started. This acquisition phase enables recovery tracking in case of sudden movements by exploiting the acquired pairings and re-initializing the block pose.

Our approach adopts the eye-in-hand visual-servo configuration, whose extrinsic camera parameters are estimated based on the end-effector design. Thus, we indicate with $s = (x, y, \log(Z/Z^*), \theta_u)$ the visual features constructed for the 2 1/2-D visual servoing tasks, where $(x, y)$ are the coordinates of the target in the camera reference frame, $Z$ and $Z^*$ are the current and desired depth of the point, respectively, and $\theta_u$ is a $6 \times 1$ vector that identifies the rotation angles (expressed in the axis–angle convention) that the camera has to follow. Moreover, we refer to $\widehat{L_s^+}$ as the approximation of the interaction matrix and to $e_s$ as the tracking error of the visual features, $s$, defined as $e_s = s - s'$, where $s'$ denotes the desired visual features.

The $6 \times 1$ vector $v$ represents the linear and angular camera velocities that are computed through the following control law:

$$v = -\lambda \widehat{L_s^+} e_s \qquad (1)$$

Hence, the above regulation control law defines the linear and angular velocities the camera has to follow to reach the target. Using forward kinematics, the vector $v \in \mathbb{R}^{6 \times 1}$ is converted into the task space and executed through a motion rate controller using inverse differential kinematics. The overall closed-loop system runs at about 9 Hz on the laptop i7-8750H CPU.

To further optimize the smoothness of the trajectory, we consider fixed constraints on the maximum joint velocities that the controller can predict. In Section 4, we show that the visual servoing pipeline we adopted increases the overall accuracy and robustness of the Jenga extraction system.

*3.5. Tactile Perception*

Tactile perception is a fundamental aspect of a human Jenga player since identifying potential removable blocks is not possible via visual analysis only. According to this, we decide to incorporate a low-cost 1D force sensor in our system, which only provides information about the perpendicular reaction force between the end-effector and the target block. Hence, while previous works adopted a 6D force sensor to adjust the direction of the end-effector while pushing [19], we take advantage of our visual-based control strategy and guarantee considerably good precision by simply pushing the block forward. The one-dimensional force sensor is mounted directly on the fingertip of the end-effector, as shown in Figure 5, and provides a digital output with a full-scale force span of 5 N. The sensor is activated when an interaction between the fingertip and the block occurs. The block can be either stuck or free to move, so the push primitive uses the reaction force to derive a binary block classification. The closed loop explained in Section 3.4 reads the measurements at 9 Hz while the Arduino microcontroller sends them at 20 Hz.

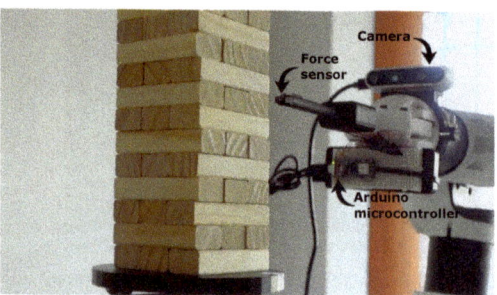

**Figure 5.** The manipulator has a human finger-inspired end-effector, which mounts a simple 1D force sensor on its tip. The force sensor controlled by an Arduino Nano board reveals the status of a block (removable or not) once it is approached and pushed. An eye-in-hand camera configuration enables the visual control pipeline to align the fingertip with the Jenga piece accurately.

A challenging aspect of the game is that the more push attempts are performed, the more unstable the tower becomes, as extracted blocks or rotations perturb its structure due to pushes and retractions. Therefore, two thresholds detect immobile or moving blocks in $thr_1 = 0.32$ N and $thr_2 = 0.18$ N. Such values are determined by combining theoretical [28] and empirical results, better depicted in Section 4.1. The threshold changes to a smaller conservative value in the second phase of the game when all the higher levels have been tested. Our choice can be defined as conservative, as it safeguards the stability of the tower rather than seeking more competitive performance.

## 4. Experiments

The experimental setup includes the anthropomorphic educational manipulator, e.DO manufactured by Comau, a depth camera Intel RealSense D435i, a MicroForce FMA piezoresistive force sensor (5 N full scale, 12-bit resolution for 0.002 N sensitivity), an Arduino Nano 33 BLE, and 3D printed components such as a rotating base, an end-effector design extension, and camera support. The software runs on a single computer with an i7-8750H CPU and 16 GB of RAM. Thus, one of our goals is to investigate and prove the effectiveness of state-of-the-art with low-cost equipment. Visual control adopts an eye-in-hand configuration with camera support, while the small force sensor is placed on top of the end-effector extension connected to the Arduino Nano.

## 4.1. Reaction Force Threshold

In this section, we first describe the experiments carried out to define the static force threshold used to check the removability of blocks. The real Jenga blocks present small differences in dimensions, generating a diverse pattern of friction forces in the tower each time it is rearranged for a new game. Although [28] tried to provide a rigorous mathematical formulation of friction forces in the tower, we prefer an experimental approach to identify the correct threshold values to detect whether or not the robot push affects the stability of the tower.

The measurements are taken with a complete tower configuration during the block extraction, starting the data collection of force reaction from when the contact between the block and end-effector starts and the force sensor detects a non-zero value. The experiment is run multiple times on different tower levels, mixing the blocks' disposition each time to test random friction conditions. The plot in Figure 6 shows the force profile of 15 blocks located at different tower positions over time.

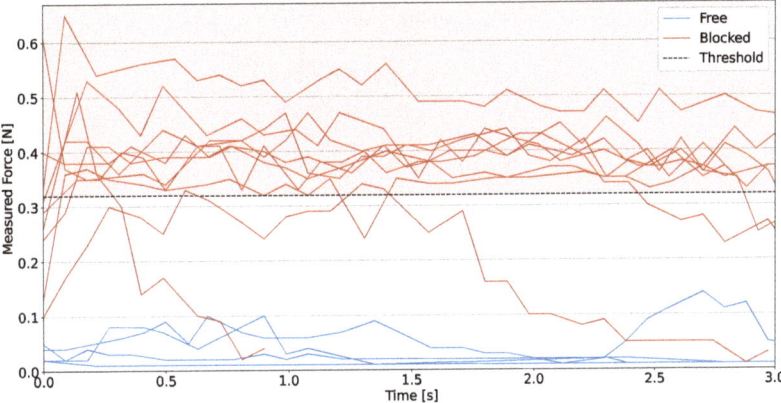

**Figure 6.** Reaction force measured over time during pushes for several blocks located in different layers and tower positions. The threshold force value of 0.32 N is also plotted to distinguish moving blocks in red from constrained ones in blue.

The choice of the threshold force values is not trivial, as it affects the system's ability to classify the block's removability state. By setting higher threshold values, more blocks can be classified as false positives, i.e., they are perceived as free to move, whereas they cannot be safely extracted. On the other hand, imposing lower threshold values affects system performance leading the system to avoid feasible block extractions.

As depicted in Figure 6, the initial threshold value is set to 0.32 N, making the system more likely to detect false positives than false negatives. In addition, this is reflected in a more aggressive strategy at the beginning of the game, enhancing performance at the cost of tower stability. After attempting to remove half of the levels, the threshold is reduced to preserve tower stability. In fact, after removing a certain number of blocks, the tower becomes increasingly unstable, and the friction forces change according to the position of the extracted blocks. Due to the static friction force, the superposition of these effects increases the probability of disrupting the tower as soon as the contact between the fingerprint and block occurs. Therefore, mechanics and empirical observations led us to lower the threshold value to 0.18 N. In general, the measurements are in the same range as the results obtained by the more extensive analysis of [18,19,28], with forces between 0 N and 1 N and thresholds $\geq 0.2$ N.

## 4.2. Instance Segmentation Experiments

In this section, we report the method and the details of the procedure to train the instance segmentation model with a carefully devised synthetic dataset, as well as the results obtained by experimental validation. The experimentation aims to assess the quality of the deep learning model and its generalization to the real pictures of our experimental environment.

### 4.2.1. Training Setup

The training of the instance segmentation model is entirely performed on a synthetic dataset crafted from a 3D model of the Jenga tower. Using the 3D modeling software Blender and its Python APIs through BlenderProc [67], hundreds of photorealistic and varied images of the tower are produced with automatic pixel-perfect annotations. This approach makes it fast and easy to obtain hundreds of samples without manually labeling real images. The training and validation datasets are composed of 800 and 80 images of size 640 × 480, respectively, rendered from a Blender synthetic scene composed of 48 cuboids arranged in a tower of 16 levels. We apply 12 different wood materials to the faces of these blocks to simulate the possible colors and wood line patterns with a realistic look. Each scene is loaded with blocks, a virtual camera, and a point light source. We design the following levels of scene randomization:

- The materials are randomly sampled and assigned to all 48 cuboids to change their look;
- A total of 6 to 24 cuboids are randomly displaced along their x and z axes by a distance between −4 mm and 4 mm;
- A point light source is positioned by randomly sampling a height of $0.1 \div 0.5$ m spanning a circular arc of 60 deg centered on the tower with a random radius between 0.4 and 0.7 m;
- A total of 2 to 9 random blocks are removed from the tower to create holes;
- Camera position is sampled on a circular arc of 20 deg with a height between 0.05 and 0.2 m and a radius between 0.25 to 0.45 m.

The camera is placed to capture two faces of the tower at the same time. For each configuration, 10 samples are acquired with the full tower and 10 with random missing blocks. The procedure is repeated for each random scene obtaining a diverse and complete dataset with different light conditions, block displacements, missing blocks, camera angles, and views of the experimental conditions. The scene's background is then filled with black, white, or gray. The render time for all the 880 images was 4 hours on an i7-9700K CPU. The test set is composed of 20 real manually-annotated pictures from our experimental setup for a total of around 800 segmented tower blocks. The images are taken in slightly different light conditions and camera positions.

The input 640 × 480 images are rescaled to 550 × 550 to be compatible with network input requirements. We adopt a ResNet-50 backbone with pre-trained weights on ImageNet. We perform 8000 training iterations (69 epochs) with a batch size of 8, SGD optimizer with a momentum of 0.9, and a weight decay of $5 \times 10^{-4}$. The initial learning rate of $10^{-3}$ is scaled down by a factor of 10 at iterations 5000, 6000, and 7000. We consider a positive intersection-over-union (IoU) value of 0.5 during training. The training is performed on a Tesla K80 GPU with Cuda 11.2.

### 4.2.2. Instance Segmentation Results

As the main metric to assess the quality of the instance segmentation, we adopt the widely used average precision at different values of intersection-over-union ($AP_{IoU}$). A predicted mask is considered a true positive (TP) if it has an IoU with the ground-truth mask over the given threshold. AP is then computed as the area under the curve of the precision–recall plot obtained varying the confidence threshold $t_c$.

Table 1 reports the AP results at different IoU values, both on the synthetic test set and the real manually-annotated dataset. As expected, increasing the IoU threshold results in a performance drop due to the stricter requirements asked of the model. Generally, we observe a certain drop in performance when considering the real-world dataset, mainly due to border effects caused by approximate hand-made annotations and a decreased recall caused by the high number of instances in a single image. However, since a high recall is not required to perform block selection and tracking effectively, we state that the obtained real-world generalization is good enough for the target application. Visual comparison of a synthetic and a real image is reported in Figure 7a,b.

**Table 1.** Mask AP results at different values of IoU for the Jenga blocks instance segmentation model on both the synthetic and the real manually-annotated test datasets.

| Dataset | $AP_{50}$ | $AP_{80}$ | $AP_{90}$ | Mean |
|---|---|---|---|---|
| Synth | 90.08 | 87.76 | 63.2 | 78.37 |
| Real | 75.98 | 53.40 | 11.53 | 53.09 |

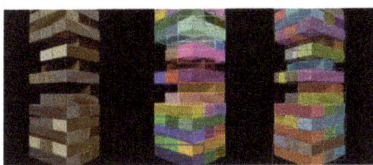 

(a) Synthetic Dataset (AP = 69.9% for 80% IoU)   (b) Real Images (AP = 88.1% for 80% IoU)

**Figure 7.** Segmentation masks on simulated (a) and real (b) images. From left to right: source image, ground truth, and predicted masks.

### 4.3. Tracking Robustness

This experiment tests the robustness of the stereo model-based tracker in two different configurations to assess its ability to keep track of the object's pose during movements. In particular, we compare our experimental visual set up with the basic functionalities of the ViSP library [66]. The main difference lies in constructing the 3D block model and the tracker initialization method. A CAD model of the single block takes only into account the target piece and requires the user to initialize the model manually. Instead, our tracking system merges different pieces around the target block according to the tower arrangement described in Section 3.4. For this test, we exploit a rotating base to automatically turn the Jenga tower around its vertical axis by 45 deg clockwise or counterclockwise at a constant speed. This rotation brings one of the faces perpendicularly to the camera axis, thus keeping the target block always in the field of view.

While the tower rotates, spanning the whole angle range, we test the tracker to follow the block moving in the images. We measure the projection error $e_{proj}$ as the difference between the tracker's estimated rotation of the block and the actual rotation of the rotating base. Fixing the maximum acceptable error $err_{thr} = 25$ deg, the failure condition is reached when $e_{proj} < e_{thr}$. For each run, we report the percentage of the total time (60 s) for which the tracker follows the block without failures, including the target block's tower level. Since this test's ultimate goal is to highlight the tracker's robustness, we keep the same threshold value used in the game. The trials are performed with two different angular velocities, $\omega_1 = 2.5$ deg/s and $\omega_2 = 8.3$ deg/s for the same target.

The results in Table 2 suggest that ViSP [66] is a scalable library that can be further optimized according to the requirements of our task. Our tracking system not only overrides the point-to-point manual initialization leveraging the segmentation mask prediction, but it also significantly improves the tracking robustness up to 7.5 times.

**Table 2.** Largest tracking time comparison between the 3D CAD models generated as a single block model and our pattern-based multi-block model.

| Level | Vel [deg/s] | Single [%] | Group [%] |
|---|---|---|---|
| 5  | $\omega_1$ | 36.9 | 100 |
| 5  | $\omega_2$ | 17.0 | 100 |
| 6  | $\omega_1$ | 13.1 | 100 |
| 6  | $\omega_2$ | 8.0  | 100 |
| 8  | $\omega_1$ | 28.9 | 100 |
| 8  | $\omega_2$ | 10.6 | 100 |
| 9  | $\omega_1$ | 19.4 | 100 |
| 9  | $\omega_2$ | 29.5 | 100 |
| 10 | $\omega_1$ | 18.4 | 100 |
| 10 | $\omega_2$ | 8.1  | 100 |
| 11 | $\omega_1$ | 24.7 | 100 |
| 11 | $\omega_2$ | 14.8 | 100 |

*4.4. Visual Servo Convergence and Accuracy*

The visual servo control law is tested, in terms of time and spatial error, by bringing the end-effector to the computed goal. To this end, the tracking system estimates the block's pose (i.e., position and orientation) to extract the visual features and compute the $6 \times 1$ velocity vector in the camera reference frame. Such velocities are then converted in the end-effector's reference frame through the extrinsic parameters, estimated on our custom fingertip's design. Then, the linear and angular velocities are translated into joint velocities and actuated accordingly.

4.4.1. Timing Convergence

The experiment begins with the robot in a default pose at the same height as the tower's first level. Then, a timer starts and automatically stops when the end-effector reaches the desired pose with a fixed tolerance on the visual servo error magnitude as *tolerance* $= 0.00002$. This test is repeated for each tower level regardless of the block configuration. In Figure 8, we report the distribution of convergence time at different tower levels. The average convergence time is roughly constant among levels, while we observe a high variance between different observations. This is caused by the fact that most of the convergence time is spent reaching the desired orientation rather than the desired position. The oscillations in the tracker estimate are higher when the camera is close to the tower, and the 3D block model degenerates to a plane face. The tracker is set to tolerate up to 25 deg of mean reprojection error before failing, so the robot performs many corrections to the orientation, using all its joints to follow the oscillations. This sometimes generates longer convergence times. However, time is not a primary constraint in Jenga, so it is acceptable to trade convergence speed for more precise end-effector alignment.

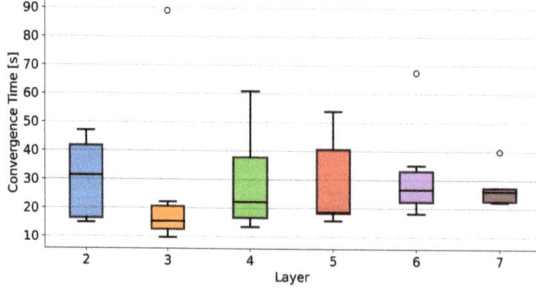

**Figure 8.** Mean and standard deviation of convergence times for blocks on the same level, where the black line is the mean, and the dispersion of the values is shown. Circles show isolated events, in which tracker's oscillations caused longer convergence times of the robot pose.

### 4.4.2. Spatial Accuracy

The second part of the test aims to assess the manipulator's accuracy in pushing blocks. It consists in manually stopping the robot as soon as the end-effector reaches the target and measuring the distance between the contact point and the center of the block. Defining $err_x$ and $err_y$ as horizontal and vertical errors, we perform a statistical analysis of the system precision. The results depicted in Figure 9 are obtained from 21 trials on different blocks. The results demonstrate the accuracy and repeatability of our system, as a mean error below 0.2 mm and a standard deviation below 1 mm are compatible with the accuracy defined in Section 3.4. Moreover, it is worth noting that the mean value for both axes is close to zero, which indicates the deep focus on estimating the extrinsic camera-to-robot and intrinsic pixel-to-mm parameters. The differences between the two axes mainly depend on the manipulator's dexterity and its mechanical tolerances.

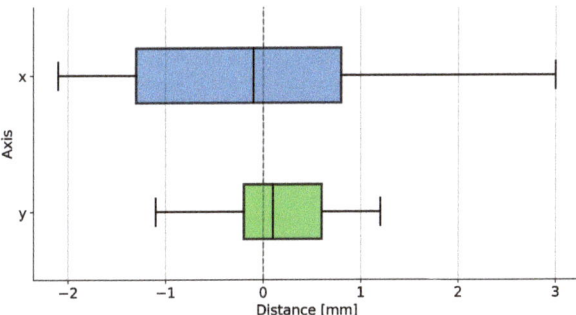

**Figure 9.** Mean and standard deviation of the distance between the contact point of the end-effector and the center of the selected block.

### 4.5. Consecutive Block Extractions

Finally, the different modules are integrated, and tests of the architecture are performed on the full system. The experiment includes 18 trials following the semi-automatic loop explained in Section 3.2. The goal is to extract as many blocks as possible without breaking the tower.

In order to properly validate our results, the chosen metric considers both correctly extracted blocks and correctly classified immobile blocks. Hence, an attempt is labeled as successful in one of the following cases:

1. A mobile block is correctly pushed without perturbing the stability of the tower;
2. The binary classification model correctly identifies the block as immobile, and the arm retracts without causing the tower to fall

After each attempt, the policy updates the block status obtained from visual and tactile measurements during the extraction process. In particular, force sensor data are used to detect the block status, while visual measurements provide feedback on the result of the extraction primitive. The latter integrates pose estimation information and the direct kinematics of the manipulator in order to verify the complete extraction of the block from the tower. The minimization of the angular tracking error in the visual servo law allows positioning the end-effector as parallel as possible to the block by pushing it out of the tower (1) along the z-axis of the block. The robotic arm then retreats the long finger back along the same axis. This open-loop push primitive runs synchronously to the force sensor at a frequency of 20 Hz to eventually abort the execution if high reaction force values are detected. If the manipulator successfully extracts the block, the policy updates the memory buffer with the current block status and verifies the presence of other removable blocks in the subspace in order to start a new attempt. Conversely, in the case of immobile blocks, the manipulator pursues to extract the remaining blocks of the same layer. In this scenario, the policy removes the block from the list of removable blocks for future extractions.

In our setup, the operator has three interactions that make the game semi-autonomous: changing the tower's position to allow the robot to switch from higher to lower layers (or vice versa), placing the extracted blocks on top to form a new level, and aborting extractions when failures occur.

Minor failures do not stop the game, as they do not cause the tower to collapse. The operator can abort an attempt when the system fails for reasons unrelated to the extraction. In this case, the loop continues on the next iteration, and the policy is not updated. Possible minor failures are:

- The robot reaches a singularity position when the target block level is at the edge of the workspace;
- The tracking is lost as either the tower is in bad light conditions or the target block lies outside the field of view;
- The initial pose estimation is not precise enough because the segmentation model does not detect all the corners of the block correctly.

To minimize these failures and prevent the perturbation of the tower's stability, we make some conservative choices. Different force thresholds are assigned depending on the game's phase, i.e., which subspace of blocks is currently tackled. This way, the policy addresses the first subspace with an aggressive force threshold of 0.32 N. After all the first subspace levels have been tested, the policy addresses the second subspace with a cautious threshold of 0.18 N. The effect of adding the extracted blocks on top is twofold: their weight contributes to changing the friction between the blocks and shifting the center of gravity of the tower. Furthermore, each new level is considered in the policy for additional attempts. This way, we obtain a minimum of 42 attempts (3 per level) and 14 block extractions (1 per level), plus 3 additional attempts and 1 extraction for every newly formed level.

The results are reported in Figure 10 with details on the distribution of complete extractions, unmovable blocks correctly classified, and attempts ended with an error. Experiments show that 80% of the extraction attempts are successful. More specifically, 45% of the attempts lead to extracting the block correctly, while 35% of them find an unmovable block. The errors (20%) are mainly caused by the system losing the tracked block or by an imprecise initial pose estimation. Each time an error occurs, the current extraction attempt is aborted, and a new one is started.

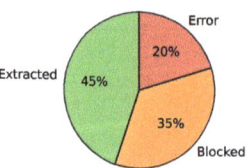 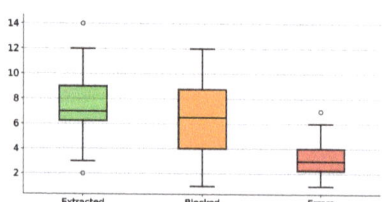

**Figure 10.** The plot on the left illustrates the average number of extracted blocks (green), the blocks correctly classified as unmovable (orange), and errors among 18 trials (red). On the right, the statistical distribution of the outcomes is plotted. Our system can perform successful extraction or correctly identify unmovable blocks in 80% of the attempts. The highest score achieved is 14 extracted blocks, accumulating more error in the long game, while the lowest score is only 2 extractions (white circles).

Each experiment ends when the tower falls. In our experiments, the fall occurred after 13.8 correct attempts on average (7.5 correct extractions and 6.3 detected unmovable blocks, respectively). In most experiments, the fall was caused by the increasing instability of the tower after multiple extractions. However, a few experiments ended earlier because of poor tracker performance in providing the position and orientation during the push movement or incorrect detection of mobile blocks (two experiments ended after only 5 extraction attempts). The highest registered score counts 14 successful extractions and 11 detected unmovable blocks. The results are comparable to those of [28], which scored 14 consecutive extractions using a 6-axis force sensor while being much higher than the 5 extractions of [27], which also had inexpensive equipment. The score is far behind both the 20 and 29 of, respectively [19], more extensive architecture and artificial intelligence, and [17], more complex hardware and simpler Jenga game setup.

## 5. Conclusions

In this paper, we propose a complete system to play the challenging game of Jenga with cost-efficient robotic hardware. The contribution of this work goes beyond the Jenga game, proposing an advanced, adaptable solution for accurate manipulator control in delicate robotic tasks. We demonstrate that a visual-based approach for perception and control can provide the robot with significant benefits in terms of scene understanding and control accuracy.

Differently from previous works, the main components of our system are a deep instance segmentation neural network used to identify each block of the Jenga tower and a visual tracking and control pipeline to continuously adjust the pose of the end-effector as it approaches the block. A low-cost 1D force sensor is integrated into the system to check the removability of the target block, drastically reducing the cost and complexity of the overall system. Our extensive experimentation shows the remarkable performance of each fundamental component of the solution. After examining the accuracy and stability of the perception and control units, we evaluate the whole system by playing Jenga and reaching a maximum of 14 successive block extractions. Future works may see the advancement of the reasoning capability of the robot: reinforcement learning agents can be investigated to optimize the planning policy for the game. Moreover, a visual-based sensorimotor agent could eventually replace the controller by directly mapping segmented and depth images to velocity commands for the end-effector.

**Author Contributions:** Conceptualization, M.C.; methodology, L.M., G.P., M.M., S.A. and F.S.; software, L.M. and G.P.; validation, L.M. and G.P.; formal analysis, M.M., S.A. and F.S.; investigation, L.M., M.M., S.A. and F.S.; data curation, L.M. and G.P.; writing—original draft preparation, L.M., G.P., M.M., S.A. and F.S.; writing—review and editing, M.M., S.A. and F.S.; visualization, L.M., G.P., M.M. and S.A.; supervision, M.C.; project administration, M.C. All authors have read and agreed to the published version of the manuscript.

**Funding:** This research received no external funding.

**Institutional Review Board Statement:** Not applicable.

**Informed Consent Statement:** Not applicable.

**Data Availability Statement:** Not applicable.

**Acknowledgments:** This work was developed with the contribution of the Politecnico di Torino Interdepartmental Centre for Service Robotics (PIC4SeR) https://pic4ser.polito.it/ (visited on 15 November 2022) and SmartData@PoliTO https://smartdata.polito.it/ (visited on 15 November 2022).

**Conflicts of Interest:** The authors declare no conflict of interest.

## References

1. Sun, X.; Zhu, X.; Wang, P.; Chen, H. A review of robot control with visual servoing. In Proceedings of the 2018 IEEE 8th Annual International Conference on CYBER Technology in Automation, Control, and Intelligent Systems (CYBER), Tianjin, China, 19–23 July 2018; pp. 116–121.
2. Zhao, Z.Q.; Zheng, P.; Xu, S.T.; Wu, X. Object detection with deep learning: A review. *IEEE Trans. Neural Netw. Learn. Syst.* **2019**, *30*, 3212–3232. [CrossRef] [PubMed]
3. Zhang, X.; Chen, Z.; Wu, Q.J.; Cai, L.; Lu, D.; Li, X. Fast semantic segmentation for scene perception. *IEEE Trans. Ind. Inform.* **2018**, *15*, 1183–1192. [CrossRef]
4. Martini, M.; Cerrato, S.; Salvetti, F.; Angarano, S.; Chiaberge, M. Position-Agnostic Autonomous Navigation in Vineyards with Deep Reinforcement Learning. In Proceedings of the 2022 IEEE 18th International Conference on Automation Science and Engineering (CASE), Mexico City, Mexico, 20–24 August 2022; pp. 477–484.
5. Salvetti, F.; Angarano, S.; Martini, M.; Cerrato, S.; Chiaberge, M. Waypoint Generation in Row-based Crops with Deep Learning and Contrastive Clustering. *arXiv* **2022**, arXiv:2206.11623.

6. Bac, C.W.; van Henten, E.J.; Hemming, J.; Edan, Y. Harvesting robots for high-value crops: State-of-the-art review and challenges ahead. *J. Field Robot.* **2014**, *31*, 888–911. [CrossRef]
7. Berenstein, R.; Shahar, O.B.; Shapiro, A.; Edan, Y. Grape clusters and foliage detection algorithms for autonomous selective vineyard sprayer. *Intell. Serv. Robot.* **2010**, *3*, 233–243. [CrossRef]
8. Kletz, S.; Schoeffmann, K.; Benois-Pineau, J.; Husslein, H. Identifying surgical instruments in laparoscopy using deep learning instance segmentation. In Proceedings of the 2019 International Conference on Content-Based Multimedia Indexing (CBMI), Dublin, Ireland, 4–6 September 2019; pp. 1–6.
9. Hasan, S.K.; Linte, C.A. U-NetPlus: A modified encoder-decoder U-Net architecture for semantic and instance segmentation of surgical instruments from laparoscopic images. In Proceedings of the 2019 41st Annual International Conference of the IEEE Engineering in Medicine and Biology Society (EMBC), Berlin, Germany, 23–27 July 2019; pp. 7205–7211.
10. Chen, X.; Guhl, J. Industrial robot control with object recognition based on deep learning. *Procedia CIRP* **2018**, *76*, 149–154. [CrossRef]
11. Domae, Y. Recent trends in the research of industrial robots and future outlook. *J. Robot. Mechatronics* **2019**, *31*, 57–62. [CrossRef]
12. Juel, W.K.; Haarslev, F.; Ramirez, E.R.; Marchetti, E.; Fischer, K.; Shaikh, D.; Manoonpong, P.; Hauch, C.; Bodenhagen, L.; Krüger, N. Smooth robot: Design for a novel modular welfare robot. *J. Intell. Robot. Syst.* **2020**, *98*, 19–37. [CrossRef]
13. Eirale, A.; Martini, M.; Tagliavini, L.; Gandini, D.; Chiaberge, M.; Quaglia, G. Marvin: An Innovative Omni-Directional Robotic Assistant for Domestic Environments. *Sensors* **2022**, *22*, 5261. [CrossRef]
14. Yu, X.; He, W.; Li, Q.; Li, Y.; Li, B. Human-robot co-carrying using visual and force sensing. *IEEE Trans. Ind. Electron.* **2020**, *68*, 8657–8666. [CrossRef]
15. Goldau, F.F.; Shastha, T.K.; Kyrarini, M.; Gräser, A. Autonomous multi-sensory robotic assistant for a drinking task. In Proceedings of the 2019 IEEE 16th International Conference on Rehabilitation Robotics (ICORR), Toronto, ON, Canada, 24–28 June 2019; pp. 210–216.
16. Dong, J.; Cong, Y.; Sun, G.; Zhang, T. Lifelong robotic visual-tactile perception learning. *Pattern Recognit.* **2022**, *121*, 108176. [CrossRef]
17. Kroger, T.; Finkemeyer, B.; Winkelbach, S.; Eble, L.O.; Molkenstruck, S.; Wahl, F.M. A manipulator plays Jenga. *IEEE Robot. Autom. Mag.* **2008**, *15*, 79–84. [CrossRef]
18. Yoshikawa, T.; Shinoda, H.; Sugiyama, S.; Koeda, M. Jenga game by a manipulator with multiarticulated fingers. In Proceedings of the 2011 IEEE/ASME International Conference on Advanced Intelligent Mechatronics (AIM), Budapest, Hungary, 3–7 July 2011; pp. 960–965.
19. Fazeli, N.; Oller, M.; Wu, J.; Wu, Z.; Tenenbaum, J.B.; Rodriguez, A. See, feel, act: Hierarchical learning for complex manipulation skills with multisensory fusion. *Sci. Robot.* **2019**, *4*, eaav3123. [CrossRef] [PubMed]
20. Bauza, S.; Castillo, J.; Nanz, A.; Kambalur, B. Deep Q-Learning Applied to a Jenga Playing Robot. *Preprint ResearchGate* 2017 Available online: https://www.researchgate.net/publication/336778754_Deep_Q-Learning_Applied_to_a_Jenga_Playing_Robot (accessed on 5 November 2022).
21. Justesen, N.; Bontrager, P.; Togelius, J.; Risi, S. Deep learning for video game playing. *IEEE Trans. Games* **2019**, *12*, 1–20. [CrossRef]
22. Mnih, V.; Kavukcuoglu, K.; Silver, D.; Graves, A.; Antonoglou, I.; Wierstra, D.; Riedmiller, M. Playing atari with deep reinforcement learning. *arXiv* **2013**, arXiv:1312.5602.
23. Silver, D.; Hubert, T.; Schrittwieser, J.; Antonoglou, I.; Lai, M.; Guez, A.; Lanctot, M.; Sifre, L.; Kumaran, D.; Graepel, T.; et al. Mastering chess and shogi by self-play with a general reinforcement learning algorithm. *arXiv* **2017**, arXiv:1712.01815.
24. Caccianiga, G.; Mariani, A.; de Paratesi, C.G.; Menciassi, A.; De Momi, E. Multi-sensory guidance and feedback for simulation-based training in robot assisted surgery: A preliminary comparison of visual, haptic, and visuo-haptic. *IEEE Robot. Autom. Lett.* **2021**, *6*, 3801–3808. [CrossRef]
25. Zheng, C.; Chen, P.; Pang, J.; Yang, X.; Chen, C.; Tu, S.; Xue, Y. A mango picking vision algorithm on instance segmentation and key point detection from RGB images in an open orchard. *Biosyst. Eng.* **2021**, *206*, 32–54. [CrossRef]
26. Zheng, W.; Xie, Y.; Zhang, B.; Zhou, J.; Zhang, J. Dexterous robotic grasping of delicate fruits aided with a multi-sensory e-glove and manual grasping analysis for damage-free manipulation. *Comput. Electron. Agric.* **2021**, *190*, 106472. [CrossRef]
27. Wang, J.; Rogers, P.; Parker, L.; Brooks, D.; Stilman, M. Robot Jenga: Autonomous and strategic block extraction. In Proceedings of the 2009 IEEE/RSJ International Conference on Intelligent Robots and Systems, St. Louis, MO, USA, 10–15 October 2009; pp. 5248–5253.
28. Kimura, S.; Watanabe, T.; Aiyama, Y. Force based manipulation of Jenga blocks. In Proceedings of the 2010 IEEE/RSJ International Conference on Intelligent Robots and Systems, Taipei, Taiwan, 18–22 October 2010; pp. 4287–4292.
29. LeCun, Y.; Bengio, Y.; Hinton, G. Deep learning. *Nature* **2015**, *521*, 436–444. [CrossRef]
30. Krizhevsky, A.; Sutskever, I.; Hinton, G.E. Imagenet classification with deep convolutional neural networks. *Commun. ACM* **2017**, *60*, 84–90. [CrossRef]
31. He, K.; Zhang, X.; Ren, S.; Sun, J. Deep residual learning for image recognition. In Proceedings of the IEEE Conference on Computer Vision and Pattern Recognition, Las Vegas, NV, USA, 27–30 June 2016; pp. 770–778.
32. Hu, J.; Shen, L.; Sun, G. Squeeze-and-excitation networks. In Proceedings of the IEEE Conference on Computer Vision and Pattern Recognition, Salt Lake City, UT, USA, 18–23 June 2018; pp. 7132–7141.

33. Woo, S.; Park, J.; Lee, J.Y.; Kweon, I.S. Cbam: Convolutional block attention module. In Proceedings of the European Conference on Computer Vision (ECCV), Munich, Germany, 8–14 September 2018; pp. 3–19.
34. Vaswani, A.; Shazeer, N.; Parmar, N.; Uszkoreit, J.; Jones, L.; Gomez, A.N.; Kaiser, Ł.; Polosukhin, I. Attention is all you need. In Proceedings of the Advances in Neural Information Processing Systems, Long Beach, CA, USA, 4–9 December 2017; Volume 30.
35. Dosovitskiy, A.; Beyer, L.; Kolesnikov, A.; Weissenborn, D.; Zhai, X.; Unterthiner, T.; Dehghani, M.; Minderer, M.; Heigold, G.; Gelly, S.; et al. An Image is Worth 16 × 16 Words: Transformers for Image Recognition at Scale. In Proceedings of the International Conference on Learning Representations, Virtual Event, 3–7 May 2021.
36. Liu, W.; Anguelov, D.; Erhan, D.; Szegedy, C.; Reed, S.; Fu, C.Y.; Berg, A.C. Ssd: Single shot multibox detector. In Proceedings of the European Conference on Computer Vision, Amsterdam, The Netherlands, 11–14 October 2016; pp. 21–37.
37. Redmon, J.; Divvala, S.; Girshick, R.; Farhadi, A. You only look once: Unified, real-time object detection. In Proceedings of the IEEE Conference on Computer Vision and Pattern Recognition, Las Vegas, NV, USA, 27–30 June 2016; pp. 779–788.
38. Redmon, J.; Farhadi, A. YOLO9000: Better, faster, stronger. In Proceedings of the IEEE Conference on Computer Vision and Pattern Recognition, Honolulu, HI, USA, 21–26 July 2017; pp. 7263–7271.
39. Redmon, J.; Farhadi, A. Yolov3: An incremental improvement. *arXiv* **2018**, arXiv:1804.02767.
40. Long, J.; Shelhamer, E.; Darrell, T. Fully convolutional networks for semantic segmentation. In Proceedings of the IEEE Conference on Computer Vision and Pattern Recognition, Boston, MA, USA, 7–12 June 2015; pp. 3431–3440.
41. Ronneberger, O.; Fischer, P.; Brox, T. U-net: Convolutional networks for biomedical image segmentation. In Proceedings of the International Conference on Medical Image Computing and Computer-Assisted Intervention, Munich, Germany, 5–9 October 2015; pp. 234–241.
42. He, K.; Gkioxari, G.; Dollár, P.; Girshick, R. Mask r-cnn. In Proceedings of the IEEE International Conference on Computer Vision, Venice, Italy, 22–29 October 2017; pp. 2961–2969.
43. Ren, S.; He, K.; Girshick, R.; Sun, J. Faster r-cnn: Towards real-time object detection with region proposal networks. In Proceedings of the Advances in Neural Information Processing Systems, Montreal, QC, Canada, 7–12 December 2015; Volume 28.
44. Li, Y.; Qi, H.; Dai, J.; Ji, X.; Wei, Y. Fully convolutional instance-aware semantic segmentation. In Proceedings of the IEEE Conference on Computer Vision and Pattern Recognition, Honolulu, HI, USA, 21–26 July 2017; pp. 2359–2367.
45. Chen, L.C.; Hermans, A.; Papandreou, G.; Schroff, F.; Wang, P.; Adam, H. Masklab: Instance segmentation by refining object detection with semantic and direction features. In Proceedings of the IEEE Conference on Computer Vision and Pattern Recognition, Salt Lake City, UT, USA, 18–22 June 2018; pp. 4013–4022.
46. Kirillov, A.; Levinkov, E.; Andres, B.; Savchynskyy, B.; Rother, C. Instancecut: From edges to instances with multicut. In Proceedings of the IEEE Conference on Computer Vision and Pattern Recognition, Honolulu, HI, USA, 21–26 July 2017; pp. 5008–5017.
47. Liang, X.; Lin, L.; Wei, Y.; Shen, X.; Yang, J.; Yan, S. Proposal-free network for instance-level object segmentation. *IEEE Trans. Pattern Anal. Mach. Intell.* **2017**, *40*, 2978–2991. [CrossRef] [PubMed]
48. Newell, A.; Huang, Z.; Deng, J. Associative embedding: End-to-end learning for joint detection and grouping. In Proceedings of the Advances in Neural Information Processing Systems, Long Beach, CA, USA, 4–9 December 2017; Volume 30.
49. Bolya, D.; Zhou, C.; Xiao, F.; Lee, Y.J. Yolact: Real-time instance segmentation. In Proceedings of the IEEE/CVF International Conference on Computer Vision, Seoul, Korea, 27 October–2 November 2019; pp. 9157–9166.
50. Bolya, D.; Zhou, C.; Xiao, F.; Lee, Y.J. YOLACT++: Better Real-time Instance Segmentation. *IEEE Trans. Pattern Anal. Mach. Intell.* **2020**, *44*, 1108–1121. [CrossRef] [PubMed]
51. Evjemo, L.D.; Gjerstad, T.; Grøtli, E.I.; Sziebig, G. Trends in smart manufacturing: Role of humans and industrial robots in smart factories. *Curr. Robot. Rep.* **2020**, *1*, 35–41. [CrossRef]
52. Hill, J.; Park, W.T. Real Time Control of a Robot with a Mobile Camera. In Proceedings of the 9th ISIR, Washington, DC, USA, 13–15 March 1979.
53. Azizian, M.; Khoshnam, M.; Najmaei, N.; Patel, R.V. Visual servoing in medical robotics: A survey. Part I: Endoscopic and direct vision imaging-techniques and applications: Visual servoing in medical robotics: A survey (Part I). *Int. J. Med. Robot. Comput. Assist. Surg.* **2014**, *10*, 263–274. [CrossRef] [PubMed]
54. Dewi, T.; Risma, P.; Oktarina, Y.; Muslimin, S. Visual Servoing Design and Control for Agriculture Robot; a Review. In Proceedings of the 2018 International Conference on Electrical Engineering and Computer Science (ICECOS), Malang, Indonesia, 2–4 October 2018; pp. 57–62.
55. Staub, C.; Osa, T.; Knoll, A.; Bauernschmitt, R. Automation of tissue piercing using circular needles and vision guidance for computer aided laparoscopic surgery. In Proceedings of the 2010 IEEE International Conference on Robotics and Automation, Anchorage, AK, USA, 3–7 May 2010; pp. 4585–4590.
56. Voros, S.; Haber, G.P.; Menudet, J.F.; Long, J.A.; Cinquin, P. ViKY Robotic Scope Holder: Initial Clinical Experience and Preliminary Results Using Instrument Tracking. *IEEE/ASME Trans. Mechatronics* **2010**, *15*, 879–886. [CrossRef]
57. Krupa, A.; Gangloff, J.; Doignon, C.; de Mathelin, M.; Morel, G.; Leroy, J.; Soler, L.; Marescaux, J. Autonomous 3D positioning of surgical instruments in robotized laparoscopic surgery using visual servoing. *IEEE Trans. Robot. Autom.* **2003**, *19*, 842–853. [CrossRef]
58. Hutchinson, S.; Hager, G.; Corke, P. A tutorial on visual servo control. *IEEE Trans. Robot. Autom.* **1996**, *12*, 651–670. [CrossRef]
59. Barth, R.; Hemming, J.; van Henten, E.J. Design of an eye-in-hand sensing and servo control framework for harvesting robotics in dense vegetation. *Biosyst. Eng.* **2016**, *146*, 71–84. [CrossRef]

60. Mehta, S.; MacKunis, W.; Burks, T. Robust visual servo control in the presence of fruit motion for robotic citrus harvesting. *Comput. Electron. Agric.* **2016**, *123*, 362–375. [CrossRef]
61. Lippiello, V.; Cacace, J.; Santamaria-Navarro, A.; Andrade-Cetto, J.; Trujillo, M.Á.; Rodríguez Esteves, Y.R.; Viguria, A. Hybrid Visual Servoing With Hierarchical Task Composition for Aerial Manipulation. *IEEE Robot. Autom. Lett.* **2016**, *1*, 259–266. [CrossRef]
62. Comport, A.; Marchand, E.; Pressigout, M.; Chaumette, F. Real-time markerless tracking for augmented reality: The virtual visual servoing framework. *IEEE Trans. Vis. Comput. Graph.* **2006**, *12*, 615–628. [CrossRef] [PubMed]
63. Lin, T.Y.; Dollár, P.; Girshick, R.; He, K.; Hariharan, B.; Belongie, S. Feature pyramid networks for object detection. In Proceedings of the IEEE Conference on Computer Vision and Pattern Recognition, Honolulu, HI, USA, 21–26 July 2017; pp. 2117–2125.
64. Collins, T.; Bartoli, A. Infinitesimal plane-based pose estimation. *Int. J. Comput. Vis.* **2014**, *109*, 252–286. [CrossRef]
65. Siciliano, B.; Sciavicco, L.; Villani, L.; Oriolo, G. *Robotics: Modelling, Planning and Control*; Advanced textbooks in control and signal processing; Springer: London, UK, 2010.
66. Marchand, E.; Spindler, F.; Chaumette, F. ViSP for visual servoing: A generic software platform with a wide class of robot control skills. *IEEE Robot. I Autom. Mag.* **2005**, *12*, 40–52. [CrossRef]
67. Denninger, M.; Sundermeyer, M.; Winkelbauer, D.; Olefir, D.; Hodan, T.; Zidan, Y.; Elbadrawy, M.; Knauer, M.; Katam, H.; Lodhi, A. BlenderProc: Reducing the Reality Gap with Photorealistic Rendering. In Proceedings of the Robotics: Science and Systems (RSS), Virtual Event, 12–16 July 2020.

**Disclaimer/Publisher's Note:** The statements, opinions and data contained in all publications are solely those of the individual author(s) and contributor(s) and not of MDPI and/or the editor(s). MDPI and/or the editor(s) disclaim responsibility for any injury to people or property resulting from any ideas, methods, instructions or products referred to in the content.

Article

# Collaborative 3D Scene Reconstruction in Large Outdoor Environments Using a Fleet of Mobile Ground Robots

John Lewis, Pedro U. Lima and Meysam Basiri *

Institute for Systems and Robotics, Instituto Superior Tecnico, University of Lisbon, 1049-001 Lisbon, Portugal
* Correspondence: meysam.basiri@tecnico.ulisboa.pt

**Abstract:** Teams of mobile robots can be employed in many outdoor applications, such as precision agriculture, search and rescue, and industrial inspection, allowing an efficient and robust exploration of large areas and enhancing the operators' situational awareness. In this context, this paper describes an active and decentralized framework for the collaborative 3D mapping of large outdoor areas using a team of mobile ground robots under limited communication range and bandwidth. A real-time method is proposed that allows the sharing and registration of individual local maps, obtained from 3D LiDAR measurements, to build a global representation of the environment. A conditional peer-to-peer communication strategy is used to share information over long-range and short-range distances while considering the bandwidth constraints. Results from both real-world and simulated experiments, executed in an actual solar power plant and in its digital twin representation, demonstrate the reliability and efficiency of the proposed decentralized framework for such large outdoor operations.

**Keywords:** scene reconstruction; cooperative mapping; point cloud registration; multi-robot system; 3D mapping; communication constraint

## 1. Introduction

The use of outdoor mobile robots for real-world applications, such as search and rescue [1,2], logistics [3], agriculture [4], industrial inspection [5], surveillance and maintenance [6,7], have increased rapidly over the past several years. This is due to the capabilities of mobile robots to assist humans in dangerous, repetitive or time-consuming tasks. A successful robot navigation for such applications relies primarily on three aspects: mapping, localization, and trajectory planning. Robotic mapping generates a map by deciphering the spatial information of the environment acquired through the robot's sensors. Commonly, mapping is carried out first to understand the environment and enhance the subsequent localization and motion planning tasks. However, for many applications, mapping must be executed frequently to continuously acquire a complete situational awareness and to support reasoning and decision making in dynamic environments.

Many outdoor robotic automation applications, such as solar farm inspection and maintenance [8–10], disaster response [11–13], agriculture [14] and city re-planning [15,16] need to cover very large areas of 1–40 acres. Traversing such expansive environments with a single mobile robot is very time-consuming or even impractical. In addition, conventional localization methods based only on GPS, odometry and IMU are not always reliable for such long-range operations. The uneven, rough, and unstructured nature of rural environments, such as in solar farms and disaster-struck regions, introduce additional localization errors. In such scenarios, a multi-robot system can be a suitable alternative to obtain full coverage of the area and execute tasks in a collaborative manner, resulting in a more complete and time-efficient solution. In regards to mapping, a multi-robot system can rapidly explore the environment in parallel and from different angles, to generate more accurate maps in less time [17] and to enhance the localization accuracy in challenging environments.

In general, a multi-robot mapping framework will require three main elements:

1. A mission-planning unit to coordinate robots to explore the environment.
2. A communication policy to share the map generated by each robot.
3. A matching and merging method to integrate individual maps into a global map.

In practice, mapping large outdoor 3D environments with a team of mobile robots is challenging due to the communication limitations and the high volume of sensor data that need to be shared and processed. High-capacity wireless communication routers commonly available to robots, such as Wi-Fi modules, typically have a limited range of about 90 m or less in open outdoor environments. On the other hand, long-range wireless devices, such as the Xbee-Pro RF module, can provide a large coverage of up to several kilometers but have a very limited bandwidth of about 200 kbs, which is not suitable for sharing high volumes of sensor data. Furthermore, the employment of 4G mobile technologies is not always possible due to the lack of coverage in many rural areas or in disaster-response scenarios. Hence, it is essential to consider such communication constraints while developing a multi-agent mapping algorithm.

In this work, we present an online, fully distributed and active framework for a team of mobile ground robots, equipped with 3D LiDAR sensors, for mapping and situational awareness in large outdoor environments. We develop our solution specifically for the application of inspection and surveillance of large multi megawatt solar plants, while considering the strict communication constraints that exist in terms of range and bandwidth in the commonly available wireless technologies. However, the proposed framework can be applied to many other outdoor exploration problems, such as search and rescue or precision agriculture. The outline of the paper is as follows: Section 2 provides the literature review on 3D mapping and localization methods and discusses the various techniques and limitations of multi-agent cooperative mapping and point cloud registration. The proposed distributed multi-agent framework is discussed in Section 3. The experiments and results are presented in Section 4. Finally, Section 5 concludes the findings and pitches possible improvements to the proposed method.

## 2. Literature Review

Over the past decade, 3D sensors have emerged as revolutionary data acquisition devices. In robotics, 3D sensory information has been used for mapping, localization, obstacle avoidance, and scene recognition. Omnidirectional LiDARs [18–21], RGBD cameras [22–25], and uni-directional LiDARs [26–28] have found applications in the field. Three-dimensional sensor-based algorithms, such as LOAM [18,19,21], have become the norm for an out-of-the-box simultaneous localization and mapping algorithm. However, the computational complexity of 3D algorithms and the size of 3D sensor data make it challenging to achieve scalability. Due to these reasons, SLAM (simultaneous localization and mapping) algorithms [29–31] are not commonly preferred with 3D sensors, especially in large areas [32,33]. Methods of point cloud compression [34] and low-cost registration [35,36] are promising endeavors but require prior training. The considerable size of 3D data further imposes constraints on a decentralized multi-agent mapping system, making it challenging to share observations continuously. Hence, it is essential to transfer only the required features.

Point clouds represent rigid body data structures, typically generated from LiDAR sensors. The process of aligning two point clouds is called point cloud registration. The process results in a rigid body transformation matrix that aligns one point cloud in the frame of another. The registration techniques [37] are categorized into local and global registration. Global registration [38–41] is ideal when the initial transformation estimate has yet to be discovered, and is perfect when the point clouds are acquired from spatially distant frames of references. When the initial transformation is known, a quick refinement can be acquired from a local registration technique. Local registration techniques, such as iterative closest point (ICP) [42,43], normal distribution transform [44], point-to-plane [45–47], color-based [48,49] or class-based methods [50,51], can be computationally expensive if the initial transformation is inaccurate. For multi-agent systems, where each agent has a different frame of reference, global point cloud alignment is refined by a local point cloud registration technique.

A multi-robot system relies on a consistent network for the exchange of observations and data. If the system is not centralized, the agents rely on peer-to-peer networking. However, these networks can be classified into two subcategories: long-range and short-range networks. Long-range networks, such as low power wide area (LPWA) [52,53] and long-term evolution machine type communication (LTE-M) [54] provide networking solutions for large areas. The rural area coverage analysis [55] for sigfox (30 km at 12 kbps), lora (15 km at 290 bps–50 kbps), LTE-M (10 km at 200 kbps–1 Mbps) showcases the constraints imposed on the data size. The Xbee-PRO RF modules [56] are commonly used for outdoor robot applications allowing a long-range radio-frequency (RF) transmission that can go up to several kilometers, with a limited bandwidth of (200 kbps). Short-range communication, such as Wi-Fi (100 m at 15 Mbps), are ideal for the transfer of large size data. Thus, for a distributed multi-agent mapping system, relying on peer-to-peer 3D sensory information transfer, covering a wide area ($\geq 1$ km$^2$) requires both short- and long-range communication systems.

Multi-agent SLAM poses many different challenges, such as inter-agent cooperation and communication [57], distributed sensor fusion [58] and collaborative planning [59]. These challenges are further enhanced when the sensors share large information packets, such as 3D data [58]. These challenges can be relaxed in a centralized system, assisted with short-range communication devices with high bandwidth [60,61]; however, this is not a realistic solution for large outdoor applications. In a communication-constrained environment [57,62–65], prior planning [66] to meet and share information can relieve stress on communication channels. However, these periodic communications can be challenging to realize when the area to be covered exceeds 1 km$^2$, especially considering the overall energy expended.

The major contribution of this article is an end-to-end active distributed homogeneous framework for the large-scale 3D mapping of environments. We incorporate a global peer-to-peer small bandwidth long-range network along with a short-range peer-to-peer network to allow a framework that can go beyond the range limits of a Wi-Fi network. The proposed approach generates a global map in each agent's frame and helps to localize agents within this map. Conditionally, the framework heavily filters point clouds to enable long-range transmission, which is then used for localization and mapping. This conditional approach ensures that only the necessary communication bandwidth and computation are used. This relative localization can also be used for improving path planning, exploration, and mapping. The framework is developed, to tackle the communication constraints, imposed in large-area mapping.

## 3. Methodology

A set of $N_a$ agents, $R$, is tasked to explore and map the environment. Each agent, $R_i$ ($\forall R_i \in R$), is equipped with a 3D LiDAR sensor, a GPS receiver, an IMU, and an odometer sensor. The LiDAR sensor has a maximum range of $L_{max}$. Considering possible GPS drifts, odometer slippage, and electromagnetic interference, each agent has an instance of an extended Kalman filter (EKF) to fuse the sensory information from GPS, IMU and odometer and obtain a more reliable estimate of the global localization $G_i$ in the geographic coordinate system. An instance of the LiDAR odometry and mapping (LOAM) [18] is used on each agent to locally map the surrounding environment from the $R_i$ perspective and to locate the agent within the map. This egocentric LOAM localization is represented in the form of an odometry message $O_i$, in the sub-map $M_i$, of agent $R_i$. Each agent is equipped with a short-range and a long-range wireless transceiver. The short-range transceiver is a Wi-Fi module that allows the peer-to-peer transfer of large map data, which is only activated when two agents are within distance $C_{max}$ of each other. The long-range transceiver is an RF module that can ensure long-range transfer of very small quantities of data, which is used only to share odometry, GPS or heavily down-sampled 3D data.

Figure 1 portrays an instance of the proposed fast decentralized multi-agent active mapping framework, executed on agent $R_i$. A separate instance of the framework is executed on each agent of R. This ensures an active decentralized framework for multi-agent 3D mapping. The framework has two modules: a continuous update module and a conditional update module. The continuous update module is executed with every new sensory update. In this module, an instance of LOAM generates the egocentric odometry $O_i$ and the map $M_i$. Added to this, an instance of extended Kalman filter fuses the sensory data from GPS, IMU and odometry to output the global localization estimate, $G_i$. A ball tree generator, as explained in Section 3.1, generates a global ball tree $B_i$ that keeps track of $O_i$ and $G_i$ throughout time and at specified distance intervals. Whenever a new tree node is added, it is also shared with all other agents using the long-range transmitter. Minimal proximity search, detailed in Section 3.2, is used to compute the proximity of an incoming tree node from an agent $R_j$ with all nodes of the ball tree $B_i$. The conditional update module is executed when the result of the minimal proximity search is true.

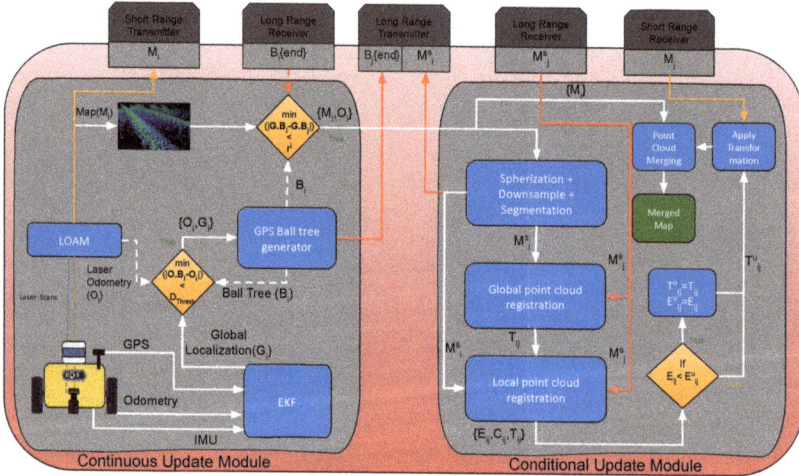

**Figure 1.** Proposed framework deployed in agent $R_i$.

Conditional update module consists of several computationally intensive processes. Spherized point cloud registration, in Section 3.3, describes down-sampling, segmentation and nearest-neighbor sampling of the $M_i$, to generate a spherized map $M^s{}_i$. $M^s{}_i$, which is considerably reduced in size but abundant in features, is then transmitted using the long-range transmitter to the respective agent $R_j$, for point cloud registration. The resultant transformation is then used for merging the complete maps once the agents are close enough to transfer the complete maps via the short-range transceiver.

*3.1. Global Ball Tree Generator*

A ball tree is a binary tree data structure, that is used for data partitioning to ensure fast data query [67]. When an agent, $R_i$, is initialized, a ball-tree, $B_i$, is instantiated with $G_i(t = 0)$ as the root node. The node $n$ of $B_i$ is represented as $B_i(n)$ and the latest node added is $B_i(end)$. The pair-wise distance used for constructing $B_i$ is the Haversine distance. The Haversine distance represents the angular distance between two points on the surface of a sphere. Ball trees with Haversine distance are shown to result in fast nearest-neighbor look-up for GPS datasets [68,69]. A node $B_i(n)$, inserted at time instant $T$, consists of the $G_i(t = T)$ and is tagged with the corresponding LOAM-odometry message $O_i(t = T)$. The framework continuously monitors the LOAM-odometry $O_i$ and iteratively calculates the distance between the current odometry $O_i(t = T)$ with that of the last node $B_i(end)$. If this distance is greater than a predefined value, $D_{thresh}$, a new node is added to $B_i$, with $G_i(t = T)$ and $O_i(t = T)$.

The process, called global ball-tree generator, is described in Algorithm 1, which is continuously run by each agent in $R$. Each node of the ball-tree has a global localization estimate $G_i$, which is mapped with the corresponding LOAM-odometry message stored at that time instant. These pair-wise data are essential to link the egocentric localization of $R_i$ with the global frame. Alternatively, we could georeference the point cloud, for each iteration of LOAM, which requires an accurate initial global localization estimate [70] and would be computationally costly [71,72]. The intermittent method proposed in this work eases the computational complexity. It also alleviates the dependence on a single initial estimate.

---

**Algorithm 1** Global ball-tree generator for agent $R_i$.

**Input:** $G_i(t), O_i(t), D_{thresh}, B_i$

**Initialize** $B_i = \text{ADD-NODE}(G_i(t=0), O_i(t=0))$

**while** mapping **do**
    Odom(Current) $= O_i(t = T)$
    Odom(Last − Node) $= B_i(end) \to O$
    **if** $\|\text{Odom}(Current) - \text{Odom}(Last - Node)\|_2 \leq D_{thresh}$ **then**
        $B_i = \text{ADD-NODE}(G_i(t=T), O_i(t=T), B_i)$
    **end if**
**end while**

**procedure** ADD-NODE($G, O, B=\text{Balltree}()$)
    $B.\text{push}(G)$
    $B(end) \to O$
    return $B$
**end procedure**

---

*3.2. Minimal Proximity Search*

In a communication-constrained environment, it is essential to ensure that the bandwidth is used for the most vital transmissions. The process explained in this section queries for possible spatial overlaps in the global frame. Minimal proximity search transmits every new node added to the ball tree $B_i(end)$ and compares it with the ball trees of other agents in $R$ for proximity within the global frame.

Every new node added to $B_i$ of agent $R_i$ is shared over the long-range transmitter, with the remaining agents. The minimal bandwidth required to transfer the node $B_i(end)$ makes it ideal for a communication-constrained environment. With no loss of generality, an agent $R_j (\in R, \forall i \neq j)$ has its own instance of LOAM, EKF and global ball-tree generator, resulting in its own sub-map $M_j$, egocentric odometry $O_j$, global localization estimate $G_j$, and global ball-tree $B_j$. Agent $R_j$ processes the incoming node information from $R_i$ by carrying out a nearest-neighbor search in $B_j$. If the global localization estimate $G_j$ entry of the resultant nearest-neighbor node, $B_j(Neighbour)$, is within a certain threshold($r^i$), of $B_i(end)$, the node $B_j(Neighbour)$ is shared with $R_i$. This is depicted in Algorithm 2.

**Algorithm 2** Minimal proximity search by $R_j$.

**Input:** $B_i(end), B_j, r^{ij}$

**while** mapping **do**
    Neighbor=$B_j$.NearestNeighborSearch($B_i(end).G$)
    **if** Neighbor.distance $\leq r^{ij}$ **then**
        return $B_j(Neighbor)$
    **end if**
**end while**

---

In an effort to minimize the effect of GPS drifts, the distance threshold, $r^i$, is a bounded dynamic distance threshold. Equation (1) ensures that $r^i$ is bounded within predefined values ($r_{min}, r_{max}$) and proportional to the uncertainty $C^i_{ekf}$ of the agent $R_i$ EKF estimate.

$$r^i = \begin{cases} r_{max} & \text{if } C^i_{ekf} \cdot r^i \geq r_{max} \\ r_{min} & \text{if } C^i_{ekf} \cdot r^i \leq r_{min} \\ C^i_{ekf} \cdot r^i & \text{else} \end{cases} \quad (1)$$

*3.3. Spherized Point Cloud Registration*

There exists a transformation, $T_{ij}$, that aligns $M_j$ with $M_i$ of agents $R_j$ and $R_i$. This transformation can be achieved by registering the map $M_j$ with the map $M_i$. However, the sizes of $M_i$ and $M_j$ are rapidly increasing as the $R_i$ and $R_j$ individually explore and map the environment, from their perspective. Sharing such large data over a long-range bandwidth-limited communication channel will lead to a high network latency and data loss. Hence, this section describes a strategy to only share small sampled subsets of the maps, and only for the regions that are expected to have sufficient overlapping features for registration.

With no loss in generality, let us assume that for two agents, $R_i$ and $R_j$, the minimal proximity search was successful. A successful minimal proximity search (Section 3.2) gives an assurance that, at two different time instances, the $R_i$ and $R_j$ are spatially close, in the global frame. The minimal proxy search results in two nodes, $node_i$ and $node_j$, of $B_i$ and $B_j$, that are globally close: $B_i(node_i).O$ and $B_j(node_j).O$ gives the egocentric odometry measurement of $R_i$ and $R_j$. For lucidity, we will refer to $B_i(node_i).O$ and $B_j(node_j).O$ as $L_i$ and $L_j$, respectively.

A Euclidean ball, of radius $r_o$, is generated in both $M_i$ and $M_j$, centered at $L_i$ and $L_j$, respectively. This method of filtering is hereby referred to as spherization. The points within the sphere are sampled and used for point cloud registration. Since they represent a fraction of the overall map, the size is considerably reduced. Added to this, the sampled map, $M^s_i$ and $M^s_j$, have features that are bound to overlap. This is because the sampled map was generated when the agents were spatially close in the global frame. Since point clouds can be considered a rigid body of particles [73], we can conclude that the $T_{ij}$ that successfully aligns $M^s_i$ with $M^s_j$ also aligns $M_i$ with $M_j$.

Spherized maps are transmitted over the long-range transmitter to the respective agents. For a seamless transmission on the constrained bandwidth channel, the spherized maps have to be less than 25 kilobytes. Thus, spherization is preceded by downsampling, ground-plane removal and outlier removal to bring down the overall size of the point cloud to the prerequisite limit. Each agent generates the spherized maps in its own frame of reference. These frames of reference will be separated by several meters, which is not ideal for a local point cloud registration algorithm. We use a global registration algorithm to align these two spherized point clouds roughly. The transformation matrix acquired from the global registration technique is then used to initialize the local point cloud registration. Local point cloud registration helps in refining the initial rough alignment.

The local point cloud registration results in the transformation, $T_{ij}$, and the RMSE, $E_{ij}$, of all inlier correspondences.

The RMSE [45], in the context of point cloud registration, refers to the root mean square value between the corresponding points of the two point clouds. For $N_c$ correspondences, between $M_i^s$ and $M_j^s$, the RMSE for transformation, $T_{ij}$, can be calculated by Equation (2). $c_i$ and $c_j$ refer to all the correspondences in $M_i^s$ and $M_j^s$, respectively. The transformation, $T_{ij}$, that minimizes $E_{ij}$, across all executions of Algorithm 3 is chosen for the full map alignment.

$$\text{RMSE} = \sqrt{\frac{\sum_{n=1}^{N_c} \|M_j^s(c_g^n) - T_{ij} * M_j^s(c_m^n)\|^2}{N_c}} \quad (2)$$

In the proposed implementation, the global registration is carried out using RANSAC (random sample consensus) [74]. The FPFH (fast point feature histograms) feature [75], a 33-dimensional vector that encapsulates the local geometric property, for each point, is calculated. RANSAC searches for these features to make a fast and approximate alignment. For local registration, we are aware that the process can be further enhanced by sharing only the features [76,77] rather than the entire point clouds and subsequently using feature-based registration methods [45,50,51]. We could also implement a semantic mapping technique [20,21] for acquiring a segmented map before spherization. However, we use point-to-plane ICP [46] to keep the overall complexity and tunable parameters to a minimum.

**Algorithm 3** Spherized point cloud registration in agent $R_i$.

**Input:** $M_i, M^s{}_j, B_i, r^s{}_{ij}$

if Minimal-Proximity-Search($B_i(end), B_j, r^s{}_{ij}$) is True then
    $M^s{}_i$ = Spherization($M_i, B_i(end).O, r^s{}_{ij}$)
    Long-range-transmission($M^s{}_i$) -> $R_j$
    $T_{ij}$ = Global-Point-Cloud-Registration($M^s{}_i, M^s{}_j$)
    $T_{ij}, E_{ij}, C_{ij}$ = Local-Point-Cloud-Registration($M^s{}_i, M^s{}_j$)
    if $E_{ij} < E^u{}_{ij}$ then
        $E^u{}_{ij} = E_{ij}$
        $T^u{}_{ij} = T_{ij}$
        return $T^u{}_{ij}$
    end if
end if

**procedure** SPHERIZATION($M, O, r$)
    $M$ = Outlier-Removal(Ground-Plane-Removal(Downsample($M$)))
    Neighbors = $M$.NearestNeighbourSearch(centre = $O$, radius = $r$)
    $M^s$ = $M$(Neighbors)
    return $M^s$
**end procedure**

## 4. Experiments and Results

### 4.1. Real World Experiments

We carried out our experiments with two Jackal robots (shown in Figure 2a) (named J1 and J2), from Clearpath Robotics, on an actual solar farm (total area approximately 1 km$^2$, depicted in Figure 3a). The two robots were equipped with a Velodyne Puck (VLP-16) sensor that has 16 layers of infra-red (IR) lasers, a horizontal field of view of 360°, a vertical field of view of 30° and a speed of up to 300,000 data points per second. A Pixhawk 2.1 cube IMU and a Here+ GPS receiver were also added to each robot.

**Figure 2.** Real-world experimental setup. (**a**) Clearpath Jackal. (**b**) Path taken.

**Figure 3.** Experimental setup: Solar farm, corresponding 3D map of solar farm obtained from LOAM [18] and digital twin. (**a**) Aerial view. (**b**) 3D point cloud. (**c**) Digital Twin.

The global path of the robots are planned beforehand to explore the regions of interest, through visiting a set of predefined GPS waypoints. The plans also include some time instances where the robots are within communication range for the sharing of map information between agents. The global paths taken by the robots are presented in Figure 2b, along with the area in which the agents were within short-range communication distance and the region that had a successful minimal proximity search outcome. The values of $(r_{max}, r_{min})$ were set to (20 m, 30 m). Figure 4 represents the various stages of the framework during the experiment. Each agent's LOAM initialization (shown in Figure 4a,d) creates an ego-centric frame of reference. Once the successful minimal proximity search is achieved, the down-sampled point cloud spheres are shared between both agents. These are then registered in the respective frames of reference, as portrayed in Figure 4b,e. Finally, when the agents are close enough for short-range communication, the latest maps generated by J1 and J2 are shared and aligned, as depicted in Figure 4c,f.

*4.2. Simulated Experiments*

The simulations were carried out on a digital twin world (a 3D Gazebo model) of the actual solar farm used in the real-world experiments, as shown in Figure 3a. The simulated environment had a total area of about 1 km$^2$. The digital twin was purely used for simulation purposes, to further test the collaborative 3D scene reconstruction framework with multiple agents, and was not linked to real-time sensory data from the actual solar farm. The complete ground truth point cloud was acquired by converting the 3D Gazebo mesh model (depicted in Figure 5b) to a 3D point cloud model (depicted in Figure 5c).

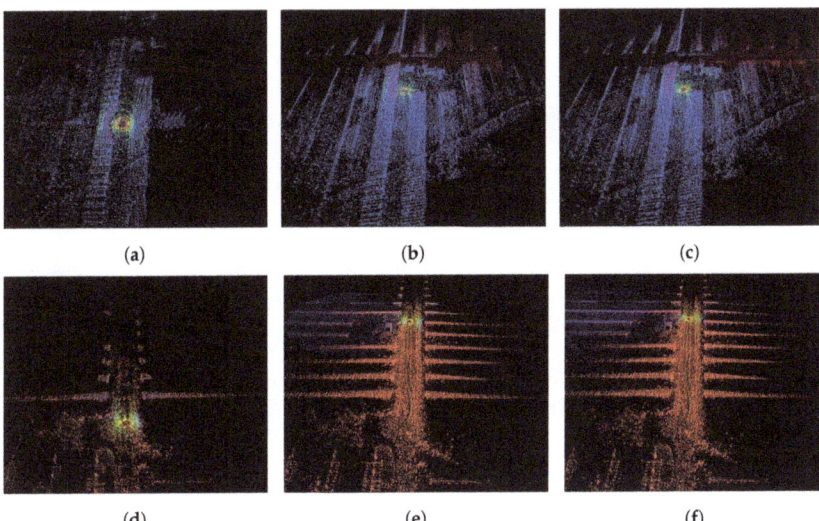

**Figure 4.** Real-world experiment—the various phases of the proposed method for agent J1 (**above**) and J2 (**below**). The blue and red points correspond to the point clouds generated by J1 and J2, respectively. (**a,d**) LOAM initialization; (**b,e**) minimal proximity search was successful, an agent receives a spherized down-sampled map from the other agent and registers this in its own map; (**c,f**) each agent receives the full map from the other agent, as they are within short communication range. The agent aligns the incoming map with its own map using the transformation acquired from the spherized registration.

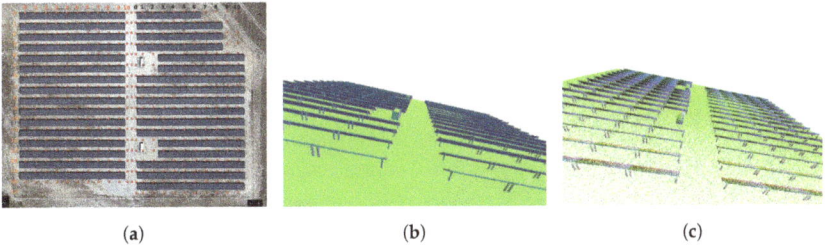

**Figure 5.** (**a**) Way points utilized for navigation, (**b**) 3D mesh of the digital twin, (**c**) 3D point cloud generated from the 3D mesh.

We initially performed a brief parameter analysis to select the values for maximum LiDAR ranges, $L_{max}$. Figure 6 details three maps, generated with three different $L_{max}$ values, by an agent following the same path in between the solar panels of the digital twin world (Figure 3c). We can note that for $L_{max} = 20$ m (in Figure 6a), LOAM is unable to properly find the correspondences that are further away. Due to this, the reconstructed panels are incorrectly curved. This issue is not seen for $L_{max} = 40$ m (in Figure 6b) and $L_{max} = 80$ m (in Figure 6c). In an effort to keep the computational load to a minimum, $L_{max}$ was selected as 40 m for the experiments.

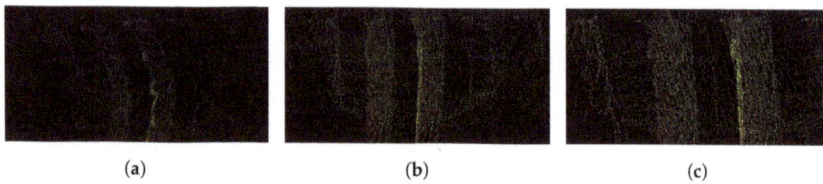

**Figure 6.** Maps, from the same viewpoint, generated with varying maximum LiDAR ranges, $L_{max}$. (**a**) $L_{max} = 20$ m. (**b**) $L_{max} = 40$ m. (**c**) $L_{max} = 80$ m.

To validate the robustness of the proposed algorithm to 3D-laser errors, we induced a Gaussian error in the simulated VLP-16 sensor. The red circles in Figure 7 map the correspondences between the ground truth and the 3D map generated by a single agent. It can be noted that, owing to LiDAR errors, there is a clear mismatch in the generated map. The resultant incorrect 3D reconstruction is evident in the encircled areas. Such reconstruction errors, across each mapping agent, is bound to make the eventual point cloud registration prone to errors. However, the cooperative framework was shown to be robust against such individual reconstruction errors and could still merge multiple maps with a good accuracy.

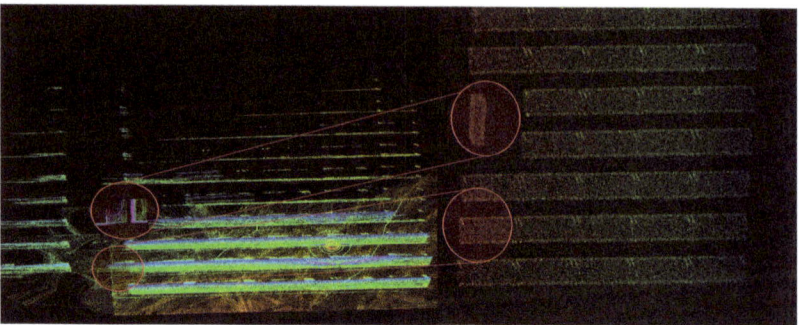

**Figure 7.** The effect of erroneous LiDAR measurements. The red circles represent the expected correspondences between the generated map and the ground truth. (**Left**): The LOAM-mapping result of an agent. (**Right**): The ground-truth 3D model.

We averaged the results over 15 simulations of varying number of UGVs ($N_a = 2$ to 5 agents). For comparing the robustness of the proposed method to the noise in the LiDAR data and resultant LOAM mapping, we executed the experiments with different LiDAR rates, $f$ of 10 hz and 5 hz. Mapping at a lower laser frequency, for the same agent speed, is relatively more error prone. Similar to the real-world experiment, navigation is carried out in between predefined waypoints, shown in Figure 5a. These waypoints are grouped as rows and divided uniformly amongst $N_a$. The selected path covered all possible communication conditions and allowed validation of the end-to-end functionality of the proposed framework. For $N_a = 3$, the distribution of agents and the path taken by each agent can be seen in Figure 8b. Each color in the point cloud, shown in Figure 8a, represents the sub-map obtained by a single agent. Note that, the ground plane was removed for ease of visualization. The errors in individual LOAM-maps can be evidently seen as artifacts in Figure 8a. The framework is robust to these errors and is able to merge the maps irrespectively.

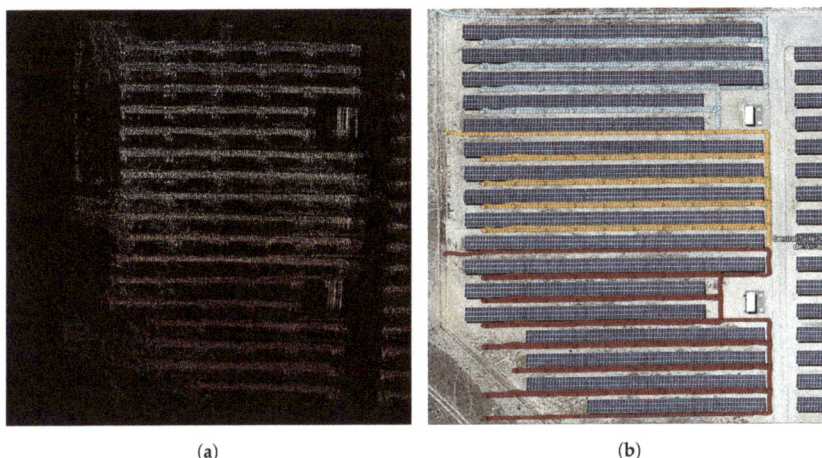

**Figure 8.** The results of an isolated iteration of simulated experiments with $N_a = 3$. (**a**) Merged map. (**b**) Path taken.

For performance analysis of the 3D scene reconstructions, we register the resultant merged map ($M_m$) from the simulated UGVs against the ground truth 3D model ($M_g$). This point cloud registration results in a transformation matrix, $T_{mg}$, and a set of corresponding points, $c_g$ and $c_m$, in $M_g$ and $M_m$, respectively.

4.2.1. RMSE Analysis

Figure 9a plots the minimum–maximum–mean RMSE values for various number of agents. We can note that there is a decline in the overall RMSE values with the increase in the number of agents, $N_a$. The mapping error, infused by the LiDAR noise, accumulates over time. This is spread across the number of agents involved and thus the overall decline in RMSE is expected, provided the map merging is accurate. This decline in RMSE implies a successful fusion of each agent's map.

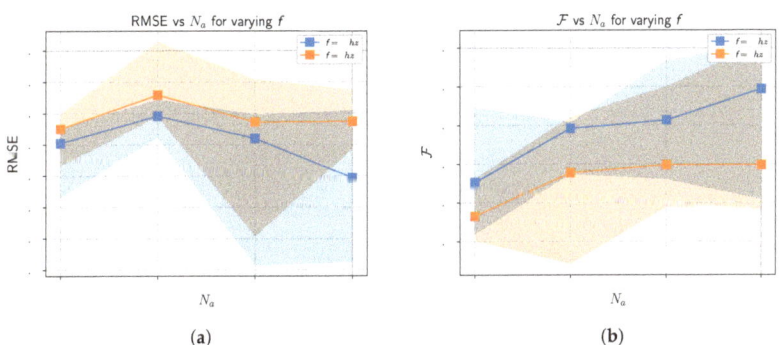

**Figure 9.** RMSE and fitness plots for varying number of agents $N_a$ across different rates. (**a**) RMSE. (**b**) Fitness.

4.2.2. Fitness Analysis

Fitness($\mathcal{F}$) property of a point cloud registration gives us the overlapping area of the two point clouds. For our scenario, where the target point cloud, $M_g$ has $N_g$ points, $\mathcal{F}$ is given by Equation (3).

$$\mathcal{F} = \frac{N_c}{N_g} \quad (3)$$

Since $M_g$ is constant throughout the analysis, a high fitness score implies an increase in the number of correspondences. In Figure 9b, we can note that with the increase in $N_a$, there is a steady increase in $\mathcal{F}$, (and thereby $N_c$), implying that the merged point clouds have more point-to-point correspondences with the $M_g$. This can be attributed to the successful blending of the map of each agent.

4.2.3. Covariance Analysis

The Fischer information matrix, $\mathcal{I}$, that is acquired as a result of the point cloud registration of $M_g$ on $M_m$, characterizes the confidence in the registration process. The inverse [78] of $\mathcal{I}$ gives the covariance matrix, $\mathcal{C}$, of the point cloud registration process [74,79,80]. $\mathcal{C}$ gives us the uncertainties involved in the 6 degrees of freedom. We utilize the determinant of $\mathcal{C}$ for our analysis of the overall uncertainty. Figure 10a showcases the healthy decline in the value of the determinant of $\mathcal{C}$. This implies that the point cloud registration is more confident in its result, with the increase in $N_a$. The reduced covariance or the increased confidence is the result of successful map merges.

The results depicted in Figures 9 and 10 further showcase the robustness of the proposed algorithm. The behavior exhibited by the agents at $f = 10$ hz is the similar to that of $f = 5$ hz. In other words, the agents showcase a decline in RMSE and $det(\mathcal{C})$ and an increase in $\mathcal{F}$ with the increase in $N_a$. Though there is a performance degradation in $f = 5$ hz with respect to $f = 10$ hz, this is expected, owing to the increased mapping error from LOAM.

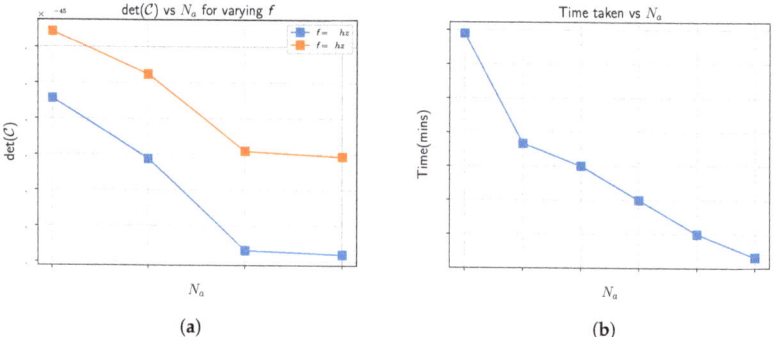

**Figure 10.** Uncertainty and time plots for varying number of agents $N_a$. (**a**) $det(\mathcal{C})$. (**b**) Time taken.

4.2.4. Time Analysis

Trivially, the time taken to map the whole map should decline linearly with $N_a$. For all runs of the simulation, the homogeneous agents have the same set of waypoints to visit. Figure 10b showcases the time taken to complete the whole map. Mapping is deemed to be complete when all agents have completed Algorithm 3, with $E^u \leq 0.4$. This threshold gives us a reliable transformation in between maps of different agents. As is evident in Figure 10b, there is an expected decline in time taken; however, this is not linear. This is because of the time taken to achieve a successful spherized registration with low $E^u$.

4.2.5. Density

The density, at a point($p$) in a point cloud, is the number of points around $p$, within a sphere of radius, $r_d$. The density at point $p$, $\mathcal{D}(p)$, is given by Equation (4). Density can be roughly considered analogous to the resolution of an image. Thus, a denser point cloud is a more detailed point cloud.

$$\mathcal{D}(p) = \frac{\text{Number of points within sphere(centre=}p\text{,radius=}r_d\text{)}}{\frac{4}{3}\pi \cdot (r_d)^3} \tag{4}$$

For analysis, we average the density of every point in the merged point cloud [81]. The spherical radius $r_d$ is chosen as 1 m. The results are shown in Figure 11a. We can see a healthy increase in the average density with $N_a$. This is attributed to the increased overlapping areas. The overlapping areas have unique points from different agents during the merging of individual point clouds. This in-turn increases the number of points per unit area. Added to this, though trivial, it is evident that the density is higher at $f = 10$ hz than $f = 5$ hz. This is due to the increase rate of the LiDAR data acquisition. Figure 11b depicts the surface density distribution across each point in the point cloud, generated with 3 agents (represented in Figure 8a). Figure 11c depicts the histogram of surface density of Figure 11b. We can note that higher surface density is rare and achievable primarily in areas of overlap.

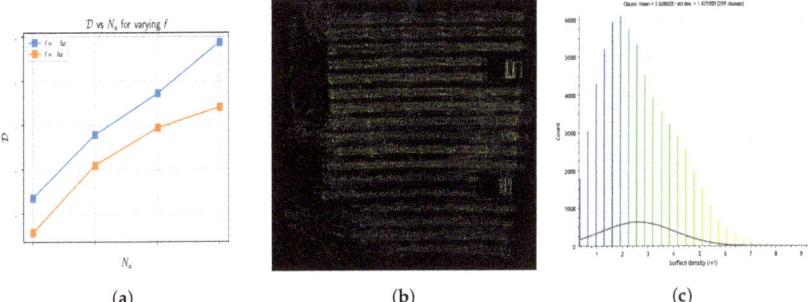

**Figure 11.** (**a**) The variation of mean density of merged map over $N_a$ across different rates, (**b**) the surface density distribution across the point cloud in Figure 8a, (**c**) The histogram of surface density distribution for the point cloud in Figure 8a.

## 5. Conclusions and Future Work

This article proposes an active, distributed, homogeneous multi-agent mapping and localization framework. The distributed framework enables conditional long-range and short-range peer-to-peer communication for small and large data. The proposed method is tested on a real-world solar farm with two UGVs and its digital twin with multiple agents. The results showcase the robustness of the proposed algorithm to independent mapping errors. However, we acknowledge that using direct point cloud registration in the framework can be error-prone, with increased LiDAR errors. Additionally, a spherized registration is robust to global localization errors up to a few meters. Thus, noisy EKF estimates, due to large GPS or magnetic interference, might lead to incorrect map merges. For future work, we aim to extend the framework to a heterogeneous team of agents with a heterogeneous set of sensors. We also plan to incorporate an optimal waypoint planning module, considering the constraints in communication and each agent's battery life. Currently, we are conducting an in-depth parameter study to understand and optimize the framework.

**Author Contributions:** Conceptualization, J.L., M.B. and P.U.L.; methodology, J.L., M.B. and P.U.L.; software, J.L.; validation, J.L.; formal analysis, J.L., M.B. and P.U.L.; investigation, J.L., M.B. and P.U.L.; resources, J.L., M.B. and P.U.L.; data curation, J.L., M.B. and P.U.L.; writing—original draft preparation, J.L., M.B. and P.U.L.; writing—review and editing, J.L., M.B. and P.U.L.; visualization, J.L.; supervision, M.B. and P.U.L.; project administration, M.B. and P.U.L.; funding acquisition, J.L., M.B. and P.U.L. All authors have read and agreed to the published version of the manuscript.

**Funding:** This work was supported by a doctoral grant from Fundação para a Ciência e a Tecnologia (FCT) UI/BD/153758/2022 and ISR/LARSyS Strategic Funding through the FCT project UIDB/50009/2020.

**Institutional Review Board Statement:** Not applicable.

**Informed Consent Statement:** Not applicable.

**Data Availability Statement:** The data presented in this study are available on request from the corresponding author.

**Conflicts of Interest:** The authors declare no conflict of interest.

## Abbreviations

The following abbreviations are used in this manuscript:

| | |
|---|---|
| EKF | Extended Kalman Filter |
| EMI | Electromagnetic Interference |
| LOAM | Lidar Odometry and Mapping in Real Time |
| IMU | Inertial Measurement Unit |

## References

1. Basiri, M.; Gonçalves, J.; Rosa, J.; Bettencourt, R.; Vale, A.; Lima, P. A multipurpose mobile manipulator for autonomous firefighting and construction of outdoor structures. *Field Robot* **2021**, *1*, 102–126. [CrossRef]
2. Karma, S.; Zorba, E.; Pallis, G.; Statheropoulos, G.; Balta, I.; Mikedi, K.; Vamvakari, J.; Pappa, A.; Chalaris, M.; Xanthopoulos, G.; et al. Use of unmanned vehicles in search and rescue operations in forest fires: Advantages and limitations observed in a field trial. *Int. J. Disaster Risk Reduct.* **2015**, *13*, 307–312. [CrossRef]
3. Limosani, R.; Esposito, R.; Manzi, A.; Teti, G.; Cavallo, F.; Dario, P. Robotic delivery service in combined outdoor–indoor environments: technical analysis and user evaluation. *Robot. Auton. Syst.* **2018**, *103*, 56–67. [CrossRef]
4. Åstrand, B.; Baerveldt, A.J. An agricultural mobile robot with vision-based perception for mechanical weed control. *Auton. Robot.* **2002**, *13*, 21–35. [CrossRef]
5. Lu, S.; Zhang, Y.; Su, J. Mobile robot for power substation inspection: A survey. *IEEE/CAA J. Autom. Sin.* **2017**, *4*, 830–847. [CrossRef]
6. Capezio, F.; Sgorbissa, A.; Zaccaria, R. GPS-based localization for a surveillance UGV in outdoor areas. In Proceedings of the Fifth International Workshop on Robot Motion and Control, Dymaczewo, Poland, 23–25 June 2005; pp. 157–162.
7. Montambault, S.; Pouliot, N. Design and validation of a mobile robot for power line inspection and maintenance. In Proceedings of the 6th International Conference on Field and Service Robotics-FSR 2007, Chamonix Mont-Blanc, France, 6–12 July 2007; Springer: Berlin/Heidelberg, Germany, 2008; pp. 495–504.
8. Akyazi, Ö.; Şahin, E.; Özsoy, T.; Algül, M. A solar panel cleaning robot design and application. *Avrupa Bilim Ve Teknoloji Dergisi* **2019**, 343–348. [CrossRef]
9. Jaradat, M.A.; Tauseef, M.; Altaf, Y.; Saab, R.; Adel, H.; Yousuf, N.; Zurigat, Y.H. A fully portable robot system for cleaning solar panels. In Proceedings of the 2015 10th International Symposium on Mechatronics and its Applications (ISMA), Sharjah, United Arab Emirates, 8–10 December 2015; pp. 1–6.
10. Kazem, H.A.; Chaichan, M.T.; Al-Waeli, A.H.; Sopian, K. A review of dust accumulation and cleaning methods for solar photovoltaic systems. *J. Clean. Prod.* **2020**, *276*, 123187. [CrossRef]
11. Schwarz, M.; Rodehutskors, T.; Droeschel, D.; Beul, M.; Schreiber, M.; Araslanov, N.; Ivanov, I.; Lenz, C.; Razlaw, J.; Schüller, S.; et al. NimbRo Rescue: Solving disaster-response tasks with the mobile manipulation robot Momaro. *J. Field Robot.* **2017**, *34*, 400–425. [CrossRef]
12. Haynes, G.C.; Stager, D.; Stentz, A.; Vande Weghe, J.M.; Zajac, B.; Herman, H.; Kelly, A.; Meyhofer, E.; Anderson, D.; Bennington, D.; et al. Developing a robust disaster response robot: CHIMP and the robotics challenge. *J. Field Robot.* **2017**, *34*, 281–304. [CrossRef]
13. Kruijff, G.J.M.; Kruijff-Korbayová, I.; Keshavdas, S.; Larochelle, B.; Janíček, M.; Colas, F.; Liu, M.; Pomerleau, F.; Siegwart, R.; Neerincx, M.A.; et al. Designing, developing, and deploying systems to support human–robot teams in disaster response. *Adv. Robot.* **2014**, *28*, 1547–1570. [CrossRef]
14. Hajjaj, S.S.H.; Sahari, K.S.M. Review of agriculture robotics: Practicality and feasibility. In Proceedings of the 2016 IEEE International Symposium on Robotics and Intelligent Sensors (IRIS), Tokyo, Japan, 17–20 December 2016; pp. 194–198.
15. Pfaff, P.; Triebel, R.; Stachniss, C.; Lamon, P.; Burgard, W.; Siegwart, R. Towards mapping of cities. In Proceedings of the 2007 IEEE International Conference on Robotics and Automation, Roma, Italy, 10–14 April 2007; pp. 4807–4813.
16. Bauer, A.; Klasing, K.; Lidoris, G.; Mühlbauer, Q.; Rohrmüller, F.; Sosnowski, S.; Xu, T.; Kühnlenz, K.; Wollherr, D.; Buss, M. The autonomous city explorer: Towards natural human-robot interaction in urban environments. *Int. J. Soc. Robot.* **2009**, *1*, 127–140. [CrossRef]
17. Simmons, R.; Apfelbaum, D.; Burgard, W.; Fox, D.; Moors, M.; Thrun, S.; Younes, H. Coordination for multi-robot exploration and mapping. In *Aaai/Iaai*; 2000; pp. 852–858. Available online: https://www.aaai.org/Papers/AAAI/2000/AAAI00-131.pdf (accessed on 9 December 2022).
18. Zhang, J.; Singh, S. LOAM: Lidar Odometry and Mapping in Real-time. In *Robotics: Science and Systems*; University of California: Berkeley, CA, USA, 2014; pp. 1–9.

19. Shan, T.; Englot, B. Lego-loam: Lightweight and ground-optimized lidar odometry and mapping on variable terrain. In Proceedings of the 2018 IEEE/RSJ International Conference on Intelligent Robots and Systems (IROS), Madrid, Spain, 1–5 October 2018; pp. 4758–4765.
20. Li, L.; Kong, X.; Zhao, X.; Li, W.; Wen, F.; Zhang, H.; Liu, Y. SA-LOAM: Semantic-aided LiDAR SLAM with loop closure. In Proceedings of the 2021 IEEE International Conference on Robotics and Automation (ICRA), Xi'an, China, 30 May–5 June 2021; pp. 7627–7634.
21. Chen, S.W.; Nardari, G.V.; Lee, E.S.; Qu, C.; Liu, X.; Romero, R.A.F.; Kumar, V. Sloam: Semantic lidar odometry and mapping for forest inventory. *IEEE Robot. Autom. Lett.* **2020**, *5*, 612–619. [CrossRef]
22. Yousif, K.; Taguchi, Y.; Ramalingam, S. MonoRGBD-SLAM: Simultaneous localization and mapping using both monocular and RGBD cameras. In Proceedings of the 2017 IEEE International Conference on Robotics and Automation (ICRA), Singapore, 29 May–3 June 2017; pp. 4495–4502.
23. Loianno, G.; Thomas, J.; Kumar, V. Cooperative localization and mapping of MAVs using RGB-D sensors. In Proceedings of the 2015 IEEE International Conference on Robotics and Automation (ICRA), Seattle, WA, USA, 26–30 May 2015; pp. 4021–4028.
24. Apriaskar, E.; Nugraha, Y.P.; Trilaksono, B.R. Simulation of simultaneous localization and mapping using hexacopter and RGBD camera. In Proceedings of the 2017 2nd International Conference on Automation, Cognitive Science, Optics, Micro Electro-Mechanical System, and Information Technology (ICACOMIT), Jakarta, Indonesia, 23–24 October 2017; pp. 48–53.
25. Paton, M.; Kosecka, J. Adaptive rgb-d localization. In Proceedings of the 2012 Ninth Conference on Computer and Robot Vision, Toronto, ON, Canada, 28–30 May 2012; pp. 24–31.
26. Lin, J.; Zhang, F. Loam livox: A fast, robust, high-precision LiDAR odometry and mapping package for LiDARs of small FoV. In Proceedings of the 2020 IEEE International Conference on Robotics and Automation (ICRA), Philadelphia, PA, USA, 23–27 May 2020; pp. 3126–3131.
27. Xu, W.; Zhang, F. Fast-lio: A fast, robust lidar-inertial odometry package by tightly-coupled iterated kalman filter. *IEEE Robot. Autom. Lett.* **2021**, *6*, 3317–3324. [CrossRef]
28. Xu, W.; Cai, Y.; He, D.; Lin, J.; Zhang, F. Fast-lio2: Fast direct lidar-inertial odometry. *IEEE Trans. Robot.* **2022**, *38*, 2053–2073. [CrossRef]
29. Durrant-Whyte, H.; Bailey, T. Simultaneous localization and mapping: Part I. *IEEE Robot. Autom. Mag.* **2006**, *13*, 99–110. [CrossRef]
30. Bailey, T.; Durrant-Whyte, H. Simultaneous localization and mapping (SLAM): Part II. *IEEE Robot. Autom. Mag.* **2006**, *13*, 108–117. [CrossRef]
31. Kim, P.; Chen, J.; Cho, Y.K. SLAM-driven robotic mapping and registration of 3D point clouds. *Autom. Constr.* **2018**, *89*, 38–48. [CrossRef]
32. Takleh, T.T.O.; Bakar, N.A.; Rahman, S.A.; Hamzah, R.; Aziz, Z. A brief survey on SLAM methods in autonomous vehicle. *Int. J. Eng. Technol.* **2018**, *7*, 38–43. [CrossRef]
33. Jiang, Z.; Zhu, J.; Lin, Z.; Li, Z.; Guo, R. 3D mapping of outdoor environments by scan matching and motion averaging. *Neurocomputing* **2020**, *372*, 17–32. [CrossRef]
34. Wiesmann, L.; Milioto, A.; Chen, X.; Stachniss, C.; Behley, J. Deep compression for dense point cloud maps. *IEEE Robot. Autom. Lett.* **2021**, *6*, 2060–2067. [CrossRef]
35. Navarrete, J.; Viejo, D.; Cazorla, M. Compression and registration of 3D point clouds using GMMs. *Pattern Recognit. Lett.* **2018**, *110*, 8–15. [CrossRef]
36. Wiesmann, L.; Guadagnino, T.; Vizzo, I.; Grisetti, G.; Behley, J.; Stachniss, C. DCPCR: Deep Compressed Point Cloud Registration in Large-Scale Outdoor Environments. *IEEE Robot. Autom. Lett.* **2022**, *7*, 6327–6334. [CrossRef]
37. Huang, X.; Mei, G.; Zhang, J.; Abbas, R. A comprehensive survey on point cloud registration. *arXiv* **2021**, arXiv:2103.02690.
38. Choy, C.; Dong, W.; Koltun, V. Deep global registration. In Proceedings of the IEEE/CVF Conference on Computer Vision and Pattern Recognition, Seattle, WA, USA, 13–19 June 2020; pp. 2514–2523.
39. Zhou, Q.Y.; Park, J.; Koltun, V. Fast global registration. In *European Conference on Computer Vision*; Springer: Cham, Switzerland, 2016; pp. 766–782.
40. Yang, H.; Shi, J.; Carlone, L. Teaser: Fast and certifiable point cloud registration. *IEEE Trans. Robot.* **2020**, *37*, 314–333. [CrossRef]
41. Lei, H.; Jiang, G.; Quan, L. Fast descriptors and correspondence propagation for robust global point cloud registration. *IEEE Trans. Image Process.* **2017**, *26*, 3614–3623. [CrossRef]
42. Besl, P.J.; McKay, N.D. Method for registration of 3-D shapes. In *Sensor Fusion IV: Control Paradigms and Data Structures*; Spie: Bellingham, WA, USA, 1992; Volume 1611, pp. 586–606.
43. Chen, Y.; Medioni, G. Object modelling by registration of multiple range images. *Image Vis. Comput.* **1992**, *10*, 145–155. [CrossRef]
44. Biber, P.; Straßer, W. The normal distributions transform: A new approach to laser scan matching. In Proceedings of the 2003 IEEE/RSJ International Conference on Intelligent Robots and Systems (IROS 2003) (Cat. No. 03CH37453), Las Vegas, NV, USA, 27–31 October 2003; Volume 3, pp. 2743–2748.
45. Rusinkiewicz, S.; Levoy, M. Efficient variants of the ICP algorithm. In Proceedings of the Third International Conference on 3-D Digital Imaging and Modeling, Quebec City, QC, Canada, 28 May–1 June 2001; pp. 145–152.
46. Low, K.L. Linear least-squares optimization for point-to-plane icp surface registration. *Chapel Hill Univ. North Carol.* **2004**, *4*, 1–3.

47. Park, S.Y.; Subbarao, M. An accurate and fast point-to-plane registration technique. *Pattern Recognit. Lett.* **2003**, *24*, 2967–2976. [CrossRef]
48. Park, J.; Zhou, Q.Y.; Koltun, V. Colored point cloud registration revisited. In Proceedings of the IEEE International Conference on Computer Vision, Venice, Italy, 22–29 October 2017; pp. 143–152.
49. Huhle, B.; Magnusson, M.; Straßer, W.; Lilienthal, A.J. Registration of colored 3D point clouds with a kernel-based extension to the normal distributions transform. In Proceedings of the 2008 IEEE International Conference on Robotics and Automation, Bangkok, Thailand, 14–17 December 2008; pp. 4025–4030.
50. Zaganidis, A.; Sun, L.; Duckett, T.; Cielniak, G. Integrating deep semantic segmentation into 3-d point cloud registration. *IEEE Robot. Autom. Lett.* **2018**, *3*, 2942–2949. [CrossRef]
51. Zaganidis, A.; Magnusson, M.; Duckett, T.; Cielniak, G. Semantic-assisted 3D normal distributions transform for scan registration in environments with limited structure. In Proceedings of the 2017 IEEE/RSJ International Conference on Intelligent Robots and Systems (IROS), Vancouver, BC, Canada, 24–28 September 2017; pp. 4064–4069.
52. Raza, U.; Kulkarni, P.; Sooriyabandara, M. Low power wide area networks: An overview. *IEEE Commun. Surv. Tutorials* **2017**, *19*, 855–873. [CrossRef]
53. Ikpehai, A.; Adebisi, B.; Rabie, K.M.; Anoh, K.; Ande, R.E.; Hammoudeh, M.; Gacanin, H.; Mbanaso, U.M. Low-power wide area network technologies for Internet-of-Things: A comparative review. *IEEE Internet Things J.* **2018**, *6*, 2225–2240. [CrossRef]
54. Vaezi, M.; Azari, A.; Khosravirad, S.R.; Shirvanimoghaddam, M.; Azari, M.M.; Chasaki, D.; Popovski, P. Cellular, wide-area, and non-terrestrial IoT: A survey on 5G advances and the road toward 6G. *IEEE Commun. Surv. Tutorials* **2022**, *24*, 1117–1174. [CrossRef]
55. Vejlgaard, B.; Lauridsen, M.; Nguyen, H.; Kovács, I.Z.; Mogensen, P.; Sorensen, M. Coverage and capacity analysis of sigfox, lora, gprs, and nb-iot. In Proceedings of the 2017 IEEE 85th Vehicular Technology Conference (VTC Spring), Sydney, Australia, 4–7 June 2017; pp. 1–5.
56. XBee RF Modules. Available online: http://www.digi.com/products/xbee-rf-solutions (accessed on 12 August 2022).
57. Corah, M.; O'Meadhra, C.; Goel, K.; Michael, N. Communication-efficient planning and mapping for multi-robot exploration in large environments. *IEEE Robot. Autom. Lett.* **2019**, *4*, 1715–1721. [CrossRef]
58. Xu, X.; Zhang, L.; Yang, J.; Cao, C.; Wang, W.; Ran, Y.; Tan, Z.; Luo, M. A review of multi-sensor fusion slam systems based on 3D LIDAR. *Remote Sens.* **2022**, *14*, 2835. [CrossRef]
59. Valencia, R.; Morta, M.; Andrade-Cetto, J.; Porta, J.M. Planning reliable paths with pose SLAM. *IEEE Trans. Robot.* **2013**, *29*, 1050–1059. [CrossRef]
60. Krinkin, K.; Filatov, A.; yom Filatov, A.; Huletski, A.; Kartashov, D. Evaluation of modern laser based indoor slam algorithms. In Proceedings of the 2018 22nd Conference of Open Innovations Association (FRUCT), Jyvaskyla, Finland, 15–18 May 2018; pp. 101–106.
61. Sayed, A.S.; Ammar, H.H.; Shalaby, R. Centralized multi-agent mobile robots SLAM and navigation for COVID-19 field hospitals. In Proceedings of the 2020 2nd Novel Intelligent and Leading Emerging Sciences Conference (NILES, Giza, Egypt, 24–26 October 2020; pp. 444–449.
62. Liu, T.M.; Lyons, D.M. Leveraging area bounds information for autonomous decentralized multi-robot exploration. *Robot. Auton. Syst.* **2015**, *74*, 66–78. [CrossRef]
63. Matignon, L.; Jeanpierre, L.; Mouaddib, A.I. Coordinated multi-robot exploration under communication constraints using decentralized markov decision processes. In Proceedings of the Twenty-Sixth AAAI Conference on Artificial Intelligence, Toronto, ON, Canada, 22–26 June 2012.
64. Arkin, R.C.; Diaz, J. Line-of-sight constrained exploration for reactive multiagent robotic teams. In Proceedings of the 7th International Workshop on Advanced Motion Control, Maribor, Slovenia, 3–5 July 2002; pp. 455–461.
65. Amigoni, F.; Banfi, J.; Basilico, N. Multirobot exploration of communication-restricted environments: A survey. *IEEE Intell. Syst.* **2017**, *32*, 48–57. [CrossRef]
66. Gao, W.; Wang, Y.; Zhong, X.; Yang, T.; Wang, M.; Xu, Z.; Wang, Y.; Xu, C.; Gao, F. Meeting-Merging-Mission: A Multi-robot Coordinate Framework for Large-Scale Communication-Limited Exploration. *arXiv* **2021**, arXiv:2109.07764.
67. Omohundro, S.M. *Five Balltree Construction Algorithms*; International Computer Science Institute: Berkeley, CA, USA, 1989.
68. Boeing, G. Clustering to reduce spatial data set size. *arXiv* **2018**, arXiv:1803.08101.
69. Bhatia, N. Survey of nearest neighbor techniques. *arXiv* **2010**, arXiv:1007.0085.
70. Hariz, F.; Souifi, H.; Leblanc, R.; Bouslimani, Y.; Ghribi, M.; Langin, E.; Mccarthy, D. Direct Georeferencing 3D Points Cloud Map Based on SLAM and Robot Operating System. In Proceedings of the 2021 IEEE International Symposium on Robotic and Sensors Environments (ROSE), Virtual Conference, 28–29 October 2021; pp. 1–6.
71. Liu, W.; Li, Z.; Li, Y.; Sun, S.; Sotelo, M.A. Using weighted total least squares and 3-D conformal coordinate transformation to improve the accuracy of mobile laser scanning. *IEEE Trans. Geosci. Remote Sens.* **2019**, *58*, 203–217. [CrossRef]
72. Janata, T.; Cajthaml, J. Georeferencing of multi-sheet maps based on least squares with constraints—First military mapping survey maps in the area of Czechia. *Appl. Sci.* **2020**, *11*, 299. [CrossRef]
73. Yang, H. A dynamical perspective on point cloud registration. *arXiv* **2020**, arXiv:2005.03190.
74. Choi, S.; Zhou, Q.Y.; Koltun, V. Robust reconstruction of indoor scenes. In Proceedings of the IEEE Conference on Computer Vision and Pattern Recognition, Boston, MA, USA, 7–12 June 2015; pp. 5556–5565.

75. Rusu, R.B.; Blodow, N.; Beetz, M. Fast point feature histograms (FPFH) for 3D registration. In Proceedings of the 2009 IEEE International Conference on Robotics and Automation, Kobe, Japan, 12–17 May 2009; pp. 3212–3217.
76. Shen, Z.; Liang, H.; Lin, L.; Wang, Z.; Huang, W.; Yu, J. Fast Ground Segmentation for 3D LiDAR Point Cloud Based on Jump-Convolution-Process. *Remote Sens.* **2021**, *13*, 3239. [CrossRef]
77. Zhang, F.; Fang, J.; Wah, B.; Torr, P. Deep fusionnet for point cloud semantic segmentation. In *European Conference on Computer Vision*; Springer: Berlin/Heidelberg, Germany, 2020; pp. 644–663.
78. Fujita, K.; Okada, K.; Katahira, K. *The Fisher Information Matrix: A Tutorial for Calculation for Decision Making Models*; 2022.
79. Pulli, K. Multiview registration for large data sets. In Proceedings of the Second International Conference on 3-d Digital Imaging and Modeling (Cat. No. pr00062), Ottawa, ON, Canada, 4–8 October 1999; pp. 160–168.
80. Barczyk, M.; Bonnabel, S.; Goulette, F. Observability, Covariance and Uncertainty of ICP Scan Matching. *arXiv* **2014**, arXiv:1410.7632.
81. Maset, E.; Scalera, L.; Beinat, A.; Visintini, D.; Gasparetto, A. Performance Investigation and Repeatability Assessment of a Mobile Robotic System for 3D Mapping. *Robotics* **2022**, *11*, 54. [CrossRef]

**Disclaimer/Publisher's Note:** The statements, opinions and data contained in all publications are solely those of the individual author(s) and contributor(s) and not of MDPI and/or the editor(s). MDPI and/or the editor(s) disclaim responsibility for any injury to people or property resulting from any ideas, methods, instructions or products referred to in the content.

Article

# Finding a Landing Site in an Urban Area: A Multi-Resolution Probabilistic Approach

Barak Pinkovich [1,*], Boaz Matalon [2], Ehud Rivlin [1] and Hector Rotstein [1]

[1] The Faculty of Computer Science, Technion Israel Institute of Technology, Haifa 3200003, Israel
[2] Rafael Advanced Defense Systems Ltd., Haifa 3102102, Israel
* Correspondence: barakp@campus.technion.ac.il

**Abstract:** This paper considers the problem of finding a landing spot for a drone in a dense urban environment. The conflicting requirements of fast exploration and high resolution are solved using a multi-resolution approach, by which visual information is collected by the drone at decreasing altitudes so that the spatial resolution of the acquired images increases monotonically. A probability distribution is used to capture the uncertainty of the decision process for each terrain patch. The distributions are updated as information from different altitudes is collected. When the confidence level for one of the patches becomes larger than a prespecified threshold, suitability for landing is declared. One of the main building blocks of the approach is a semantic segmentation algorithm that attaches probabilities to each pixel of a single view. The decision algorithm combines these probabilities with a priori data and previous measurements to obtain the best estimates. Feasibility is illustrated by presenting several examples generated by a realistic closed-loop simulator.

**Keywords:** unmanned aerial vehicles; search theory; perception; semantic segmentation

## 1. Introduction

Unlike conventional aircraft that takeoff and land on designated and controlled areas outside city limits, future commercial drones are expected to operate smoothly in crowded urban environments. Consequently, the well-defined zones delimited for landing, takeoff and safe flight will be replaced by dynamic and opportunistic areas within cities. To deal with this challenge, future delivery and transportation drones must be able to solve the "last-mile" problem (this refers to the last part in the delivery process of a product, namely, the section of transport from the last distribution center to the end customer), and find a place for landing with the following characteristics:

1. Relatively close to the intended destination. These places cannot be limited to predetermined areas like heliports, sports fields, or similar places
2. Appropriate for the drone's size and weight.
3. Appropriate for landing under harsh (or relatively harsh) flight conditions compatible with the drone's flying capabilities.
4. Pose no safety concerns to itself, other vehicles, or living beings in the environment.

This paper describes a multi-resolution probabilistic approach to search for a landing place in a dense urban environment such as the one illustrated in Figure 1. In this context, multi-resolution is achieved by iteratively decreasing the altitude from which the visual sensor observes the urban environment, hence generating a sequence of images with monotonically increasing spatial resolution. Notice that the purpose is not to generate super-resolution images but rather to improve the level of confidence of the exploration. The result is probabilistic in the sense that confidence levels of different regions of the urban scenario are computed based on a priori knowledge and the result of observations. Somewhat related ideas for the different problem of performing an energy-efficient close inspection in an agricultural field were recently considered in [1].

Citation: Pinkovich, B.; Matalon, B.; Rivlin, E.; Rotstein, H. Finding a Landing Site in an Urban Area: A Multi-Resolution Probabilistic Approach. Sensors 2022, 22, 9807. https://doi.org/10.3390/s22249807

Academic Editor: José María Martínez-Otzeta

Received: 8 November 2022
Accepted: 9 December 2022
Published: 14 December 2022

**Copyright:** © 2022 by the authors. Licensee MDPI, Basel, Switzerland. This article is an open access article distributed under the terms and conditions of the Creative Commons Attribution (CC BY) license (https://creativecommons.org/licenses/by/4.0/).

**Figure 1.** Simulated urban environment in which the drone attempts to land.

## 2. Related Work

Although to the best of the author's knowledge the problem considered in this paper is new, the probabilistic viewpoint provides connections with extensive existing literature, including search theory and Bayes-based decision-making. For example, Ref. [2] provides a survey on the usage of Bayesian networks usage for intelligent autonomous vehicle decision-making with no focus on specific missions. Similarly, Ref. [3] describes spacecraft autonomy challenges for future space missions, in which real-time autonomous decision-making and human-robotic cooperation must be considered. In a related autonomous spacecraft mission, Ref. [4] studies the selection of an appropriate landing site for an autonomous spacecraft on an exploration mission. The problem is formulated so that three main variables are defined on which to select the landing site: terrain safety, engineering factors (spacecraft's descending trajectory, velocity, and available fuel), and the site preselected by using the available a priori information. The approach was tested using a dynamics and spacecraft simulator for entry, descent, and landing.

The problem considered here is also somewhat related to *forced landing*. This is because a UAV may need to decide the most suitable forced landing sites autonomously, usually from a list of known candidates [5]. In that work, references were made to the specifications for a forced landing system laid out in a NASA technical report essentially consisting of three main criteria: risk to the civilian population, reachability, and probability of a safe landing. The emphasis is on public safety, where human life and property are more important than the UAV airframe and payload. Specifications were included in a multi-criteria decision-making (MCDM) Bayesian network. See [6] for an application of the model to a real-life example. The initial design for UAVs' autonomous decision systems for selecting emergency landing sites in a vehicle fault scenario are also considered in [6]. The overall design consists of two main components: preplanning and real-time optimization. In the preplanning component, the system uses offline information, such as geographical and population data, to generate landing loss maps over the operating environment. In the real-time component, onboard sensor data are used to update a probabilistic risk assessment for potential landing areas.

Another related field of interest is *search and rescue*, a challenging task, as it usually involves a large variety of scenarios that require a high level of autonomy and versatile decision-making capabilities. A formal framework casting the search problem as a decision between hypotheses using current knowledge was introduced in [7]. The search task was defined as follows: given a detector model (i.e., detection error probabilities for false alarms and missed detections) and the prior belief that the target exists in the search region, determine the evolution of the belief that the target is present in the search region as a function of the observations made until time $t$. The belief evolution is computed via a recursive Bayesian expression that provides a compact and efficient way to update the belief function at every time step as the searcher observes a sequence of unexplored and/or previously visited cells in the search region. After generating a method for computing the belief evolution for a sequence of imperfect observations, the authors investigated the search control policy/strategy. In the context of an area search problem, [1] investigated the uncertainty associated with a typical search problem to answer a crucial question: how many image frames would be required by a camera onboard a vehicle to classify a target as "detected" or "undetected" with uncertain prior information on the target's existence. The paper presents a formulation incorporating uncertainty using beta distributions to create robust search actions. As shown below, these ideas are highly related to our approach.

Finally, ideas somewhat related to the ones considered here were used to pose a path-planning algorithm for performing an energy-efficient close inspection on selected areas in agricultural fields [8].

## 3. Problem Formulation

Suppose a drone needs to find a landing place in an urban area $\mathcal{A}$. For simplicity, consider $\mathcal{A}$ to be planar, with an attached coordinate system $\{W\}$ such that $\mathcal{A}$ lies on the $x - y$ plane, measuring the altitude along the $z$-direction. Buildings and other constructions are modeled as occupied volumes over $\mathcal{A}$. For example, the area considered in this paper will be 1 km by 1 km square. The drone can fly at different altitudes $h$ while collecting measurements using a monocular camera and a visual sensor with range capabilities. Examples of the latter are an RGB-D sensor, a Lidar, or a couple of stereo cameras. Let $\mathcal{A}_h$ be the plane parallel to $\mathcal{A}$ at an altitude $h$ onto which $\mathcal{A}$'s relevant characteristics can be mapped. For instance, a no-fly zone $\mathcal{U}$ in $\mathcal{A}$ (e.g., the base of a building) will be mapped onto the corresponding subset $\mathcal{U}_h$ in $\mathcal{A}_h$ (at least if $h$ is smaller than the corresponding building's altitude).

The mechanical structure of the drone, the size of its propellers, and the flying conditions (e.g., wind) constrain the minimum dimensions of the landing site on which the drone can safely land. Conservatively, $\mathcal{A}$ will be divided into a grid of identical cells $c_{ij}$, so that $\cup_{ij} c_{ij} = \mathcal{A}$, and each $c_{ij}$ is in principle a landing site candidate. Note that this division is in-line with search theory (see, e.g., [9]) but stems from a different motivation: it is not a unit area being explored but the smallest area of interest. As discussed next, this will impact our development in several ways.

The images taken by a camera on the drone at time instant $t$ will be a function of the pose $p_t$ and the camera's field-of-view. Note that the camera's orientation can differ from the drone's by a relative rotation between the two. Assuming that the FOV is fixed and known, let $I_t(h)$ be the camera's image at time $t$ and $F_t(h)$ be the camera's corresponding footprint, namely the 3D structure mapped onto the image plane. At the time $t$, the drone will have a unique pose $p_t$, but the dependence on the altitude is specifically considered in the notation; this is because the altitude scales the resolution and the footprint: for smaller $h$, one obtains a better resolution at the cost of a smaller footprint. The structure $F_t(h)$ is built on a collection of cells $C_t(h) = \cup_{\{j,k\} \in \{J_t(h), K_t(h)\}} c_{ij} \subset \mathcal{A}$.

Without additional constraints, the drone could fly over $\mathcal{A}$ at a relatively low altitude $h_{min}$, searching for an appropriate cell $c_{ij}$ on which to land. However, at this altitude, the camera's *footprint* $C_t(h_{min})$ will include a relatively small number of cells $c_{ij}$ and hence the drone would spend a potentially prohibitive amount of time/energy exploring the whole $\mathcal{A}$. On the other hand, the altitude can be selected to be the maximum allowable by regulations, say $h_{max}$, resulting in as large as possible footprints. In an extreme case, $C_t(h_{max}) = \mathcal{A}$. This maximizes the area subtended by a single image and minimizes the exploration time, but will give rise to an image resolution that cannot guarantee the safety of landing, i.e., it will not resolve relatively small obstacles. The trade-off between altitude and resolution is solved in this work by considering a multi-resolution approach: the exploration will start at high altitude, say $h_1$, looking for the largest possible subset of $\mathcal{A}$ that appears to be *feasible* for landing, say $L_1$. Subsequently, the drone will reduce its altitude to $h_2$ and re-explore $L_1$ with the higher resolution resulting from $h_2 < h_1$. This process results in a sequence of $\mathcal{A} \supset L_1 \supset \cdots \supset L_N$ that will eventually converge to a collection of one or more safe landing places.

Figures 1–3 illustrate this scenario.

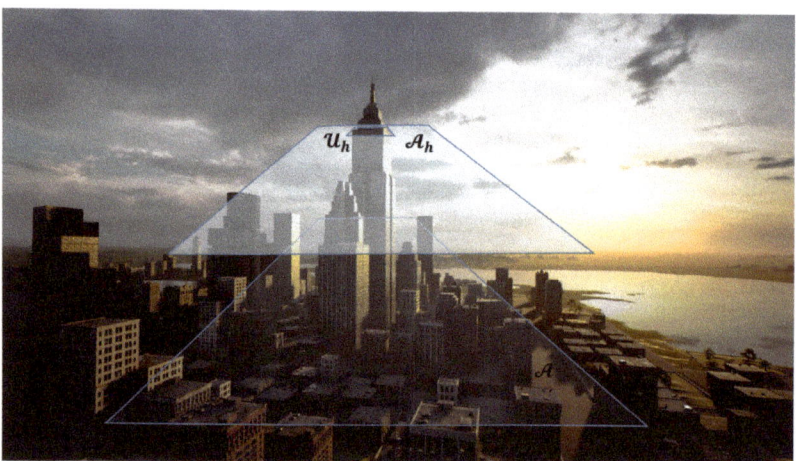

**Figure 2.** The area of interest $\mathcal{A}$, the plane $\mathcal{A}_h$ at an altitude $h$, and the mapping of an obstacle $\mathcal{U}_h$.

**Figure 3.** The plane at a given altitude is divided into small cells.

*3.1. A Probabilistic Model*

In classical search theory [9], finding a target is often formulated as a decision problem by defining a set of binary variables:

$$\mathcal{H}_{ij} \doteq \begin{cases} 1 & \text{if } c_{ij} \text{ has a target} \\ 0 & \text{otherwise} \end{cases} \tag{1}$$

This can be extended to the case of interest by defining:

$$\mathcal{H}_{ij} \doteq \begin{cases} 1 & \text{if } c_{ij} \text{ is appropriate for landing} \\ 0 & \text{otherwise} \end{cases} \quad (2)$$

In the presence of uncertainty, each cell is likely to be suitable for landing with some probability $K_{ij}$ referred to as the *fitness for landing*. Clearly, if a cell is appropriate for landing, then $K_{ij} = 1$, and if it is not, then $K_{ij} = 0$. In real-world scenarios, prior knowledge about the fitness for landing can be based on using types of maps or 3D models that can be imprecise or outdated. Consequently, uncertainty in $K_{ij}$ needs to be incorporated into the model. Probably the simplest way to model the decision problem would be to introduce a binary distribution and say that the probability of $K_{ij} = 1$ is $p$ and $K_{ij} = 0$ is $1 - p$. However, as observed by [1], this model fails to capture the *uncertainty* of the information, and instead, it is preferable to use a beta distribution for describing the prior knowledge together with its underlying uncertainty. Note that the beta distribution and the binomial and Bernoulli distributions form a *conjugate pair*, so that if a Bernoulli distribution can model the sensor, then the observation of new data changes only the parameters of the prior. At the same time, the conjugacy property ensures that the posterior is in the same class (i.e., beta). This is critical when propagating the belief in the fitness for landing in a Bayesian framework. Using these models simplifies the decision problem into a binary outcome as defined in Equation (2).

The beta distribution is defined as,

$$Pr(K_{ij}|\alpha, \beta) = \frac{\Gamma(\alpha + \beta)}{\Gamma(\alpha)\Gamma(\beta)} K_{ij}^{\alpha-1}(1 - K_{ij})^{\beta-1} \quad (3)$$

where $0 < K_{ij} < 1$ and $\Gamma(\alpha)$ is the gamma function defined as,

$$\Gamma(\alpha) = \int_0^\infty x^{\alpha-1} e^{-x} dx \quad (4)$$

When $\alpha$ is an integer, $\Gamma(\alpha) = (\alpha - 1)!$. The parameters $\alpha$ and $\beta$ can be considered prior "successes" and "failures". The beta distribution is somewhat similar to the binomial distribution. The main difference is that the random variable is $\mathcal{H}_{ij}$ and the parameter is $K_{ij}$ in the binomial distribution, whereas the random variable is $K_{ij}$ and the parameters are $\alpha$ and $\beta$ in the beta distribution.

Bayes theorem is often used in search theory [1,7,9] to update the aggregated belief (e.g., posterior distribution), which is proportional to the *likelihood* function times the prior distribution:

$$Pr\left(K_{ij} \big| S_{ij}^N, \alpha, \beta\right) \propto Pr\left(S_{ij}^N \big| , K_{ij}\right) Pr(K_{ij}|\alpha, \beta) \quad (5)$$

Here, $S_{ij}^N = \sum_{n=1}^N \mathcal{H}_{ij}^n$ is the number of successes in $N$ Bernoulli trials. $Pr\left(K_{ij}\big|S_{ij}^N, \alpha, \beta\right)$ is the posterior distribution for $K_{ij}$ given the number of successes. $Pr\left(S_{ij}^N\big|K_{ij}\right)$ is the *likelihood function* and $Pr(K_{ij}|\alpha, \beta)$ is the prior distribution for $K_{ij}$. In [7], the likelihood distribution is given by a binomial distribution series of $N$ observations:

$$Pr\left(S_{ij}^N \big| K_{ij}\right) = \binom{S_{ij}^N}{N} K_{ij}^{S_{ij}^N} (1 - K_{ij})^{N - S_{ij}^N} \quad (6)$$

Note that the underlying assumption is that the Bernoulli distribution provides an appropriate statistical model for the sensor used. This assumption is more or less natural when considering a series of noisy images. In our case, simple image processing algorithms are replaced by a more complex *meta*-sensor: fitness is computed by a semantic segmentation algorithm that associates for each pixel on a given image the suitability for landing on the corresponding cell on the ground. An appropriate model for this process is discussed next.

### 3.2. A Correlated Detection Model

The probability framework is motivated by the sensors' limitations for establishing whether a given cell is appropriate for landing. As mentioned above, the main limitations are the camera resolution, the environmental conditions limiting visibility, and possibly scene dynamics. The probability estimated by a sensor that a given cell is appropriate for landing may vary according to altitude, with lower altitudes having higher confidence levels (up to a certain point depending on the sensor).

In [1], the authors assumed independent Bernoulli trials when detecting a target in recurrent visits. Independent trials are acceptable when the condition of the experiment does not change. However, in this research, our multi-resolution approach implies that when observing cell $c_{ij}$ at different altitude levels, one cannot assume uncorrelated measurements between different levels. At each level, the experiment's condition changes (e.g., different resolution), and if a landing place exists, then it is expected that the rate of success will depend on previous trials and will increase when the level of details increases while descending toward cell $c_{ij}$.

Generalizing the binomial distribution typically involves modifying either the assumption of constant "success" probability and/or the assumption of independence between trials in the underlying Bernoulli process. The approach to generalizing the binomial distribution in this research follows the generalized Bernoulli distribution (GBD) model [10] by relaxing the assumption of independence between trials. The GBD model was further considered in statistics literature [11–14]) aiming to obtain its central limit theorems, including the strong law of large numbers and the law of the iterated logarithm for partial sums.

Consider a Bernoulli process $\{\mathcal{H}_{ij}^n, n \geq 1\}$ in which the random variables $\mathcal{H}_{ij}^n$ are correlated so that the success probability for the trial conditional on all the previous trials depends on the total number of successes achieved to that point. More precisely, for some $0 < K_{ij} < 1$,

$$Pr\left(\mathcal{H}_{ij}^{n+1} \big| \mathcal{F}_{ij}^n\right) = (1 - \theta_{ij}^n) K_{ij} + \theta_{ij}^n \frac{S_{ij}^n}{n} \quad (7)$$

where $0 \leq \theta_{ij}^n \leq 1$ are dependence parameters, $S_{ij}^N = \sum_{n=1}^{N} \mathcal{H}_{ij}^n$ for $N \geq 1$ and $\mathcal{F}_{ij}^N = \sigma(\mathcal{H}_{ij}^1, \cdots, \mathcal{H}_{ij}^N)$. If $\mathcal{H}_{ij}^1$ has a Bernoulli distribution with parameter $K_{ij}$, it follows that $\mathcal{H}_{ij}^1, \mathcal{H}_{ij}^2, \cdots$ are identically distributed Bernoulli random variables.

By replacing the binomial distribution in Equation (5) with the GBD at each altitude $h_n$, the aggregated belief that a cell $c_{ij}$ is suitable for landing given the number of successes will be proportional to the product of the prior distribution and the altitude correlation-based distribution.

### 4. Testing Environment

To develop and test the probability multiple-resolution approach, a simulation environment was created using AirSim [15], a drone and car simulator built on the Unreal Engine [16]. AirSim is an open-source, cross-platform simulator for physically and visually realistic simulations. It is developed as an Unreal plugin that can be integrated into any Unreal environment. Within AirSim, a drone can be controlled using a Python/C++ API; for our project's requirements, the drone can be configured similarly to a real drone in terms of dynamics, sensor data, and computer interface. The drone can be flown in the simulation environment from one waypoint to another while acquiring data from the sensors defined in the platform. The simulator computes images taken by a downwards-facing camera and a GPS/inertial navigation system for the current configuration.

The simulator has two primary purposes:

1. To test the overall multi-resolution approach as the unit under test (UUT). The simulator functions as the hardware in the loop (HIL) tester's data generator in this mode. The data generated by the simulator are streamed into the drone's mission compute, and the system processes the data and computes the next coordinate to which the drone flies. Note that in this case, the simulator drives the real-time functioning of the closed-loop system.
2. To generate offline data for the search algorithm. As mentioned above, the drone can be flown using the API around the map at different scenarios and heights while generating data at predetermined rates. Typical data consist of RGB images, segmented images, and inertial navigation data, forming a probabilistic model analysis and validation data set.

The 3D model used in the simulator was the Brushify Urban Buildings Pack [17], purchased from the Unreal marketplace. Figures 4 and 5 show a simple example of RGB and corresponding segmentation images for the cameras simulated on the drone. The images highlight the observation that objects that occupy a cell are hardly detectable from a high altitude (e.g., a phone booth), while when descending, the gathered information allows the algorithm to detect these objects and decide that the drone cannot land in these specific cells.

**Figure 4.** High-altitude urban scene captured with a downward-looking camera. On the left is a simulated image. On the right is the corresponding segmented scene.

 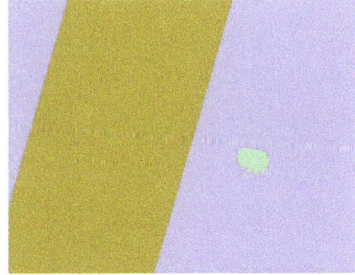

**Figure 5.** Low-altitude urban scene captured with a downward-looking camera. On the left is a simulated image. On the right is the corresponding segmented scene.

## 5. Analysis and Preliminary Results

Obtaining prior knowledge about the urban scene is necessary for a probability model. For this purpose, a labeled 3D digital surface model (DSM) was generated by the simulation using a $2 \times 2$ [m] cell resolution.

The DSM, shown in Figure 6, allows for choosing the parameters of the prior distribution for each cell given the label of that cell. The labels chosen to be represented with initial probabilities were such that they were visible from a high altitude and may be considered appropriate for landing or not when descending. Figure 7 shows the beta distribution's parameters for each label. These parameters were chosen to give some knowledge on an appropriate (or not) place to land. Still, there is sufficient uncertainty in the prior's belief to allow some degree of freedom to change the values and the new belief with new observations.

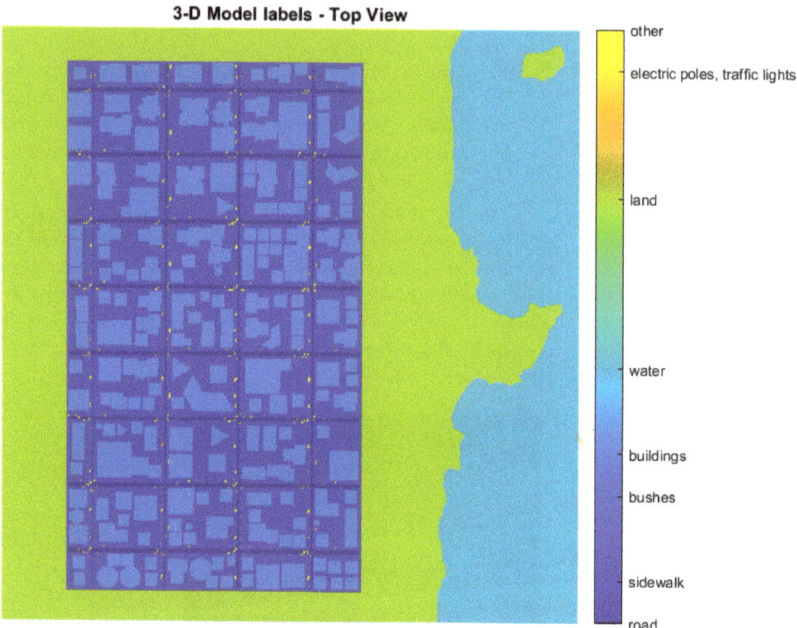

**Figure 6.** Top view of the digital surface model with chosen labels.

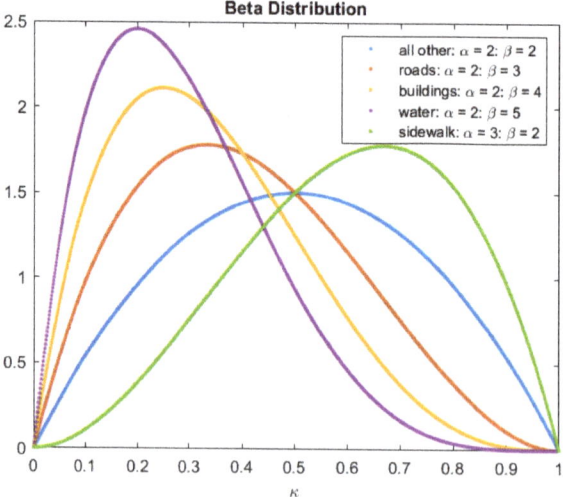

**Figure 7.** Prior beta distribution for chosen labels.

The $\alpha$ and $\beta$ parameters in Equation (3) are stored for each cell $c_{ij}$ in a 2D model of the urban scene, as illustrated in Figure 8. To determine if cell $c_{ij}$ contains a place to land, the probability $Pr(K_{ij} > \kappa)$ is calculated:

$$Pr(K_{ij} > \kappa) = \int_{\kappa}^{1} Pr(K_{ij})\, dK \qquad (8)$$

For instance, Figure 9 shows the prior belief that a landing place exists for $Pr(K_{ij} > 0.5)$. Only sidewalks are somewhat appropriate for landing, even for $\kappa = 0.5$, and the belief can easily be changed when new observations are obtained. Notice that cells in both Figures 8 and 9 contain values for $\alpha$, $\beta$, and prior probabilities.

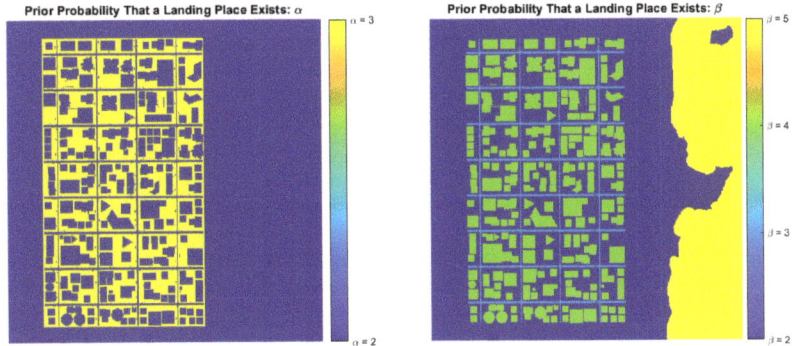

**Figure 8.** Prior $\alpha$ and $\beta$ parameters for each cell in the urban scene.

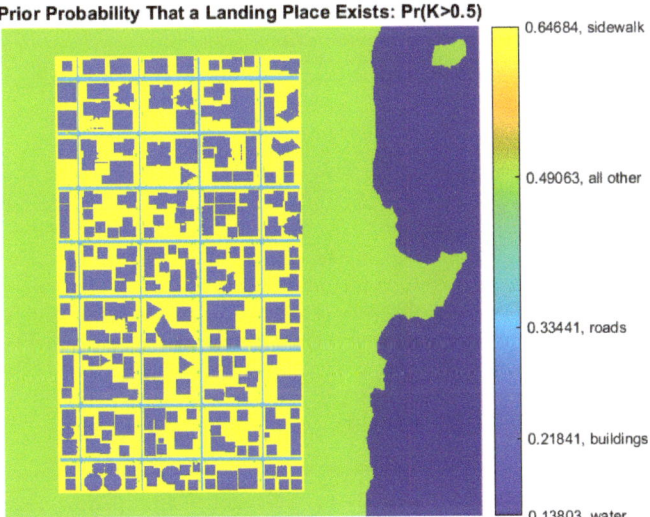

**Figure 9.** Prior landing probability for $\kappa = 0.5$.

The downwards-looking camera mounted on the drone provides color images that need to be converted into information on how suitable each cell is for landing. Clearly, this relation may be highly complex. In recent years, deep neural networks have shown to be successful in various computer-vision applications, including the type of semantic segmentation problems relevant to our purpose. Consequently, we chose to employ the semantic segmentation network BiSeNet [18], which fuses two information paths: context path and spatial path. The context allows information from distant pixels to affect a pixel's classification at the cost of reduced spatial resolution. In contrast, the spatial path maintains fine details by limiting the number of down-sampling operations. This net also provides a reasonable compromise between segmentation accuracy and compute requirements.

Other network architectures, potentially more complex and accurate, are currently under investigation, including DeepLabV3+ [19] (a state-of-the-art convolutional net that combines an atrous spatial pyramid with an encoder–decoder structure) and Segformer [20] (a relatively efficient transformer-based segmentation model). Preliminary results when using these semantic segmentation architectures on real-life data are discussed in a forthcoming paper currently under review.

BiSeNet was trained and validated on images taken by the camera while flying in the urban environment at different altitudes. The model uses the labels in the digital surface model in Figure 6 for training. During inference, the model predicts probability scores (summing to 1) for the different categories using captured images. Each category is also assigned an a priori weight representing how suitable this category is for landing (e.g., weight[sidewalk] = 0.8 and weight[building] = 0). We take a weighted average of the categories' probabilities using the predefined weights to obtain a final score $p_{mn}$ for each image pixel. $p_{mn}$, which can vary between 0 to 1, which describes the probability that a landing site exists based on the observed data. Using the 6-DOF of the drone, the image footprint, i.e., pixels coordinates projected on the ground, was transformed to a world coordinate system to be associated with each cell $c_{ij}$ in the grid. The outcome of a Bernoulli trial for success or failure is given by counting $N_p$, the number of pixels associated with cell $c_{ij}$ that pass $p_{mn} > 0.5$ and are relative to $N_{pc}$, the total number of pixels associated with $c_{ij}$. If the relative amount is greater than 0.99, then $c_{ij}$ holds a successful trial.

$$N_p = \sum_{m,n \in i,j}^{N_{pc}} \{p_{mn} > 0.5\} \qquad (9)$$

$$\mathcal{H}_{ij} \doteq \begin{cases} 1, & \frac{N_p}{N_{pc}} > 0.99 \\ 0, & \text{otherwise} \end{cases} \qquad (10)$$

### 5.1. A Single-Altitude Bayesian Update

The Bayesian update was tested for several flight scenarios. Suppose now that the drone flies and takes images at a constant altitude and that the semantic segmentation network analyzes images. Each cell $c_{ij}$ belonging to an image footprint is associated with the corresponding projected pixels, and a Bernoulli trial is performed according to Equation (10). The trial is performed on each cell only once to prevent added correlation effects at that altitude. The outcome of the single trial is added to the $\alpha$ and $\beta$ values previously selected as the prior (shown in Figures 7 and 8) and then integrated numerically according to Equation (8). Figures 10 and 11 show the updating stage at different time instances and altitudes.

### 5.2. An Altitude-Based Bernoulli Trial Distribution

In order to study the GBD model, an experiment with Bernoulli trials at different altitudes was performed. Several objects are placed on the ground at different locations around the urban scene. The experiment was planned so that a single object was selected for the drone to descend upon at different locations. At each location, images are taken as input for the semantic segmentation network. On each output of the network, a Bernoulli trial is performed on a single cell according to Equation (10), so that at each altitude that an image is taken, there is a single success or failure output on a given cell $c_{ij}$. Figures 12 and 13 show the input and output at selected altitudes in a single location. A phone booth was selected for the drone to descend upon. There are 31 locations around the urban scene with the phone booth placed on the sidewalk. A cell in the world coordinate system was selected at each location so that the phone booth occupies partially or the entire cell. The expected outcome is that the phone booth would be partially detected at high altitudes, and the selected cell would be detected as fit for landing. In contrast, more details will be detected when descending, and the cell will be detected as unfit for landing.

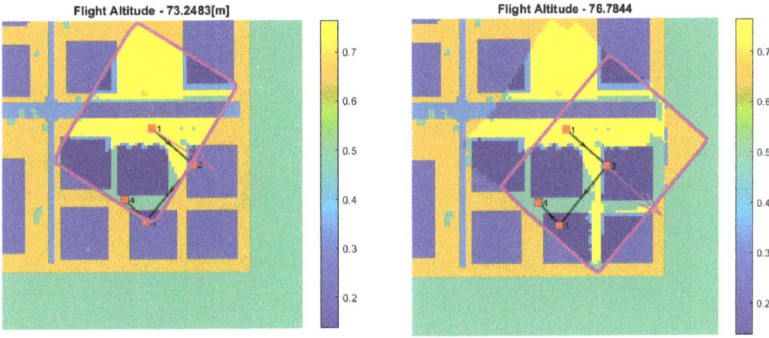

**Figure 10.** Bernoulli trial update: each cell is updated only once.

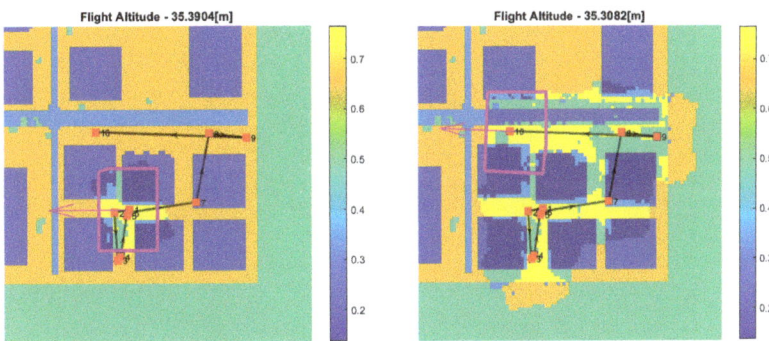

**Figure 11.** Bernoulli trial update: each cell is updated only once.

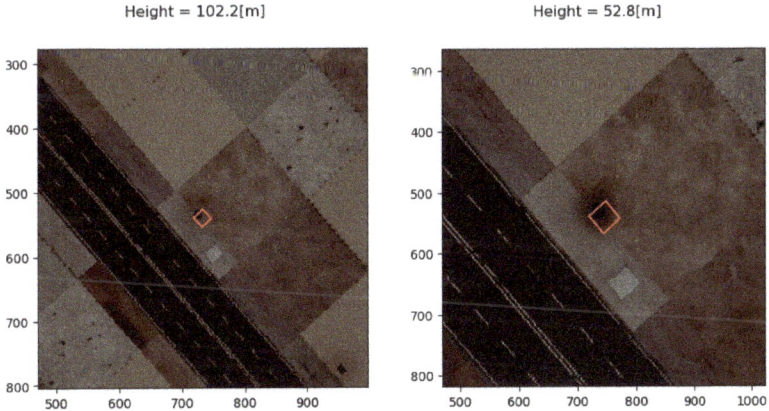

**Figure 12.** Semantic segmentation input: telephone booth partially occupies the cell.

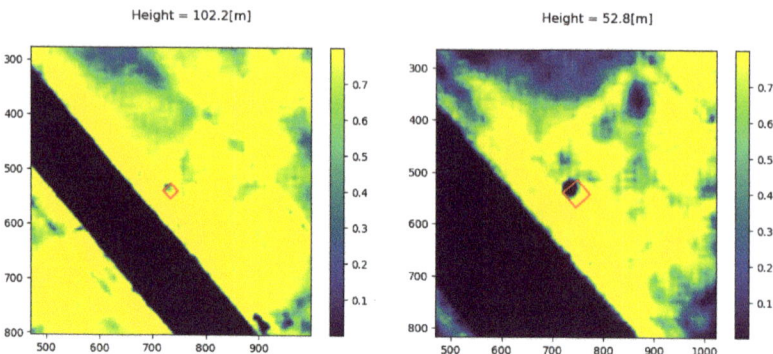

**Figure 13.** Semantic segmentation output: telephone booth partially occupies the cell.

Figure 14 shows the histogram for the 31 locations with altitude-based trials. We can see that the aforementioned expected behavior is indeed observed under approximately 160 m. Peculiarly, above this altitude, the success frequency diminishes markedly. This may be explained by the fact that there were no images from these altitudes in the training set, making the net's prediction unreliable. It should be noted that even in lower altitudes, its prediction can be noisy for various reasons, mainly because the training set is not diverse enough. We expect that more diversity in the training set will yield greater reliability of the net, which in turn, will result in a better fit for the model.

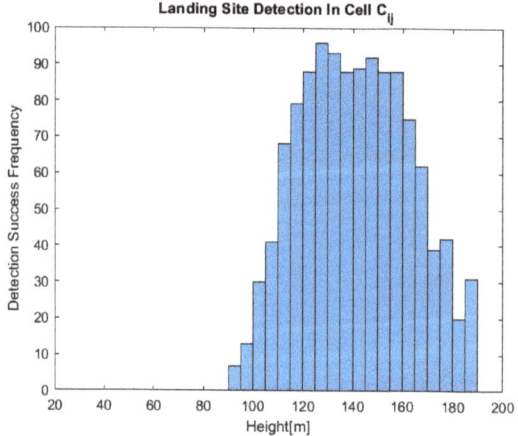

**Figure 14.** An altitude-based Bernoulli trial histogram.

## 6. Conclusions and Further Work

This paper has presented a multi-resolution probabilistic approach to finding an appropriate landing site for drones in a dense urban environment. The approach uses a priori data (e.g., a map or a DSM) to estimate the fitness for landing probability distribution for each cell on which the environment is divided. Distribution and not probabilities are used in an attempt to model the uncertainty of the data. Subsequently, the data collected by a visual sensor and processed by a semantic segmentation neural net are used to update the distribution using Bayesian networks. In order to do that, the probability of success is factored into the results obtained by the net. Images are captured at different altitudes in an attempt to solve the trade-off between image quality, including spatial resolution, context, and others. After presenting theoretical aspects, the simulation environment in which the approach was developed and tested is detailed, and the experiments conducted for validation are described. The overall approach is shown to produce the desired results, at least for the simulation environment in which it was tested.

Further work is currently underway in three main directions. Firstly, we would like to establish some success criteria for the procedure. For instance, we would like to develop bounds to enable more accurate ways of analyzing our results. Secondly, we would like to generate more realistic images on which semantic segmentation can be trained and tested. Lastly, we would like to test the approach on actual data and conduct a flight test to achieve real-life validation.

**Author Contributions:** Conceptualization, B.P., E.R. and H.R.; methodology, E.R. and H.R.; software, B.P. and B.M.; validation, B.P. and B.M.; formal analysis, B.P.; investigation, B.P., B.M. and H.R.; resources, E.R.; data curation, B.M.; writing—original draft preparation, B.P. and H.R.; writing—review and editing, B.P. and H.R.; supervision, E.R. All authors have read and agreed to the published version of the manuscript.

**Funding:** This research was partially funded by HYUNDAI NGV.

**Institutional Review Board Statement:** Not applicable.

**Informed Consent Statement:** Not applicable

**Data Availability Statement:** Not applicable.

**Acknowledgments:** We would like to thank Daniel Weihs, head of the Technion Autonomous System Program (TASP), for his continuous support.

**Conflicts of Interest:** The authors declare no conflict of interest.

# References

1. Bertuccelli, L.F.; How, J. Robust UAV search for environments with imprecise probability maps. In Proceedings of the 44th IEEE Conference on Decision and Control, Seville, Spain, 12–15 December 2005; pp. 5680–5685.
2. Torres, R.D.d.L.; Molina, M.; Campoy, P. Survey of Bayesian Networks Applications to Intelligent Autonomous Vehicles. *arXiv* **2019**, arXiv:1901.05517.
3. Starek, J.A.; Açıkmeşe, B.; Nesnas, I.A.; Pavone, M. Spacecraft autonomy challenges for next-generation space missions. In *Advances in Control System Technology for Aerospace Applications*; Springer: Berlin/Heidelberg, Germany, 2016; pp. 1–48.
4. Serrano, N. A bayesian framework for landing site selection during autonomous spacecraft descent. In Proceedings of the 2006 IEEE/RSJ International Conference on Intelligent Robots and Systems, Beijing, China, 9–15 October 2006; pp. 5112–5117.
5. Coombes, M.; Chen, W.H.; Render, P. Site selection during unmanned aerial system forced landings using decision-making bayesian networks. *J. Aerosp. Inf. Syst.* **2016**, *13*, 491–495. [CrossRef]
6. Ding, J.; Tomlin, C.J.; Hook, L.R.; Fuller, J. Initial designs for an automatic forced landing system for safer inclusion of small unmanned air vehicles into the national airspace. In Proceedings of the 2016 IEEE/AIAA 35th Digital Avionics Systems Conference (DASC), Sacramento, CA, USA, 25–29 September 2016; pp. 1–12.
7. Chung, T.H.; Burdick, J.W. A decision-making framework for control strategies in probabilistic search. In Proceedings of the 2007 IEEE International Conference on Robotics and Automation, Roma, Italy, 10–14 April 2007; pp. 4386–4393.
8. Stache, F.; Westheider, J.; Magistri, F.; Stachniss, C.; Popović, M. Adaptive path planning for UAVs for multi-resolution semantic segmentation. *Robot. Auton. Syst.* **2023**, *159*, 104288. [CrossRef]
9. Stone, L.D. *Theory of Optimal Search*; Elsevier: Amsterdam, The Netherlands, 1976.
10. Drezner, Z.; Farnum, N. A generalized binomial distribution. *Commun. -Stat.-Theory Methods* **1993**, *22*, 3051–3063. [CrossRef]
11. Drezner, Z. On the limit of the generalized binomial distribution. *Commun. Stat. Methods* **2006**, *35*, 209–221. [CrossRef]
12. James, B.; James, K.; Qi, Y. Limit theorems for correlated Bernoulli random variables. *Stat. Probab. Lett.* **2008**, *78*, 2339–2345. [CrossRef]
13. Wu, L.; Qi, Y.; Yang, J. Asymptotics for dependent Bernoulli random variables. *Stat. Probab. Lett.* **2012**, *82*, 455–463. [CrossRef]
14. Zhang, Y.; Zhang, L. Limit theorems for dependent Bernoulli variables with statistical inference. *Commun. -Stat.-Theory Methods* **2017**, *46*, 1551–1559. [CrossRef]
15. Shah, S.; Dey, D.; Lovett, C.; Kapoor, A. Airsim: High-fidelity visual and physical simulation for autonomous vehicles. In *Proceedings of the Field and Service Robotics*; Springer: Berlin/Heidelberg, Germany, 2018; pp. 621–635.
16. Games, E. Unreal Engine. 2021. Available online: https://www.unrealengine.com (accessed on 9 March 2021).
17. Garth, J. UBrushify—Urban Buildings Pack. 2020. Available online: https://www.unrealengine.com/marketplace/en-US/product/brushify-urban-buildings-pack (accessed on 9 March 2021).
18. Yu, C.; Wang, J.; Peng, C.; Gao, C.; Yu, G.; Sang, N. Bisenet: Bilateral segmentation network for real-time semantic segmentation. In Proceedings of the European Conference on Computer Vision (ECCV), Munich, Germany, 8–14 September 2018; pp. 325–341.

19. Chen, L.C.; Zhu, Y.; Papandreou, G.; Schroff, F.; Adam, H. Encoder-decoder with atrous separable convolution for semantic image segmentation. In Proceedings of the European Conference on Computer Vision (ECCV), Munich, Germany, 8–14 September 2018; pp. 801–818.
20. Xie, E.; Wang, W.; Yu, Z.; Anandkumar, A.; Alvarez, J.M.; Luo, P. SegFormer: Simple and efficient design for semantic segmentation with transformers. *Adv. Neural Inf. Process. Syst.* **2021**, *34*, 12077–12090.

*Article*

# Vision-Based Detection and Classification of Used Electronic Parts

**Praneel Chand [1,\*] and Sunil Lal [2]**

[1] Centre for Engineering and Industrial Design (CEID), Waikato Institute of Technology, Hamilton 3200, New Zealand
[2] School of Mathematical and Computational Sciences, Massey University, Palmerston North 4410, New Zealand
\* Correspondence: praneelchand10@yahoo.co.nz or praneel.chand@wintec.ac.nz

**Abstract:** Economic and environmental sustainability is becoming increasingly important in today's world. Electronic waste (e-waste) is on the rise and options to reuse parts should be explored. Hence, this paper presents the development of vision-based methods for the detection and classification of used electronics parts. In particular, the problem of classifying commonly used and relatively expensive electronic project parts such as capacitors, potentiometers, and voltage regulator ICs is investigated. A multiple object workspace scenario with an overhead camera is investigated. A customized object detection algorithm determines regions of interest and extracts data for classification. Three classification methods are explored: (a) shallow neural networks (SNNs), (b) support vector machines (SVMs), and (c) deep learning with convolutional neural networks (CNNs). All three methods utilize 30 × 30-pixel grayscale image inputs. Shallow neural networks achieved the lowest overall accuracy of 85.6%. The SVM implementation produced its best results using a cubic kernel and principal component analysis (PCA) with 20 features. An overall accuracy of 95.2% was achieved with this setting. The deep learning CNN model has three convolution layers, two pooling layers, one fully connected layer, softmax, and a classification layer. The convolution layer filter size was set to four and adjusting the number of filters produced little variation in accuracy. An overall accuracy of 98.1% was achieved with the CNN model.

**Keywords:** vision system; object detection; object classification; shallow neural networks (SNNs); support vector machines (SVMs); deep learning; convolutional neural networks (CNNs)

## 1. Introduction

One of the key principles of a circular economy [1] is the elimination of waste and pollution. This facilitates a robust system that is beneficial for businesses, humans, and the environment. Recycling and reusing products should be emphasized in every part of the economy. In educational environments where resourcing can be constrained, equipment and consumables used in projects can be recycled or reused [2].

Higher education institutions that provide training for engineers often place high emphasis on practical activities and assessments. Courses in fields such as electrical and electronic engineering often rely on hardware components such as resistors, capacitors, inductors, voltage regulators, and diodes for project work. As an example, students are required to construct an electrotechnology product in the Electrical and Electronics Applications course at Waikato Institute of Technology [3]. The construction could be on a printed circuit board (PCB), Veroboard, or breadboard. After project work, the constructed PCBs are left in storage or thrown away (Figure 1). Used components are often discarded instead of being reused. In the circular economy concept, components on these circuit boards could be removed as part of soldering practice lessons. Since sorting parts manually is mundane, this could be achieved using an intelligent automated sorting system. Thus, this research proposes that a vision-based system be used to detect and classify parts.

**Figure 1.** Circuit boards discarded after project work.

According to Mathworks [4], identifying objects in images or videos is a computer vision technique known as object recognition. A variety of artificial intelligence methods can be used for object recognition. Techniques in machine learning and deep learning have become popular recently [5–7]. Object detection is similar to object recognition but varies in execution. In objection detection, instances of objects are identified and also located in an image. This enables many objects to be located and identified in an image.

Machine learning [8] is a sub-class of artificial intelligence and deep learning [9] is a sub-class of machine learning. Traditional machine learning approaches have interconnected steps such as segmentation, feature extraction and classification. Conventional traditional machine learning classification methods for object recognition include shallow neural networks (SNNs) and support vector machines (SVMs) [10,11]. Deep learning primarily utilizes deep neural networks that consist of multiple hidden layers. Feature extraction and classification is learned by the deep neural network. This provides superior flexibility because the framework can be re-trained using a custom dataset for transfer learning. Deep learning can also achieve better classification than traditional machine learning. However, it achieves this at the expense of requiring high-end computing power, larger training datasets, and longer training time. A comparison of traditional machine learning and deep learning applied to image recognition showed an increase in accuracy of less than 5% [12].

A common application of vision-based detection of electronic components is inspecting the integrity and quality of PCBs [13–15]. Image classification techniques based on deep neural networks have been used to detect integrated circuit (IC) components and verify their correct placement on the finished PCB product in [13]. Verification is similar to classification and a best accuracy of 92.31% was achieved. Machine learning is used to inspect components prior to assembly in [14]. The purpose of prior inspection is to reduce the number of defective components mounted and reduce falsely rejected components. Scale-invariant feature transform (SIFT) parameters are extracted from raw images and used with an artificial neural network (ANN) or an SVM for classification. Classification accuracies of up to approximately 97% were achieved. Tiny surface mount electronic components on PCBs are recognized using machine learning and deep learning in [15]. Machine learning with SVM+ principal component analysis (PCA) achieved an overall true positive rate (TPR) of 93.29%. The TPR was further improved to 99.999% with the deep learning-based Faster SqueezeNet.

Some recent methods to classify loose electrical and electronic components are based on deep learning models [16–18]. In [16], a customized CNN architecture is developed to classify three types of components: resistors, diodes, and capacitors. The developed system's performance is benchmarked against pre-trained AlexNet, GoogleNet, ShuffleNet, and SqueezeNet deep learning architectures. While the accuracy of the pre-trained models ranged from 92.95% to 96.67%, the proposed CNN model achieved 98.99% accuracy. Post-training evaluation in a real-world setting was not conducted. In [17] and [18], variations of the 'you only look once' (YOLO) deep learning model [19] are utilized. The speed and accuracy of real-time object detection makes YOLO a popular choice. It is capable of directly outputting the position and category of an object through its neural network. Four electronic components (three types of capacitors and an inductor) are classified using YOLO-V3 and Mobilenet in [18]. A mean average precision (mAP) of 0.9521 was achieved. The YOLOv4-tiny network is combined with a multiscale attention module (MAM) and used to classify twenty types of electronic components in [17]. This improves the accuracy of the original algorithm from 93.74% to 98.6%. A potential deep learning model for detecting and classifying parts is Faster R-CNN [20]. However, the drawback of using Faster R-CNN for classifying electronic parts is explained in [13]. It achieves very poor results and according to the authors Faster R-CNN is not designed for small, relatively featureless objects such as ICs.

A non-deep learning based machine learning method for classifying electrical and electronic parts is presented in [21]. In this implementation, a K nearest neighbor (KNN) classification algorithm is used to classify capacitors, diodes, resistors, and transistors. Classification is performed based on physical appearances such as length, width, number of legs, shape (roundness of objects), and correlation of input images with standard database images. Full results and analysis are not presented, and accuracy is not quantified. While KNNs are simple and easy to implement, they can become significantly slower as the volume of data increases.

Recently, weakly supervised learning (WSL) has become popular in the computer vision community. A survey of various methods for object localization and detection is provided in [22]. An advantage of WSL is that it can perform object localization and detection at image level speeds of conventional fully supervised learning tasks. Typically, weakly labelled training images can be input to either machine learning methods (e.g, SVMs), or off-the-shelf deep models (e.g., AlexNet or R-CNN), or novel deep WSL frameworks. WSL is applied to video salient object detection in [23]. Co-salient object detection distinguishes common and salient objects in a group of relevant images. A summarize and search method that employs dynamic convolution to distinguish salient objects is presented in [24]. The current literature search did not determine any suitable implementations of WSL and video salient object detection for sorting electronic parts.

A machine learning method that utilizes an SNN classifier to identify capacitors within a scene of scattered electronic components is presented in [25]. A feature extraction algorithm detects objects and converts them to a 20 × 20-pixel grayscale image for the SNN. An overall accuracy of 82.7% is achieved. This method is further extended to a three class problem for classifying capacitors, potentiometers, and voltage regulators in [26]. By increasing the size of the grayscale image to 30 × 30 pixels and correspondingly adjusting the size of the hidden neuron layer, an overall accuracy of 85.6% is achieved. Capacitor classification achieves an accuracy of 91.4%.

Unlike the other reviewed methods [25,26], utilize lower resolution grayscale images for classification. This reduces the complexity of the classifier and requires lower computational power (processor and memory use). However, the accuracy is also reduced. Hence, this paper investigates the use of alternative methods based on SVM and CNN to improve classification using the low-resolution grayscale images.

## 2. Materials and Methods

*2.1. Conceptual Framework*

Figure 2 shows a visualization of the object sorting system. An overhead camera coupled with a Niryo Ned robotic arm [27] is used to detect, classify, and shift objects within a pre-defined workspace. This workspace has a size of 194 mm horizontally (h, x) by 194 mm vertically (v, y) and its boundaries are marked by one origin marker, top left (TL), and three edge markers, top right (TR), bottom left (BL), and bottom right (BR). The Niryo Ned robotic system has been selected because it features the open-source Robot Operating System (ROS) platform [28] and supports Matlab integration via the ROS Toolbox [29]. A graphical user interface (GUI)-based controller has been developed in Matlab to communicate commands to the robot and perform image acquisition [30]. Another feature is the relatively low cost of the hardware which is approximately US $3299 [30].

**Figure 2.** Visualization of object sorting system.

A Logitech HD C270 web camera is mounted in the center of the workspace. It is positioned at a height of approximately 0.37 m. Using a camera resolution of 960 × 720 pixels, the four boundary markers are clearly visible near the limits of the camera image at this height. Figure 3 illustrates a sample camera image at the height of 0.37 m. The camera height is adjustable since the four boundary markers are also used to automatically calibrate pixel distances (1), (2). The TL marker is the origin marker and is used to compute pixel and physical distances in the workspace. It also translates workspace distances to the robot's reference frame.

$$x_{cal} = 194/(0.5(h_{TR} - h_{TL} + h_{BR} - h_{BL})), \quad (1)$$

$$y_{cal} = 194/(0.5(v_{BL} - v_{TL} + v_{BR} - v_{TR})), \quad (2)$$

where

$x_{cal}$ is the x-axis calibrated pixel distance scale factor in mm/pixel,
$y_{cal}$ is the y-axis calibrated pixel distance scale factor in mm/pixel.
H and v are horizontal and vertical pixel numbers, respectively.

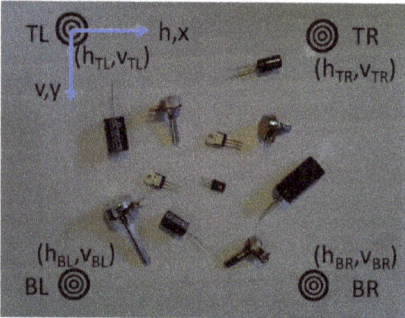

**Figure 3.** Sample camera image at the height of 0.37 m.

The overall general framework of the vision-based classification system designed in this paper is illustrated in Figure 4. First, an image of the workspace is captured using the web camera via Matlab. Following this, the acquired image is processed for object detection. Bounding boxes are placed around detected objects and the center of the bounding boxes represents the location (position) of the objects. After determining the bounding boxes, the partial image inside each bounding box is considered a region of interest (ROI) for the classifier and is resized according to the classifier requirements. Once the ROI is resized, the classifier uses it to match the image to an object class it has been trained to recognize. The classified object can then be moved by the robotic arm to a target location.

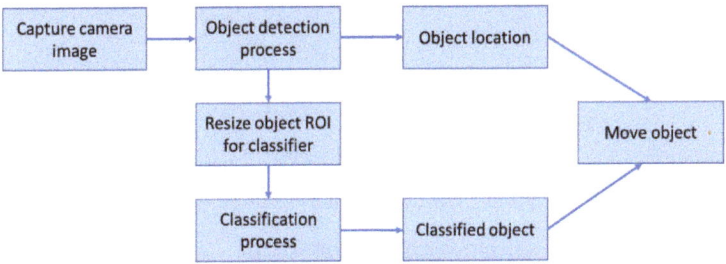

**Figure 4.** General framework of the vision-based classification system.

*2.2. Object (Component) Detection*

The major parts of the object (component) detection process are highlighted in Figure 5. Various image processing algorithms are applied to extract ROIs containing unclassified objects. Figure 6 shows representative images of the various stages of the object detection process.

To reduce the complexity of the object detection and classification process, grayscale images are used. Hence, the first part of the process is to convert the RGB color image to grayscale using the weighted method (3). Following this, edge detection algorithms can be applied to determine the boundaries (outlines) of objects within images [31]. Of the available algorithms in Matlab (Sobel, Canny, Prewitt, and Roberts), Canny performed the best in detecting shape outlines (Figure 6a). Canny uses two thresholds which makes it less likely to be fooled by noise and more likely to detect true weak edges. The values for the high and low thresholds are 0.1 and 0.04, respectively.

$$gray = 0.299R + 0.587G + 0.114B. \qquad (3)$$

The output of the Canny edge detection algorithm is a binary image which is then dilated to further improve connectivity between the edges. This is achieved by applying a rectangular structuring element that enlarges the edges of the binary image (Figure 6b). Edge connectivity is important as the next stage involves flood-filling the binary image to form filled (solid) shapes representing the detected objects (Figure 6c). After flood-filling, the binary image is further processed by measuring the properties of the image regions. The "BoundingBox" property argument returns a set of positions and sizes of the smallest boxes, i, containing each detected object (Figure 6d) (4). This represents the ROIs. The green crosses in Figure 6d mark the bounding box centers (BBC) that represent the location of the objects in the workspace (5) and (6).

$$BB_i = [ho_i, vo_i, hw_i, vh_i], \qquad (4)$$

where
$BB_i$ is the ith bounding box,
$ho_i$ is the horizontal pixel number of the top left corner,
$vo_i$ is the the vertical pixel number of the top left corner,
$hw_i$ is the horizontal width in pixels,
$vh_i$ is the vertical height in pixels.

$$BBC(h_i, v_i) = (ho_i + 0.5 \times hw_i, vo_i, + 0.5 \times vh_i), \qquad (5)$$

$$BBC(x_i, y_i) = ((ho_i + 0.5 \times hw_i - h_{TL}) \times x_{cal}, (vo_i + 0.5 \times vh_i - v_{TL}) \times y_{cal}). \quad (6)$$

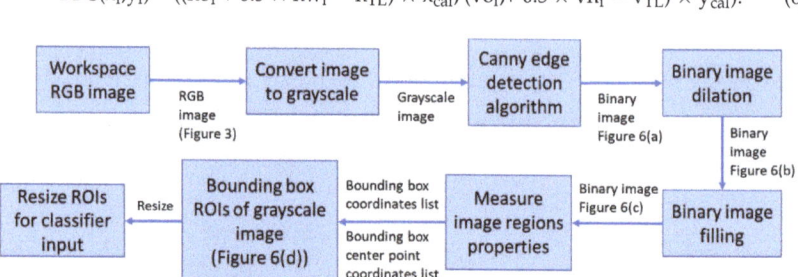

**Figure 5.** Main parts of the object detection process (Figures 3 and 6).

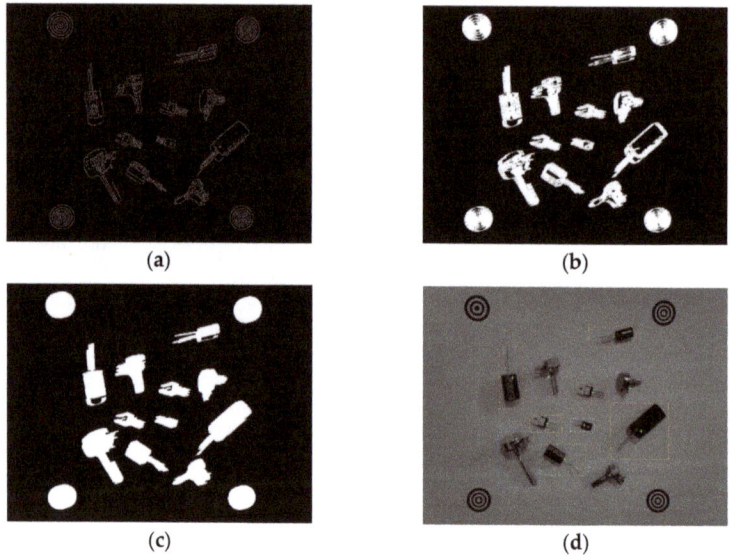

**Figure 6.** Sample images of various stages of the object detection process. (**a**) Canny edge detection binary image; (**b**) Binary image dilation; (**c**) Filled binary image; (**d**) Grayscale image with bounding boxes and center point coordinates.

After detecting bounding boxes (ROIs), the size of each bounding box is checked against an estimated size threshold representing the dimensions of the smallest component to be detected. This eliminates small boxes that may have been erroneously detected due to noise or tiny holes in components such as voltage regulators. The pick and place task assumes that objects are physically separated and do not overlap.

The final stage before input to the component classifier involves standardizing the size of the ROI images. The ROIs of the grayscale image inside the bounding boxes are rescaled to 30 × 30 pixels. This has been arbitrarily selected to reduce complexity of the classifier and represents 900 inputs.

*2.3. Component Classification*

Component classification determines which class or category the detected component belongs to. Several methods of doing this are outlined in Section 1. Three techniques utilized in this research are described below.

2.3.1. Shallow Neural Network (SNN)

The SNN classifier has 900 inputs, one hidden layer, and three outputs representing the components (capacitor, potentiometer, and regulator) as shown in Figure 7. It is designed and implemented using the Neural Pattern Recognition tool (nprtool) in Matlab 2021a. The classifier is a feedforward neural network that is backpropagation trained using the scaled conjugate gradient method [32]. The performance function is the Cross-Entropy method (7) which generates batches of episodes and removes bad episodes in a batch to train the network on better ones. The tansig function is utilized in the hidden layer while the softmax function is employed in the output layer. These are the default settings of the nprtool. The main variable adjusted in the SNN is the number of neurons in the hidden layer.

$$J = -\frac{1}{M} \sum_{m=1}^{M} \sum_{i=1}^{K} t_m^i \ln\left(y_m^i\right) \qquad (7)$$

where

J is the cost,
M is the number of training data,
K is the number of output classes,
y is the output (contains K values, one for each class).

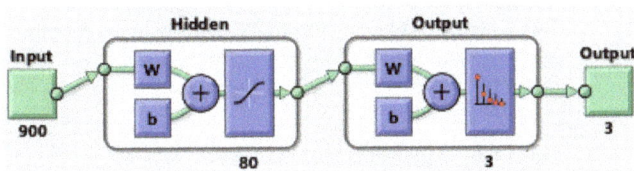

**Figure 7.** SNN architecture.

2.3.2. Support Vector Machine (SVM) and Principal Component Analysis (PCA)

The SVM classifier also has 900 inputs and three outputs. It is designed using the Matlab Classification Learner App. Error-correcting output codes (ECOC) [33] are used to train the classifier which works by solving for a hyperplane that separates two class data with maximal margin [34]. The support vectors are the points which lie near the separating hyperplane. The SVM is trained for a 3-class problem on a one vs all approach. Since the original training data is not linearly separable, four different kernel functions $K(x_i, x)$ (linear (8), quadratic (9), cubic (10), and Gaussian (11)) are applied to the classifier. These transform the original input space into vectors of a highly dimensional feature space for the SVM to classify. The general structure of the SVM is shown in Figure 8

$$K(x_i, x) = (x_i . x) \qquad (8)$$

$$K(x_i, x) = (x_i . x + 1)^2 \qquad (9)$$

$$K(x_i, x) = (x_i . x + 1)^3, \qquad (10)$$

$$K(x_i, x) = e^{-\|x_i - x\|^2 / 2\sigma^2}. \qquad (11)$$

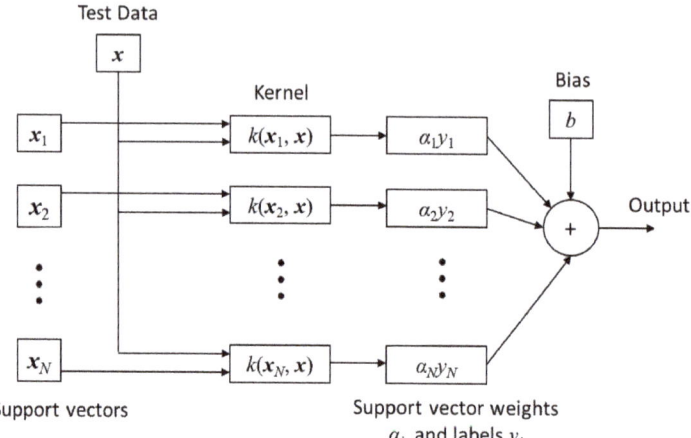

**Figure 8.** SVM architecture.

Using 900 predictors in the input space can impair computational time. Hence, PCA [35] is used to determine the principal components for feature optimization [36,37]. This singular value decomposition reduces the data dimensionality and projects it to a lower dimensional environment for the SVM. This naturally comes at the expense of accuracy. Hence, it is important to compare the SVM + PCA classifier accuracy with the SVM only classifier.

2.3.3. Convolutional Neural Network (CNN)

Like the SNN, the CNN classifier is also a feedforward neural network with 900 inputs (30 × 30-pixel image). It can extract features from the two-dimensional image and optimize parameters using backpropagation. The high performance of CNNs makes them a preferred deep learning architecture as outlined in Section 1 and in [38]. The basic structure of a CNN is shown in Figure 9. The hidden layers consist of a series of convolution, rectified linear unit (ReLU), and pooling layers. In the convolution layer, the image is examined by applying a filter smaller than the original image to determine its properties. Following this, the ReLU layer removes negative values from the output of the convolution layer. The pooling layer reduces the original size of the image by retaining important features and ignoring unnecessary features in the image. The fully connected (FC) layer converts the matrix image into a flat vector for the SoftMax function to determine the output classification.

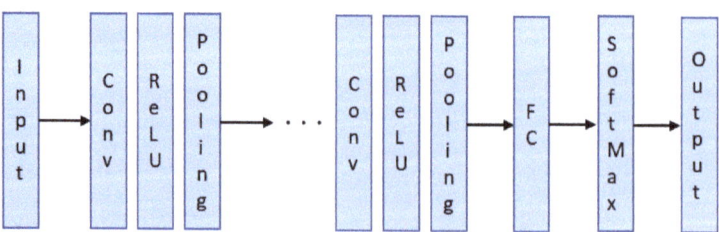

**Figure 9.** Basic structure of a CNN.

The architecture of the proposed CNN inspired by [39] has three convolution layers, two pooling layers, one fully connected layer, softmax, and an output classification layer as shown in Figure 10. The filter size for all three convolution layers is set to 4 × 4 with a stride of 1. A filter size of 3 × 3 is utilized for the two pooling layers and the stride is set to 3.

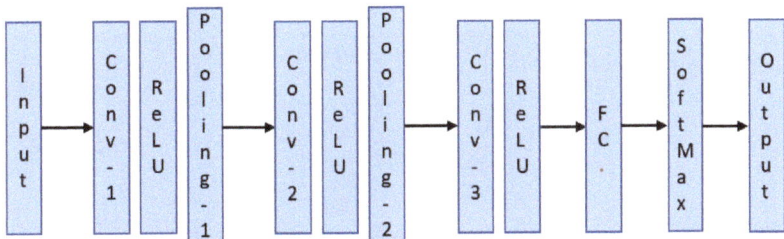

**Figure 10.** Architecture of the proposed CNN.

## 3. Results

### 3.1. Datasets and Configuration

The dataset used in this research consisted of a total of 1734 images extracted via the object detection process described in Section 2.2. Each class (capacitor, potentiometer, and regulator) had 578 images. A sample of the dataset images derived from object detection process is shown in Figure 11. Further details of the dataset are available in [40]. The dataset was randomly divided into 70% training (1214 images), 15% validation (260 images) and 15% test (260 images). Five-fold cross-validation was used in the training process.

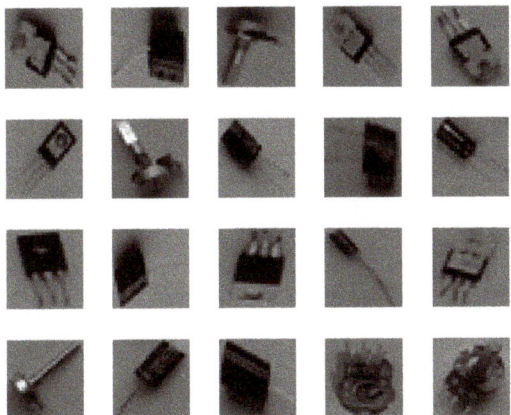

**Figure 11.** Sample of 25 images from the database.

A Windows 10 HP ProBook 450 G7 laptop running Matlab 2021a was used to implement the various classifiers. The hardware configuration had an Intel i7-10510U processor and 16 GB RAM.

After training the classifiers and testing them on the dataset, the best classifiers for SNN, SVM, and CNN were put to test in the real world. This was done with new independent data generated from the evaluation of ten multi-object scenes with a total of 104 objects.

### 3.2. SNN Classifier Accuracy

The SNN classifier model was tested with a variety of hidden layer neurons ranging from 10 to 120. When the number of hidden neurons was below 40 (10, 20, or 30) the test accuracies were all below 90%. Details of the test accuracies when the number of hidden layer neurons varied between 40 and 120 is shown in Table 1. Good classification is possible with any of the classifiers with 40, 60, 80, or 100 neurons. The model with 80 hidden neurons was selected since it had the best overall accuracy. Figure 12 illustrates the test confusion matrix and the confusion matrix of the real-world test with 104 new objects.

**Table 1.** Accuracies of the tested SNN models.

| Hidden Neurons | Test Accuracy % |
|---|---|
| 40 | 93.1 |
| 60 | 92.3 |
| 80 | 93.5 |
| 100 | 92.3 |
| 120 | 87.8 |

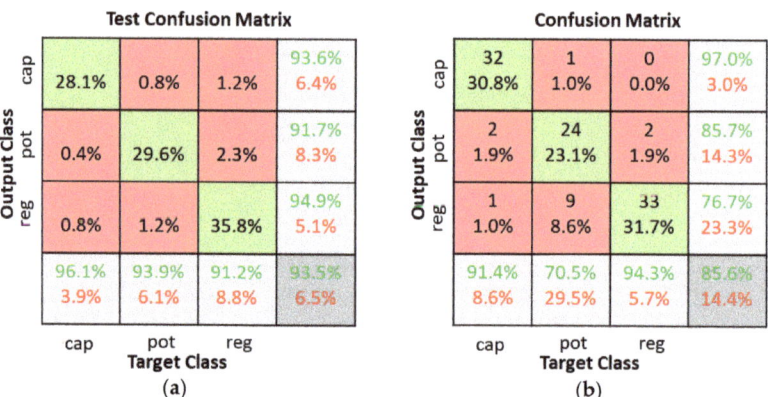

**Figure 12.** SNN classifiers with 80 hidden neurons confusion matrices. (**a**) Dataset test; (**b**) Real world test with 104 new objects.

*3.3. SVM + PCA Classifier Accuracy*

In the SVM + PCA classifier experiments, the number of components in PCA was varied between 10 and 50. Four kernel functions (linear, quadratic, cubic, and medium Gaussian) were also tested. The results of the various combinations tested are summarized in Figure 13. The horizontal lines without markers in Figure 13 represent the accuracy of SVM classifiers using the various kernel functions without PCA. Without PCA, the SVM classifiers achieved accuracies of 78.2%, 93.9%, 94.9%, and 92.4% with the linear, quadratic, cubic, and medium Gaussian kernels, respectively. Using PCA with the linear and medium Gaussian kernels degraded accuracies to below 70%. The quadratic and cubic kernels achieved low reduction in accuracy when the number of PCA components was between 20 and 30. The SVMs with cubic kernel function were the best overall achieving accuracies of 94.9% without PCA and 94.6% with 20 component PCA. Figure 14 illustrates the test confusion matrix and the confusion matrix of the real-world test with 104 new objects for the SVM classifiers with cubic kernel function. The real-word test achieved the same results with the SVM and SVM + PCA with 20-component classifiers.

*3.4. CNN Classifier Accuracy*

The CNN classifier model was tested with a 4 × 4 filter size for all convolution layers and a 3 × 3 filter size for the pooling layers. The stride in the convolution and pooling layers was set to one and three, respectively. The number of filters in the convolution layers was varied as shown in Table 2. Table 3 shows the CNN model training parameters. As shown in Figure 15, there was little change in overall accuracy when the number of filters in the convolution layers varied. Hence, Configuration 1 was selected since it has the least number of filters. Figure 16 illustrates the test confusion matrix and the confusion matrix of the real-world test with 104 new objects for the CNN classifiers using Configuration 1.

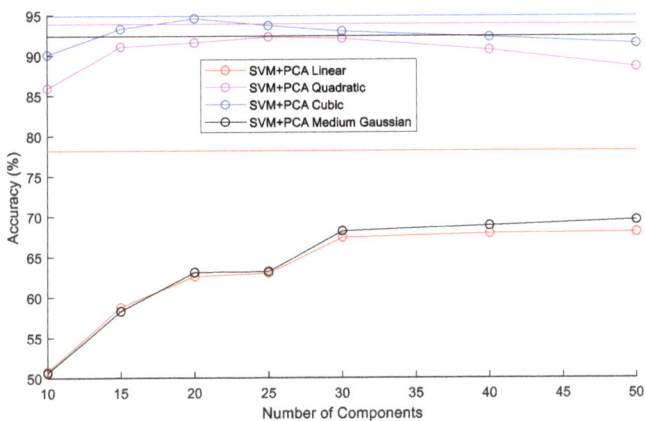

**Figure 13.** Comparison of SVM classifier accuracies with various kernels and PCA component numbers.

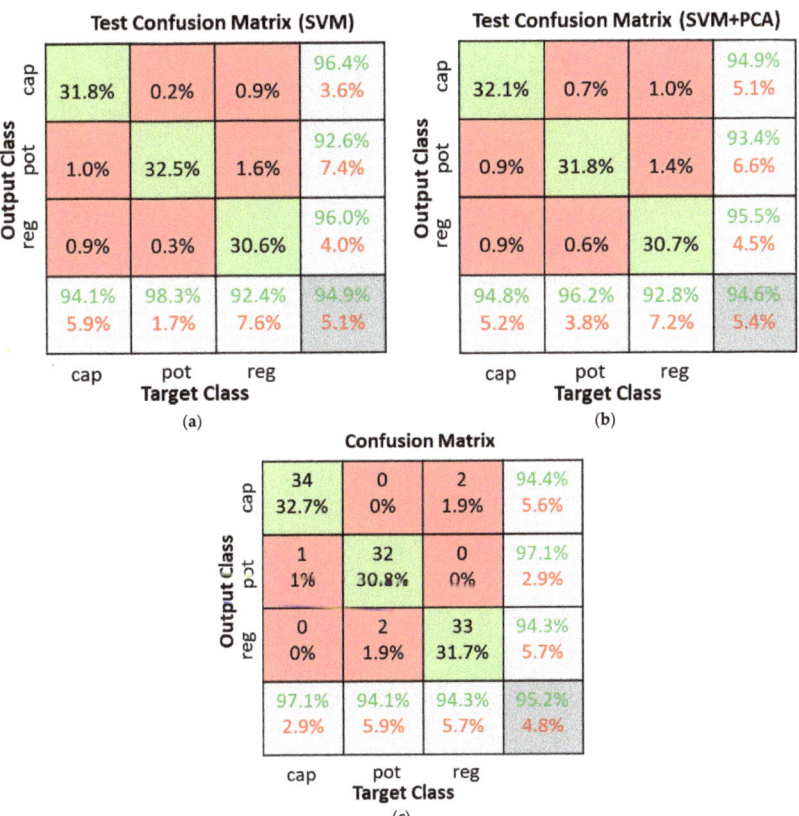

**Figure 14.** SVM classifiers with cubic kernel function confusion matrices. (**a**) SVM dataset test; (**b**) SVM + PCA with 20 components dataset test; (**c**) Real world test with 104 new objects.

**Table 2.** Filter numbers in convolution layers.

| Configuration Number | Value [Conv-1 Conv-2 Conv-3] |
|---|---|
| 1 | [10 20 40] |
| 2 | [12 24 48] |
| 3 | [15 30 60] |

**Table 3.** CNN model training parameters.

| Parameters | Value |
|---|---|
| Optimize method | stochastic gradient descent with momentum (sgdm) |
| Initial learning rate | 0.02 |
| Maximum epochs | 7 |
| Validation frequency | 20 |

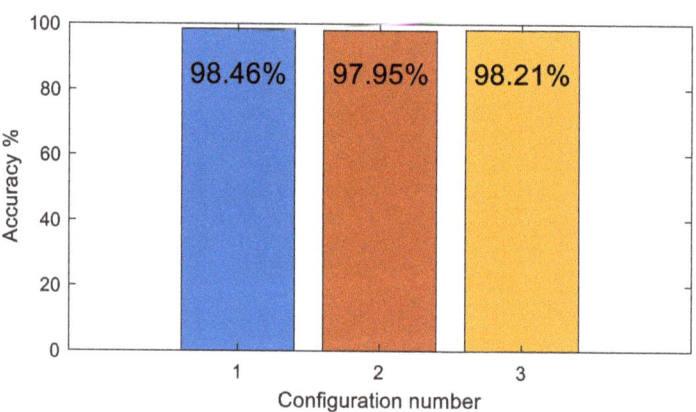

**Figure 15.** CNN classifier accuracy for various convolution filter configurations.

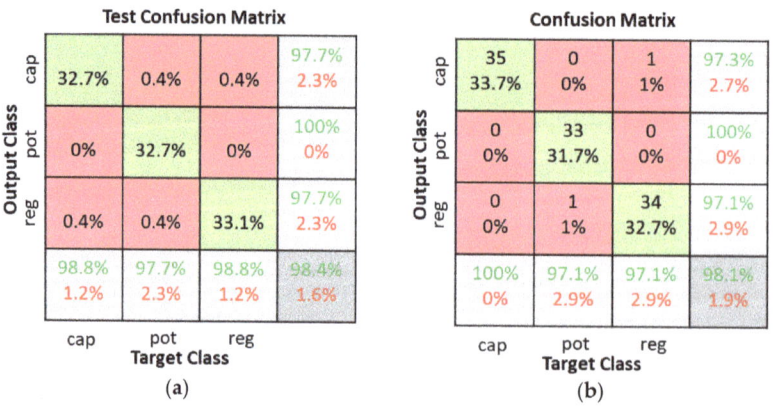

**Figure 16.** CNN classifiers with Configuration 1 confusion matrices. (**a**) Dataset test; (**b**) Real world test with 104 new objects.

## 4. Discussion

### 4.1. Overall Comparison of the Three Classifiers

The receiver operating characteristic (ROC) curves for the SNN with 80-hidden neuron classifier, SVM with cubic kernel and 20 PCA-component classifier, and CNN Configuration 1 classifier are shown in Figure 17. It is clearly visible that the CNN classifier has a superior ROC curve and performs the best for all object classes. Figure 18 compares the key performance criteria metrics of the classifiers based on the real-world test with 104 new objects. The CNN classifier has the best sensitivity and precision across all component classes. It also achieved the best accuracy of 98.1%. The SVM + PCA classifier can produce good results which are close to the CNN.

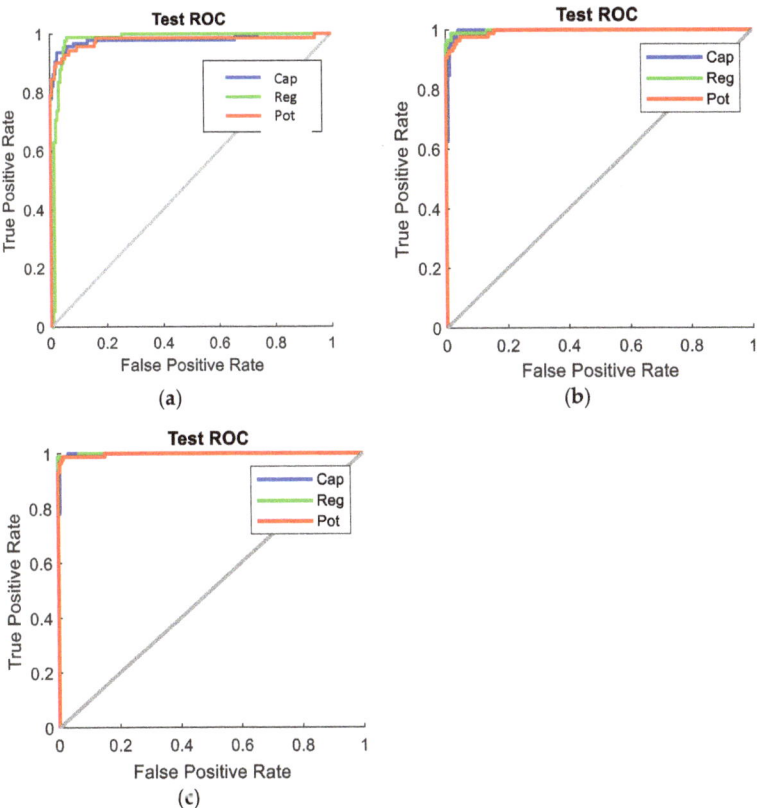

**Figure 17.** ROC curves of the best classifiers. (**a**) SNN with 80 hidden neurons; (**b**) SVM + PCA with 20 components; (**c**) CNN with Configuration 1.

### 4.2. Comparison with Accuracy of Other Classifiers

The classifiers developed in this paper utilize low-resolution grayscale images. Other methods reviewed in Section 1 use higher resolution and color images. Therefore, these other classifiers are inherently more complex and require heavier computational power. Table 4 compares the CNN classifier presented in this paper with the properties of other representative deep learning models from Section 1. Model complexity excludes the ReLU layers for all models. A direct comparison of computation volume and speed is not possible due to variations such as image resolution and object class numbers. Therefore, an approximate comparison based on image input size and network complexity is made in Table 4. The key feature of our method is that it can perform on a standard laptop computer.

The accuracy level of the developed CNN classifier is comparable with the other methods despite it using low resolution (30 × 30-pixel) grayscale images. However, the classifiers developed in [15,17] are capable of detecting a much wider range of electronic components. The training dataset employed in this research is small but sufficient for the three types of parts as there is not a large variation in physical properties of the items in each class. This is validated based on the classification results. The dataset can be expanded to include a larger variety of project parts if needed. For example, if ceramic and electrolytic capacitors need to be classified, then a new or expanded dataset can be utilized. The method presented in this paper is like YOLO as it has the ability to detect and classify electronic components with a single image of the entire workspace.

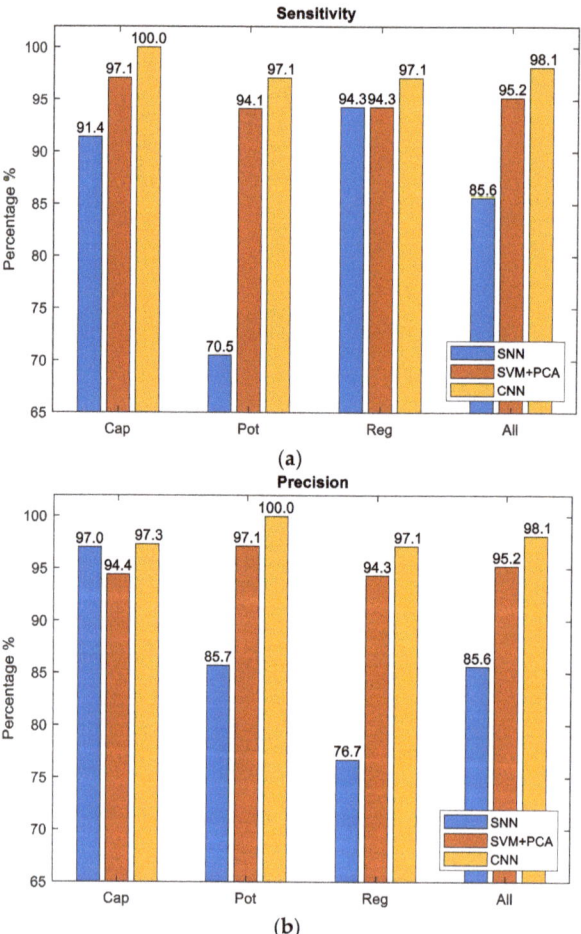

**Figure 18.** Key performance criteria metrics. (**a**) Sensitivity; (**b**) Precision.

Table 4. Comparison of the proposed CNN classifier with other deep learning methods.

| Reference | Dataset Properties | Classes | Model Complexity | Accuracy |
|---|---|---|---|---|
| Atik (2022) [16] | Color, 227 × 227 × 3 pixels, 5332 images | 3 | Custom CNN with 13 layers | 98.99% |
| Xu et al. (2020) [15] | Color, 112 × 112 × 3 pixels, 40000 images | 22 | Faster SqueezeNet with 23 layers | 99.999% TPR when FPR = $10^{-6}$ |
| Huang et al. (2019) [18] | Color, 416 × 416 × 3 pixels, 43,160 images | 4 | YOLO-V3-Mobilenet with 30 layers | 95.21% mAP |
| Guo et al. (2021) [17] | Color, 608 × 608 × 3 pixels, 12,000 images | 20 | YOLOv4-tiny + MAM with 24 layers | 98.6% mAP |
| Proposed CNN classifier | Grayscale, 30 × 30 pixels, 1734 images | 3 | Custom CNN with 7 layers | 98.4% test, 98.1% real world test |

## 5. Conclusions

This paper presented the development of vision-based methods for the detection and classification of used electronic parts. Three classes of components were considered: capacitors, potentiometers, and voltage regulator ICs. A customized method for detecting multiple objects in a workspace and extracting data for classifier input was developed. Low resolution (30 × 30-pixel) grayscale images are input into the classifiers. This reduces the complexity of the classifiers and inherently requires lower computational power (processor and memory use). Three types of classifiers were investigated: SNN, SVM + PCA, and CNN. After training and testing the classifiers on the dataset, the best classifiers were put to test in the real world. As expected, the SNN classifier achieved lowest overall accuracy (93.5% in dataset and 85.6% in real word). This was followed by the SVM + PCA classifier with 20 components (94.6% in dataset and 95.2% in real world). The best accuracy was achieved with the CNN classifier (98.4% in dataset and 98.1% in real world). The accuracy of the CNN classifier is comparable to other relevant deep learning models.

Future work will involve extending this detection and classification method to other electronic parts. This will require increasing the dataset size for each component. The size of the input image to the classifier is currently limited by the resolution of the camera (960 × 720 pixels). In addition to this, the pick and place of objects detected via the developed object detection algorithm is being implemented.

**Author Contributions:** Conceptualization, P.C. and S.L.; methodology, P.C.; software, P.C.; validation, P.C., and S.L.; formal analysis, P.C.; investigation, P.C.; resources, P.C.; data curation, P.C.; writing—original draft preparation, P.C.; writing—review and editing, P.C. and S.L.; visualization, P.C.; project administration, P.C.; funding acquisition, P.C. All authors have read and agreed to the published version of the manuscript.

**Funding:** This research received partial funding from the Waikato Institute of Technology Contestable Fund.

**Data Availability Statement:** Datasets used in this study are available from Praneel Chand via email: praneelchand10@yahoo.co.nz.

**Acknowledgments:** The author greatly appreciates the assistance of Niryo and The Brainary for their technical support.

**Conflicts of Interest:** The authors declare no conflict of interest.

## References

1. Arruda, E.H.; Melatto, R.A.P.B.; Levy, W.; Conti, D.d.M. Circular economy: A brief literature review (2015–2020). *Sustain. Oper. Comput.* **2021**, *2*, 79–86. [CrossRef]
2. Chand, P.; Sepulveda, J. Automating a Festo Manufacturing Machine with an Allen-Bradley PLC. *J. Mechatron. Robot.* **2021**, *5*, 23–32. [CrossRef]

3. Chand, P.; Foulkes, M.; Kumar, A.; Ariyarathna, T. Using Simulated Work-Integrated Learning in Mechatronics Courses. In Proceedings of the 2021 IEEE International Conference on Engineering, Technology & Education (TALE), Wuhan, China, 5–8 December 2021; pp. 1–6.
4. MathWorks. Object Recognition—3 Things You Need to Know. Available online: https://au.mathworks.com/solutions/image-video-processing/object-recognition.html (accessed on 9 September 2022).
5. Janiesch, C.; Zschech, P.; Heinrich, K. Machine learning and deep learning. *Electron. Mark.* **2021**, *31*, 685–695. [CrossRef]
6. Sharma, N.; Sharma, R.; Jindal, N. Machine Learning and Deep Learning Applications-A Vision. *Glob. Transit. Proc.* **2021**, *2*, 24–28. [CrossRef]
7. Sarker, I.H. Deep Learning: A Comprehensive Overview on Techniques, Taxonomy, Applications and Research Directions. *SN Comput. Sci.* **2021**, *2*, 420. [CrossRef]
8. Shalev-Shwartz, S.; Ben-David, S. *Understanding Machine Learning: From Theory to Algorithms*; Cambridge University Press: New York, NY, USA, 2014; p. 410.
9. Goodfellow, I.; Bengio, Y.; Courville, A. *Deep Learning*; MIT Press: Cambridge, MA, USA, 2016; p. 800.
10. Hegde, R.B.; Prasad, K.; Hebbar, H.; Singh, B.M.K. Comparison of traditional image processing and deep learning approaches for classification of white blood cells in peripheral blood smear images. *Biocybern. Biomed. Eng.* **2019**, *39*, 382–392. [CrossRef]
11. Khan, S.; Sajjad, M.; Hussain, T.; Ullah, A.; Imran, A.S. A Review on Traditional Machine Learning and Deep Learning Models for WBCs Classification in Blood Smear Images. *IEEE Access* **2021**, *9*, 10657–10673. [CrossRef]
12. Lai, Y. A Comparison of Traditional Machine Learning and Deep Learning in Image Recognition. *J. Phys. Conf. Ser.* **2019**, *1314*, 012148. [CrossRef]
13. Reza, M.A.; Chen, Z.; Crandall, D.J. Deep Neural Network–Based Detection and Verification of Microelectronic Images. *J. Hardw. Syst. Secur.* **2020**, *4*, 44–54. [CrossRef]
14. Goobar, L. *Machine Learning Based Image Classification of Electronic Components*; KTH: Stockholm, Sweden, 2013.
15. Xu, Y.; Yang, G.; Luo, J.; He, J. An Electronic Component Recognition Algorithm Based on Deep Learning with a Faster SqueezeNet. *Math. Probl. Eng.* **2020**, *2020*, 2940286. [CrossRef]
16. Atik, I. Classification of Electronic Components Based on Convolutional Neural Network Architecture. *Energies* **2022**, *15*, 2347. [CrossRef]
17. Guo, C.; Lv, X.-l.; Zhang, Y.; Zhang, M.-l. Improved YOLOv4-tiny network for real-time electronic component detection. *Sci. Rep.* **2021**, *11*, 22744. [CrossRef]
18. Huang, R.; Gu, J.; Sun, X.; Hou, Y.; Uddin, S. A Rapid Recognition Method for Electronic Components Based on the Improved YOLO-V3 Network. *Electronics* **2019**, *8*, 825. [CrossRef]
19. Jiang, P.; Ergu, D.; Liu, F.; Cai, Y.; Ma, B. A Review of Yolo Algorithm Developments. *Procedia Comput. Sci.* **2022**, *199*, 1066–1073. [CrossRef]
20. Ren, S.; He, K.; Girshick, R.; Sun, J. Faster R-CNN: Towards Real-Time Object Detection with Region Proposal Networks. In Proceedings of the Advances in Neural Information Processing Systems 28 (NIPS 2015), Cambridge, MA, USA, 7–12 December 2015; pp. 1–9.
21. Chigateri, M.K.; Manjuvani, K.M.; Manjunath, K.M.; Moinuddin, K. The Detection of Electrical and Electronics Components using K Nearest Neighbour (KNN) Classification Algorithm. *Int. Res. J. Eng. Technol. (IRJET)* **2016**, *3*, 169–175.
22. Zhang, D.; Han, J.; Cheng, G.; Yang, M.-H. Weakly Supervised Object Localization and Detection: A Survey. *IEEE Trans. Pattern Anal. Mach. Intell.* **2022**, *44*, 5866–5885. [CrossRef]
23. Zhao, W.; Zhang, J.; Li, L.; Barnes, N.; Liu, N.; Han, J. Weakly Supervised Video Salient Object Detection. In Proceedings of the 2021 IEEE/CVF Conference on Computer Vision and Pattern Recognition (CVPR), Nashville, TN, USA, 20–25 June 2021; pp. 16821–16830.
24. Zhang, N.; Han, J.; Liu, N.; Shao, L. Summarize and Search: Learning Consensus-aware Dynamic Convolution for Co-Saliency Detection. In Proceedings of the 2021 IEEE/CVF International Conference on Computer Vision (ICCV), Montreal, QC, Canada, 10–17 October 2021; pp. 4147–4156.
25. Kumar, R.; Lal, S.; Kumar, S.; Chand, P. Object detection and recognition for a pick and place Robot. In Proceedings of the Asia-Pacific World Congress on Computer Science and Engineering, Nadi, Fiji, 4–5 November 2014; pp. 1–7.
26. Chand, P. Investigating Vision Based Sorting of Used Items. In Proceedings of the 2022 IEEE International Conference on Artificial Intelligence in Engineering and Technology (IICAIET), Kota Kinabalu, Malaysia, 13–15 September 2022; pp. 1–5.
27. Niryo. NED User Manual. Available online: https://docs.niryo.com/product/ned/v4.0.0/en/index.html (accessed on 30 June 2022).
28. Open-Robotics. ROS Melodic Morenia. Available online: https://wiki.ros.org/melodic (accessed on 19 September 2022).
29. Mathworks. *ROS Toolbox User's Guide*; The Mathworks, Inc.: Natick, MA, USA, 2022; p. 624.
30. Chand, P. Developing a Matlab Controller for Niryo Ned Robot. In Proceedings of the 2022 International Conference on Technology Innovation and Its Applications (ICTIIA), Tangerang, Indonesia, 23–25 September 2022; pp. 1–5.
31. Nixon, M.S.; Aguado, A.S. *Feature Extraction and Image Processing for Computer Vision*, 4th ed.; Elsevier Academic Press: London, UK, 2019; p. 650.
32. Møller, M.F. A scaled conjugate gradient algorithm for fast supervised learning. *Neural Netw.* **1993**, *6*, 525–533. [CrossRef]

33. Dietterich, T.G.; Bakiri, G. Solving Multiclass Learning Problems via Error-Correcting Output Codes. *J. Artif. Intell. Res.* **1995**, *2*, 263–286. [CrossRef]
34. Cortes, C.; Vapnik, V. Support-vector networks. *Mach. Learn.* **1995**, *20*, 273–297. [CrossRef]
35. Wold, S.; Esbensen, K.; Geladi, P. Principal component analysis. *Chemom. Intell. Lab. Syst.* **1987**, *2*, 37–52. [CrossRef]
36. Guo, Q.; Wu, W.; Massart, D.L.; Boucon, C.; de Jong, S. Feature selection in principal component analysis of analytical data. *Chemom. Intell. Lab. Syst.* **2002**, *61*, 123–132. [CrossRef]
37. Song, F.; Guo, Z.; Mei, D. Feature Selection Using Principal Component Analysis. In Proceedings of the 2010 International Conference on System Science, Engineering Design and Manufacturing Informatization, Yichang, China, 12–14 November 2010; pp. 27–30.
38. Alzubaidi, L.; Zhang, J.; Humaidi, A.J.; Al-Dujaili, A.; Duan, Y.; Al-Shamma, O.; Santamaría, J.; Fadhel, M.A.; Al-Amidie, M.; Farhan, L. Review of deep learning: Concepts, CNN architectures, challenges, applications, future directions. *J. Big Data* **2021**, *8*, 53. [CrossRef]
39. Mathworks. Create Simple Deep Learning Network for Classification. Available online: https://au.mathworks.com/help/deeplearning/ug/create-simple-deep-learning-network-for-classification.html (accessed on 5 October 2022).
40. Chand, P. Low Resolution Used Electronics Parts Image Dataset for Sorting Application. *Data*, 2022; *under review*.

Article

# AIDM-Strat: Augmented Illegal Dumping Monitoring Strategy through Deep Neural Network-Based Spatial Separation Attention of Garbage

Yeji Kim and Jeongho Cho *

Department of Electrical Engineering, Soonchunhyang University, Asan 31538, Republic of Korea
* Correspondence: jcho@sch.ac.kr

**Abstract:** Economic and social progress in the Republic of Korea resulted in an increased standard of living, which subsequently produced more waste. The Korean government implemented a volume-based trash disposal system that may modify waste disposal characteristics to handle vast volumes of waste efficiently. However, the inconvenience of having to purchase standard garbage bags on one's own led to passive participation by citizens and instances of illegally dumping waste in non-standard plastic bags. As a result, there is a need for the development of automatic detection and reporting of illegal acts of garbage dumping. To achieve this, we suggest a system for tracking unlawful rubbish disposal that is based on deep neural networks. The proposed monitoring approach obtains the articulation points (joints) of a dumper through OpenPose and identifies the type of garbage bag through the object detection model, You Only Look Once (YOLO), to determine the distance of the dumper's wrist to the garbage bag and decide whether it is illegal dumping. Additionally, we introduced a method of tracking the IDs issued to the waste bags using the multi-object tracking (MOT) model to reduce the false detection of illegal dumping. To evaluate the efficacy of the proposed illegal dumping monitoring system, we compared it with the other systems based on behavior recognition. As a result, it was validated that the suggested approach had a higher degree of accuracy and a lower percentage of false alarms, making it useful for a variety of upcoming applications.

**Keywords:** waste disposal; object detection; multi-object tracking; articular point; garbage bag

**Citation:** Kim, Y.; Cho, J. AIDM-Strat: Augmented Illegal Dumping Monitoring Strategy through Deep Neural Network-Based Spatial Separation Attention of Garbage. *Sensors* **2022**, *22*, 8819. https://doi.org/10.3390/s22228819

Academic Editor: José María Martínez-Otzeta

Received: 27 October 2022
Accepted: 13 November 2022
Published: 15 November 2022

**Copyright:** © 2022 by the authors. Licensee MDPI, Basel, Switzerland. This article is an open access article distributed under the terms and conditions of the Creative Commons Attribution (CC BY) license (https://creativecommons.org/licenses/by/4.0/).

## 1. Introduction

Economic and social progress in the Republic of Korea resulted in an enhanced standard of living, which subsequently led to enormous amounts of waste from enriching consumer goods. A significant societal issue is created by this rise in garbage levels, which also harms the environment [1]. Additionally, used-up household items, garbage, and construction waste produce foul odors and pollutants, ruining the urban landscape and threatening citizens' health. To address this issue and develop a clean, garbage-less environment, the government implemented a volume-rate waste disposal system in 1995. The new program has a pricing model that enables people to bear a volume-rate cost from their garbage to voluntarily reduce waste and maximize the separate disposal of recyclable items, in contrast to the existing program that imposed incremental fees based on the sizes of houses or the rate of property tax [2].

Waste eligible for volume-rate disposal corresponds to municipal waste generated by households and small enterprises. Standardized volume-based bags must be purchased to dispose of waste. As a motivation for minimizing a pollutant's effect on health and the environment and an economic incentive to improve optimal waste disposal and increase knowledge of the citizens, the volume-based garbage disposal system aims to convey a need for the reduction of illegal garbage dumping and the cooperation and participation [3]. The method can lessen the burden and cost associated with gathering, moving, and processing waste. However, regular instances of illegal rubbish dumping are caused by the bother of having to purchase conventional garbage bags on one's own and the challenging process of handling enormous waste. The uncovered cases of illegal garbage dumping in Seoul went from 99,098 in 2014 to 128,144 in 2020, revealing a year-on-year increase, and it is one of the numerous social problems that must be overcome [4]. Notable in particular are the rising instances of unlawful rubbish disposal in non-standard bags, such as white disposable delivery plastic bags or black disposable plastic bags, as more take-out food deliveries take place. Such illicit dumping is steadily increasing in the absence of aggressive prosecution, necessitating different measures.

Watchpersons or government officials patrol to find illegal dumping situations occasionally, but such efforts need a larger labor force in wide areas. The recently installed closed-circuit television (CCTV) in locations with a concentration of unlawful dumping contains video recordings. However, the lack of manpower to conduct ongoing surveillance or analyze every single film makes it difficult to bring charges for illegal dumping [5]. Another comparable technique employs CCTV and human body identification sensors to send out an audio warning to onlookers to promote awareness, but the alert does not reveal illegal dumping; it causes noise disturbances due to the frequent pointless broadcasts. This approach may temporarily frighten illegal dumpers psychologically but has limited impacts in ending illegal dumping. Figure 1 depicts the illegal dumping monitoring system that is now in use with the CCTV and audio broadcasts as being surrounded by various forms of unlawfully placed rubbish. This demonstrates the limitations of the current illegal dumping monitoring system despite significant initial investment in the system.

**Figure 1.** Limitations of the current illegal dumping monitoring system.

Recently proposed methods combine deep-learning object detection technology widely in use with camera-based monitoring to monitor illegal dumping. The new approach can address the limitations of the existing methods requiring significant manpower and have the benefit of reducing unnecessary noise by enhancing false alarm rates. Min and Lee [6] proposed a way of catching illegal garbage dumping using a deep neural network trained on the joints of persons that are collected by image processing. By separating dumping postures from the other non-dumping postures, their system determines whether dumping is legal or illegal. Bae et al. [7] used the real-time object detection model, You Only Look Once (YOLO), to learn about the illegal dumping operation itself and to create zones for observation and non-observation in order to lower the system's false alert rate. The trained model detects an act of dumping and then identifies it as illegal only when the coordinates of the activities are within the observation zone. Jeong et al. [8] used the Gaussian Mixture Model to examine object changes that are based on histogram differences. Their suggested approach is based on the idea that at the point of dumping, there is a divide between the dumper and the trash. Kim et al. [9] proposed a system that detects illegal dumping using probabilistic analysis of the object trajectory.

As a result, several techniques exist to track unlawful dumping using object detection and video analysis technologies based on convolutional neural networks (CNNs), as well as detecting sensors. Nevertheless, Refs. [6,7] consider an act of dumping as illegal when a non-dumping posture is similar to a dumping posture, even in the absence of garbage in hand, thus raising frequent false alarms. Therefore, Refs. [7,9] designated an observation zone for illegal dumping. As a result, their system cannot detect illegal dumping when it occurs outside of the surveillance zone and is susceptible to numerous missed detections. Therefore, Refs. [6–9] merely identifies characteristic changes of a dumper or only differentiates standard or non-standard garbage bags, which may raise a false alarm even when garbage is in a standard bag, all of which are issues still to be addressed. As a result, a more comprehensive monitoring system for unlawful dumping is required, one that goes beyond the dumping acts itself or isolated, small surveillance zones.

This study suggests a strategy of augmented illegal dumping monitoring (AIDM) that determines the distance between the dumper's wrist and the garbage bag. To estimate the dumper's wrist joint, Single Person Pose Estimation, which is a method for estimating spatial dependence combinations between body parts, is required and is largely divided into a tree-structured graphical model [10,11] and a non-tree model [12,13]. Afterward, CNN was applied to increase the reliability of joint estimation [14,15]. However, when two people are detected on one screen, the precise joint of each person cannot be extracted, so research on Multi-Person Pose Estimation [16,17] has been actively conducted. Among them, the OpenPose [18] model has been used in many fields and introduced in this study because it extracts joint points at a relatively high speed, and the amount of computation does not increase significantly even if the number of people increases.

The proposed method uses the OpenPose model [18] that can determine the articulation points of a person to extract the wrist joint and then uses the YOLO method [19] to classify four types of garbage bags. Additionally, to reduce errors from the unwarranted calculation of the distance of the wrist joint to the already dumped garbage bag or the issue of not identifying the same garbage due to the change in frames, we implement a Simple Online Realtime Tracking with A Deep Association Metric (DeepSORT) [20] that can keep track of multiple objects for tracking the garbage bag identifiers (IDs). We suggest an algorithm that can identify illegal dumping by keeping track of garbage bags that have already been dumped and those that are still to be dumped separately and deciding when the distance between the dumper's wrist and the bag of trash is more than a certain threshold. The test findings demonstrate that our method of determining illegal dumping based on the distance of the actual dumper's wrist to the garbage bag has better efficacy than other recently published methods that are based on behavior recognition or dumping zone designation. This research has the following contributions:

- With improved detection performance, the proposed monitoring system for illegal dumping can reduce noises caused by unnecessary audio guidance due to the inaccuracies of the existing illegal dumping broadcasting system;
- Using the object detection model, YOLO can differentiate the standard bags that are legal for garbage dumping and the other non-standard bags. Also, the proposed technique can minimize errors of falsely recognizing dumping-like behavior as illegal dumping through OpenPose, which can extract the articulation points;
- Our suggested method tracks the objects throughout the entire video without the use of specifically designated observation zones to evaluate whether illegal dumping happened;
- By introducing the object tracking model DeepSORT, we give IDs to already dumped garbage and garbage held in a dumper's hand and track the objects to detect illegal dumping, thus lowering the missed detection rate.

In this Section, we discussed the need for an illegal dumping monitoring system and the goal of the study. In Section 2, we introduce the components of our illegal dumping monitoring system. In Section 3, we describe the design process of the proposed system. In Section 4, we describe the experimental conditions, testing, and results for the evaluation of the proposed system's performance. In the last section, Section 5, we conclude our research.

## 2. Materials and Methods

The monitoring system for illegal dumping that is presently in operation cannot decide on a dumping act itself for the illegality, and thus the impact is not as high as expected concerning the investment for the system implementation. Additionally, the systems that were recently designed using research on illicit dumping practices as the subject are highly susceptible to the probability of mistakenly associating suspicious behavior with unlawful dumping. As a result, we propose an improved monitoring system that identifies illegal dumping by classifying the types of garbage bags and estimating the distance of the dumper's wrist to the garbage bag, as shown schematically in Figure 2. The object detector recognizes and classifies the rubbish bag while concurrently extracting a person's joints from the input image. The object detector then begins tracking the garbage-classified object. Then, it continuously calculates the distance of the extracted wrist joint to the object detected as the non-standard bag. The dumping is considered unlawful if the distance exceeds a certain level.

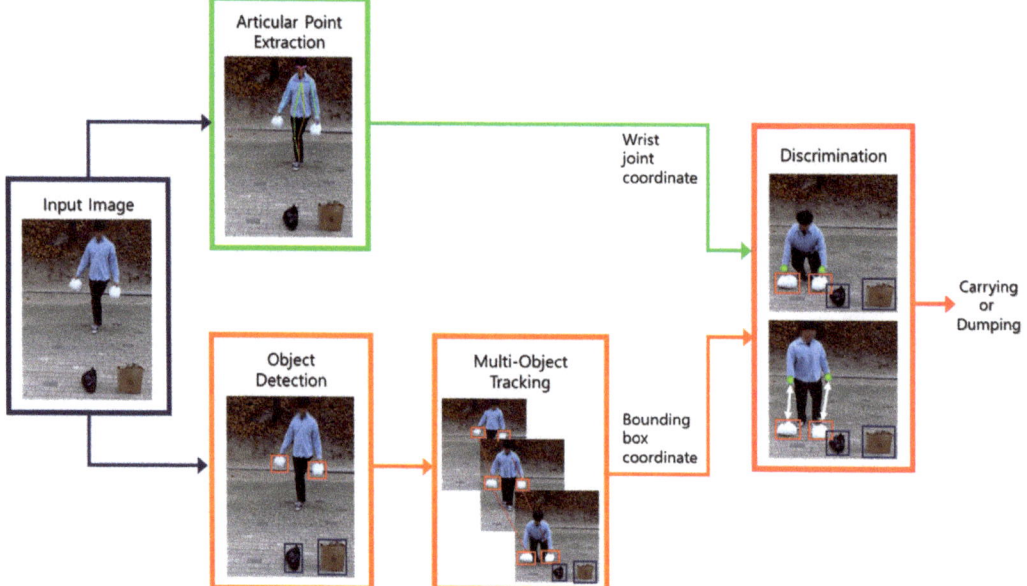

**Figure 2.** Schematic diagram of the proposed monitoring system for illegal dumping.

### 2.1. Articular Point Extraction

Deep-learning-based posture estimation is done in two ways: the top-down technique for first finding an area with a person in it and then determining the posture in the area, and the bottom-up method for estimating the posture from the characteristic points of a human body without finding a person. To predict the joints, we employed the bottom-up OpenPose model in this study. Fast joint extraction is possible with OpenPose, and if more individuals are added, the calculation volume does not considerably rise, making it appropriate for crowded areas [21].

Based on a CNN, OpenPose infers characteristic points as joints and delineates them and uses VGGNet to enhance learning efficacy by extracting features of a wider area with fewer parameters. VGGNet creates a feature map $F$, which goes through a multilayered convolution branch $\alpha$ to create a confidence map $\gamma$ representing the positions of the joints and goes through another branch $\beta$ to create an affinity field $\delta$ indicating associations (location and direction) between body parts. The model first trains $\delta$ to obtain optimal $\delta$ predictions, which are then used to train $\gamma$. $\delta^t$ and $\gamma^t$ in the $t$th step are iterated by the respective branch $\beta^t$ up to the $T_A$ step, and then iterated by the branch $\alpha^t$ up to the $T_A + T_B$ Step, as summarized below [22]:

$$\delta^t = \beta^t\left(F, \delta^{t-1}\right), 2 \leq t \leq T_A \qquad (1)$$

$$\gamma^t = \alpha^t\left(F, \delta^{T_A}, \gamma^{t-1}\right), T_A < t \leq T_A + T_B \qquad (2)$$

$\delta$ and $\gamma$ obtained in each step are used to match a person's arms and legs. The points finally determined as the person's arms and legs are connected to extract the joints of the body. OpenPose learns through a loss function $f$ composed of an objective function $f_\delta^t$ for the joint associations and another objective function $f_\gamma^t$ for the locations of the articular points. The object functions are as follows:

$$f_\delta^t = \sum_{v=1}^{V} \sum_{P} W(P) \|\delta_v^t(P) - \delta_v^*(P)\|_2^2 \qquad (3)$$

$$f_\gamma^t = \sum_{r=1}^{R} \sum_{P} W(P) \|\gamma_r^t(P) - \gamma_r^*(P)\|_2^2 \qquad (4)$$

where $\delta_v^*$ is the ground truth (GT) of the affinity field and $\gamma_r^*$ is the GT of the confidence map; $R$ is the number of confidence maps corresponding to the number of the joints, and $V$ is the number of the two joints connected; $W$ is a binary mask for the GT and set as zero (0) when the pixel $P$ has no GT for the joint to avoid adverse effect on true positive predictions. The loss function $f$ is the sum of $\delta$ losses incurred from the first step to $T_A$ step and the sum of $\gamma$ losses incurred from $T_A + 1$ step to $T_A + T_B$ step, as shown below:

$$f = \sum_{t=1}^{T_A} f_\delta^t + \sum_{t=T_A+1}^{T_A+T_B} f_\gamma^t \qquad (5)$$

Finally, the model outputs $\gamma$ that contains the location of the articular point. If $\gamma$ has multiple similar peak values around the articular point, the non-maximum suppression [23] is used to identify the highest peak value at the articular point.

*2.2. Object Detection*

One of the areas of study in computer vision is object detection technology, which is employed to automatically operate and adjust particular devices. The detection involves classification and localization. In classification, a single object in the image is classified with class probabilities, and localization is a process of determining the location of the object. Object detection methods are largely divided into two-stage detectors and one-stage detectors. The two-stage detector conducts the localization and classification sequentially to obtain the results. In the first stage, the area where an object is likely to present is inferred quickly through the regional proposal. In the second stage, the classification identifies the type of object. The major models include Regions with CNN (R-CNN) [24], Fast R-CNN [25], Faster R-CNN [26], and Mask R-CNN [27]. The two-stage detectors generally have higher accuracy but slow speed.

Unlike the two-stage detector, which performs two processes sequentially, the one-stage detector produces results faster by conducting localization and classification concurrently. The main models include YOLO [19] and Single Shot Multibox Detector (SSD) [28]. YOLO, in particular, significantly enhances the speed of two-stage detectors and can estimate the class probability and the bounding box simultaneously, making it frequently utilized in real-time processing. Furthermore, the training process traverses the full image, learning not just the characteristics of individual objects but also the overall context of the image, resulting in exceptional performance when extending to additional locations. The following steps are taken during the training [19,29,30]:

After dividing the input image into $S \times S$ grid areas, its characteristics are extracted using the convolutional layer, and the prediction tensor is created through the fully connected layer. Each grid cell is represented by the B number of the bounding boxes, each of which has the corresponding confidence score (CS). The bounding box has information about

$(x, y, w, h, CS)$, with $(x, y)$ being the centroid coordinates of the bounding box and $(w, h)$ being its width and height. $CS$ is the probability of the object being within the bounding box and shows whether the class is correctly predicted, as shown below:

$$CS = \Pr(Obj) \times \text{IoU}_{PB}^{GT} \tag{6}$$

where $\Pr(Obj)$ is the probability of the object being within the bounding box; $\text{IoU}_{PB}^{GT}$, the intersection over the union (IoU), shows the extent to which GT matches the box (PB) determined by the model and corresponds to the overlapping area of the actual value and the predicted value, as shown below:

$$\text{IoU}_{PB}^{GT} = \frac{PB \cap GT}{PB \cup GT} \tag{7}$$

The conditional probability $P_{Class}$ indicating which class multiple objects included in the bounding box belong to and the Class-specific Confidence Score (CCS) indicating the probability that the object is contained within the bounding box area, and it matches with the actual value of the classified object are expressed as follows:

$$P_{Class} = \Pr(Class_i | Obj) \tag{8}$$

$$CCS = CS \times P_{Class} = \Pr(Class) \times \text{IoU}_{PB}^{GT} \tag{9}$$

As shown above, the bounding box with the highest $CCS$ is finally chosen as the bounding box for the given object among the B number of the bounding boxes predicted.

### 2.3. Object Tracking

Multi-object tracking (MOT) [31–33] is a technique for tracking the locations of numerous objects in a video in real time. It first assigns a unique identifier (ID) to each identified object to track its movement by comparing the previous frame and the current frame. Major MOT methods include Simple Online and Realtime Tracking (SORT) [34] and DeepSORT [20]. SORT is a tracking method to analyze only the degree of similarity of the association between objects using only the information of the objects detected in the current frame and the previous frame of the image. However, if the item is obstructed by a barrier during the object tracking, it cannot be identified as the same object indefinitely and thus obtains a new ID that differs from the ID previously assigned. Furthermore, the movement of multiple objects instead of one causes frequent ID-switching, which hinders smooth tracking [20,35].

As an extension of SORT, DeepSORT has object detection, the Kalman filter-based estimation as well as the matching cascade that uses a deep-learning feature Re-ID, and thus addresses the drawbacks of SORT, that is, unstable to occlusion or ID-switching [20]. The Kalman filter is used to update the identified object by estimating its location in the future frame using information from the previous frame. Then, to match the identified object, DeepSORT utilizes the Mahalanobis distance, which gives an object's location based on the movement effective for short-term prediction, and the cosine distance that uses the object's appearance for the long-term signaling block followed by the recovery of its identity. We determine the Cost Matrix $D_{CM}$ as the weighted mean of the Mahalanobis distance $D_{MA}$ and the cosine distance $D_{Cos}$ for the calculation of the similarity matrix.

$$D_{CM} = \rho \times D_{MA} + (1 - \rho) \times D_{Cos} \tag{10}$$

where $\rho$ is a hyperparameter used to control the matrix impact; when the camera motion is large, it is set $\rho = 0$, using $D_{Cos}$ only. Then, the IoU matching is performed on the tracks and detections that are not related. The IoU matching process uses three states to obtain information about continuous tracking: matched tracks for objects being tracked continuously, unmatched detections for designating a recently appeared object as the final object, and unmatched tracks for designating an object's temporary status when the tracked object cannot be found, and the tracking cannot continue.

## 3. Proposed Architecture Design

In this section, we describe the detailed procedure for designing the proposed monitoring system that detects illegal dumping based on the distance between the potential dumper's wrist joints found using OpenPose and the garbage bag location obtained through YOLO and DeepSORT. The block diagram in Figure 3 shows the system schematically.

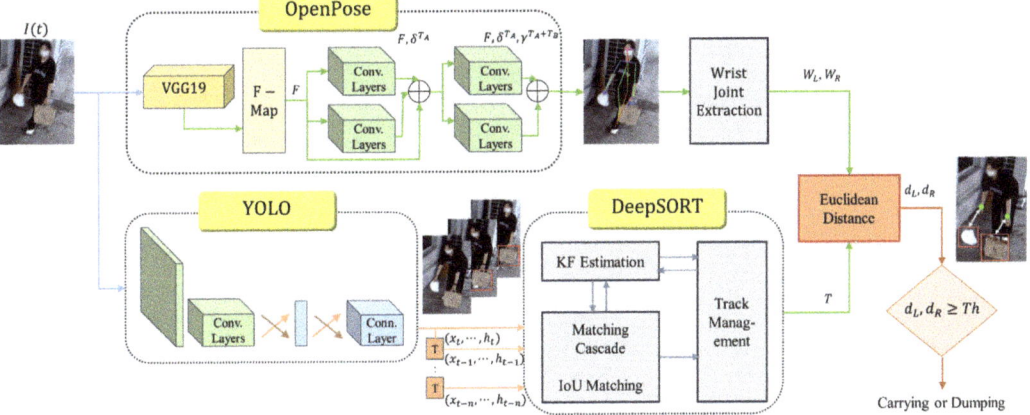

**Figure 3.** Block diagram of the proposed illegal dumping monitoring system.

### 3.1. Extraction of the Articular Points of the Wrist Using OpenPose

The articular points of the person's wrist are retrieved from the video $I(t)$ of a possible dumper walking into the observation zone while holding the trash. To accomplish this, we input the given image to VGG-19 in OpenPose to generate a feature map, which is then used to generate a confidence map $\gamma$ for displaying the locations of the joints and an affinity field $\delta$ for demonstrating the correlation between the body parts. As we detect illegal dumping based on the point in time when a part of the extracted joints separates from the garbage, in the case of the finger closest to the trash, the next closest wrist joint is selected because the joint coordinates cannot be extracted when the finger is often obscured by other objects. As a result, of the 18 joint coordinates that are retrieved, we only use the elbow and shoulder that are connected to the wrist, and we disregard the remaining 12 coordinates that are beyond the area of interest. The three joints of the shoulder, elbow, and wrist are displayed on the screen in a state where the left arm and the right arm are separated. Then, the joint coordinates of the left wrist $W_L$ and the joint coordinates of the right wrist $W_R$ are finally estimated.

### 3.2. Tracking the Garbage Bag Using YOLO and DeepSORT

To identify the garbage bag held by the potential dumper, we employ the real-time object detection model YOLO to obtain the bounding box $(x, y, w, h)$ of the garbage bag as the identified object. Then, from the bounding box, we extract the top centroid $T(t_1, t_2)$, which can be expressed as $t_1 = x + \frac{w}{2}$, $t_2 = y$. Furthermore, to identify illegal dumping in real time, we employ DeepSORT to determine whether the object in the previous frame $I(t-1)$ and the object in the current frame $I(t)$ are the same. Here, the Kalman filter, the matching cascade, and the IoU matching [20] are conducted recursively to determine the similarity between each object. Using three states, the matched tracks for the objects being tracked continuously, the unmatched detections for designating a newly discovered object as the final object, and the unmatched tracks for designating a temporary status to the object when the tracked object is not found and the tracking cannot continue, the IoU matching finally defines an ID to the object. Here, the ID contains the types of detected objects and the order (Class name, Class number) that the objects are detected. This enables the continuous recognition of the same garbage bag even when it is occluded by other obstacles. Moreover, it is possible to suppress the ID switching that may occur due to the movement of multiple garbage bags instead of one garbage bag. Accordingly, even if the detected garbage bag is dumped, it can be made to have the same ID, making a judgment on illegal dumping possible.

*3.3. Discriminator for the Determination of Illegal Dumping*

As described above, to determine the illegality of the garbage bag held by the potential dumper, we compute the Euclidean distance between the wrist joint coordinates $W_L(w_{L,1}, w_{L,2})$ and $W_R(w_{R,1}, w_{R,2})$ obtained from OpenPose and the top centroid $T(t_1, t_2)$ of the bounding box obtained from YOLO, as shown below:

$$d_L(T, W_L) = \sqrt{(t_1 - w_{L,1})^2 + (t_2 - w_{L,2})^2} \qquad (11)$$

$$d_R(T, W_R) = \sqrt{(t_1 - w_{R,1})^2 + (t_2 - w_{R,2})^2} \qquad (12)$$

As the final step, we check if the $d_L$ or $d_R$ that are calculated per frame exceeds the pre-defined threshold $Th$ to evaluate whether the garbage bag being tracked is dumped illegally. When $d_L$ and $d_R$ are below the threshold, we set the object ID to 1 to indicate that the potential dumper has the garbage bag. The ID remains 1 while every frame is examined until the point of garbage bag dumping. By contrast, for the garbage bags that are dumped already, $d_L$ and $d_R$ both surpass the threshold. As a result, we set the object ID to zero (0) to indicate that the garbage bag is not held by the dumper. Thus, immediately after the garbage bag is dumped, that is, when $d_L$ or $d_R > Th$, a judgment is made that the object is dumped, the ID changes from 1 to 0, and the alarm goes off. Furthermore, as the already dumped garbage bags are detected and set to 0, they are not falsely identified as those being held by the dumper even when the dumper's wrist gets close to the garbage bag.

## 4. Experimental Results

To assess the performance of the proposed illegal dumping monitoring method, we took into account eight scenarios that were similar to actual instances of illegal dumping, including garbage dumping by one hand, dumping by both hands, garbage dumping without bending the waist, and dumping yet to have occurred with the garbage in the dumper's hand. We then gathered the data for these cases. Furthermore, to determine the performance against the existing garbage dumping monitoring techniques, we included the approach [7] that learns the dumping postures to decide on illegal dumping and the method, Post+det, that learns the dumping postures as well as the garbage bags. There were a total of eight situations included in the performance test.

*4.1. Experimental Environment*

The proposed illegal garbage dumping monitoring system was implemented by NVIDIA GeForce GTX 1060 Ti and Intel Core i7-8700 CPU. To train YOLOv4 for real-time object detection, we collected illegal dumping films for each situation using a Logitech C920 PRO HD. The dataset includes videos of the simulation of actual illegal dumping scenes, with 30 videos of about 10 s for each scenario.

Commonly dumped garbage includes black plastic bags, white plastic bags, and paper bags containing general garbage, as well as volume-based bags that are recommended to be used. We selected four types of bags that are dumped the most, as shown in Figure 4a, to simulate actual dumping scenes under the environment in Figure 4b. We labeled the black plastic bag trashBLK, the white plastic bag trashWHT, the paper bag trashPBG, and the standard bag trashAUT. For the YOLOv4 training, we utilized a total of 12,891 images, with the image size set to 608 × 608, the batch size to 8, and the maximum number of batch learning to 15,000. There may be several items in a single photograph. There are 13,186, 16,147, 15,611 and 11,711 trashBLK, trashWHT, trashPBG, and trashAUT in all of the photos, respectively.

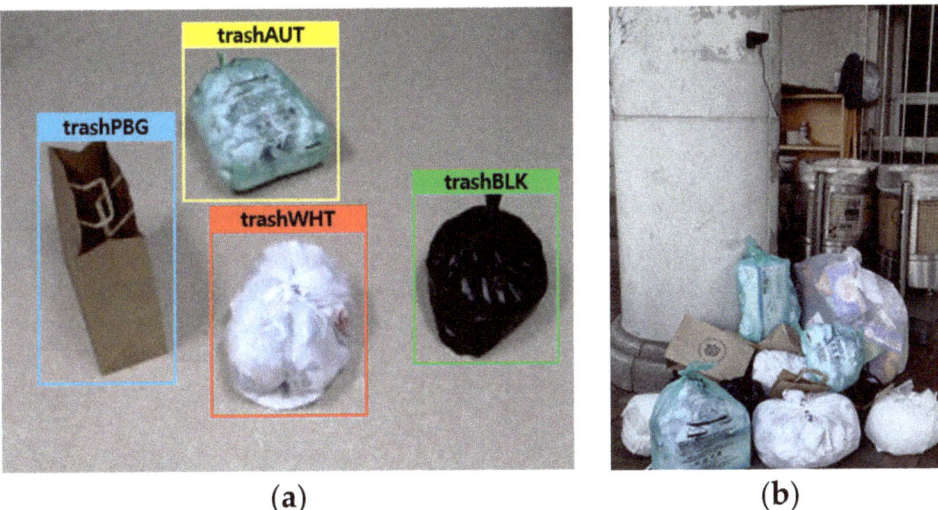

**Figure 4.** The environment for the collection of illegal dumping data and scenario-based evaluation; (a) standard and non-standard garbage bags used in the training and the evaluation; (b) data collection environment.

*4.2. Evaluation of Object Detection Performance*

We used the average precision (AP) as a performance indicator for assessing the performance of the object detection model YOLOv4, which is trained on the different types of collected garbage bags. To denote the model's performance as a single numerical value, we utilized the precision-recall curve and the accuracy to evaluate the confidence of the object identified by the model. Precision is the rate of the correctly detected objects among the detected objects, recall is the rate of the detected objects among all the objects that should be detected, and accuracy is the rate of the correctly detected objects among all the objects, as demonstrated below [6]:

$$\text{Precision} = \frac{TP}{TP + FP} \quad (13)$$

$$\text{Recall} = \frac{TP}{TP + FN} \quad (14)$$

$$\text{Accuracy} = \frac{TP + TN}{TP + FP + FN + TN} \quad (15)$$

where the True Positive (TP) means the object that should be identified is correctly detected, the False Positive (FP) means the object that should not be detected is wrongly detected, the False Negative (FN) means the object that should be detected is not detected, and the True Negative (TN) means the object that should not be detected is not detected. Seven hundred and ninety-eight images were used to determine object detection, and the results are shown in Table 1.

**Table 1.** Performance of identifying the four types of garbage bags using YOLOv4.

| Class ID | Object | AP | TP | FP |
|---|---|---|---|---|
| 0 | trashBLK | 99.77% | 685 | 36 |
| 1 | trashWHT | 99.53% | 706 | 7 |
| 2 | trashPBG | 98.96% | 823 | 7 |
| 3 | trashAUT | 99.24% | 712 | 8 |

As illustrated in the table, when the IoU is 0.5, the detection performance indicator, AP, for each class is mostly above 99%, while the average indicator, meanAP (mAP), for all classes is 99.38%, indicating that the model can classify all four objects with high accuracy. However, trashBLK indicates a lower precision than the other types of garbage bags due to the occasional false recognition of a person's black hair or shoes.

*4.3. Evaluation of the Illegal Dumping Monitoring Performance*

The data gathered for the evaluation has a total of four types of garbage bags previously described. As shown in Figure 5, we developed eight different dumping scenarios, S1 through S8, which are comparable to real garbage dumps.

**Figure 5.** Eight types of scenarios (S1~S8) for the performance evaluation of the illegal dumping monitoring system; (**a**) S1—a bending posture with no garbage bag; (**b**) S2—a dumping scenario with the non-standard bag in one hand; (**c**) S3—a dumping position with the non-standard garbage bags on both hands; (**d**) S4—a dumping posture with the legal standard garbage bag in hand; (**e**) S5—a dumping scenario with legal standard garbage bags in both hands; (**f**) S6—a dumping posture with the standard bag in one hand and the non-standard bag in the other hand; (**g**) S7—a dumping posture without bending the waist with the non-standard bags in both hands; and (**h**) S8—a bending position without dumping with the non-standard bag.

The proposed AIDM determines illegal dumping based on the distance ($d_L$, $d_R$) between the wrist joints of a dumper and the detected object, not the dumping posture. To achieve this, we established a threshold ($Th$) to 90 cm, taking into account the installation angle and the distance between the camera and the visible object. To verify the utility of the proposed method, we performed a comparison against the existing monitoring techniques: the technique [7] that determines whether illegal dumping has occurred solely based on a dumping posture with the body bent forward, and the technique, Post+det, that monitors illicit dumping through the detection of garbage and dumping postures. The test results are reported in Table 2 in terms of the reliability of the determination of illegality at the site of dumping using the scenarios S1 to S8.

As can be seen from the comparison, [7] recorded a lower accuracy in the scenarios S1, S4, S5, and S7 because it determines whether dumping is legal by learning the shapes of the dumpers rather than the garbage bags, in contrast to the Post+det and the ADIM, which can identify the standard bags that can be legally dumped. Furthermore, the Post+det appears to demonstrate a higher detection performance overall than [7]. However, it occasionally failed to detect suspicious dumping actions, leading to lower accuracy in scenarios S2, S3, and S6. Particularly for S7, it failed to detect anything since the garbage dumping occurred without bending the body. In contrast, the proposed model demonstrated at least

93% accuracy in identifying illegal dumping in all the scenarios, demonstrating that it is a stable illegal dumping monitoring system. On the whole, the average accuracy of [7], the Post+det, and the AIDM for detecting illegal dumping are 0.43, 0.63, and 0.97, respectively. Therefore, it can be said that the proposed AIDM has a more robust and improved detection performance than the existing method.

Table 2. Performance comparison of dumping monitoring models per scenario.

| Test Scenario | Dumping Monitoring Model | | | | | | | | | | | | | | |
|---|---|---|---|---|---|---|---|---|---|---|---|---|---|---|---|
| | [7] | | | | | Post+det | | | | | Proposed AIDM | | | | |
| | TP | TN | FP | FN | Acc. | TP | TN | FP | FN | Acc. | TP | TN | FP | FN | Acc. |
| S1 | 0 | 1 | 29 | 0 | 0.03 | 0 | 30 | 0 | 0 | 1.00 | 0 | 30 | 0 | 0 | 1.00 |
| S2 | 25 | 0 | 0 | 5 | 0.83 | 17 | 0 | 0 | 13 | 0.57 | 28 | 0 | 0 | 2 | 0.93 |
| S3 | 28 | 0 | 0 | 2 | 0.93 | 7 | 0 | 0 | 23 | 0.23 | 30 | 0 | 0 | 0 | 1.00 |
| S4 | 0 | 7 | 23 | 0 | 0.23 | 0 | 28 | 2 | 0 | 0.93 | 0 | 30 | 0 | 0 | 1.00 |
| S5 | 0 | 1 | 29 | 0 | 0.03 | 0 | 25 | 5 | 0 | 0.83 | 0 | 30 | 0 | 0 | 1.00 |
| S6 | 28 | 0 | 0 | 2 | 0.93 | 17 | 0 | 0 | 13 | 0.57 | 29 | 0 | 0 | 1 | 0.97 |
| S7 | 2 | 0 | 0 | 28 | 0.07 | 0 | 0 | 0 | 30 | 0.00 | 29 | 0 | 0 | 1 | 0.97 |
| S8 | 0 | 12 | 18 | 0 | 0.40 | 0 | 28 | 2 | 0 | 0.93 | 0 | 28 | 2 | 0 | 0.93 |
| Average Acc. | | | 0.43 | | | | | 0.63 | | | | | 0.97 | | |

Figure 6 shows the test results for scenario S4, where a legal volume-based waste bag is thrown on one hand. From top to bottom, the results are taken from each time point of T/4-, T/2-, 3T/4-, and T-seconds. At $T/4 \sim T/2$ s, the dumper is shown walking with the garbage in hand to the designated dumping site. In Figure 6a, there is no change since the dumper has to bend his body for the dumping to be detected as such. In Figure 6b, the system found the legal standard bag trashAUT, and in Figure 6c, it concurrently located the person's joints and detected trashAUT. The dumper dropped the trash bag at the 3T/4-s point. [7] detected the dumping posture only and not the type of garbage bag, identifying it as illegal and indicating the red alarm. On the other hand, the Post+det and the AIDM can differentiate the standard bag, showing the green alarm after detecting the dumping action and deeming it legal. The T-second mark is the moment right before the dumper departs the site after dumping the garbage. The alarm was no longer displayed in [7] and the Post+det for garbage dumping as the dumper stopped bending their body, whereas the AIDM kept the green alarm as the garbage bag discarded by the dumper had a unique ID.

Figure 7 additionally demonstrates the test results for scenario S7, where the dumper dumps the non-standard garbage bags without bending their body. Similar to the above instances, at $T/4 \sim T/2$ s, [7] did not identify anything, while the Post+det detected three types of garbage bags, trashBLK, trashWHT, and trashPBG. The AIDM found the person's articular points and, like the Post+det, detected all three types of garbage bags. At the 3T/4-second mark, in which the garbage is dumped, [7] the Post+det failed to detect a dumping action as the dumper did not bend his body. On the other hand, the AIDM identified the non-standard garbage bag and determined that the distance from the wrist to the bag was above the threshold, thus deeming it unlawful and showing the red alarm.

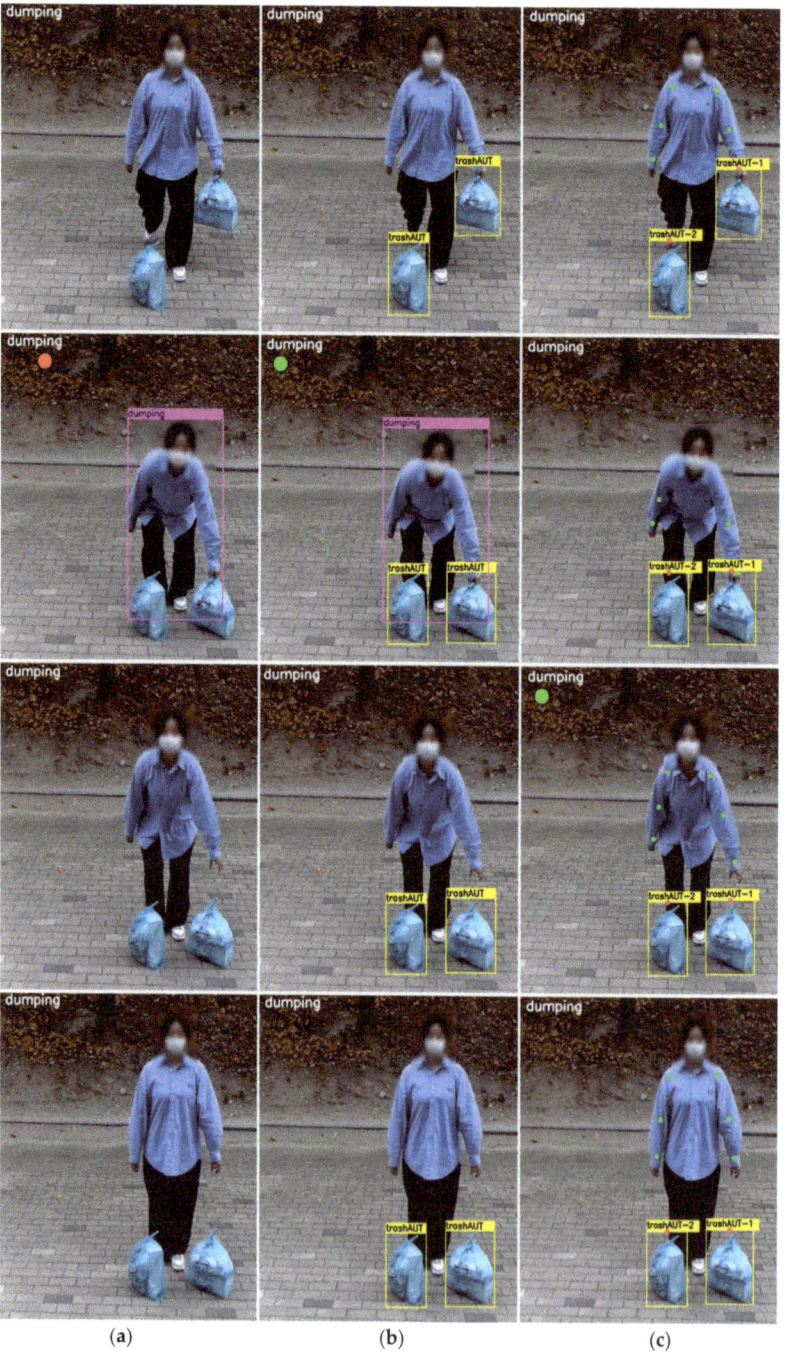

**Figure 6.** Illustrations of the outcomes from the monitoring models for the disposal of the legal standard garbage bag in one hand; (**a**) [7], (**b**) Post+det, and (**c**) Proposed AIDM.

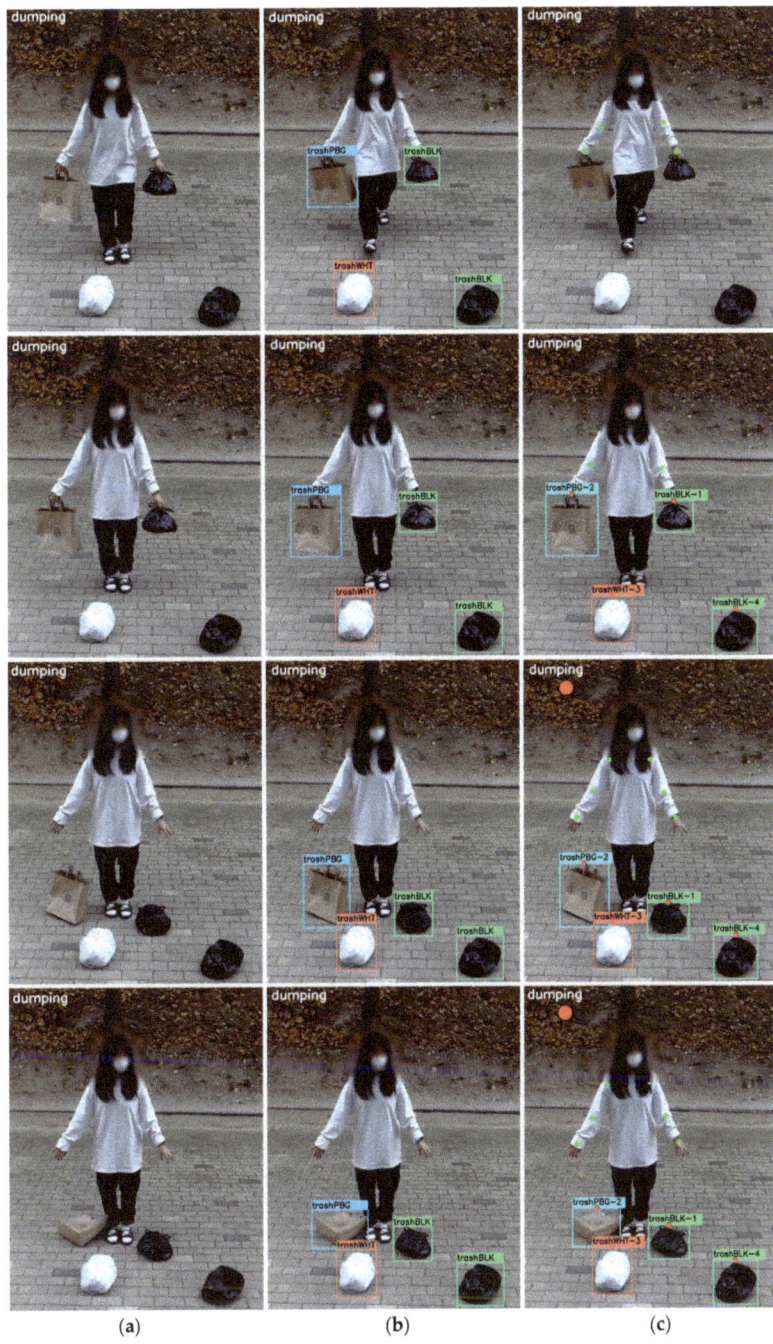

**Figure 7.** Examples of the findings from the monitoring models for rubbish disposal without body bending; (**a**) [7], (**b**) Post+det, and (**c**) Proposed AIDM.

## 5. Conclusions

The government of the Republic of Korea has implemented a volume-based waste disposal system that can change the disposal features to efficiently handle massive amounts of waste. However, illegal dumping often occurs as people dump garbage in disposable black plastic bags or white plastic bags used for food deliveries. Recently, methods have been implemented in areas where illegal garbage dumping occurs to control such behavior by installing closed-circuit television (CCTV) and the transmission of audio warnings using human body detection sensors. Nevertheless, the effect is limited. As a result, numerous actions are necessary since unlawful dumping is constantly growing in the absence of strict prosecution. Therefore, this study suggested a deep neural network-based illegal dumping monitoring technique that can determine the distance between the dumper's wrist and the garbage bag. The proposed technique retrieves the articular points of a dumper using OpenPose and identifies the type of garbage bag through the object detection model YOLO. Furthermore, to reduce false detection of illegal dumping, we introduced a method of tracking the IDs issued to the waste bags using the MOT model. The test results demonstrate that our approach of determining illegal dumping based on the distance of the actual dumper's wrist to the garbage bag has better performance than other recently published methods based on behavior recognition or dumping zone designation. We expect the proposed method to be widely utilized in the future.

**Author Contributions:** Y.K. and J.C. took part in the discussion of the work described in this paper. All authors have read and agreed to the published version of the manuscript.

**Funding:** This research was funded by a National Research Foundation of Korea (NRF) grant funded by the Korean government (MOE) (No. 2021R1I1A3055973) and the Soonchunhyang University Research Fund.

**Institutional Review Board Statement:** Not applicable.

**Informed Consent Statement:** Not applicable.

**Data Availability Statement:** The data presented in this study are available on request from the corresponding author.

**Conflicts of Interest:** The authors declare that they have no competing interests.

## References

1. Park, J. An Evaluation of Volume-Based Waste Collection Fee System. Master's Thesis, Kyunghee University, Seoul, Korea, 2000.
2. Kim, D. A Study on Improvement Plan for Trash Specific Duty. Master's Thesis, Chosun University, Gwangju, Korea, 2006.
3. Kim, J. A Study on the Estimation for Improvement of the Volume based Waste Fee System. Master's Thesis, Chung-Ang University, Seoul, Korea, 2003.
4. Seoul Information Communication Plaza. Performance of Cracking Down on Illegal Dumping of Garbage. Available online: https://opengov.seoul.go.kr/ (accessed on 1 September 2022).
5. Mu, J. A Study on Improving Household Waste Collection Systems. Master's Thesis, Chung-Ang University, Seoul, Korea, 2016.
6. Min, H.; Lee, H. Garbage Dumping Detection System using Articular point Deep Learning. *J. Korea Multimed. Soc.* **2021**, 24, 1508–1517.
7. Bae, C.; Kim, H.; Yeo, J.; Jeong, J.; Yun, T. Development of Monitoring System for Detecting Illegal Dumping Using Deep Learning. In Proceedings of the Korean Society of Computer Information, Jeju, Korea, 16–18 July 2020.
8. Jeong, J.; Kwon, S.; Kim, Y.; Hong, S.; Kim, Y. Development of Illegal Dumping System using Image Processing. In Proceedings of the Korean Institute of Information Scientists and Engineers, Jeju, Korea, 18–20 June 2017.
9. Kim, J.; Kim, H.; Kim, P.; Lee, Y. The Design of Intelligent System for Statistically Determining Illegal Garbage Dumping through Trajectory Analysis. In Proceedings of the Korean Institute of Information Scientists and Engineers, Jeju, Korea, 18–20 June 2017.
10. Ramanan, D.; Forsyth, D.; Zisserman, A. Tracking people and recognizing their activities. In Proceedings of the IEEE Conference on Computer Vision and Pattern Recognition, San Diego, CA, USA, 20–25 June 2005.
11. Yang, Y.; Ramanan, D. Articulated Human Detection with Flexible Mixtures of Parts. *IEEE Trans. Pattern Anal. Mach. Intell.* **2013**, 35, 2878–2890. [CrossRef] [PubMed]
12. Lan, X.; Huttenlocher, D. Beyond trees: Common-factor models for 2D human pose recovery. In Proceedings of the IEEE International Conference on Computer Vision, Beijing, China, 17–21 October 2005.

13. Dantone, M.; Gall, J.; Leistner, C.; Van Gool, L. Human Pose Estimation Using Body Parts Dependent Joint Regressors. In Proceedings of the IEEE Conference on Computer Vision and Pattern Recognition, Portland, OR, USA, 23–28 June 2013.
14. Tompson, J.; Goroshin, R.; Jain, A.; LeCun, Y.; Bregler, C. Efficient Object Localization using Convolutional Networks. In Proceedings of the IEEE Conference on Computer Vision and Pattern Recognition, Boston, MA, USA, 7–12 June 2015.
15. Chen, Y.; Shen, C.; Wei, X.-S.; Liu, L.; Yang, J. Adversarial PoseNet: A Structure-Aware Convolutional Network for Human Pose Estimation. In Proceedings of the IEEE International Conference on Computer Vision, Venice, Italy, 22–29 October 2017.
16. Sun, M.; Savarese, M. Articulated part-based model for joint object detection and pose estimation. In Proceedings of the IEEE International Conference on Computer Vision, Barcelona, Spain, 6–13 November 2011.
17. Fang, H.-S.; Xie, S.; Tai, Y.-W.; Lu, C. RMPE: Regional Multi-person Pose Estimation. In Proceedings of the IEEE International Conference on Computer Vision, Venice, Italy, 22–29 October 2017.
18. Cao, Z.; Hidalgo, G.; Simon, T.; Wei, S.E.; Sheikh, Y. OpenPose: Realtime multi-person 2D pose estimation using part affinity fields. *IEEE Trans. Pattern Anal. Mach. Intell.* **2021**, *43*, 172–186. [CrossRef] [PubMed]
19. Redmon, J.; Divvala, S.; Girshick, R.; Farhadi, A. You only look once: Unified, real-time object detection. In Proceedings of the IEEE Conference on Computer Vision and Pattern Recognition, Las Vegas, NV, USA, 27–30 June 2016.
20. Wojke, N.; Bewley, A.; Paulus, D. Simple Online Realtime Tracking with a Deep Association Metric. In Proceedings of the IEEE Conference on Image Processing, Beijing, China, 17—20 September 2017.
21. Badave, H.; Kuber, M. Evaluation of Person Recognition Accuracy based on OpenPose Parameters. In Proceedings of the International Conference on Intelligent Computing and Control Systems, Madurai, India, 6–8 May 2021.
22. Simonyan, K.; Zisserman, A. Very Deep Convolutional Networks for Large-Scale Image Recognition. *arXiv* **2015**, arXiv:1409.1556.
23. Hosang, J.; Benenson, R.; Schiele, B. Learning Non-maximum Suppression. In Proceedings of the IEEE Conference on Computer Vision and Pattern Recognition, Honolulu, HI, USA, 21–26 July 2017.
24. Girshick, R.; Donahue, J.; Darrell, T.; Malik, J. Rich Feature Hierarchies for Accurate Object Detection and Semantic Segmentation. In Proceedings of the IEEE Conference on Computer Vision and Pattern Recognition, Columbus, OH, USA, 23–28 June 2014.
25. Girshick, R. Fast R-CNN. In Proceedings of the IEEE International Conference on Computer Vision, Santiago, Chile, 7–13 December 2015.
26. Ren, S.; He, K.; Girshick, R.; Sun, J. Faster R-CNN: Towards real-time object detection with region proposal networks. *IEEE Trans. Pattern Anal. Mach. Intell.* **2015**, *39*, 1137–1149. [CrossRef] [PubMed]
27. He, K.; Georgia, G.; Dollar, P.; Girshick, R. Mask R-CNN. In Proceedings of the IEEE International Conference on Computer Vision, Venice, Italy, 22–29 October 2017.
28. Liu, W.; Anguelov, D.; Erhan, D.; Szegedy, C.; Reed, S.; Fu, C.Y.; Berg, A.C. SSD: Single shot multibox detector. In Proceedings of the European Conference on Computer Vision, Amsterdam, The Netherlands, 11–14 October 2016.
29. Bochkovskiy, A.; Wang, C.Y.; Liao, H.Y.M. Yolov4: Optimal speed and accuracy of object detection. *arXiv* **2020**, arXiv:2004.10934.
30. Akyol, G.; Kantarcı, A.; Çelik, A.; Cihan Ak, A. Deep Learning Based, Real-Time Object Detection for Autonomous Driving. In Proceedings of the IEEE Conference on Signal Processing and Communications Applications, Gaziantep, Turkey, 5–7 October 2020.
31. Teknomo, K.; Takeyama, Y.; Inaura, H. Frame-based tracing of multiple objects. In Proceedings of the IEEE Workshop on Multi-Object Tracking, Vancouver, BC, Canada, 8 July 2001.
32. Luo, W.; Xing, J.; Milan, A.; Zhang, X.; Liu, W.; Kim, T.K. Multiple Object Tracking: A Literature Review. *arXiv* **2017**, arXiv:1409.7618. [CrossRef]
33. Wang, Z.; Zheng, L.; Liu, Y.; Li, Y.; Wang, S. Towards Real-Time Multi-Object Tracking. In Proceedings of the European Conference on Computer Vision, Glasgow, UK, 23–28 August 2020.
34. Bewley, A.; Ge, Z.; Ott, L.; Ramos, F.; Upcroft, B. Simple Online and Realtime Tracking. In Proceedings of the IEEE International Conference on Image Processing, Phoenix, AZ, USA, 25–28 September 2016.
35. Pereira, R.; Carvalho, G.; Garrote, L.; Nunes, U.J. Sort and Deep SORT Based Multi-Object Tracking for Mobile Robotics: Evaluation with New Data Association Metrics. *Appl. Sci.* **2022**, *12*, 1319. [CrossRef]

*Article*

# Application of Smoothing Spline in Determining the Unmanned Ground Vehicles Route Based on Ultra-Wideband Distance Measurements

Łukasz Rykała [1,*], Andrzej Typiak [1], Rafał Typiak [1] and Magdalena Rykała [2]

1. Faculty of Mechanical Engineering, Military University of Technology, 00-908 Warsaw, Poland
2. Faculty of Security, Logistics and Management, Military University of Technology, 00-908 Warsaw, Poland
* Correspondence: lukasz.rykala@wat.edu.pl

**Abstract:** Unmanned ground vehicles (UGVs) are technically complex machines to operate in difficult or dangerous environmental conditions. In recent years, there has been an increase in research on so called "following vehicles". The said concept introduces a guide—an object that sets the route the platform should follow. Afterwards, the role of the UGV is to reproduce the mentioned path. The article is based on the field test results of an outdoor localization subsystem using ultra-wideband technology. It focuses on determining the guide's route using a smoothing spline for constructing a UGV's path planning subsystem, which is one of the stages for implementing a "follow-me" system. It has been shown that the use of a smoothing spline, due to the implemented mathematical model, allows for recreating the guide's path in the event of data decay lasting up to a several seconds. The innovation of this article originates from influencing studies on the smoothing parameter of the estimation errors of the guide's location.

**Keywords:** UGV; ultra-wideband; UWB; smoothing spline; nonparametric regression; path planning; follow-me; smoothing parameter

**Citation:** Rykała, Ł.; Typiak, A.; Typiak, R.; Rykała, M. Application of Smoothing Spline in Determining the Unmanned Ground Vehicles Route Based on Ultra-Wideband Distance Measurements. *Sensors* **2022**, *22*, 8334. https://doi.org/10.3390/s22218334

Academic Editors: José María Martínez-Otzeta and Andrey V. Savkin

Received: 3 October 2022
Accepted: 28 October 2022
Published: 30 October 2022

**Copyright:** © 2022 by the authors. Licensee MDPI, Basel, Switzerland. This article is an open access article distributed under the terms and conditions of the Creative Commons Attribution (CC BY) license (https://creativecommons.org/licenses/by/4.0/).

## 1. Introduction

The term unmanned ground vehicles (UGVs) refers to robots that can travel on land without human operators [1]. In some cases, UGVs can operate autonomously, while in others, operators can control them remotely [2]. In the so called "follow-me" mode, the operator does not have to manually control the platform. This mode allows the vehicle to follow the route set by the guide [3]. Navigating UGVs in "follow-me" mode requires the precise location of the guide to be determined. Guides are responsible for creating paths for UGVs, as mentioned earlier. Maintaining a set distance from the guide and keeping the platform's heading are the most important aspects of this mode [4]. UGVs should be able to follow the guide in a smooth motion, but if there is an emergency, the guide may stop during movement [5]. It is possible to implement these functionalities using the components of the "follow-me" system, including the guide's observation subsystem, the path planning subsystem, and the control subsystem [6]. "Follow-me" systems can be divided, inter alia, because of the mode of interaction and degree of autonomy [3]. Mode of interaction refers to the way that the platform interacts with the guide and it can be explicit or implicit. If a human does not directly command the platform, the mentioned mode is explicit. On the other hand, in the case of the degree of autonomy, the most common variant is partial autonomy. Fully autonomous systems use multiple technologies simultaneously and are extremely expensive. The use of "follow-me" systems in dangerous terrain means that the UGV relies heavily on the guide's movement (implicit mode of interaction and partial autonomy). The platform also does not have to follow the guide in real-time, if it is not necessary in a given situation. To do this, the guide moves first, marking a certain path, then stops and waits for the UGV to reach it. In this case, determining the exact location of the guide is crucial.

Knowing the guide's location is the basis of the "follow-me" system, and this task is performed within the guide's observation subsystem [7]. This article is based on the field test results of an outdoor localization subsystem based on ultra-wideband (UWB) technology [8] constituting one of the elements of the "follow-me" system. In order to locate objects using UWB technology, very short data packets are sent wirelessly with a very low power spectral density using the radio energy scattering technique (time of flight). It provides the bandwidth needed to transfer the required amounts of data [9,10].

The mentioned outdoor localization subsystem consists of a total of five UWB modules. Four of them (receivers) were deployed on an existing UGV, while the fifth module (transmitter) was attached to the guide as a part of the developed subsystem. The described subsystem estimates the relative operator's position based on distance measurements [11,12]. Various technologies are used in commercial "follow-me" system solutions, including UWB [13,14]. Because the main task of the platform in the above-mentioned cases is to keep following the guide (not necessarily along the path indicated by him), a smaller number of receivers is used (usually two), which results in a lower accuracy of the guide's localization.

Using the guide's location subsystem, the UGV's desired route is determined based on the above premise. In the path planning subsystem, successfully calculated guide positions are used to create a route [15,16]. Finally, the planning subsystem aims to provide input signals to the control systems that facilitate the execution of the planned path [17].

The problem of determining the route of a guide's movement requires solving the problem of fitting a continuous function to a discrete set of the guide's locations. The numerical methods used in solving the data fitting problem are interpolation and approximation [18]. Interpolation is rarely used in relation to data from experimental measurements because of the presence of disturbances in devices (the interpolation function must pass through the given points) [19]. Approximation, in turn, allows for smoothing and simplifying the course of the analyzed data sets [20]. In addition, this method can be used for large data sets as opposed to interpolation.

The approximation is the problem of describing a data set using approximating functions f(x). When using the approximation methods, a certain set of base functions is assumed, from which the approximating function is defined and the method of its use is determined. The most frequently used form of approximating functions is the so-called general polynomials. The approximation task consists of finding the coefficient values of the generalized polynomial so that the approximating function minimizes the adopted criterion, e.g., the sum of the squared differences. The concept of approximation is also closely related to the concept of regression, which is a solution to the problem of point approximation for a data set, but its final result is, apart from the sought function coefficients, also a function model. Regression allows for determining the symbolic form of the function, which, meeting the adopted criteria, reflects the individual values of the dependent variable for the previously defined set of independent variable values [18,21].

Regression can be divided into two main types: parametric and non-parametric [22]. Some sources also distinguish semiparametric methods, which are rarely used [22]. Mentioned regression models are chosen according to the prior knowledge of the functional form and the random error distribution. The most important criterion, however, is the knowledge of the functional form. If it is known, parametric regression will be able to fit the data. The parametric approach requires knowledge of a mathematical model, which can be simple (e.g., linear regression), and its parameters are assumed directly [23]. Because the guide can move in any direction, marking the platform's path, it is impossible to make any assumptions about its route. Therefore, the application of the parametric regression approach in the analyzed case becomes very difficult to implement.

There are also non-parametric regression methods in which the form of the model is not clearly defined and their parameters are not taken directly [24,25]. Nonparametric regression methods, including Kernel regression [26,27], LOWESS (locally weighted scatterplot smoothing) [22], and smoothing spline [28–31], have an extensive form of a mathematical model. Nonparametric regression models are much more flexible and computationally complex compared with parametric models. In addition, they avoid erroneous fitting results when the wrong model is used. The result of the application of the above-mentioned methods is not a mathematical relationship; therefore, the mentioned results are also difficult to export [22].

To determine the guide's route, the desired method should be characterized by a moderate computational complexity, have the ability to parametrically shape the smoothing of the coordinates of the guide's location, and be able to estimate the missing coordinates based on the present values.

Kernel regression smoothing is a technique that uses kernel functions as a weighing function for developing a non-parametric regression model. It can be applied to high-dimensional data sets and it can be used for fitting the data without making any distributional assumptions about it. It is more flexible than other non-parametric approaches, but it does not have any direct smoothing parameter [32,33]. In turn, the LOWESS method is based on the simplicity of linear least squares regression, which makes it highly exposed to the effects of outliers in the data set [34]. Similar to the mentioned Kernel regression method, LOWESS does not have a direct smoothing parameter. Additionally, the Kernel method is much more computationally complex than the LOWESS method.

However, the only analyzed non-parametric method that meets the mentioned criteria is a smoothing spline. Among other methods of non-parametric regression, it is distinguished by a lower computational complexity and the presence of a direct parameter smoothing the given waveform. Moreover, it is not exposed to the effects of outliers in the data set. Therefore, a smoothing spline was chosen as the method for calculating the guide's route.

Researchers have focussed on the study of path planning algorithms of autonomous robots (which can also work under the "follow-me" system) using various modern methods, including smoothing splines [35]. No studies were found on the use of the smoothing spline method in the context of planning the movement of the UGV (or robot in general) as part of the "follow-me" system, hence the article is an innovation in the field. Moreover, no studies were found on the influence of the smoothing parameter on the estimation on the guide's path.

The aim of the article is to determine the route of the guide using a smoothing spline based on the designated locations using UWB technology. In order to generate a smoothing spline, it is necessary to specify a value for the smoothing parameter. Because of this, it is necessary to conduct research on the influence of the aforementioned parameter on the estimation of the guide's route and select a value that meets the selected evaluation criterion, e.g., minimization of the sum of errors.

## 2. Materials and Methods

A smoothing spline (so-called polynomial spline or polynomial smoothing curve) is a k-th degree piecewise polynomial that has k−1 continuous derivatives. The mentioned curve is most often used to approximate a data set of points with cubic polynomials (3rd order, two continuous derivatives). The advantage of using the mentioned curve is the possibility of reaching a compromise between two opposing aims:

- fitting the value of the dependent variable to the set of independent variable values,
- smoothing the course of the value of the dependent variable (minimizing the curvature of the trajectory and its acceleration) [22].

In order to describe the mathematical model of a polynomial curve, the first step is to define the vectors of the dependent variables q and the independent variables t:

$$q = [q_0, q_1, q_2, \ldots, q_n]^T \qquad (1)$$

$$t = [t_0, t_1, t_2, \ldots, t_n]^T \qquad (2)$$

Then, the parameters of the aforementioned curve s(t) are obtained by minimizing the dependence S:

$$S = \lambda \sum_{k=1}^{n} [s(t_k) - q_k(t_k)]^2 + (1-\lambda) \int_{t_0}^{t_n} \ddot{s}(t)^2 dt \qquad (3)$$

where $\lambda \in [0,1]$—the so-called smoothing parameter, $s_i(t_i) = [s_1(t_1), \ldots, s_n(t_n)]^T$—smoothing spline function parameters [22].

The curve parameters are determined for each node, while the nature of its estimation is determined by the smoothing parameter $\lambda$, which takes values in the range of [0, 1]. In extreme cases, when $\lambda = 0$, a linear approximation is obtained using the least squares method, and for $\lambda = 1$, interpolation using a cubic polynomial is obtained. Thus, when $\lambda$ tends to zero, the smoothing effect of the course is maximized, while when $\lambda$ tends to 1, the fidelity of the mapping of the set of points is maximized [36].

*2.1. Determination of the Value of the Smoothing Parameter*

The present article is a direct extension of the research carried out in [8]. Moreover, the results of the mentioned research form the basis of the article.

A Decawave TREK1000 evaluation kit [37], which consists of five UWB modules, was used in the research. The developed system consists of five modules: four receivers called anchors and a transmitter called a tag, carrying out continuous distance measurements with a frequency of 10 Hz. UWB modules were placed on the UGV (anchors) and the guide (tag). The accuracy of a single anchor–tag measurement is approximately 10 cm using the two-way ranging time-of-flight (TOF) technique. The UWB system provides information about the distance from the individual anchors to the tag.

For the research, it was assumed that the human guide moves along seven rectilinear paths inclined at an angle of 0°, 30°, 60°, 90°, 120°, 150°, and 180°, respectively, to the x-axis of the xy coordinate system in the area satisfying the following inequalities: −10 m < x < 10 m and 0 m < y < 20 m (Figure 1).

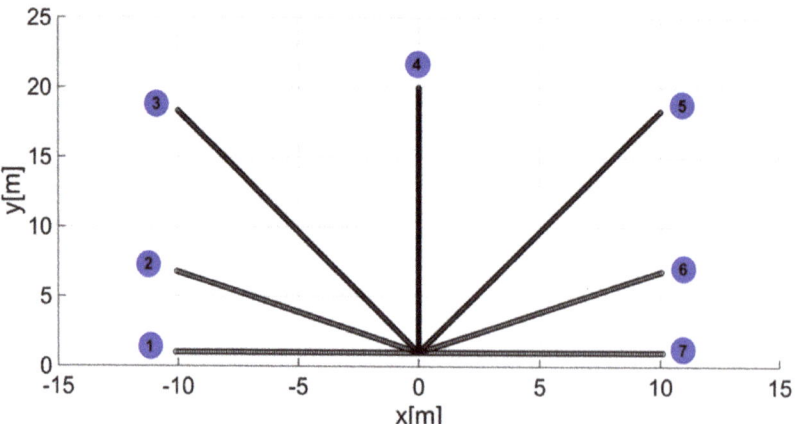

**Figure 1.** Guide paths 1–7 with the adopted xy coordinate system [6].

Next, guide paths no. 1–7 (Figure 1) were recreated with the assumption that the UGV remains stationary. Moreover, the guide was supposed to turn 180 around its axis after reaching the turning point and then return along the same track to the starting point. The arrangement of the anchors on the UGV is shown in Figure 2 (spatial configuration of the anchors for the correct operation of the location subsystem).

**Figure 2.** Configuration of UWB modules on a UGV.

During the movement, the guide carried a specially made frame with the necessary equipment (mobile location kit): a UWB tag, a GPS module, and a power supply system (Figure 3).

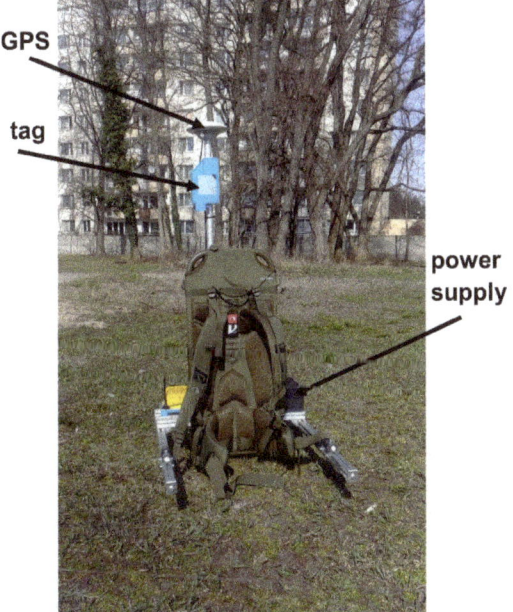

**Figure 3.** Mobile UWB guide location kit.

In order to determine the location errors of the results obtained with the UWB technology, SwiftNav DURO satellite receivers operating in the RTK mode were used (error: 1 cm horizontally and 1.5 cm vertically) [38].

The starting point of the article is the final results of research on the described location system based on UWB technology using the nonlinear programming (NLP) method based on the Levenberg–Marquardt (LM) algorithm [8]. The mentioned results (Figure 4) are the basis for the further determination of the guide's path.

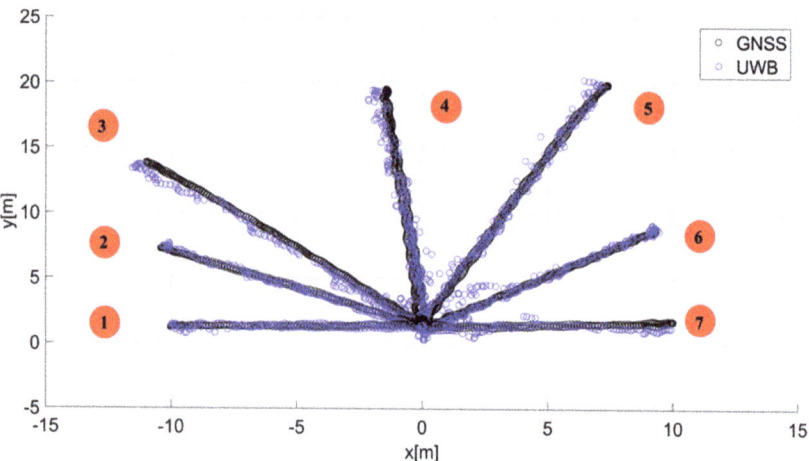

**Figure 4.** The results of the guide's location for routes no. 1–7 using UWB technology along with the reference positions obtained using the GNSS module. Own elaboration based on [8].

In order to calculate the guide's route, first, the smoothing parameter value should be specified. Then, after determining said parameter, it becomes possible to implement a polynomial curve for the results of the experimental research of the location subsystem.

The following values were adopted to evaluate the obtained results:

- total error

$$e_c(t) = \sqrt{e_x(t)^2 + e_y(t)^2} \tag{4}$$

where $e_x(t)$ is the error of mapping the guide's location on the $x$-axis of the coordinate system at time t, $e_y(t)$ is the error of mapping the guide's location on the $y$-axis of the coordinate system at time t.

- quality indicator

$$Q = \sum e_c(t) \tag{5}$$

- mean value of the quality indicator

$$Q_{av} = \frac{\sum e_c(t)}{l} \tag{6}$$

where l is the number of distance measurements [6].

Most often, the value of the smoothing parameter is determined using the following relationship:

$$\lambda_p = \frac{1}{1 + \frac{p^3}{6}} \tag{7}$$

where p is the average spacing of data points [36].

The smoothing parameter calculated according to dependence 7 (p = 0.1) is approximately $\lambda_p = 0.99$. Such a high value indicates the maximization of data fidelity, which, due to the presence of disturbances resulting in localization errors, is not always the most recommended solution. Therefore, the influence of the smoothing parameter on the guide's route estimation is determined in the article. In order to achieve the above-mentioned purpose and to select the final value of the smoothing parameter smoothing splines were calculated for the smoothing parameters λ ∈ [0.05; 0.1; 0.15, ... , 0.95] and routes no. 1–7 (Figure 4). Then, for each value of the smoothing parameter, the quality indicator Q and finally the average value of quality indicator $Q_{av}$ were determined. Additionally, for each case, the mean square value of the acceleration $a_{RMS}$ (the second derivative of the dependent variable) was also determined. Based on the minimization of the average quality indicator, the final value of the smoothing parameter was selected. The knowledge of a chosen smoothing parameter made it possible to determine the estimation of guide routes no. 1–7 using a smoothing spline. Matlab/Simulink software with the Curve Fitting Toolbox was used in the research.

## 2.2. The Influence of the Smoothing Parameter on the Guide's Path Estimation

The results of the research on the influence of the smoothing parameter on the values of the Q quality indicator and the average square acceleration values for guide routes no. 1–7 are shown in Figures 5–11. Figure 12 shows the values of the average quality indicator and the mean square root acceleration values for all of the considered results.

The values of certain quality indicators Q (Figures 5–11) obtain values ranging from about 79 m (traffic path no. 2) to about 375 m (traffic path no. 3). Determination of the average value of the quality indicator $Q_{av}$ allows for generalizing the obtained results and for determining the final value of the smoothing parameter minimizing the indicator, which is the sum of the total errors (dependence 5). On the other hand, the courses of the mean square values of acceleration $a_{RMS}$ for all of the considered paths (Figures 5–11) show an increasing trend obtaining the minimum quality indicator Q for the value of λ = 0.05 and the maximum for the value of λ = 0.95.

The $Q_{av}$ indicator obtained the minimum value for the smoothing parameter λ = 0.15 (Figure 12), which was adopted in the further part of the research as the smoothing parameter of the smoothing spline.

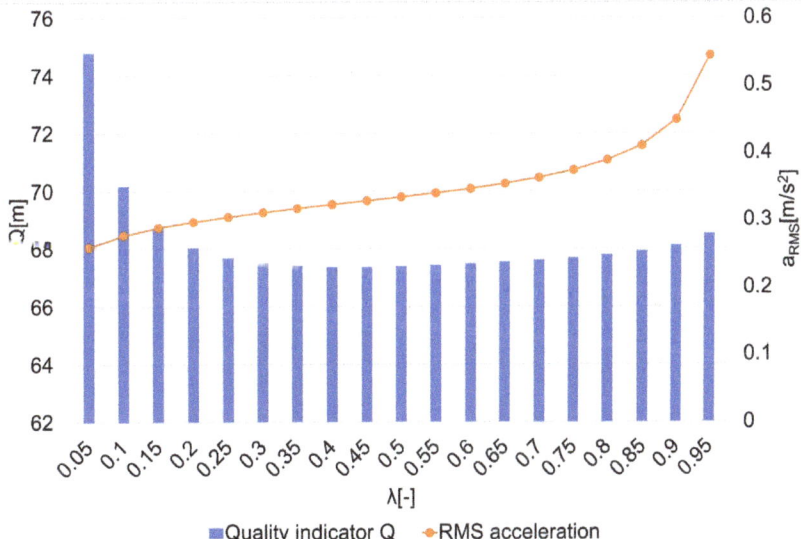

**Figure 5.** The values of the quality indicator Q and the mean square value of acceleration $a_{RMS}$ for the smoothing parameters in the case of path no. 1 [6].

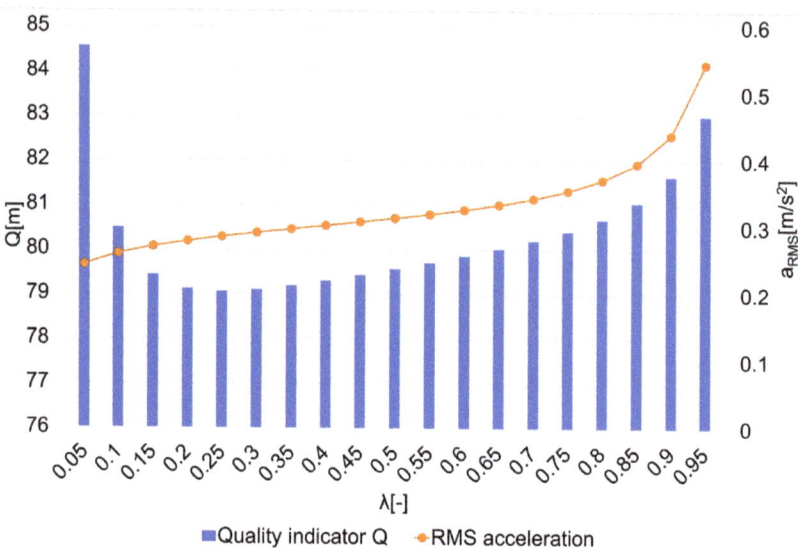

**Figure 6.** The values of the quality indicator Q and the mean square value of acceleration $a_{RMS}$ for the smoothing parameters in the case of path no. 2 [6].

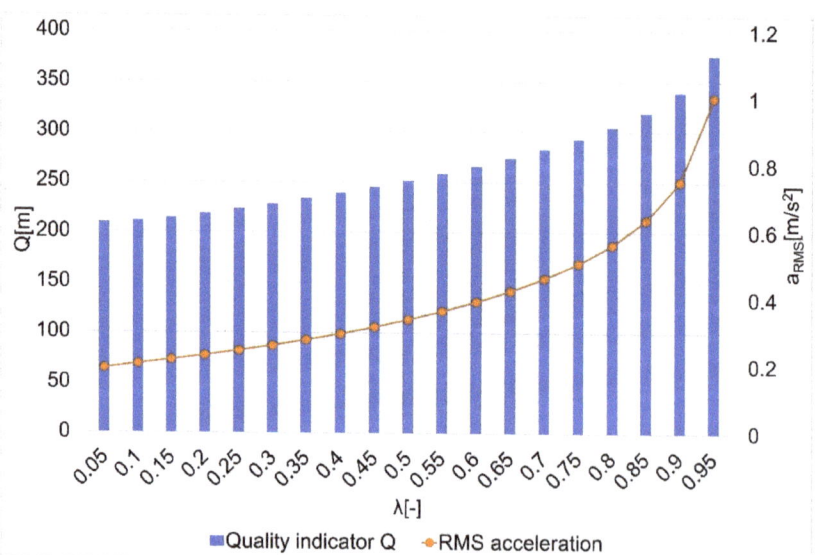

**Figure 7.** The values of the quality indicator Q and the mean square value of acceleration $a_{RMS}$ for the smoothing parameters in the case of path no. 3 [6].

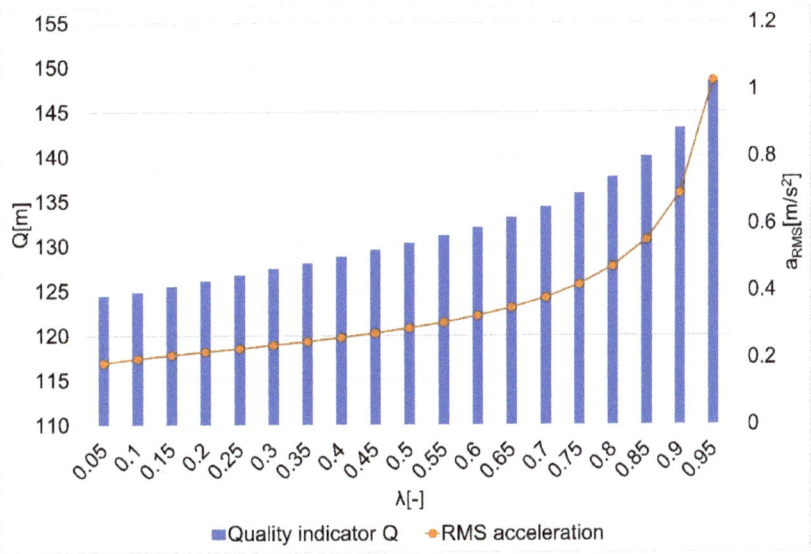

**Figure 8.** The values of the quality indicator Q and the mean square value of acceleration $a_{RMS}$ for the smoothing parameters in the case of path no. 4 [6].

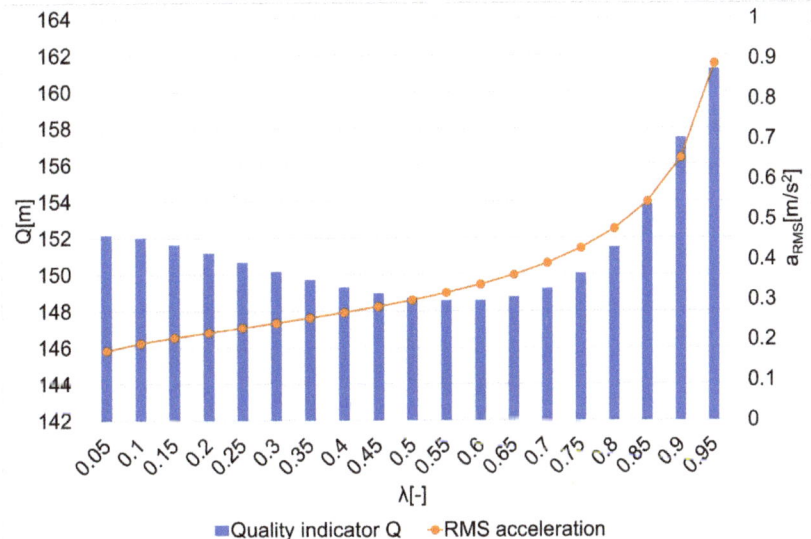

**Figure 9.** The values of the quality indicator Q and the mean square value of acceleration $a_{RMS}$ for the smoothing parameters in the case of path no. 5 [6].

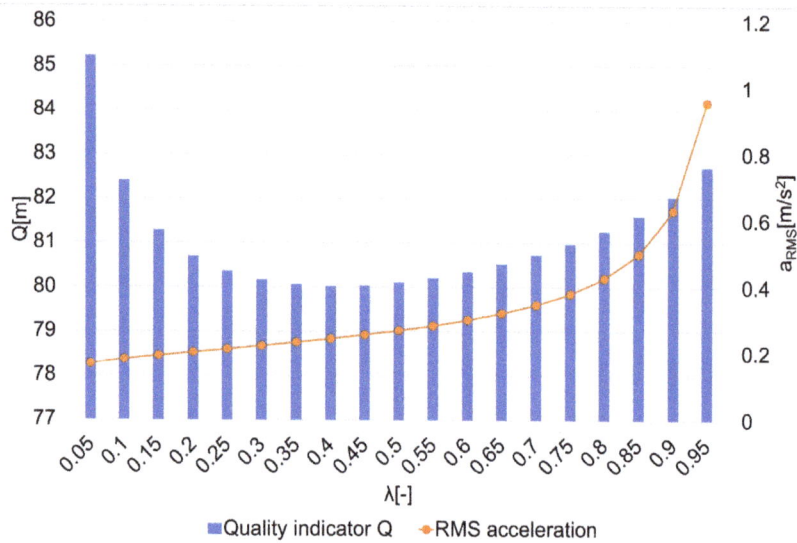

**Figure 10.** The values of the quality indicator Q and the mean square value of acceleration $a_{RMS}$ for the smoothing parameters in the case of path no. 6 [6].

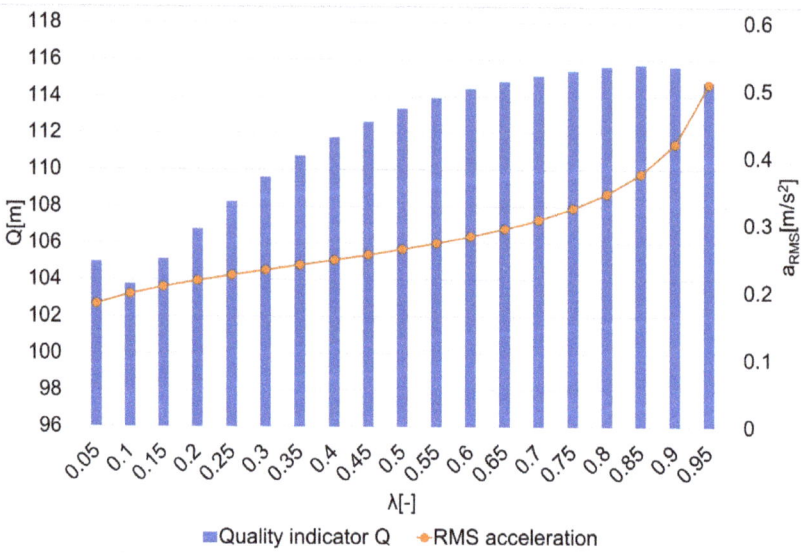

**Figure 11.** The values of the quality indicator Q and the mean square value of acceleration $a_{RMS}$ for the smoothing parameters in the case of path no. 7 [6].

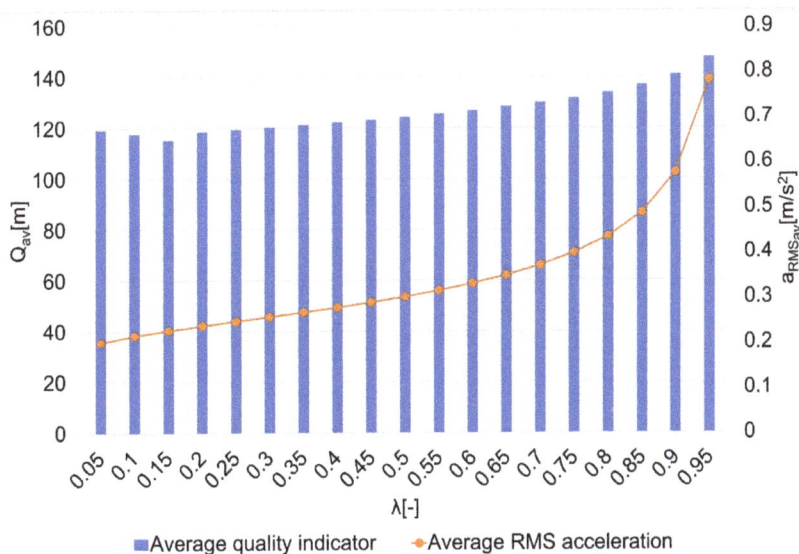

**Figure 12.** The values of the average quality indicator Q and the average mean square value of acceleration $a_{RMS}$ for the smoothing parameters in the case of paths no. 1–7 [6].

## 3. Results

After determining the value of the smoothing parameter λ (λ = 0.15), the guide's route for paths 1–7 (Figure 1) was determined using a smoothing spline. Figures 13a–19a show the x(t) and y(t) coordinates of the guide's location along with their estimates for the considered cases. Errors in determining the guide's route using the mentioned method are shown graphically in Figures 13b–19b. Moreover, Figures 13c–19c show the guide's location along with the estimation of its route with the use of a smoothing spline concerning all of the analyzed paths.

The courses of the estimated coordinates of the guide's position at x(t), y(t) are presented as a function dependent of time in Figures 13a–19a, while in Figures 13c–19c they are presented as a function independent of time in the form of y(t) = f(x(t)). In turn, location errors on the $x$ and $y$ axes of the xy coordinate system and the total errors are shown in Figures 13b–19b. In all of the cases, the decay of signals can be noticed (Figures 13a–19a), which increases the total errors of the estimated path. Basic descriptive statistics of the total errors for all of the considered paths are shown in Figure 20.

The error values do not exceed the following values: minimum 0.07 m, mean 0.57 m, RMS 0.76 m, and maximum 2.03 m (Figure 20). The largest decay of location signals is noticeable in the case of path no. 3 (decay lasting approx. 10 s, Figure 15a), and it translates into the above-mentioned maximum values of the total errors (Figures 15b and 20). However, even in this case, the estimated trajectory retains the shape of the reference trajectory (Figure 15c).

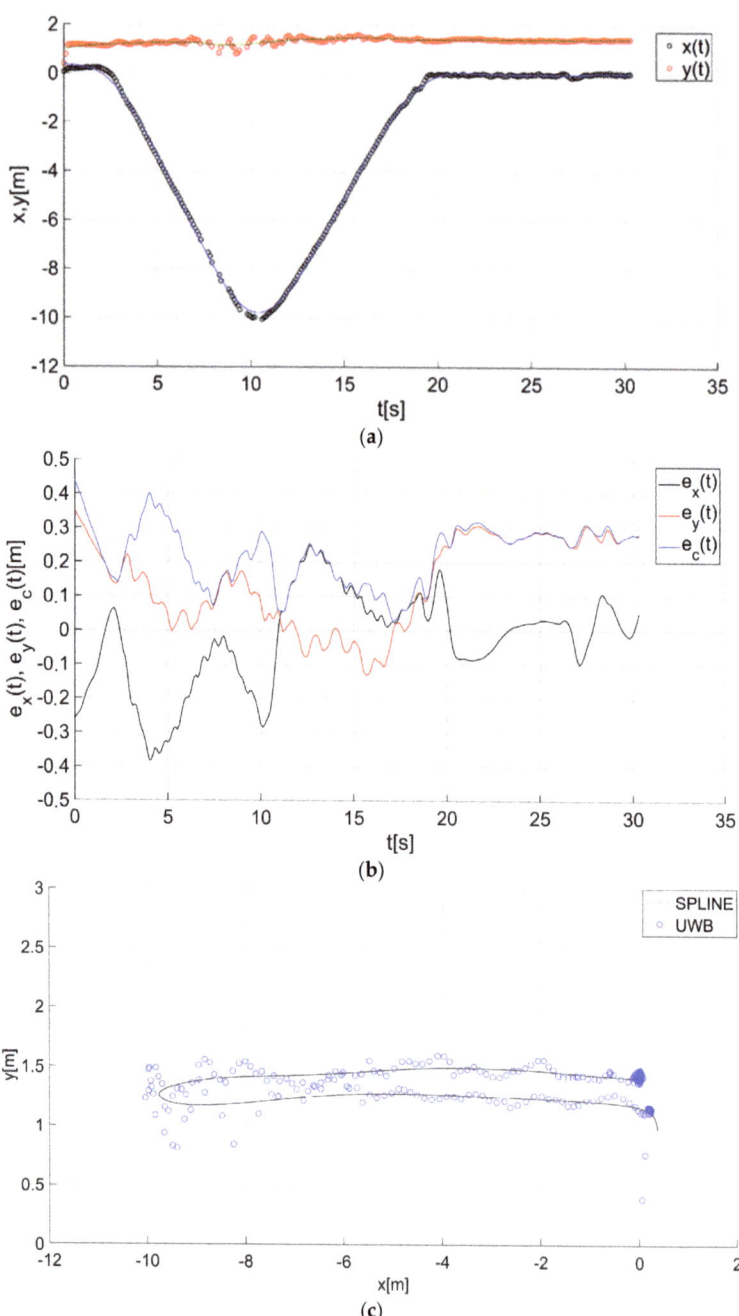

**Figure 13.** Results for the estimation of guide path no. 1 with the use of the smoothing spline: (**a**) the course of the guide's location coordinates x(t), y(t) with their continuous estimates, (**b**) the course of the estimated location errors $e_x(t)$, $e_y(t)$, $e_c(t)$, (**c**) guide's location along with the path estimation [6].

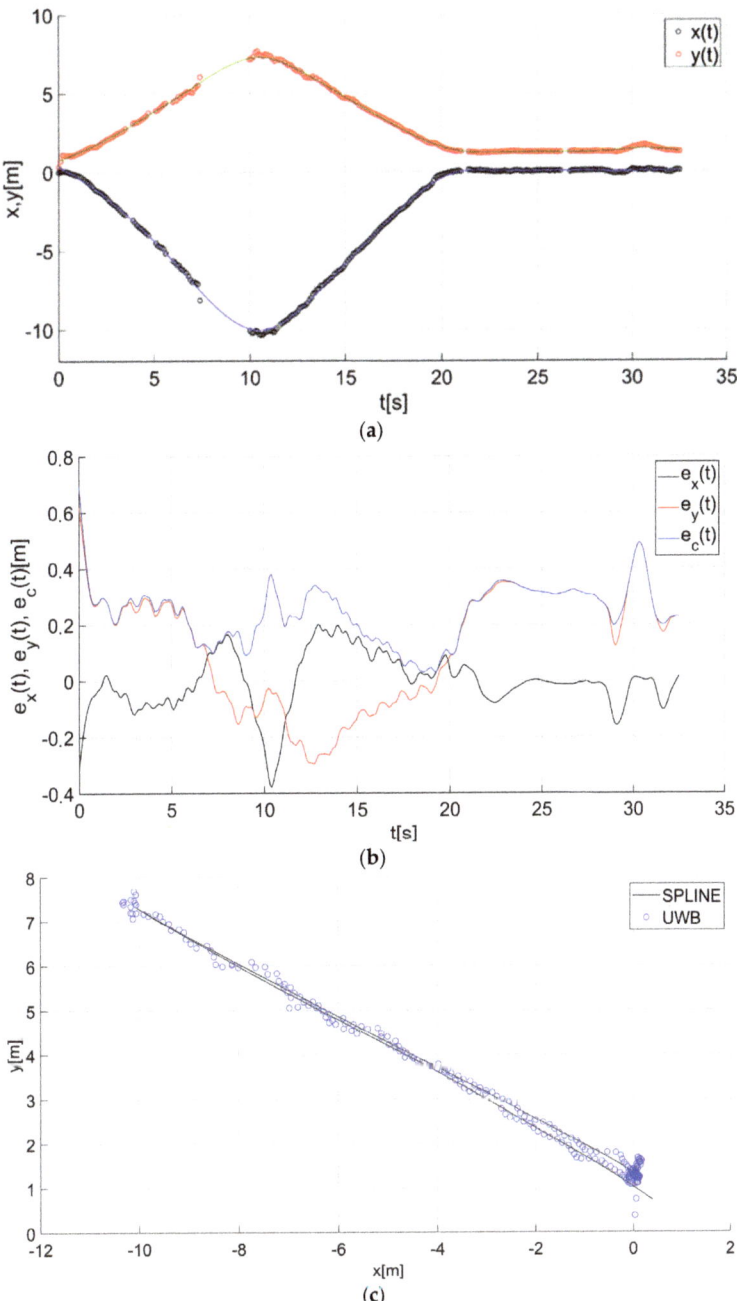

**Figure 14.** Results for the estimation of guide path no. 2 with the use of the smoothing spline: (**a**) the course of the guide's location coordinates x(t), y(t) with their continuous estimates, (**b**) the course of the estimated location errors $e_x(t)$, $e_y(t)$, $e_c(t)$, (**c**) guide's location along with the path estimation [6].

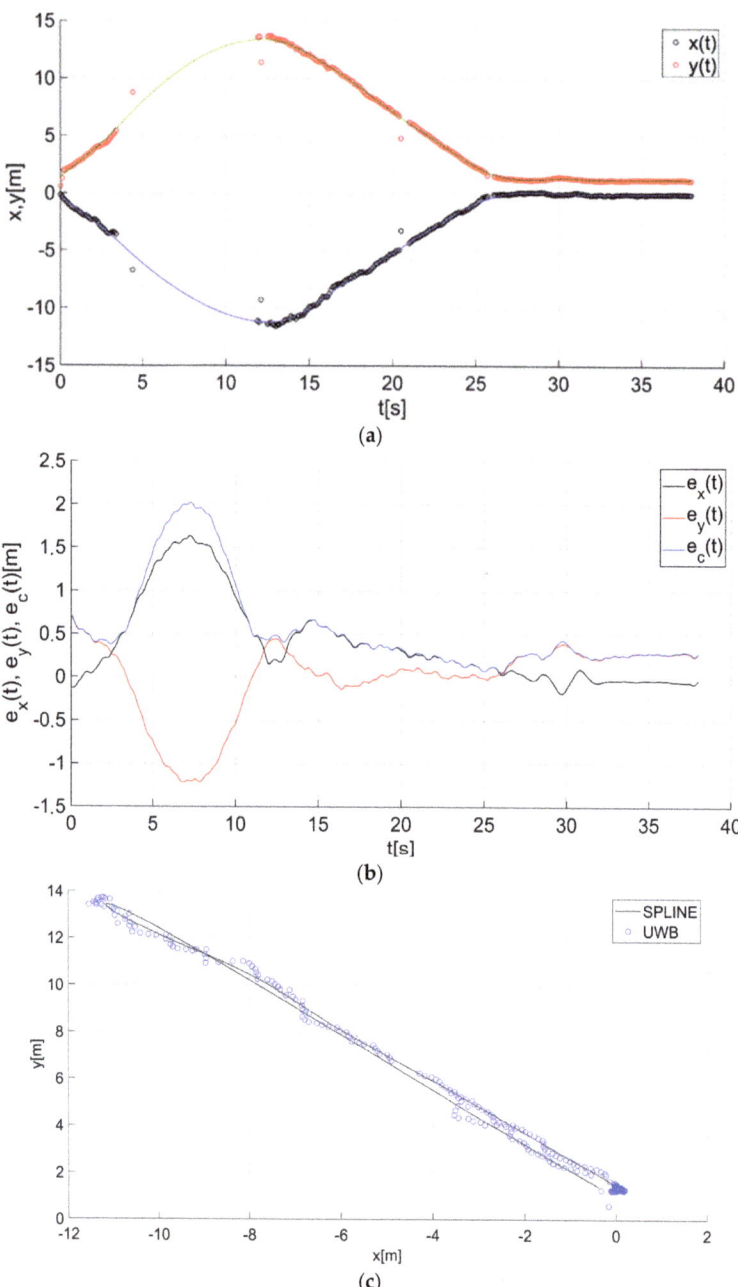

**Figure 15.** Results for the estimation of guide path no. 3 with the use of the smoothing spline: (**a**) the course of the guide's location coordinates x(t), y(t) with their continuous estimates, (**b**) the course of the estimated location errors $e_x(t)$, $e_y(t)$, $e_c(t)$, (**c**) guide's location along with the path estimation [6].

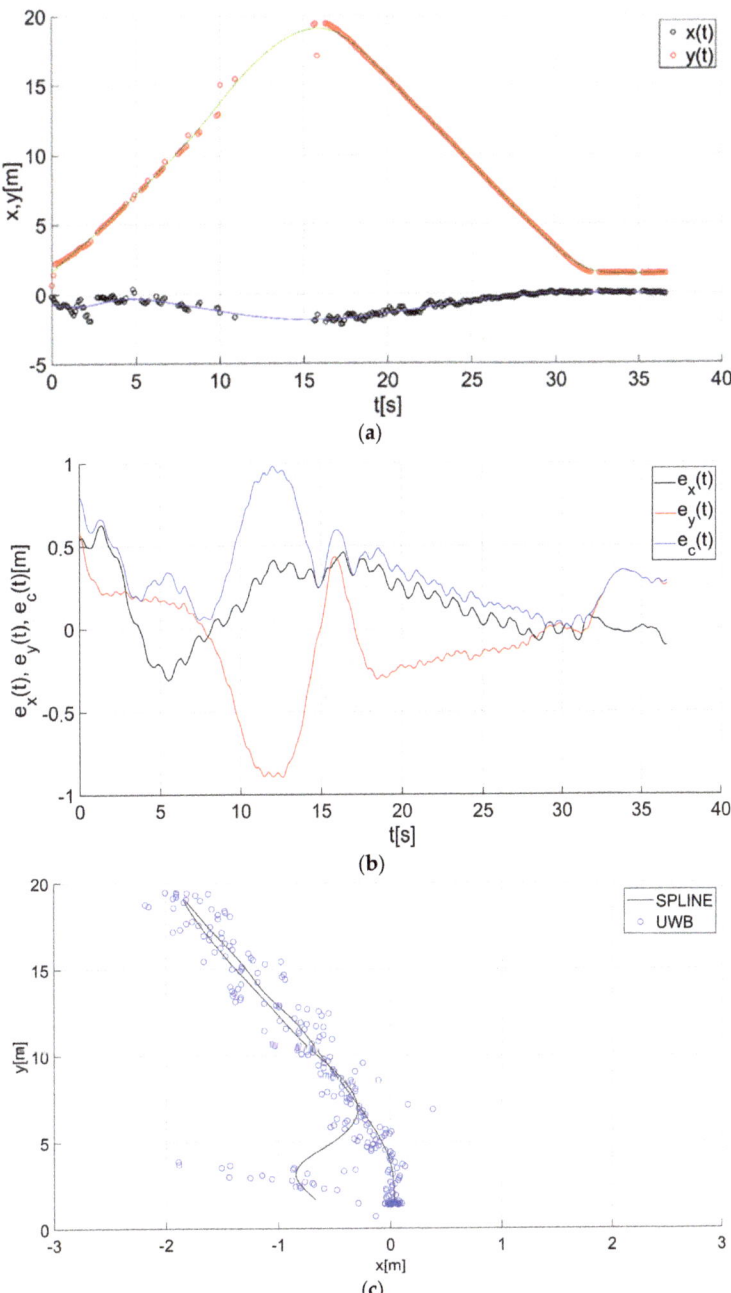

**Figure 16.** Results for the estimation of guide path no. 4 with the use of the smoothing spline: (**a**) the course of the guide's location coordinates x(t), y(t) with their continuous estimates, (**b**) the course of the estimated location errors $e_x(t)$, $e_y(t)$, $e_c(t)$, (**c**) guide's location along with the path estimation [6].

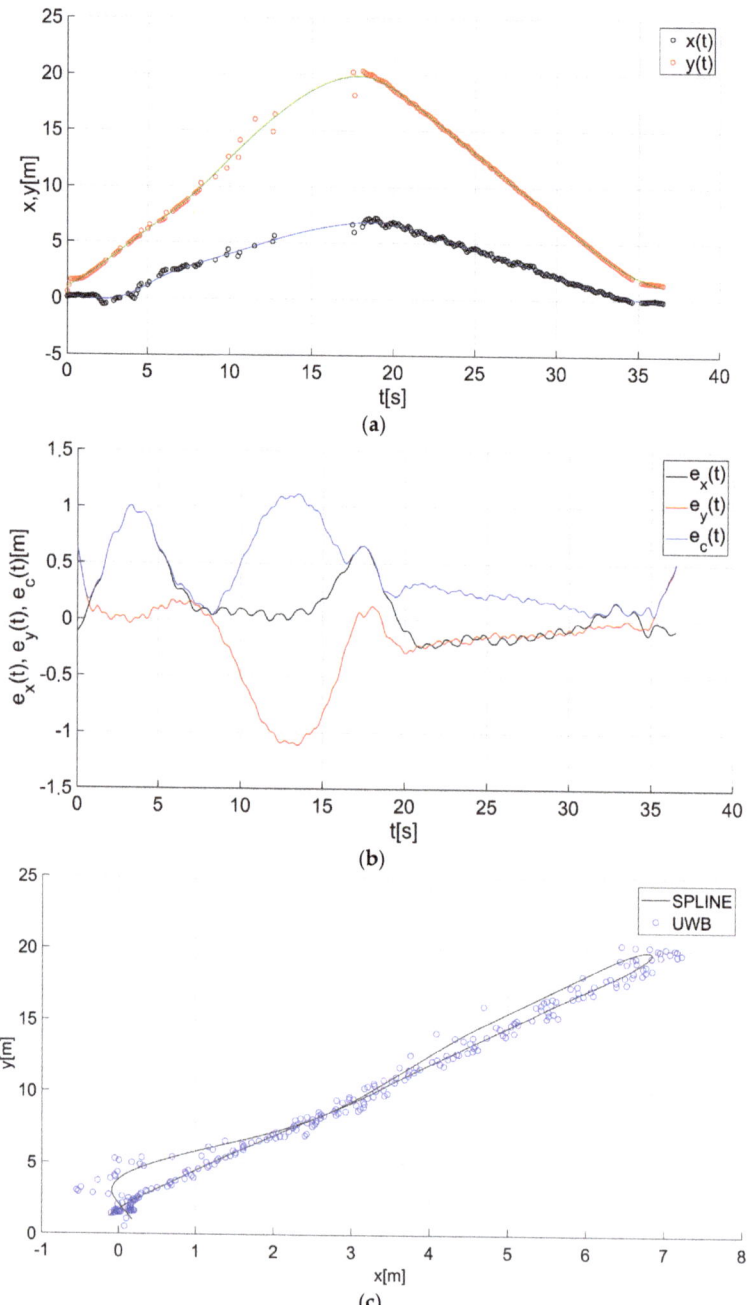

**Figure 17.** Results for the estimation of guide path no. 5 with the use of the smoothing spline: (**a**) the course of the guide's location coordinates x(t), y(t) with their continuous estimates, (**b**) the course of the estimated location errors $e_x(t)$, $e_y(t)$, $e_c(t)$, (**c**) guide's location along with the path estimation [6].

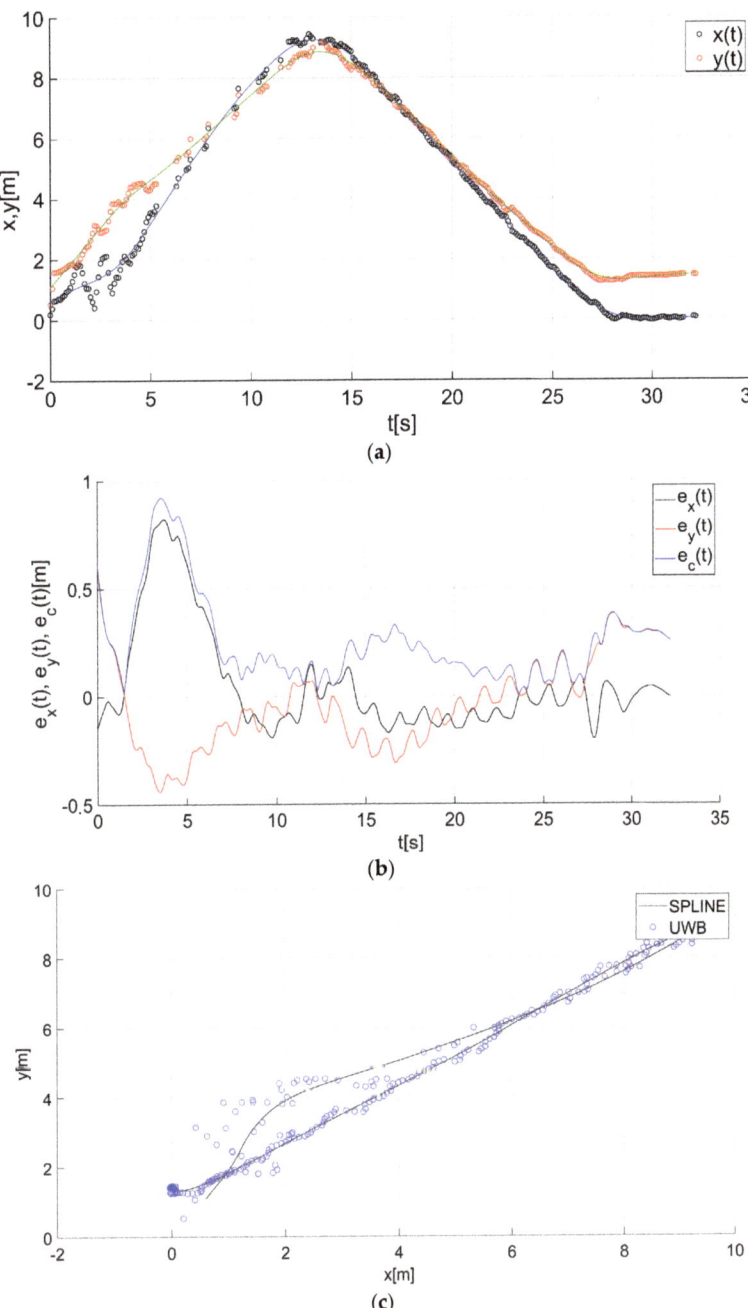

**Figure 18.** Results for the estimation of guide path no. 6 with the use of smoothing spline: (**a**) the course of the guide's location coordinates x(t), y(t) with their continuous estimates, (**b**) the course of the estimated location errors $e_x(t)$, $e_y(t)$, $e_c(t)$, (**c**) guide's location along with the path estimation [6].

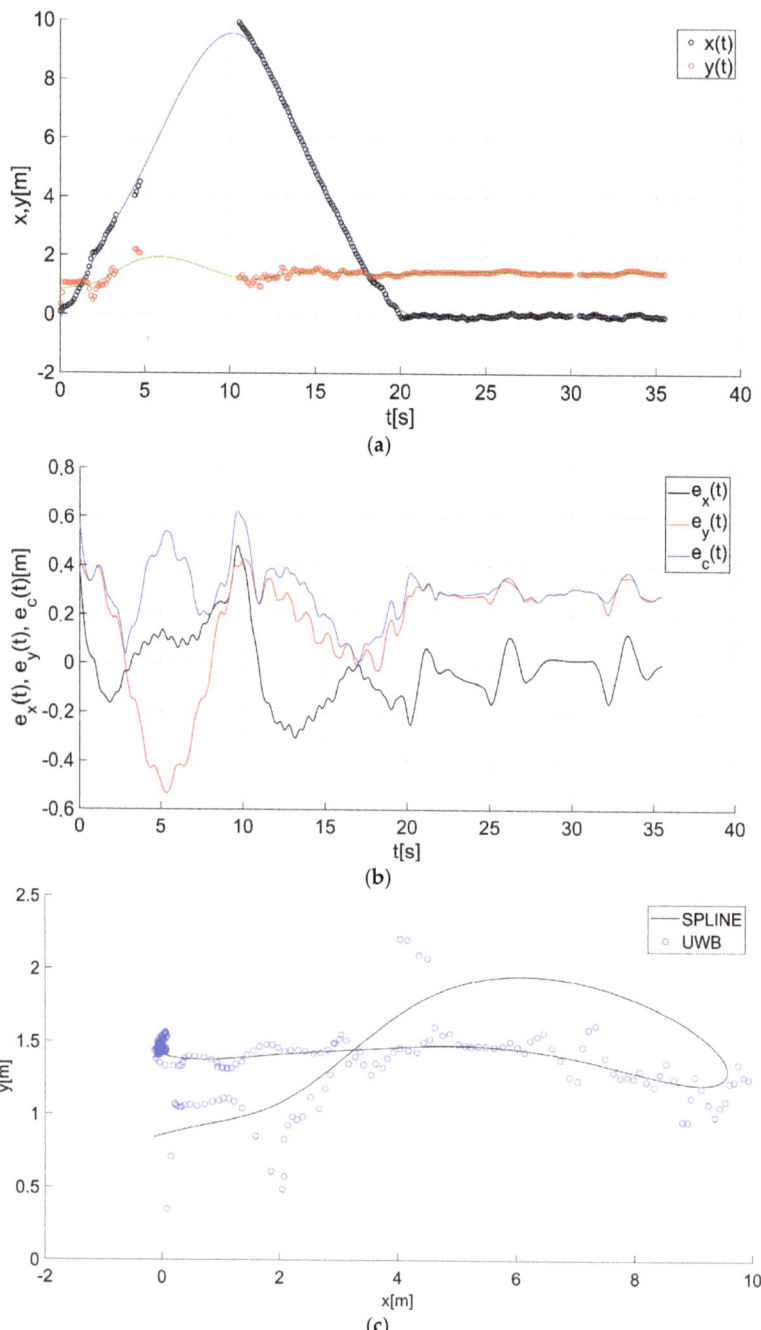

**Figure 19.** Results for the estimation of guide's path no. 7 with the use of smoothing spline: (**a**) the course of the guide's location coordinates x(t), y(t) with their continuous estimates, (**b**) the course of the estimated location errors $e_x(t)$, $e_y(t)$, $e_c(t)$, (**c**) guide's location along with the path estimation [6].

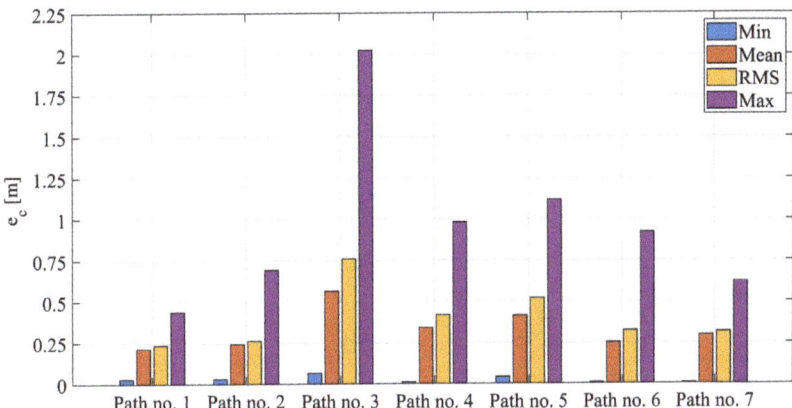

**Figure 20.** Basic descriptive statistics: minimum, mean, root mean square, and maximum of total errors for paths no. 1–7.

## 4. Conclusions

The method of determining the guide's route using a smoothing spline was discussed in the article. In the context of the "follow-me" systems, there has been no systematic study of the application of the smoothing spline method for planning the movement of the UGV, so the article represents an innovation in the field.

The influence of the smoothing parameter on the estimation of the guide's route was also determined in the article. In addition, no studies were found that examined the impact of the described parameter on path estimation.

As a result of the implemented mathematical model, it has been shown that the smoothing spline can recreate the path of the guide after a 10 s period of decay of the guide's localization results. The occurrence of the aforementioned guide's location decays increased the total errors for estimating the guide's route.

The value of the smoothing parameter affects the guide's route estimation. The choice of the final value of the smoothing parameter requires additional experimental studies. The dependence existing in the literature [36] that allows for automatically determining the value of the smoothing parameter for any data is not universal and it is only a preliminary estimate of the parameter value. It always has to be adapted to the application under consideration.

The value of the smoothing parameter also affects the estimation of the linear acceleration of the guide. An increase in the value of the smoothing parameter increases the mean square value of the linear acceleration of the guide.

**Author Contributions:** Conceptualization, Ł.R., A.T., R.T. and M.R.; data curation, Ł.R., R.T. and M.R.; formal analysis, Ł.R. and A.T.; funding acquisition, Ł.R.; investigation, Ł.R., A.T. and R.T.; methodology, Ł.R., A.T., R.T. and M.R.; project administration, A.T. and R.T.; resources, Ł.R.; software, Ł.R.; supervision, A.T. and R.T.; validation, Ł.R., A.T., R.T. and M.R.; visualization, Ł.R.; writing—original draft, Ł.R., A.T., R.T. and M.R.; writing—review and editing, Ł.R., A.T., R.T. and M.R. All authors have read and agreed to the published version of the manuscript.

**Funding:** This research received no external funding.

**Institutional Review Board Statement:** Not applicable.

**Informed Consent Statement:** Not applicable.

**Data Availability Statement:** The data presented in this study are available on request from the corresponding author.

**Conflicts of Interest:** The authors declare no conflict of interest.

## References

1. Guastella, D.C.; Muscato, G. Learning-Based Methods of Perception and Navigation for Ground Vehicles in Unstructured Environments: A Review. *Sensors* **2021**, *21*, 73. [CrossRef] [PubMed]
2. Liu, O.; Yuan, S.; Li, Z.A. Survey on Sensor Technologies for Unmanned Ground Vehicles. In Proceedings of the 3rd International Conference on Unmanned Systems (ICUS), Harbin, China, 27 November 2020; pp. 638–645. [CrossRef]
3. Islam, M.J.; Hong, J.; Sattar, J. Person-following by autonomous robots: A categorical overview. *Int. J. Robot. Res.* **2019**, *38*, 1581–1618. [CrossRef]
4. Malon, K.; Łopatka, J.; Łopatka, M. UWB based follow-me system for unmanned ground vehicle operator. In Proceedings of the 12th International Scientific Conference Intelligent Technologies in Logistics and Mechatronics Systems (ITELMS 2018), Penevezys, Lithuania, 27 April 2018; pp. 187–194.
5. Bonci, A.; Cen Cheng, P.D.; Indri, M.; Nabissi, G.; Sibona, F. Human-Robot Perception in Industrial Environments: A Survey. *Sensors* **2021**, *21*, 1571. [CrossRef] [PubMed]
6. Rykala, Ł. Shaping the Guide Location System in Terms of Determining the Route of the Unmanned Ground Vehicle. Ph.D. Thesis, Military University of Technology, Warsaw, Poland, 2022.
7. Feng, T.; Yu, T.; Wu, L.; Bai, Y.; Xiao, Z.; Lu, Z. A Human-Tracking Robot Using Ultra Wideband Technology. *IEEE Access* **2018**, *6*, 42541–42550. [CrossRef]
8. Rykała, Ł.; Typiak, A.; Typiak, R. Research on Developing an Outdoor Location System Based on the Ultra-Wideband Technology. *Sensors* **2020**, *20*, 6171. [CrossRef] [PubMed]
9. Alarifi, A.; Al-Salman, A.; Alsaleh, M.; Alnafessah, A.; Al-Hadhrami, S.; Al-Ammar, M.A.; Al-Khalifa, H.S. Ultra Wideband Indoor Positioning Technologies: Analysis and Recent Advances. *Sensors* **2016**, *16*, 707. [CrossRef] [PubMed]
10. Vitanov, R.I.; Nikolov, D.N. A State-of-the-Art Review of Ultra-Wideband Localization. In Proceedings of the 57th International Scientific Conference on Information, Communication and Energy Systems and Technologies (ICEST), Ohrid, North Macedonia, 16–18 June 2022; pp. 1–4. [CrossRef]
11. Zafari, F.; Gkelias, A.; Leung, K.K. A Survey of Indoor Localization Systems and Technologies. *IEEE Commun. Surv. Tutor.* **2019**, *21*, 2568–2599. [CrossRef]
12. Typiak, R.; Rykała, Ł.; Typiak, A. Configuring a UWB Based Location System for a UGV Operating in a Follow-Me Scenario. *Energies* **2021**, *14*, 5517. [CrossRef]
13. Terabee Follow-Me System for Mobile Robots. Available online: https://www.terabee.com/introducing-a-new-relative-positioning-system-for-mobile-platforms/ (accessed on 25 October 2022).
14. Weston Robot—UWB Follow-Me. Available online: https://www.westonrobot.com/uwb-follow-me (accessed on 25 October 2022).
15. Zhang, H.-Y.; Lin, W.-M.; Chen, A.-X. Path Planning for the Mobile Robot: A Review. *Symmetry* **2018**, *10*, 450. [CrossRef]
16. Koubaa, A.; Bennaceur, H.; Chaari, I.; Trigui, S.; Ammar, A.; Mohamed-Foued, S.; Alajlan, M.; Cheikhrouhou, O.; Javed, Y. Introduction to Mobile Robot Path Planning. Robot Path Planning and Cooperation. In *Studies in Computational Intelligence*; Springer: Cham, Switzerland, 2018; Volume 772.
17. Tzafestas, S.G. *Introduction to Mobile Robot Control*; Elsevier: Amsterdam, The Netherlands, 2014.
18. Boehm, W. *Numerical Methods*; CRC Press: Boca Raton, FL, USA, 2021.
19. Lunardi, A. *Interpolation Theory*; CRM Series; Edizioni della Normale: Pisa, Italy, 2018; Volume 16.
20. Hoffman, J.D.; Frankel, S. *Numerical Methods for Engineers and Scientists*; CRC Press: Boca Raton, FL, USA, 2018.
21. Young, D.S. *Handbook of Regression Methods*; Chapman and Hall/CRC: London, UK, 2017.
22. Mahmoud, H.F. Parametric Versus Semi and Nonparametric Regression Models. *Int. J. Stat. Probab.* **2021**, *10*, 1–90. [CrossRef]
23. Montgomery, D.C.; Peck, E.A.; Vining, G.G. *Introduction to Linear Regression Analysis*; John Wiley & Sons: New Jersey, NJ, USA, 2021.
24. Čížek, P.; Sadıkoğlu, S. Robust nonparametric regression: A review. *Wiley Interdiscip. Rev. Comput. Stat.* **2020**, *12*, e1492. [CrossRef]
25. Gorji, F.; Aminghafari, M. Robust nonparametric regression for heavy-tailed data. *J. Agric. Biol. Environ. Stat.* **2020**, *25*, 277–291. [CrossRef]
26. Rahmawati, D.P.; Budiantara, I.N.; Prastyo, D.D.; Octavanny, M.A.D. Mixture Spline Smoothing and Kernel Estimator in Multi-Response Nonparametric Regression. *IAENG Int. J. Appl. Math.* **2021**, *51*, 1–12.
27. Sahoo, D.; Hoi, S.C.; Li, B. Large scale online multiple kernel regression with application to time-series prediction. *ACM Trans. Knowl. Discov. Data* **2019**, *13*, 1–33. [CrossRef]
28. Widyastuti, D.A.; Fernandes, A.A.R.; Pramoedyo, H. Spline estimation method in nonparametric regression using truncated spline approach. *J. Phys. Conf. Ser.* **2021**, *1872*, 012027. [CrossRef]
29. Zhang, X.; Datta, G.S.; Ma, P.; Zhong, W. Bayesian spline smoothing with ambiguous penalties. *Can. J. Stat.* **2022**, *50*, 20–35. [CrossRef]
30. Mariati, N.P.A.M.; Budiantara, I.N.; Ratnasari, V. The Application of Mixed Smoothing Spline and Fourier Series Model in Nonparametric Regression. *Symmetry* **2021**, *13*, 2094. [CrossRef]
31. Dontchev, A.L.; Kolmanovsky, I.V.; Tran, T.B. Constrained data smoothing via optimal control. *Optim. Control Appl. Methods* **2022**, *43*, 1257–1269. [CrossRef]
32. Kernel Regression. Available online: https://bowtiedraptor.substack.com/p/kernel-regression (accessed on 25 October 2022).

33. Kernel Regression—With Example and Code. Available online: https://medium.com/towards-data-science/kernel-regression-made-easy-to-understand-86caf2d2b844 (accessed on 25 October 2022).
34. LOESS. Available online: https://towardsdatascience.com/loess-373d43b03564 (accessed on 25 October 2022).
35. Ravankar, A.; Ravankar, A.A.; Kobayashi, Y.; Hoshino, Y.; Peng, C.-C. Path Smoothing Techniques in Robot Navigation: State-of-the-Art, Current and Future Challenges. *Sensors* **2018**, *18*, 3170. [CrossRef] [PubMed]
36. Curve Fitting Toolbox™ User's Guide. Available online: https://www.mathworks.com/help/pdf_doc/curvefit/curvefit.pdf (accessed on 25 October 2022).
37. Decawave TREK1000. Available online: https://www.digikey.pl/pl/product-highlight/d/decawave/trek1000-twr-rtls-ic-evaluation-kit (accessed on 25 October 2022).
38. Duro User Manual—SwiftNav. Available online: https://www.swiftnav.com/latest/duro-user-manual (accessed on 25 October 2022).

Article

# Improved Monitoring of Wildlife Invasion through Data Augmentation by Extract–Append of a Segmented Entity

Jaekwang Lee, Kangmin Lim and Jeongho Cho *

Department of Electrical Engineering, Soonchunhyang University, Asan 31538, Korea
* Correspondence: jcho@sch.ac.kr

**Abstract:** Owing to the continuous increase in the damage to farms due to wild animals' destruction of crops in South Korea, various methods have been proposed to resolve these issues, such as installing electric fences and using warning lamps or ultrasonic waves. Recently, new methods have been attempted by applying deep learning-based object-detection techniques to a robot. However, for effective training of a deep learning-based object-detection model, overfitting or biased training should be avoided; furthermore, a huge number of datasets are required. In particular, establishing a training dataset for specific wild animals requires considerable time and labor. Therefore, this study proposes an Extract–Append data augmentation method where specific objects are extracted from a limited number of images via semantic segmentation and corresponding objects are appended to numerous arbitrary background images. Thus, the study aimed to improve the model's detection performance by generating a rich dataset on wild animals with various background images, particularly images of water deer and wild boar, which are currently causing the most problematic social issues. The comparison between the object detector trained using the proposed Extract–Append technique and that trained using the existing data augmentation techniques showed that the mean Average Precision (mAP) improved by $\geq 2.2\%$. Moreover, further improvement in detection performance of the deep learning-based object-detection model can be expected as the proposed technique can solve the issue of the lack of specific data that are difficult to obtain.

**Keywords:** object detection; surveillance; semantic segmentation; data augmentation

Citation: Lee, J.; Lim, K.; Cho, J. Improved Monitoring of Wildlife Invasion through Data Augmentation by Extract–Append of a Segmented Entity. *Sensors* **2022**, *22*, 7383. https://doi.org/10.3390/s22197383

Academic Editor: José María Martínez-Otzeta

Received: 4 September 2022
Accepted: 26 September 2022
Published: 28 September 2022

**Copyright:** © 2022 by the authors. Licensee MDPI, Basel, Switzerland. This article is an open access article distributed under the terms and conditions of the Creative Commons Attribution (CC BY) license (https:// creativecommons.org/licenses/by/ 4.0/).

## 1. Introduction

Damage to crops due to attacks by wild animals is one of the primary reasons for a reduction in crop yield. As with indiscriminate logging and the expansion of urban environments, including roads and buildings, incidents of crop attacks by wild animals have increased as they have lost their habitats. According to the Ministry of Environment in South Korea, the amount of damage to crops by wild animals between 2014 and 2018 was ~57 billion KRW, which is 11.4 billion KRW annually; the damage by wild boars and water deer is the largest [1]. Water deer are listed as endangered in the International Union for Conservation of Nature Red List of Threatened Species. Wild boars usually inhabit deep mountains and areas with broad-leaved trees, but during the mating season or preparation for winter, they often come down to urban areas in search of food. Particularly, the ecosystems near urban areas do not have the predators of wild boar; therefore, their population increases. There are several incidents where water deer and wild boars, having had a huge increase in population, destroy crops and appear in residential areas, causing damage to people's life or properties [1] (Figure 1).

(a)　　　　　　　　　　　　(b)

**Figure 1.** Examples of the threats and damage by wild animals: (**a**) appearance of wild animals in urban areas; (**b**) destruction of crops by wild animals.

To mitigate such damages, farms have attempted to dispel animals by installing electrical fences or using sound and light via warning lamps and explosive ultrasound. However, electrical fences may lead to casualties and, if damaged, may cause high maintenance costs. Additionally, warning lamps or explosive ultrasound can become less effective in the future as animals get accustomed to them. Recently, methods that prevent the invasion of wild animals have been proposed, which use robots equipped with deep learning-based object-detection technology to detect wild animals and use LEDs and alarms only when they detect objects in real-time monitoring [2]. However, deep learning-based object-detection technology requires sufficient data to train the deep learning model. Currently, training data are collected either by directly taking pictures of objects, extracting images from video recordings, or web crawling. However, there are certain limitations humans have in acquiring images of wild animals, such as access challenges, leading to challenges in model training. Overfitting can also be an issue while training a model in such a case [3–7].

To overcome the aforementioned data-collection issues, a large amount of training data can be generated via data augmentation [8,9]. Data augmentation is a technique that increases the amount of limited data artificially by increasing the number of images through applying different types of transformation to an original image. Although various data augmentation techniques have been proposed, there are still many limitations in performing data augmentation with a limited amount of data. Therefore, this study proposes an Extract–Append data augmentation method where only the objects of interest, specifically wild boars and water deer, are extracted from a minimal number of images via semantic segmentation, and corresponding objects are appended automatically to numerous background images. Masks, the shapes of the objects to be extracted from the segmentation network, are acquired, and the segmented objects are produced by the binarization and synthesis process. Later, the augmented training data are acquired from the inverse binarization and the synthesis of various background images. This study compared and evaluated the object-detection performance of the proposed and existing data augmentation methods to verify the usefulness of the proposed method. The contribution of this study is as follows:

1. It proposes the Extract–Append data augmentation method, which automatically generates a large amount of diverse data by extracting the masks of the objects of interest from segmentation and synthesizing them with countless arbitrary backgrounds.
2. It enables the synthesis of the object with various backgrounds without losing the original object shape by suggesting a data-processing method, which synthesizes the extracted object with the background image after creating a space so that the extracted object shape can be maintained as accurately as possible on the arbitrary background image.
3. It provides a method that could extract the mask of an object to facilitate additional training automatically, even if new background images were acquired later based on the previously trained model on a specific object.

The rest of the article comprises the following sections: in Section 2, the related research on data augmentation is described; in Section 3, the proposed Extract–Append data augmentation technique is explained; in Section 4, the test process and results are presented; and, finally, in Section 5, the present study is concluded.

## 2. Related Works

Generally, data augmentation uses spatial-level transformation and pixel-level transformation. The former involves applying spatial changes to an object. For example, it includes flipping, rotating, and cropping [10–13]. In contrast, the latter involves pixel-level image transformation and includes contrast, which adjusts the ratio of contrast in an image, and the addition of random noise to increase the adaptability of the data under various environments [14,15]. Other methods have been proposed, including cutout, which removes a part of the image by randomly masking it with squares [16], or mixup [17], which generates new data by mixing up two images by a certain ratio. However, if data augmentation is performed with a minimum number of images, only the images with limited backgrounds (environments) are produced, which makes it difficult to expect an improvement in the performance of the detection model, and in this case, data augmentation via cutout or mixup can instead play the role of noise [18,19].

Various data augmentation methods have been proposed to resolve these issues. D. Yorioka et al. [20] attempted to solve the lack of data by generating a significant number of fake images based on GAN; however, GAN training requires tremendous time, and it is difficult to train a GAN model effectively with a minimal number of data. V. Olsson et al. [21] proposed ClassMix, which increased the amount of data by synthesizing the backgrounds and objects extracted from the segmentation. However, this requires training of objects and environments, and in synthesizing the extracted objects and environments, some information can be lost. S. Bang et al. [22] proposed a method to extract the objects in an image by masking and generating backgrounds from the masked space via GAN; however, under the condition where only a limited number of data could be used, the GAN-based background-generation process may result in distorted backgrounds and a long training time. G. Ghiasi et al. [23] suggested a method that arbitrarily selects two images and, after random scaling, attaches the object to another image. However, even this method cannot overcome the issue of diversity if a small number of images limits it, and it cannot avoid the degradation of image resolution during the random scaling process. Table 1 summarizes the strengths and weaknesses of existing and proposed augmentation techniques.

**Table 1.** Comparison of the strengths and weaknesses of existing and proposed augmentation techniques.

| Augmentation Method | Strengths | Weaknesses |
|---|---|---|
| Conventional | - We can create additional images by changing the direction and angle of the object based on the acquired image.<br>- We can obtain additional images by adjusting the contrast ratio of the acquired image or adding noise to the image. | - Because it is augmented using only the collected images, there is a limit to the diversity of the object's environment.<br>- Regardless of the object, every pixel within the image may be transformed, changing the object's unique characteristics. |
| Proposed | - Objects in the collected image can be combined with various random backgrounds to create an unlimited variety of data.<br>- The mask for the object is extracted through segmentation and combined with a random background, so it is very unlikely to act as noise. | - The object's mask quality is determined by its segmentation performance.<br>- There is a slight sense of heterogeneity because the object is pasted on a random background after extraction. |

## 3. Methodology

The existing data augmentation techniques can enable augmentation only for acquired image data. Therefore, they are limited in diversity and in the number of images that can be augmented. This study proposes the Extract–Append technique that can generate a large amount of diverse data by extracting objects using masks obtained through segmentation and synthesizing them with arbitrary backgrounds to solve these problems. The acquired limited image produces the mask of an object through a segmentation network. Subsequently, the binarization process transforms it into a binary mask, which is synthesized with an input image to extract the concerned object. The binary mask is again transformed into a mask to secure a space in the object's shape, which is to be added to an arbitrary background through the inverse binarization process. Synthesis of the transformed mask and the new background image produces a background image with an object-shaped space, and then it is added to the extracted object to create a new image. The augmented image data are used in training the detection network. Figure 2 gives an overview of the object-detection system, including the proposed Extract–Append technique.

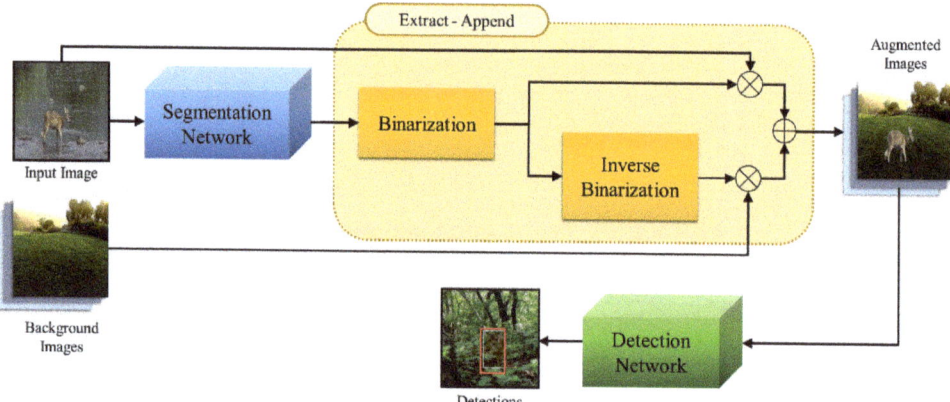

**Figure 2.** Block diagram of the object-detection system, including the proposed Extract–Append technique.

### 3.1. Semantic Segmentation

One of the most important application areas in image processing is segmentation, which categorizes and classifies images into similar regions in terms of semantic or cognitive perspectives. Here, semantic segmentation is a technique that can discern objects not by simple boundaries but by semantic regions and aims to classify objects by determining what each object signifies in an image that contains various objects, including cars, people, animals, and trees. When classifying an object, all pixels are grouped and categorized according to similar colors; through this classification, the mask of an object is extracted. To extract a more accurate mask for an object, manual photoshopping or GrabCut could be used; however, the study considered semantic segmentation to automatically extract the mask of specific objects universally. Generally, semantic segmentation is in an encoder–decoder structure. The encoder gradually performs downsampling to reduce the amount of calculation based on the size of an input image and improve the calculation speed to extract and compress the features of the object information to be extracted. However, the decoder performs upsampling to recover the lost spatial information due to reducing the spatial dimension in the encoder, and gradually attempts to recover clear object boundaries. In this way, semantic segmentation extracts the mask containing the object's information [24–26].

### 3.2. Extract–Append for Data Augmentation

The proposed Extract–Append data augmentation process is summarized in Algorithm 1, in which the shape of an object is extracted using the mask of the said extracted object once the segmentation network training is completed, and the object is appended to various arbitrary backgrounds.

---

**Algorithm 1.** Extract–Append Algorithm

---

**Require:** Pretrained semantic segmentation model $\Phi$
**Input:** Input image containing an object $I_{obj}$, Background image $I_{back}$
**Output:** Create new image $A_{obj}$

1: $M_S \leftarrow \Phi(I_{obj})$ ▷ Extract the mask of an object
2: $\hat{M}_S = \begin{cases} 1, \text{object} \\ 0, \text{background} \end{cases}$ ← Binarization of $M_S$
3: **for** each iteration do
4: $E_{obj} \leftarrow I_{obj} \odot \hat{M}_S$ ▷ Extract an object from $I_{obj}$
5: $C_{back} \leftarrow I_{obj} \odot |1 - \hat{M}_S|$ ▷ Making room for object insertion in $I_{back}$
6: $A_{obj} \leftarrow E_{obj} + C_{back}$
7: **end for**

---

From the image $I_{obj}$, acquired from web crawling and video frames that include the concerned object, the RGB 3-channel mask $M_S$ of the object is extracted via the segmentation network $\Phi$. This is transformed into a 1-channel binary mask $\widehat{M_S}$ that has either a 0 or 1 value via the binarization process, and the object $E_{obj}$ extracted by using this mask can be derived as in the following equation:

$$E_{obj} = I_{obj} \odot \widehat{M_S} \qquad (1)$$

Here, $\odot$ refers to a dot product. In other words, since the object has a value of 1 in the binary mask, the dot product of the binary mask and the input image results in a black background, and only the object retains its original color. In this way, the object is solely extracted. Later, to synthesize the extracted object with an arbitrary background $I_{back}$, an inversion of the binary mask is again performed so that the background is 1 and the object is 0. The dot product of the transformed mask and an arbitrary background results in an arbitrary background $C_{back}$ that has a value of 0 in the space of the object's shape to be appended, as shown in the following equation:

$$C_{back} = I_{back} \odot |1 - \widehat{M_S}| \qquad (2)$$

If $E_{obj}$ with only object information is added to this, a new image $A_{obj}$, which is an arbitrary background with the appended object, is created.

$$A_{obj} = E_{obj} + C_{back} \qquad (3)$$

A detailed block diagram of the proposed Extract–Append data augmentation process is illustrated in Figure 3.

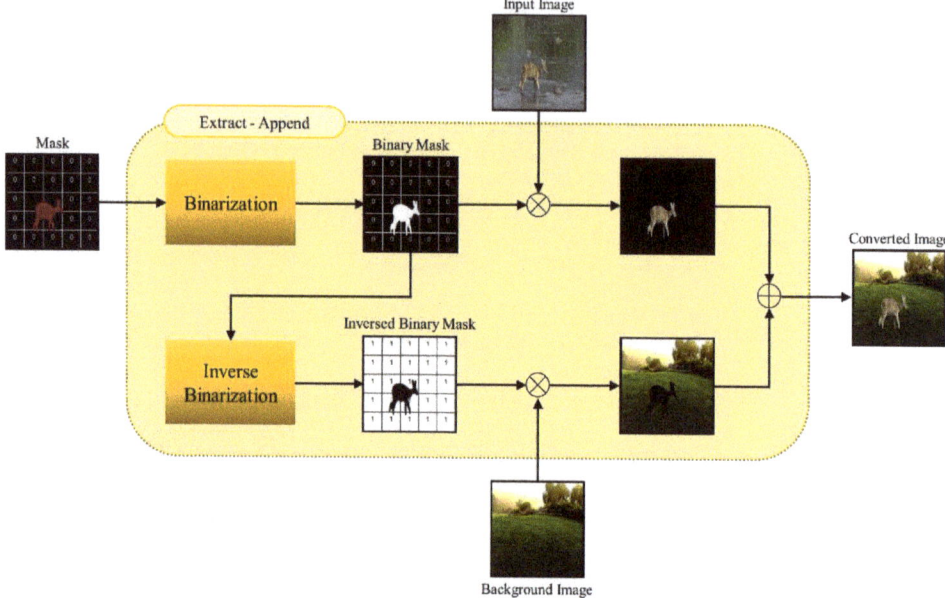

**Figure 3.** Block diagram of the proposed Extract–Append data augmentation process.

*3.3. Object Detection*

A large amount of augmented data generated by the proposed technique are used in the deep learning-based object-detection model for the surveillance of wild animals—the ultimate aim of this study. The detection model allowed for real-time processing and considered the You Only Look Once (YOLO) network, a one-stage detection method that performs classification and localization simultaneously. YOLO categorizes the input image into grids of S × S size, and each grid cell estimates B number of bounding boxes and the bounding box's confidence score (CS). Here, a bounding box has five pieces of information (x, y, w, h, and C). x and y are the box's central coordinates, corresponding to the boundary of the grid cell, and w and h also refer to the width and height, corresponding to the grid cell. Finally, C refers to the probability that the bounding box is included in a specific object. CS is the multiplication of the probability that the bounding box is included in the object $PR_{obj}$ and the Intersection over Union (IoU), the width of the overlapping region between the estimated and real values, and refers to the degree of confidence that an object exists within the bounding box as shown below.

$$CS = PR_{obj} \times IoU \tag{4}$$

Each grid cell estimates the CS of N number of classes, and Conditional Class Probability (CCP)—which is the probability that if an object exists in a cell, it will be the $k$ st class—is defined as shown below.

$$CCP = PR(Class(k)|Object) \tag{5}$$

Therefore, the class-specific CS (CCS), which refers to how identical the probability that a specific object exists in each bounding box is to the actual value, can be summarized as below. The bounding box with the highest CCS among B number of bounding boxes that each grid cell estimated for an object is determined to be the bounding box for the said object [27].

$$CSS = PR_{obj} \times IoU \times CCP \tag{6}$$

## 4. Experimental Results

To realize a model for monitoring wild animals, such as water deer and wild boars, through the proposed data augmentation structure, the mask of objects should be first extracted, and to this end, the study used the segmentation model DeepLabv3+ [24]. The study trained the object-detection model using augmented data after embodying an Extract–Append processor based on the extracted mask of the object and attempted to validate the usefulness of the proposed method by evaluating the model's detection performance. As a model for evaluating the object-detection performance, the study considered YOLOV4-tiny, and the training was performed with a NVIDIA RTX 3060 and Intel Core i7-1200F CPU. The reason for choosing YOLOV4-tiny among the various YOLO models is because it allows for an easy realization of an onboard embedded system and real-time processing. Its processing speed is relatively much faster than more recent models while its performance is slightly poor. A high-performance computer needs to be used to realize object detection using YOLOv4 in an actual farm, but this is unrealistic to carry out. In contrast, YOLOV4-tiny can allow for real-time object recognition on an embedded single board computer, such as Jetson Nano.

The resolution of the input images was 416 × 416, and to compare the performance of the proposed Extract–Append technique with the existing data augmentation-based object-detection performance, the study categorized the dataset used in training into five types. Dataset D1 used in the evaluation test was created assuming that only 60 images per class were acquired by the image extraction from the videos or web crawling of wild boars and water deer, according to the aim of augmenting the data with a minimal number of images acquired limitedly. Dataset D2 was created with 480 images per class by adding the data transformed from D1 via spatial-level transformation. Augmentation by spatial-level transformation is one of the most widely used data augmentation techniques, and thus, it was included in all dataset constructions, except for D1. Dataset D3 was created with 540 images per class by image-contrast augmentation, one of the pixel-level transformation techniques. Cut-and-paste augmentation was used to create Dataset D4, with 1080 images per class, which is similar to the Extract–Append technique proposed in this study. Finally, Dataset D5 was created with 1480 images per class by Extract–Append augmentation, which conveniently allows for synthesizing objects with unlimited, arbitrary background images. The object-detection performance evaluation via the proposed data augmentation used 100 images per class. Figure 4 shows examples of the results from the data augmentation technique used in the training of the object-detection model.

(a)　　　　　　(b)　　　　　　(c)　　　　　　(d)　　　　　　(e)

**Figure 4.** Example of water deer and wild boar images transformed by data augmentation: (**a**) original image (D1); (**b**) spatial-level transformation (D2); (**c**) pixel-level transformation (D3); (**d**) cut-and-paste (D4); (**e**) Extract–Append (D5).

The evaluation index for validating the performance of the model used the mean Average Precision (mAP), with Precision and Recall defined as follows:

$$\text{Precision} = \frac{\text{TP}}{\text{TP} + \text{FP}} \qquad (7)$$

$$\text{Recall} = \frac{\text{TP}}{\text{TP} + \text{FN}} \qquad (8)$$

Here, TP refers to True Positive, meaning that an object that needs to be detected was detected; FP refers to False Positive, meaning that an object that should not be detected was detected; FN refers to False Negative, meaning that an object that should not be detected was not detected; Precision, meaning accuracy, refers to the ratio of the objects detected by the model that was correctly detected; and Recall, meaning reproduction, refers to the ratio of the objects that should be detected and were correctly detected by the model. A PR curve is the cumulation of Precision and Recall from the highest CS, which is a value that expresses how accurately a model detects an object. The x-axis is Recall, and the y-axis is Precision. While the PR curve can determine the Precision value by the change in Recall, it is inconvenient to compare the performance of each technique quantitatively. To solve this inconvenience, the Average Precision, the area below the PR curve, is used, and the performance of each technique is evaluated by mAP, which is the Average Precision of each object if there are multiple objects divided by their numbers.

The study performed an evaluation test by changing IoU to examine the performance level of the object detector trained with the data generated by the proposed Extract–Append technique; the result is presented in Table 2. With IoU at 0.3, the performance of the object detector trained with the proposed D5 improved by 0.6% and 2.6% at minimum and maximum, respectively, based on mAP compared with that of the object detectors trained with D1 to D4. In addition, even if IoU increased to 0.5, the object detector trained with D5 showed a higher performance of 0.8% and 3.7% at minimum and maximum, respectively, than the object detectors trained with D1 to D4. Furthermore, when IoU is 0.7, the performance of the object detector based on the proposed technique improved by up to 34.8%, showing a 2.1% higher improvement from the object detector trained with D4, which is similar to the Extract–Append approach. It should be noted here that since D2 augments only a limited number of images, it is restricted in the number of images to be augmented as it faces the issue of the diversity of data. As discussed earlier, D3 can rather degrade the performance of an object detector as it adds noise to the limited number of images, thus showing the smallest performance improvement among all the data augmentation techniques used in the evaluation test. D4 extracts objects from existing images and fills the extracted space with RGB, similar to the background, using GAN. However, at this point, GAN takes a long time to be trained, and, as shown in Figure 4d, noise is added to the generated image, leading to limited performance improvement. Furthermore, it uses the existing background again; thus, as in D2 discussed above, it is limited with respect to the diversity of data. In contrast, the proposed D5 uses semantic segmentation to extract and synthesize an object to an arbitrary, intact background image. Therefore, the training time is much shorter than the method based on GAN, and since the synthesis uses various arbitrary background images, it can solve the data diversity issue. Furthermore, there is no limitation in the number of data to be augmented, resulting in a better performance of the object detector than the one trained with D1 to D4.

Figure 5 illustrates the examples of the results of the object-detection model trained with data augmented by each technique, with the IoU at 0.7. The blue box in the figure is the result where the model identified the object correctly, the red color indicates that the model incorrectly identified the object as another object, and the white color shows the ground-truth.

Table 2. Comparison of the object-detection performance by data augmentation techniques.

| Data | Data Augmentation | | | | Class | AP0.3 | AP0.5 | AP0.7 | mAP0.3 | mAP0.5 | mAP0.7 |
|---|---|---|---|---|---|---|---|---|---|---|---|
| | Spatial-Lev. Trans. | Pixel-Lev. Trans. | Cut-Paste | Extract-Append | | | | | | | |
| D1 | | | | | WaterDeer | 93.0 | 91.7 | 51.3 | 93.9 | 92.4 | 57.8 |
| | | | | | WildBoar | 93.9 | 93.0 | 63.8 | | | |
| D2 | O | | | | WaterDeer | 95.8 | 95.5 | 89.4 | 94.8 | 94.4 | 89.0 |
| | | | | | WildBoar | 93.7 | 93.7 | 88.7 | | | |
| D3 | O | O | | | WaterDeer | 93.6 | 93.4 | 65.1 | 93.5 | 93.0 | 68.3 |
| | | | | | WildBoar | 93.5 | 92.6 | 71.6 | | | |
| D4 | O | | O | | WaterDeer | 97.0 | 96.9 | 92.1 | 95.5 | 95.3 | 90.5 |
| | | | | | WildBoar | 94.8 | 93.7 | 88.9 | | | |
| D5 | O | | | O | WaterDeer | 97.2 | 97.2 | 94.3 | 96.1 | 96.1 | 92.6 |
| | | | | | WildBoar | 95.1 | 95.1 | 91.0 | | | |

**Figure 5.** Example of water deer and wild boar detection results by data augmentation techniques: (**a**) data augmentation not applied (D1); (**b**) spatial-level transformation (D2); (**c**) pixel-level transformation (D3); (**d**) cut-and-paste (D4); (**e**) Extract–Append (D5).

## 5. Conclusions

An unbiased, diverse, and significant amount of data is necessary when training a deep learning-based object-detection model. Notably, building a training dataset for specific objects requires considerable time and labor, and generally, this is resolved through data augmentation. However, existing data augmentation techniques rely on spatial- or pixel-level transformation of images, which is limited in augmenting data based on a minimal number of images, resulting in a degraded model performance and the problem of diversity of training images. Therefore, this study proposed an Extract–Append data augmentation technique to resolve the issue of a lack of specific data and promote the performance improvement of a deep learning-based object-detection model. The proposed data augmentation technique extracts only specific objects through semantic segmentation, generates a diverse and vast amount of augmented training data from synthesis with varying arbitrary background images, and synthesizes the data without changing the shape of the extracted objects. The study conducted a performance comparison test between the object detector based on the proposed Extract–Append technique and the others based on the existing data augmentation techniques, which demonstrated that the object detector trained with the proposed approach showed a detection performance improvement of up to 34.8%. In addition, compared to the cut-and-paste technique, the proposed technique improved the detection performance by 2.1%. Following these results, it is anticipated that the proposed data augmentation technique can solve the issues of lack of data and diversity to enhance the performance of various deep learning-based rare object-detection models. In the future, we will conduct additional training on other rare objects besides water deer and wild boars. We aim to generalize the proposed data augmentation technique by synthesizing these objects with various background images. Furthermore, we will also continue to complement the mask-extraction technique, which lacks data in the segmentation process.

**Author Contributions:** J.L., K.L. and J.C. took part in the discussion of the work described in this paper. All authors have read and agreed to the published version of the manuscript.

**Funding:** This research was funded by a National Research Foundation of Korea (NRF) grant funded by the Korean government (MOE) (No.2021R1I1A3055973) and the Soonchunhyang University Research Fund.

**Institutional Review Board Statement:** Not applicable.

**Informed Consent Statement:** Not applicable.

**Data Availability Statement:** The data presented in this study are available on request from the corresponding author.

**Conflicts of Interest:** The authors declare that they have no competing interests.

## References

1. Ministry of Environment. Current Status of Farm Damage Caused by Harmful Wild Animals by Year. Available online: http://me.go.kr/home/web/policy_data/read.do?pagerOffset=0&maxPageItems=10&maxIndexPages=10&searchKey=&searchValue=&menuId=10261&orgCd=&condition.code=A2&condition.deleteYn=N&seq=7009 (accessed on 1 August 2022).
2. Kim, D.; Yoo, S.; Park, S.; Kim, D.; Lee, J. Design and Implementation of Real-Time Monitoring Platform for Preventing Wild Animals. *J. Korean Inst. Commun. Inf. Sci.* **2021**, *46*, 1294–1300. [CrossRef]
3. Rao, J.; Zhang, J. Cut and Paste: Generate Artificial Labels for Object Detection. In Proceedings of the International Conference on Video and Image Processing, New York, NY, USA, 27 December 2017.
4. Marcus, D.B.; Christof, S.; Andreas, H. Augmentor: An Image Augmentation Library for Machine Learning. *arXiv* **2017**, arXiv:1708.04680v1.
5. Perez, L.; Wang, J. The Effectiveness of Data Augmentation in Image Classification using Deep Learning. *J. Open Source Softw.* **2017**, *2*, 432.
6. Mahajan, D.; Girshick, R.; Ramanathan, V.; He, K.; Paluri, M.; Li, Y.; Bharambe, A.; van der Maaten, L. Exploring the Limits of Weakly Supervised Pretraining. In Proceedings of the European Conference on Computer Vision, Munich, Germany, 8–14 September 2018.
7. Maharana, K.; Mondal, S.; Nemade, B. A review: Data pre-processing and data augmentation techniques. *Glob. Transit. Proc.* **2022**, *3*, 91–99. [CrossRef]

8. Mikoajczyk, A.; Grochowski, M. Data augmentation for improving deep learning in image classification problem. In Proceedings of the International Interdisciplinary PhD Workshop, Świnoujście, Poland, 9–12 May 2018.
9. Wong, S.; Gatt, A.; Stamatescu, V.; McDonnell, M. Understanding Data Augmentation for Classification: When to Warp? In Proceedings of the International Conference on Digital Image Computing: Techniques and Applications, Gold Coast, QLD, Australia, 30 November–2 December 2016.
10. Shijie, J.; Ping, W.; Peiyi, J.; Siping, H. Research on data augmentation for image classification based on convolution neural networks. In Proceedings of the Chinese Automation Congress, Jinan, China, 20–22 October 2017.
11. Krizhevsky, A.; Sutskever, I.; Hinton, G. ImageNet Classification with Deep Convolutional Neural Networks. *Commun. ACM* **2017**, *60*, 84–90. [CrossRef]
12. Yun, S.; Han, D.; Oh, S.; Chun, S.; Choe, J.; Yoo, Y. CutMix: Regularization Strategy to Train Strong Classifiers with Localizable Features. In Proceedings of the IEEE/CVF International Conference on Computer Vision, Seoul, Korea, 27 October–2 November 2019.
13. Xu, D.; Lee, M.; Hsu, W. Patch-Level Regularizer for Convolutional Neural Network. In Proceedings of the IEEE International Conference on Image, Taipei, Taiwan, 22–25 September 2019.
14. Noh, H.; You, T.; Mun, J.; Han, B. Regularizing Deep Neural Networks by Noise: Its Interpretation and Optimization. In Proceedings of the International Conference on Neural Information Processing Systems, Red Hook, NY, USA, 4–9 December 2017.
15. Jin, J.; Dundar, A.; Culurciello, E. Robust Convolutional Neural Networks under Adversarial Noise. *arXiv* **2015**, arXiv:1511.06306v1.
16. DeVries, T.; Taylor, G. Improved Regularization of Convolutional Neural Networks with Cutout. *arXiv* **2017**, arXiv:1708.04552v2.
17. Zhang, H.; Cisse, M.; Dauphin, Y.; Lopez-Paz, D. mixup: Beyond Empirical Risk Minimization. *arXiv* **2017**, arXiv:1710.09412v2.
18. Yang, S.; Xiao, W.; Zhang, M.; Guo, S.; Zhao, J.; Shen, F. Image Data Augmentation for Deep Learning: A Survey. *arXiv* **2022**, arXiv:2204.08610v1.
19. Shorten, C.; Khoshgoftaar, T. A survey on Image Data Augmentation for Deep Learning. *J. Big Data* **2019**, *6*, 1–48. [CrossRef]
20. Yorioka, D.; Kang, H.; Iwamura, K. Data Augmentation For Deep Learning Using Generative Adversarial Networks. In Proceedings of the IEEE Global Conference on Consumer Electronics, Kobe, Japan, 13–16 October 2020.
21. Olsson, V.; Tranheden, W.; Pinto, J.; Svensson, L. ClassMix: Segmentation-Based Data Augmentation for Semi-Supervised Learning. In Proceedings of the IEEE Conference on Applications of Computer Vision, Waikoloa, HI, USA, 3–8 January 2021.
22. Bang, S.; Baek, F.; Park, S.; Kim, W.; Kim, H. Image augmentation to improve construction resource detection using generative adversarial networks, cut-and-paste, and image transformation techniques. *Autom. Constr.* **2020**, *115*, 103198. [CrossRef]
23. Ghiasi, G.; Cui, Y.; Srinivas, A.; Qian, R.; Lin, T.; Cubuk, E.; Le, Q.; Zoph, B. Simple Copy-Paste is a Strong Data Augmentation Method for Instance Segmentation. In Proceedings of the IEEE/CVF Conference on Computer Vision and Pattern Recognition, Nashville, TN, USA, 20–25 June 2021.
24. Chen, L.; Papandreou, G.; Kokkinos, L.; Murphy, K.; YFuille, A. DeepLab: Semantic Image Segmentation with Deep Convolutional Nets, Atrous Convolution, and Fully Connected CRFs. *IEEE Trans. Pattern Anal. Mach. Intell.* **2018**, *40*, 834–848. [CrossRef] [PubMed]
25. Long, J.; Shelhamer, E.; Darrell, T. Fully Convolutional Networks for Semantic Segmentation. *IEEE Trans. Pattern Anal. Mach. Intell.* **2017**, *39*, 640–651.
26. Chen, L.; Zhu, Y.; Papandreou, G.; Schroff, F.; Adam, H. Encoder-Decoder with Atrous Separable Convolution for Semantic Image Segmentation. In Proceedings of the European Conference on Computer Vision, Munich, Germany, 8–14 September 2018.
27. Redmon, J.; Divvala, S.; Girshick, R.; Farhadi, A. You Only Look Once: Unified, Real-Time Object Detection. In Proceedings of the IEEE Conference on Computer Vision and Pattern Recognition, Las Vegas, NV, USA, 27–30 June 2016.

Article

# A Bio-Inspired Endogenous Attention-Based Architecture for a Social Robot

Sara Marques-Villarroya *, Jose Carlos Castillo *, Juan José Gamboa-Montero, Javier Sevilla-Salcedo and Miguel Angel Salichs

RoboticsLab, Universidad Carlos III de Madrid, 28911 Leganés, Spain; jgamboa@ing.uc3m.es (J.J.G.-M.); javier.sevilla@ing.uc3m.es (J.S.-S.); salichs@ing.uc3m.es (M.A.S.)
* Correspondence: smarques@ing.uc3m.es (S.M.-V.); jocastil@ing.uc3m.es (J.C.C.)

**Abstract:** A robust perception system is crucial for natural human–robot interaction. An essential capability of these systems is to provide a rich representation of the robot's environment, typically using multiple sensory sources. Moreover, this information allows the robot to react to both external stimuli and user responses. The novel contribution of this paper is the development of a perception architecture, which was based on the bio-inspired concept of *endogenous attention* being integrated into a real social robot. In this paper, the architecture is defined at a theoretical level to provide insights into the underlying bio-inspired mechanisms and at a practical level to integrate and test the architecture within the complete architecture of a robot. We also defined mechanisms to establish the most salient stimulus for the detection or task in question. Furthermore, the attention-based architecture uses information from the robot's decision-making system to produce user responses and robot decisions. Finally, this paper also presents the preliminary test results from the integration of this architecture into a real social robot.

**Keywords:** perception; social robots; bio-inspired attention; human–robot interaction

## 1. Introduction

In human–robot interaction (HRI), it is essential to select the most relevant stimuli to achieve a natural experience. In some cases, this interaction encounters constraints from the perception capabilities of the robot, especially considering the computational resources that are needed by state-of-the-art perception techniques, such as those that are based on deep learning. Therefore, to create an agile interaction with a high level of detail, a compromise is required regarding the number of detectors that can run simultaneously, in most cases. Detectors are associated with the number of stimuli and the quality and delay of the detections. Consequently, mechanisms that select the most salient stimuli should play an essential role in HRI; however, these selection mechanisms are often omitted. These mechanisms are similar to those in animals and are related to a key concept: attention.

There are different definitions of the term *attention*. Talsma et al. defined it as a multisensory cognitive function that allows humans and animals to continuously and dynamically select a particular stimulus from all of the available information in their environment [1]. Similarly, Broadbent characterised attention as the selective filtering of input stimuli to make the amount of data to be processed more manageable [2]. The common concept is that attention is a mechanism for selecting the most relevant stimulus in the environment. This selection means that some detectors (e.g., selectors that are associated with specific tasks) can be active only when required and can otherwise be idle or run with fewer resources when the interaction does not require them.

Typically, attention is a multisensory process, although vision is the most studied modality. According to Stein et al., the multisensory response occurs when inputs from different modalities (i.e., sight, touch and hearing) elicit a single response [3]. This result is usually amplified when the inputs appear in the same space and are synchronised in time, meaning that single-sensor stimuli are weaker.

Another interesting concept is the *focus of attention* (FOA). The FOA is the stimulus that was selected by the attentional mechanism and is then processed in detail. Information from regions outside of the FOA is either stored in the short-term memory or, in most cases, ignored, which causes a phenomenon called *change blindness* in the peripheral areas [4]. This event improves perceptual efficiency by only processing the most relevant information from the environment and allowing an increased processing and resolution load to be focused on the FOA and not on stimuli within the periphery [1].

When selecting the most relevant stimulus in a scene, we have to consider that humans detect objects within a visual location better when we know some of the characteristics beforehand (colour, movement, etc.). This type of known information is called the *perceptual set* [5]. According to Rosenbaum et al., the response time to stimuli is shorter when the user knows what kind of movement to expect in advance. This information is known as the *motor set* [6]. The perceptual and motor sets together comprise the *attentional set*, which is defined as the representations that are involved in selecting the relevant stimuli and responses to perform a task correctly.

The selection of the FOA can be either endogenous or exogenous. *Endogenous attention*, also called top-down or voluntary attention, is goal-driven and directed towards events or stimuli that are consciously selected by the individual [7]. This attention can optimise performance according to the task demands and may be maintained at a location for extended periods. In contrast, *exogenous attention*, also called bottom-up or involuntary attention, is driven by the importance of the stimuli and is an adaptive tool that enables the detection and processing of salient events that appear outside of the FOA. Exogenous attention can be understood as a momentary interruption of endogenous attention or the reorientation of attention towards a different stimulus. According to Corbetta et al., involuntary attention mechanisms, although primarily stimulus-driven, are modulated by goal-directed influences through attentional groups as they impose task completion as the priority measure [8]. Therefore, most attention-grabbing stimuli are related to the task at hand [9]. To ensure that a social robot can handle multiple stimuli while focusing on its current task, a multisensory endogenous attention-based system is necessary to filter out irrelevant stimuli. In addition, these systems allow for better resource management because they give priority to the detectors that are associated with the current task.

The main contribution of this work is the definition and development of a novel perception architecture that integrates bio-inspired concepts from biological attention and focuses on endogenous components. Our multisensory architecture sorts the detected stimuli in order of importance, considering the tasks that are performed by a social robot and focusing on HRI since the architecture is proposed for use in a social robot. We extend the existing endogenous attention systems by integrating stimuli from different modalities (vision, sound and touch) into a single system. In addition, we introduce concepts such as sustained and punctual attention to prioritise HRI and achieve natural interactions. Finally, we tested our system using different applications of a real robotic platform and achieved competitive responses in real time.

The rest of this article is structured as follows. Section 2 offers a selection of the most relevant works within the field of attention and studies the main mechanisms that govern these processes. Artificial attention models are also explored to consider their advantages and limitations and the most common types of sensory sources. Section 3 introduces the robotic platform that was employed in this study and the integration of our software architecture. Section 4 presents the main contribution of this work: the bio-inspired perception architecture that was based on endogenous attention. Section 5 introduces the experimental methodology that was followed to assess the proposed architecture. Section 6 presents the main results, which are organised into three case studies, and Section 7 analyses these results and explores the limitations of the system in its current state. Finally, the main conclusions that were drawn from this work are presented in Section 8.

## 2. Background

The selection of predominant stimuli is an innate mechanism in the animal world that offers significant advantages for survival, for example, hearing sounds that indicate danger or particular patterns in the environment [10]. Attention is, therefore, a multisensory process that involves many sources of information. Animals complement visual clues with those from other senses, such as hearing, smell or touch. The following sections review the mechanisms that drive the attention processes in living organisms, focusing on those that are related to endogenous attention.

### 2.1. Natural Attention Models

Several studies have highlighted the different mechanisms that underlie the attention process. For example, when selecting the most relevant features of a scene, one of the most influential human attention theories is *feature integration theory* [11]. This theory suggests that our perception of the environment occurs in two phases: an early stage (pre-attention), in which the individual features of an object (colour, shape, motion, etc.) are processed, and a late phase, which focuses on finding the FOA by fusing those particular features into a single conjunction of properties (i.e., an object). The *Simon effect* measures the response time to visual stimuli and demonstrates that the position of a stimulus is directly related to our ability to react to it [12]. Herbranson explored the hierarchical components within attention mechanisms, through which users can shift from focusing on a specific feature to a global analysis of the situation [10].

In some cases, humans can divide attention and allow different relevant stimuli to be focused on simultaneously, although accuracy is reduced compared to analysing features separately. For example, Cherry et al. studied the *cocktail party effect* phenomenon using a listening-based task [13]. Participants listened to two conversations at the same time. Punctual attention allowed them to focus on one of the two conversations while filtering out the other, so this type of attention allows the ability to focus a specific stimulus or activity in the presence of other distracting stimuli. In this experiment, the authors demonstrated that it was impossible to focus on both conversations simultaneously. This phenomenon also occurs with visual stimuli [14].

Endogenous attention voluntarily processes task-related stimuli and features within the environment. In the literature, works such as [15–17] have divided this attention mechanism into sustained attention and punctual attention (which is also referred to as selective attention in the literature). Sohlberg et al. defined sustained attention as the ability to maintain a consistent behavioural response during continuous or repetitive activity and punctual attention as the ability to maintain a cognitive set that requires the activation and inhibition of responses, depending on the discrimination of stimuli. Fisher described sustained attention as the ability to maintain sensitivity to incoming stimuli over time and selective attention as the ability to process part of the sensory input while excluding others.

Shulman et al. showed that these processes occur in specific regions of the brain [18]. For example, areas that are sensitive to movement react poorly to changes in colour. Therefore, we can establish a parallelism between certain regions in the brain that are specialised for certain detection tasks and detectors in software that tend to process a single kind of stimuli or detections of the same nature (e.g., an object detector is specialised for that kind of stimuli and is blind to movement). A change in task also usually causes a modification in the neural response. For example, changing from sorting numbers (odd and even) to sorting letters (vowels and consonants) produces a momentary decrease in the performance of the second task, which is known as *change cost* [19].

### 2.2. Artificial Attention Models

Over the years, there has been an increase in interest around the design of artificial attention models. These models select the most relevant parts of the environment and filter out irrelevant information to enable the faster processing of the important stimuli [20,21].

The field of robotics has significantly benefited from these systems as they have allowed robots to become more autonomous. Therefore, mechanisms that are capable of selecting relevant and helpful information are essential for robotics applications, especially when processing high-resolution and high-frequency sensory information, such as images.

Computational attention models usually process the different stimuli in the environment according to the task that is to be performed [22]. These models can react to a stimulus or task in a manner that is comparable to humans or other animals. The processing in such models can be divided into three phases [23]:

- *Orientation* selects where or what feature to focus on next (for example, in a white room with a red dot on a wall, that dot would be the relevant stimuli);
- *Selection* establishes how to focus on the feature (for example, the system chooses whether to pay attention to auditory or visual stimuli);
- *Amplification* decides how to process the selected stimulus, which is different from how it processes non-relevant features (for example, when we see a red dot in a white room, amplification selects the best method to analyse that object but in the rest of the room, object recognition may not be used).

Over recent years, there have been many examples of artificial attention systems in the literature, which have mainly focused on vision. Along this line, Meibodi et al. [24] presented an attention model that uses exogenous attention, endogenous information and memory to complete the attention process. Another example is the work of Zhu et al. [25], who predicted the head and eye movements of their participants using attention studies. In that work, they incorporated concepts such as visual uncertainty and balance to achieve good results in their predictions. Another example that is worth mentioning is the work that was developed by Yang et al. [26], who presented an inverse reinforcement learning model to predict the next endogenous focus of attention in humans. Along the same line, Fang et al. [27] revised the role of top-down modelling in salient object detection and designed a novel densely nested top-down flow (DNTDF)-based framework. They compared their results to other datasets to verify the correctness of their model. Finally, Adeli et al. [28] proposed an attention system that was based on using neural networks to predict the next focus of attention in real scenarios. They used bio-inspired concepts, such as the ventral, frontal and visual areas of the human brain.

Endogenous Attention Models

The literature offers different approaches for endogenous models; although in most cases, these are descriptive and are not supported by concrete developments [23]. This has been caused by a lack of generality in the algorithms. For example, it is crucial to evaluate the placement of an object within a room to understand the relevance of that object for a given task. Therefore, generic object recognition algorithms are not valid for these models [29]. Nevertheless, we can find approaches that implement attention models, such as the work of Wang and Shen [30], and use exogenous information to predict endogenous FOAs in complex situations. Another important concept is feature-based attention. This allows for the precise location of the property of interest in time, e.g., identifying upward movements when the task is to control a lifting movement [31]. In these models, the effects of endogenous knowledge indicate how exogenous stimuli should be processed.

Theses endogenous feature-based attention models emerged first [32,33]. Soon after, architectures such as the *guided search theory* [34] and the *feature gate model* [35] appeared. These architectures attempt to answer how feature gains should be adapted to achieve optimal task performance. All of these models have a supervised learning phase to compute property gains using manually labelled examples [36]. A common characteristic of all of the works that have been presented in this section is that they tend to output a region in the selected space (or image) as the following FOA. Based on these models, Beuter et al. [37] proposed fusing the exogenous and endogenous information within an image to determine the most salient areas in order to control a robot's navigation. In the same way, Yu et al. [38] presented an artificial visual attention manager that uses the features of an object to guide the perception of the robot to the next FOA, according to the current task, context and learned knowledge.

A more recent example that uses probabilistic reasoning and inference tools is the model that was proposed by Borji and Itti [39]. This work introduced an architecture for modelling endogenous visual attention that was based on reasoning in a task-dependent manner. Their model analyses the semantic values of objects within a scene and tracks the user's eye movements, then combines both with a probabilistic approach to predict the next object that is needed by the user to perform their task.

The works that have been presented in this section use endogenous attention as a complement to exogenous information and place more importance on salient stimuli that are related to the task that is to be performed. However, to the extent of our knowledge, no work in the literature has proposed a purely endogenous architecture that reacts in real time. Moreover, works on attention management that is applied to robotics are scarce and they have usually focused on mobile robots that need attention to navigate correctly. In contrast, this paper presents an endogenous attention manager for social robots that prioritises natural human–robot interaction over other stimuli and real-time reactions. This system analyses stimuli from different sources (sight, sound and touch) to decide which is the most relevant stimulus, taking into account all of the information in the environment (not just visual information as in the literature). This feature allows the robot to interact with its environment and perform more complex tasks that include multimodal detectors.

## 3. The Robotic Platform

To test our approach, we used the social robot Mini, which utilises the perception architecture that was presented in Salichs et al. [40]. Mini was developed at RoboticsLab and is a desktop robot that is 59 cm tall and can share its emotional state through expressive eyes, an LED heartbeat, cheeks and the movement of its arms, head and base. The aim of this robot is to assist and entertain older adults who suffer from mild cognitive problems. In terms of the robot's hardware, Mini has LEDs in its heart and cheeks, a VU meter as a mouth and two OLED screens that serve as eyes. Between the head and the body, Mini has a neck with two degrees of freedom that allows head movements in the vertical and horizontal axes. It also has two arms with one degree of freedom and a base, which are all controlled by servomotors. These elements allow Mini to achieve a natural vivacity. Mini also includes capacitive sensors to detect tactile stimuli and an external tablet to enhance its interaction with users by displaying multimedia content, such as videos, images or buttons, for games and exercises. The base holds a computer that has an Arduino microcontroller and a small battery for safe shutdown. Finally, the robot has an RGB-D camera at its base (Realsense D435i camera: https://www.intelrealsense.com/depth-camera-d435i/ (accessed on 10 July 2022), see Figure 1). For particular skills (e.g., to play tangram), we installed a USB fisheye camera (US-BFHD04H 1080P H.264 camera: http://www.elpcctv.com/elp-free-driver-1080p-full-hd-h264-usb-webcam-camera-module-for-car-bus-plane-video-surveillance-p-350.html (accessed on 10 July 2022)) with a 180° field of view, which provided images with a resolution of 1080p at 30 frames per second, on the robot's chest. That camera focused on the table between the robot and the user to see the game pieces.

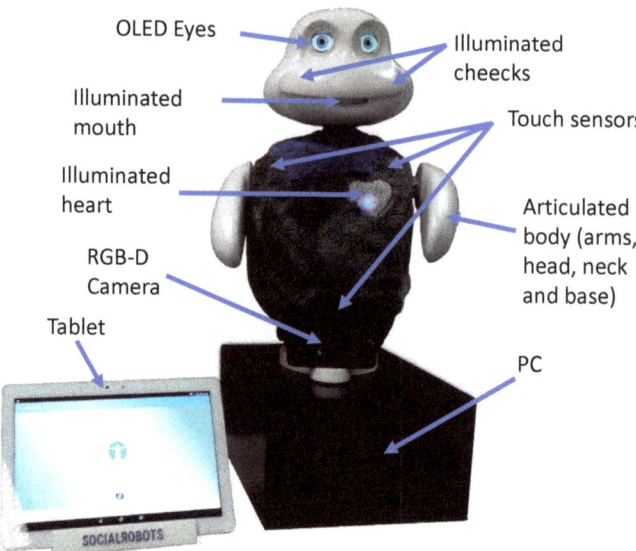

**Figure 1.** The social robot Mini.

## 3.1. The Software Architecture

The robot's software architecture is divided into blocks, as shown in Figure 2. Our work focused on the attention manager and detectors that run the robot. The detector and actuator blocks are directly connected to the physical devices of the robot and acquire information from the sensors or by commanding the actuators, respectively.

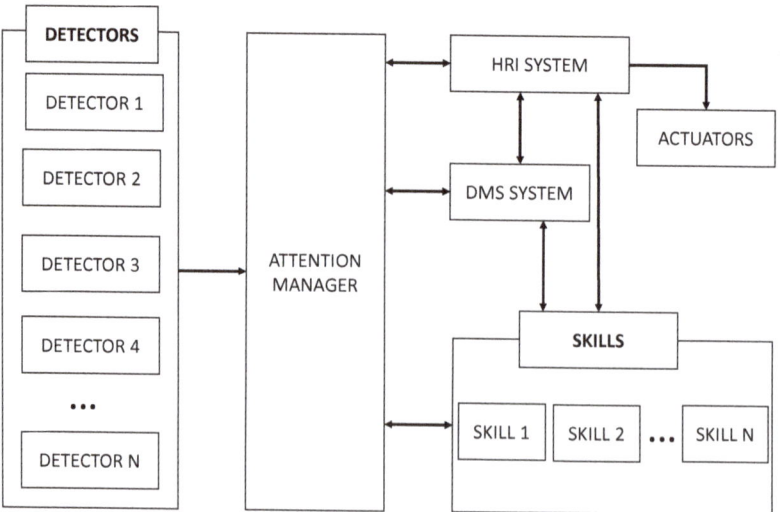

**Figure 2.** A general overview of the main software components of the Mini architecture and the connections between them. This work added the attention manager block, which communicated with the HRI and DMS systems and the robot's skills.

The *skills* correspond to the main functionalities of the robot and each skill uses perceptual information to achieve its goal. The skills also use HRI capabilities to generate dialogues and interact with the user. The robot integrates various categories of skills, such as the following:

- **Games** contains all of the games that Mini can play, such as bingo, akinator [41], tangram [42] and quiz games, which were chosen to stimulate elderly people following the recommendations of doctors or psychologists;
- **Multimedia** includes all of the multimedia capabilities that the robot can use (e.g., showing pictures or videos, playing music, audiobooks or movies and telling jokes);
- **Information** contains the information that the robot can provide, such as the news or weather [43], as well as an introduction for itself and instructions on how to interact with it;
- **Cognitive stimulation** incorporates all of the cognitive stimulation exercises (e.g., memory, attention, planning and comprehension exercises);
- **General** contains the essential communication elements for the user and the robot, such as reminders or relevant information data;
- **Sleep** includes the robot's configuration to simulate sleep (i.e., eyes closed and a neutral position of the arms, base and head).

The *human–robot interaction* (HRI) system manages the dialogue between the user and the robot. This part of the architecture processes and analyses perceptual information to decide whether it is relevant to the interaction; for example, when a skill needs the robot to provide information or the robot is waiting for an answer from the user. The robot only communicates with the user using voice or the menus on the tablet, but it also uses visual cues on the tablet and different gestures to improve communication. Finally, the *decision-making system* (DMS) controls the activation and deactivation of the robot's functionalities. This component uses the robot's internal information (e.g., motivation) and environment information (e.g., the user requesting a specific skill) to proactively decide the robot's next action.

To test the operation of the attention manager, we explored three case studies. In the first case study, we used a multimedia skill, specifically the telling jokes skill (Section 5.2). Next, we tested a cognitive stimulation skill, specifically a memory game that is based on recognising famous monuments around the world (Section 5.3). Finally, we tested a game skill, specifically the tangram game, by adding vision detectors to the attention manager (Section 5.4).

## 4. The Attention Manager

This paper proposes a bio-inspired endogenous artificial attention manager for social robots. We took inspiration from psychological and neuroscientific works that considered how human attention works within social interactions. It is important to note that the main purpose of social robots is to provide human–robot interaction as naturally as possible. To identify the requirements for developing this bio-inspired system, we studied how endogenous attention works in biological systems and its application to HRI, which led us to choose the following specifications for the attention manager:

- To decide on the optimal FOA considering the available stimuli and the current task;
- To process information from multiple sensory modalities (e.g., vision, touch and sound);
- To respond in real time, avoiding lengthy processing and response times;
- To select the correct FOA to achieve natural human–robot interaction;
- To include a voluntary attention process that detects the relevant information for a goal in a sustained manner;
- To include a mechanism that allows the robot to focus on a stimulus punctually, even when it is not the most salient stimulus, in a sustained manner;
- To take into account the current task of the robot (i.e., the system needs to communicate with the DMS to know what task the robot is currently performing and its next action

and then, with this information, the system can sort the stimuli according to their relevance to the task);
- To filter out irrelevant stimuli for the current task;
- To consider scalability and modularity (i.e., allow the inclusion of new sources of information);
- To manage resources efficiently (i.e., inhibit or disable detectors depending on the task).

According to these requirements, we developed the architecture of the bio-inspired endogenous attention manager, as shown in Figure 3. In the following sections, we explain each of the blocks that make up the proposed architecture in more detail.

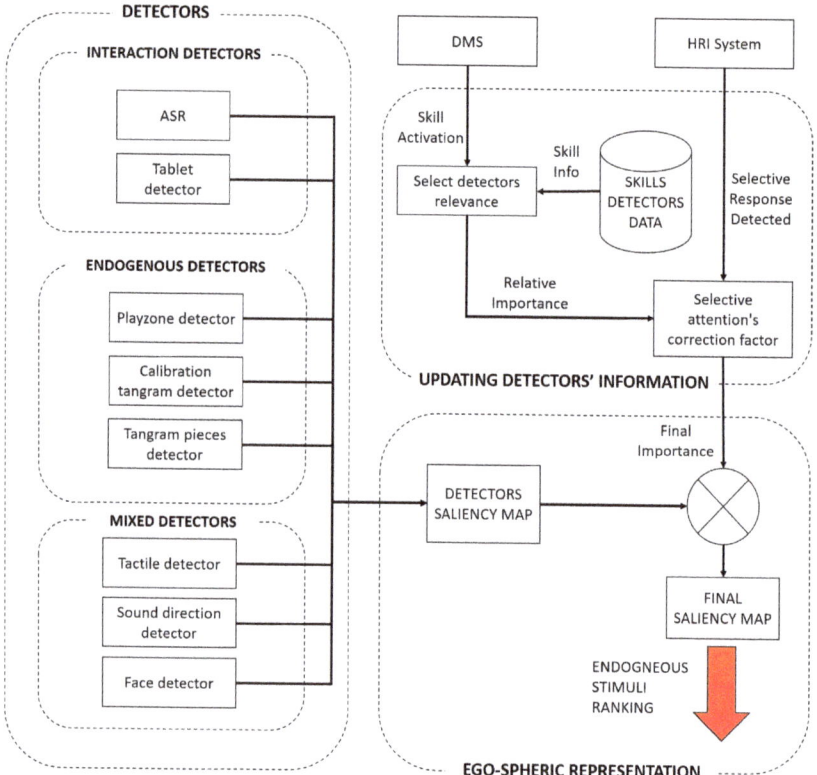

Figure 3. A schematic diagram of the endogenous attention manager.

## 4.1. The Detectors

The detectors lie at the base of the architecture. They are responsible for extracting information from the robot's environment and generating a rich representation. To optimise the computational load of the robot, the architecture can activate the detectors only when they are necessary.

According to their modality, we classified the detectors into visual, auditory and tactile. This modality is related to how the attention manager analyses the specific relative importance of each stimulus, as explained in Section 4.2. Additionally, the detectors could be organised by type: interaction, endogenous and mixed.

- Mini uses *interaction detectors* to achieve bidirectional communication with the user. In our case, the robot had two types of interaction detectors: automatic speech recognition (ASR) [44] for voice communication and the tablet [45], which the user used for tactile feedback. As explained in Section 3.1, the HRI manager controls the robot's dialogue. Therefore, it handles both the questions that are asked by the robot and the answers that are given by the user. The attention manager receives the user's answers from the detectors and through which channel they were produced using the information from this software module (see [46] for more details on how the HRI manager works in the Mini). The interaction detections are then positioned in the space around the robot. In the case of the tablet, we assumed that the device was always in a fixed place in front of the robot. When the robot uses the ASR, the attention manager tries to locate the user's face to select it as the FOA. Our architecture integrated face detection and localisation system called face-detection-retail-004, which relies on Squeezenet and an SSD network (face detector network: https://docs.openvinotoolkit.org/latest/_models_intel_face_detection_retail_0004_description_face_detection_retail_0004.html (accessed on 10 July 2022)). This detector worked well with a frontal view of the user's face, achieving an average accuracy of 83%. For each detection, the algorithm provides a unique user ID, the confidence level for the predicted class and the coordinates of the upper left and the lower right corners of the face's bounding box.

The attention manager uses this information to fix the FOA in the centre of the detection box when the robot has to pay attention to the user (for example, when the robot is talking or waiting for an answer from the user).

In contrast, when the user is not sitting in front of the robot, the attention manager uses the Mini's omnidirectional microphone to locate the direction of the user's voice. For this, the architecture uses the Open embeddeD audition System (ODAS) library because it allows for the 3D localisation, tracking, separation and filtering of sound sources [47].

Additionally, when the robot is waiting for a voice response from the user and it does not detect anybody, the attention manager tries to find the direction of the sound to locate the user, even when the user is behind the robot. Conversely, when the face detection system finds a user, the system prioritises the face over the voice stimuli.

- *Endogenous detectors* are specifically developed to provide information for a certain skill. Therefore, the robot only activates these sensors when it needs them to perform the relevant skill.

  The attention manager fully controls the activation and deactivation of these detectors. Focusing on a specific skill that was used in the case studies, the tangram game skill integrates three endogenous vision-based detectors: the first is in charge of detecting the play zone, the second is responsible for performing the calibration that is necessary to play correctly and the third is used to recognise and locate the game pieces. We assumed that the game board was in a fixed position between the user and the robot to place these stimuli. This assumption meant that all of the detectors that were operating in the playing area had the same FOA, which was the centre of the board.

- *Mixed detectors* can detect endogenous and exogenous information in the environment. For example, considering the face detector, when the robot is idle and a person enters the room, the face that is detected captures the robot's attention exogenously. In contrast, when the robot is waiting for a user response during a game, the user's face captures the robot's attention endogenously. These two behaviours result in the detector being considered as a mixed detector. These detectors are always active, but the attention manager only uses their data when they are required for the current skill. Apart from the face detector, the Mini system uses other three mixed detectors: the first detects tactile interactions, the second locates the user's face and the third recognises the direction of sound. To detect and identify tactile events, the robot uses an algorithm that is capable of detecting touch in three areas (the belly and the left and right arms) using capacitive sensors. When the attention manager detects a tactile event, the attention focuses on the relative position of the sensor that was activated with respect to the centre of the robot.

## 4.2. Updating Detector Information

Endogenous attention is a voluntary process; therefore, it is directed by the task that the robot is performing. To add this feature into the attention manager, the system needs to communicate with the DMS that is in charge of deciding which task the robot performs at a given time. In our case, the DMS informed the attention manager when there was a change in the robot's skill and whether the change involved the activation of a new ability or the deactivation of the current skill. This communication has to be asynchronous and the architecture needs to adapt quickly to the requirements of the new skill. When the robot initiates a new skill, the attention manager uses the skill detectors database (see the *Skills Detectors Database* in Figure 3) to load the associated information, such as the detectors that are associated with that skill and their relative importance (weight) in terms of attention.

As stated at the beginning of Section 4, the architecture has to process information from different sensory modalities. Our robot could react to different visual, auditive and tactile stimuli and each sensory modality had a relative importance that depended on the task that the robot was performing and whether the stimulus was relevant to the HRI. However, these modalities do not represent the same information. In our case, the stimuli were different, so we did not need a fusion step to merge information from different sensors that were detecting similar stimuli at the same location.

Visual stimuli were associated with the task itself and were not used by the HRI mechanisms. For example, when the robot played tangram, the detector returned the position of the pieces on the playing area. Equation (1) characterised the weight or importance of this type of detector:

$$\omega_{vision} = \begin{cases} 1.0, & if \quad n_{vision\ detectors} > 0 \\ 0.0, & otherwise \end{cases} \quad (1)$$

Voice interaction was also considered. For example, when the robot asked a question and waited for an answer. Considering the importance of HRI for these robots, sound stimuli were more critical than visual stimuli and, therefore, had a higher relative importance value (see Equation (2)):

$$\omega_{audio} = \begin{cases} 2.0, & if \quad n_{audio\ detectors} > 0 \\ 0.0, & otherwise \end{cases} \quad (2)$$

In our robot, tactile stimuli were considered as an interruption to the task. For example, when the robot told a joke, the user could touch its belly to stop the activity. Similarly, when the robot was playing the tangram game, the user could touch its arm to request a hint. Since different studies have demonstrated that humans pay more attention to tactile stimuli, the attention manager placed more importance on these detections. Equation (3) defined the weight of this type of stimuli:

$$\omega_{tactile} = \begin{cases} 3.0, & if \quad n_{tactile\ detectors} > 0 \\ 0.0, & otherwise \end{cases} \quad (3)$$

This system not only focuses on events that continuously appeared over time (sustained attention), but it also reacts to punctual stimuli at specific times (punctual attention). In this last case, a stimulus may not be as salient as the sustained a priori stimuli, but to calculate their relevance, the temporal factor has to also be considered. This mechanism is especially useful when the robot maintains a dialogue with the user via different channels (e.g., voice or tablet). In these cases, sustained attention tends to pay attention to the user's face, but the system must correct this issue to allow for variations between all of the possible stimuli. For this reason, the attention manager introduces a correction factor to the sustained and selective stimuli to modify their weights at each time step. This correction factor multiplies the relative importance of a specific stimulus by a factor of three. This way, the robot is able to focus its attention on the response channel with the correction factor.

The attention manager communicates with the HRI system, which controls the dialogues, and informs the general architecture about which input method the user selected to respond (see the upper right box in Figure 3).

In the current version of the attention manager, we assumed that the stimuli that could trigger punctual attention just came from the interaction detectors, specifically the automatic speech recognition and the tablet detector. Equation (4) shows how the system calculated the final importance of the interaction detectors:

$$\begin{pmatrix} \omega_{FinalASR} \\ \omega_{FinalTablet} \end{pmatrix} = \begin{pmatrix} \omega_{ASR} \\ \omega_{Tablet} \end{pmatrix} * \begin{pmatrix} \omega_{SA1} & 0.0 \\ 0.0 & \omega_{SA2} \end{pmatrix} \quad (4)$$

where $\omega_{ASR}$ and $\omega_{Tablet}$ are the a priori weight of the detectors at each time step and $\omega_{SA1}$ and $\omega_{SA2}$ are the correction factors that corresponded to punctual attention for the two detectors (the $\omega_{SA1}$ and $\omega_{SA2}$ values were always either 1.0 or 3.0, depending on the detector's attention type (sustained or punctual); for example, when the detector produced punctual attention because the user interacted with the robot using that channel, the correction factor was 3, in the other cases, the value was 1).

### 4.3. Creation of the Ego-Spheric Representation

The system requires the integration of stimuli from different sensory sources. In human attention, the brain performs a sensory integration process that allows us to perceive information coherently [1]. Our architecture integrates this process and enables the generation of a unique representation for all endogenous stimuli.

In our work, we used an ego-spheric representation to achieve multisensory aggregation. We took inspiration from the work of Bodiroza et al. [48] and the representation allowed the rendering of any stimuli around the robot, not only those in front of the camera. The sphere represented a multimodal egocentric map, in which the system recreated the areas of the salient stimuli that were located by the robot. In the ego-spheric representation, the robot's centre corresponded to the sphere's centre and the system showed the salient data as small spheres around the Mini. The size of each small sphere in the representation depended on the importance of that stimulus. Moreover, the position of each stimulus was relative to the centre of the robot. Equation (5) was used to calculate the final saliency map, where $\omega$ is the relative importance depending on the sensory modality that was tuned-up (as shown in Section 4.2), $F$ is the required detector, $d$ is the total modalities and $c$ is the final combined map:

$$c_t = \sum_{j=1}^{d} \omega_{jt} \cdot F_{jt} \quad (5)$$

where $c_t =$ is the final combined map in instant $t$
$j \in \{1, 2, ..., number\ of\ detectors\ activated\}$
$\omega_{jt} =$ is the relative importance
$F_{jt} =$ is the needed detector
$d =$ is the total modalities

The classical approach of saliency maps is to only represent the stimuli in front of the camera and to not take depth into account [23]. Therefore, they usually only include visual stimuli. However, in our work, we included multisensory information in the representation that could appear at any place around the robot, including tactile, auditive and visual stimuli. In the Figure 4, the yellow mark represents the centre of the robot and the pink sphere is the space around it. Note that the green dot is bigger than the others, meaning that the stimulus was more significant than the others.

In our architecture, the most salient stimulus is not the only output. Instead, the endogenous attention processing generates a sorted list that includes all of the stimuli that exceed the salience threshold.

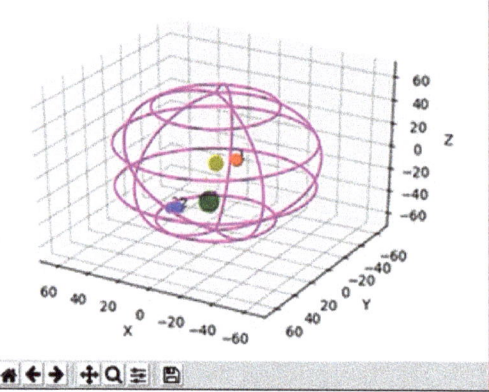

**Figure 4.** An example of an ego-spheric representation with simultaneous tactile (green circle), auditive (red circle) and visual (blue circle) stimuli.

## 5. Methodology

In Section 3.1, we explained that the Mini can perform a wide range of skills. In this section, we describe how the attention manager works and how it exchanges information with the other software architecture blocks using three realistic case studies. The first case study involved the system performing an entertainment skill: joke telling. The second case study evaluated the attention manager when performing a cognitive stimulation skill, more precisely a memory and attention exercise that consisted of recognising famous world monuments. Finally, the third case study showed how the system works when performing a game skill: the tangram game. For each of these skills, we describe how the different blocks of the software architecture communicated during the execution of the skill and the response of the endogenous attention manager. We focused on where the robot placed the FOA and the differences between its sustained and punctual attention. Table 1 summarises the case studies that are presented in this work. It indicates the type of skill, the objective of the test and the types of detectors that were involved. The objectives were cumulative, i.e., in case study 2, we aimed to meet the goals of case study 1 as well as those of case study 2.

**Table 1.** A summary of the three case studies that are presented in this work.

|  | Skill | Goal | Detectors Involved |
|---|---|---|---|
| Case Study 1 | Entertainment | - System responds to multiple activated detectors<br>- System responds to sustained or punctual attention<br>- Communication with other software blocks<br>- Sorted list according to salience as the output | - Interaction detectors<br>- Mixed detectors |
| Case Study 2 | Cognitive Stimulation | - System responds to changes in communication channels | - Interaction detectors<br>- Mixed detectors |
| Case Study 3 | Game | - System responds to stimuli from different sensory modalities | - Interaction detectors<br>- Mixed detectors<br>- Endogenous detectors |

## 5.1. Experimental Setup

In the case studies, the user sat in front of the robot (see Figure 5). In this case, four kinds of sensors provided the information about the robot's environment: an omnidirectional microphone *ReSpeaker Mic Array v2.0* (ReSpeaker Mic Array v2.0: https://wiki.seeedstudio.com/ReSpeaker_Mic_Array_v2.0/ (accessed on 10 July 2022)), an RGB-D camera (Realsense D435i: https://www.intelrealsense.com/depth-camera-d435i/ (accessed on 10 July 2022)), three capacitive touch sensors that were placed on the arms and belly of the robot and a tablet (Samsung Galaxy Tab A: https://www.samsung.com/es/tablets/galaxy-tab-a/galaxy-tab-a-10-1-inch-white-32gb-lte-sm-t585nzwephe/ (accessed on 10 July 2022)), which was placed between the robot and the user.

As well as from the sensors, the robot integrated a GPU USB *Intel Neural Compute Stick* (Intel NCS2: https://ark.intel.com/content/www/us/en/ark/products/140109/intel-neural-compute-stick-2.html (accessed on 10 July 2022)) to extend the robot's processing capabilities. Finally, the third case study included additional accessories for the skill, such as a play zone, which was also placed between the robot and the user, and a set of tangram pieces. All of the software modules were connected using ROS [49] and the system operated at 3 fps for visual stimuli, which still allowed the robot to have a human-like reaction time [50] while controlling the computational load.

To test the functioning of the system, the research team performed ten repetitions of each case study and the times that are presented in Section 6 are those that were obtained in the last repetition. In addition, the system was stressed by performing the different skills consecutively and triggering all of the possible options in each skill that was selected for the case studies. The average interaction time was 80.3 s in case study 1, 269.1 s in case study 2 and 148.7 s in case study 3.

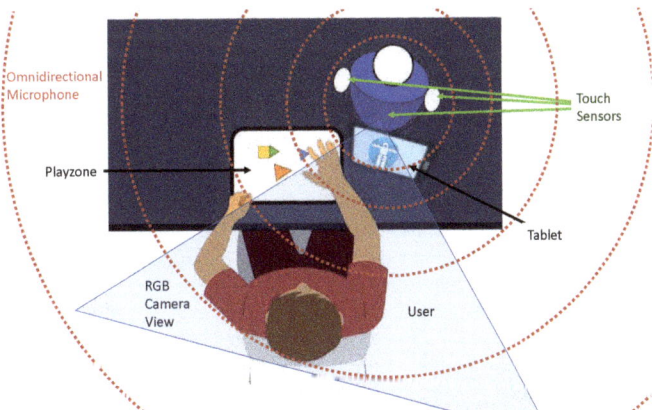

**Figure 5.** An illustration of the environmental setup of the case studies. We used the same colour coding as that in the 3D representation of the environment (blue for the camera, green for the touch sensors and red for the speaker).

## 5.2. Case Study 1: Attention during an Entertainment Skill

In this case study, we used an entertainment skill that tells jokes to the user [43]. The user could choose the type and subject of the jokes. The execution was divided into two states. The first state was selection of the skill, for which we assumed that the robot was initially awake and wanted to interact with the user. Then, the robot asked the user what they wanted to do using different communication channels (voice and the tablet). The user answered using one of these channels. After the selection, the joke telling skill initiated and the robot told three jokes to the user using voice communication.

In this case, we checked the system response when several detectors were active simultaneously and the different results that were obtained depending on whether the attention was sustained or punctual. Another important aspect was the communication between our system and the other software blocks within the robot (the HRI system, DMS and skills). Finally, the case study also checked the correct functioning of the output stimuli list, which was sorted according to salience. Section 6.1 shows the step-by-step operation of the Mini software and the behaviour of the attention manager during the case study in detail.

*5.3. Case Study 2: Attention during a Cognitive Stimulation Skill*

This case study involved a cognitive stimulation exercise that was based on memory and attention [51]. During the exercise, the Mini displayed a well-known monument, such as the Eiffel Tower or the Coliseum, and asked the user for their relative city, giving three choices.

As in the previous case study, there were two stages. The first stage consisted of the selection of the skill. To test a different set of inputs, in this case, the user asked the robot for the exercise directly by saying "I want to play to the monuments game".

In the second part of the case study, we evaluated the operation of the attention manager while the user was completing the stimulation exercise. During the execution, the Mini requested information using its text-to-speech functionality and activated the ASR as the default communication channel. However, when the ASR detected a communication problem (recognition failure or no response from the user), the Mini continued the communication using the tablet as the input. The game had seven different questions and for the first three, the attention manager received the user voice as the input, while for the last four questions, it acquired information from the tablet.

In this case study, we tested the attention manager with a more complex skill and in a dynamic setting, in which we intended to check that robot produced the correct output when interacting with the user and when changing communication channels. Section 6.2 presents the results from the communication between the software modules in the Mini and the attention manager output during the exercise.

*5.4. Case Study 3: Attention during a Game*

In this case, the user played the tangram game and the robot controlled the game development and helped the user by providing hints. In this case, the robot used an additional camera. This device focused on the table between the robot and the user to detect the play zone of the game, in which the user could freely move and place the tangram pieces (see Figure 5). To simplify the case study, this case did not include the selection state as it was similar to those of the previous case studies. During the game, the robot used endogenous detectors that were explicitly developed for this skill, such as the play zone detector, calibration detector or tangram piece detector, as described in [42]. Furthermore, the robot also included touch detection to provide hints to the user. Finally, the robot asked the user questions using the different communication channels (voice and the tablet) during the game.

For this skill, the robot used multiple kinds of detectors as it needed endogenous, interaction and mixed detectors to function correctly for the game. Therefore, in this case study, we tested the system attention output in a complex scenario using stimuli from different sensory modalities (sight, sound and touch). Section 6.3 details the connections between the different software blocks and the output of the endogenous attention manager.

## 6. Results

This section presents the results that were obtained by the complete endogenous attention manager in the different types of robot activities. Based on the attention manager specifications that were described in Section 4, the obtained results helped to verify that the attention manager satisfied the following requirements:

1. Able to identify the most salient stimulus for a given task;
2. Able to handle multimodal stimuli (visual, auditory and tactile);
3. Real-time responses;
4. Able to select the most relevant information to achieve natural human–robot interaction;
5. Able to react to sustained and punctual attention;
6. The correct management of the activation and deactivation of detectors to reduce computational load.

### 6.1. Case Study 1: Attention during an Entertainment Skill

In this case study, we considered that the user wanted to interact with the robot and that they wanted to listen to short, robot-related jokes. Figure 6 shows a sequence diagram with the messages that were exchanged among the different Mini software modules during the execution of the joke telling skill. Additionally, Figure 7 summarises the results of this case study, showing the relative importance of each detector during sustained attention and punctual attention.

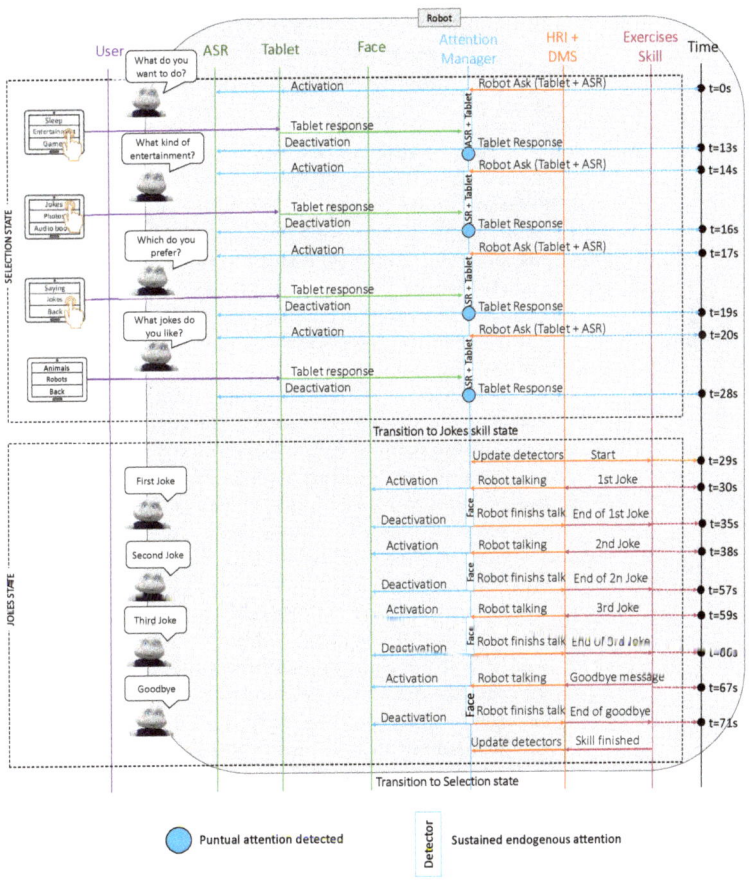

**Figure 6.** A sequence diagram of the connections between the different software blocks in the Mini during the execution of the joke telling skill. The blue boxes in the attention manager row display sustained attention from the detectors. The blue dots show the punctual attention that was produced by the user's responses.

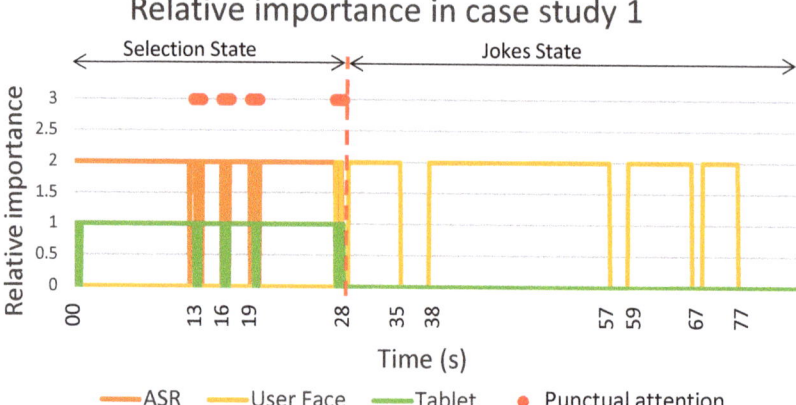

**Figure 7.** The relative importance of the detectors during sustained and punctual attention in case study 1. The orange, yellow and green lines show the sustained attention of the ASR, the user's face and the tablet, respectively. The red dots display the punctual attention that was detected during the case study.

This case study consisted of two parts: the selection state and the jokes state. Initially, the robot was awake and initiated the interaction by asking the user what they wanted to do (at $t = 0$ s). This question caused the attention manager to activate the available communication channels, initiating the tablet and ASR detectors. At this point, the system paid attention to the user's face and the tablet in a sustained manner.

As explained in Section 4.1, when the robot expected a voice response, it also tried to locate the face of the closest user, assuming that this user was the person who responded. Furthermore, using the weights that were discussed in Section 4.2, the attention manager placed more importance to the user's face, despite both detectors being active.

After a few seconds, the user answered using the tablet; therefore, there was a punctual change in the attention manager to place more importance on the device for a few seconds (see the blue point in Figure 6 and the red point in Figure 7 at $t = 13$ s). Next, the robot prompted three more questions (from $t = 14$ s to $t = 16$ s, from $t = 17$ s to $t = 19$ s and from $t = 20$ s to $t = 28$ s in Figure 6) to select which entertainment skill the user wanted and which topic they liked jokes about. The output of the attention manager for these three questions was identical to that described for the first question: the user answered using the tablet.

Once the user selected the skill, the case study continued in the jokes state (at $t = 29$ s). This state described the behaviour of the system during the execution of the joke telling skill. This skill was the simplest among the three case studies since the robot only interacted with the user through voice. Note that when the robot was talking, the sustained attention manager located the user's face to look at this point when the Mini was talking to achieve a natural interaction. Therefore, the Mini paid attention to the user's face while telling each joke and in the goodbye message (from $t = 28$ s to $t = 35$ s, from $t = 38$ s to $t = 57$ s, from $t = 59$ s to $t = 67$ s and from $t = 68$ s to $t = 77$ s in Figure 7).

### 6.2. Case Study 2: Attention during a Cognitive Stimulation Skill

In this case study, we considered that the user wanted to perform a specific cognitive stimulation exercise: the monuments game. In this game, the robot displayed pictures of famous monuments on the tablet and asked the user for their relative city, giving three possible answers. This exercise had seven different questions and during the case study, the user answered the first three questions using voice. During the fourth question, a communication error appeared and forced a change of communication channel and the user responded to the last four questions using the tablet.

The Figure 8 shows a sequence diagram with the messages that were passing among the different Mini software modules during the case study. Complementarily, Figure 9 summarises the results of this test, showing the relative importance of each detector during sustained attention and the points at which there was a punctual change in the robot's attention.

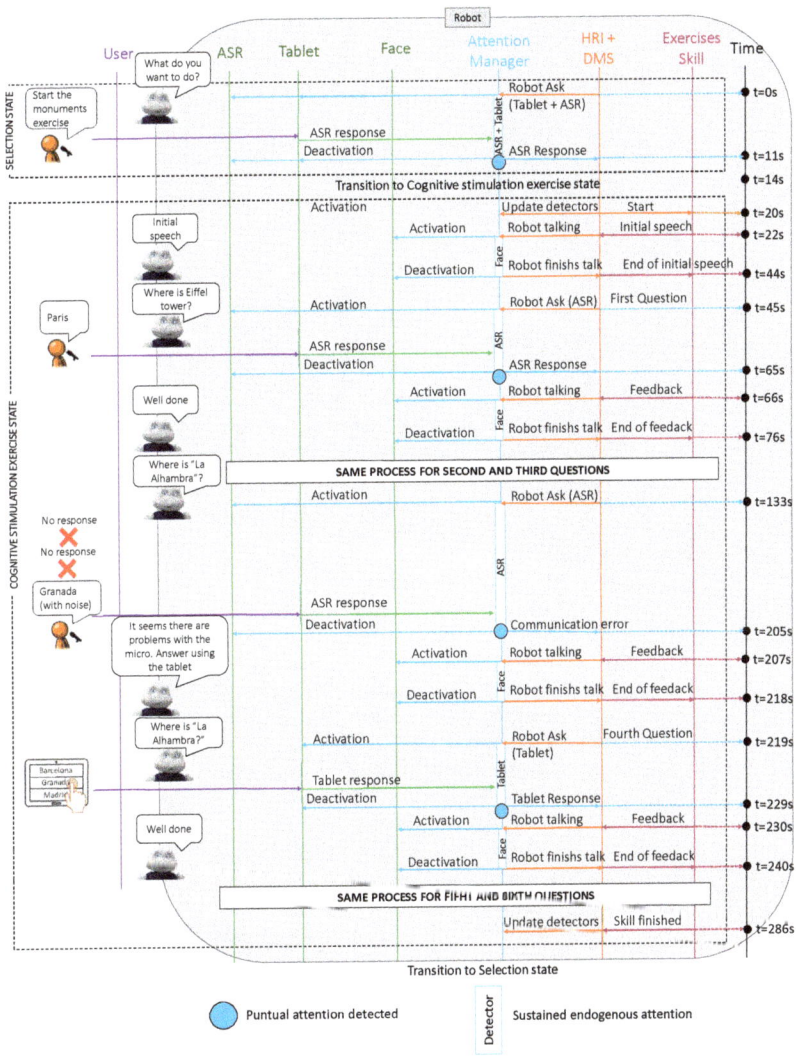

**Figure 8.** A sequence diagram of the connections between the different software blocks in the Mini during the execution of the cognitive stimulation exercise.

As in case study 1, this case study was divided into two parts: the selection state and the exercise state. The robot was initially awake and started the interaction by asking the user what they wanted to do (at $t = 0$ s). In this case, the user selected a cognitive stimulation exercise using voice (at $t = 11$ s in Figure 9). The behaviour during this first part of the case study (selection state) was similar to that in case study 1. When the robot asked a question, the attention manager activated the communication channels to receive the user's answer (ASR and tablet). Despite having both detectors active, the sustained attention placed more importance on the user's face due to the weights that were discussed previously. In this case, since the user answered using the voice, the most relevant stimulus for the attention manager corresponded to the punctual attention (see the red point in Figure 9).

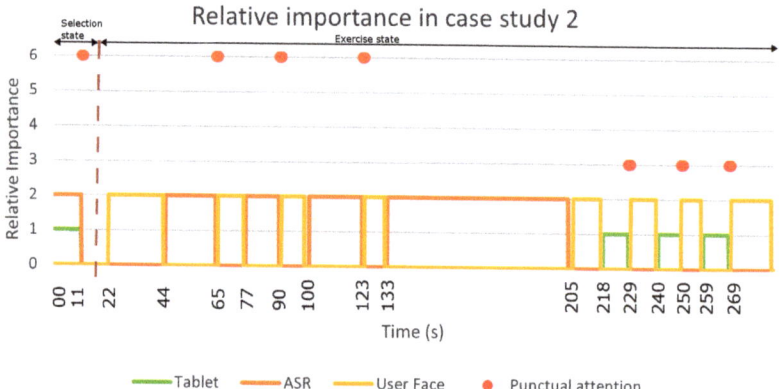

**Figure 9.** The relative importance of the detectors during sustained and punctual attention in case study 2. The orange, yellow and green lines show the sustained attention of the ASR, the user's face and the tablet, respectively. The red dots display the punctual attention that was detected during the case study. The user's answers produced the first three dots using ASR, while the tablet responses produced the last four dots.

Once the user selected the skill (at $t = 20$ s in Figure 8), the robot briefly introduced the exercise and explained what the user had to do. During this speech, the attention manager activated the face detector to pay attention to the user (from $t = 22$ s to $t = 44$ s in Figure 9). Then, the Mini asked for the Eiffel Tower's location, giving different options for the answer (Paris, Berlin or Rome). In this case, the user could reply using ASR, so the attention manager activated this channel (at $t = 45$ s) and the user responded after a few seconds (at $t = 65$ s).

As this was a correct answer, the robot responded with positive feedback for the user. During this interaction, the attention manager activated the face detector again because the user was the most salient stimuli (from $t = 66$ s to $t = 76$ s in Figure 8). The game continued in this way for two more questions (from $t = 77$ s to $t = 90$ s and from $t = 100$ s to $t = 123$ s in Figure 9). Next, the robot asked where the Alhambra is, so the attention manager again activated the ASR (at $t = 133$ s in Figure 8). When a user does not answer within a predetermined time or the speech recognition fails, the detector reports a communication error between the robot and the user. In this case, the robot prompted the user to use the tablet to respond (at $t = 207$ s in Figure 8). As in the previous case, the attention manager activated the face detector to look at the user while talking. Then, the robot repeated the question; However, in this case, it displayed the options on the tablet so that the user could answer. At this point, the attention manager activated the tablet's detector (from $t = 218$ s to $t = 229$ s in Figure 9) and waited for the user's response (see the red point in Figure 9 at $t = 229$ s). Finally, when the robot received the answer, the attention manager activated the face detector again to provide the feedback to the user (from $t = 230$ s to $t = 240$ s in Figure 8).

In summary, the exercise had seven questions and the user used voice to answer the first three (see the red points in Figure 9 at $t = 65$ s, $t = 90$ s and $t = 123$ s). During the fourth question, we simulated a communication problem with the ASR (from $t = 133$ s to $t = 205$ s) and the user responded using the tablet (see the red point at $t = 229$ s in

Figure 9). The user also answered the last three questions using the tablet (see the red points at $t = 250$ s and $t = 269$ s in Figure 9).

*6.3. Case Study 3: Attention during a Game*

This case study tested the operation of the attention manager when a user played the tangram game with the robot. In this case, the user played with a physical tangram and the robot controlled the development of the game using computer vision to detect the play zone and pieces and helped the user when they needed help. This last case was the most complex of the three case studies, as it included seven detectors and more complex interactions. As in the previous cases, Figure 10 shows a sequence diagram describing the messages that were passed among the different Mini software blocks during the execution of the game and Figure 11 presents the relative importance of each detector during sustained and punctual attention.

In this case study, we omitted the game selection phase to focus on how the FOA evolved during a complex game. The dynamics of this discarded phase were identical to those of the previous case studies. Therefore, the robot started the interaction by welcoming the user to the game. During the welcome message, the attention manager activated the face detector because the robot was talking to the user, so the most salient stimulus was the user's face ($t = 0$ s to $t = 9$ s in Figure 11). Next, the robot explored the area in front of it to find the play zone ($t = 10$ s to $t = 14$ s in Figure 10). When the robot could not see the play zone, the Mini informed the user that the game could not start. The next step involved a calibration to enhance the tangram detector accuracy; hence, the robot provided some instructions to the user ($t = 15$ s to $t = 35$ s in Figure 11).

During this time, the robot focused its attention on the user's face. Before starting the calibration process, the robot asked the user to confirm that they had followed the instructions (at $t = 38$ s in Figure 11). In the same way as in the cognitive stimulation case study, when the user did not respond or the speech recognition failed (see the red point in Figure 11 at $t = 90$ s), the Mini repeated the question but displayed the options on the tablet (at $t = 91$ s in Figure 10). Due to the change in the communication channel, the attention manager deactivated the ASR and activated the tablet detector. When the user responded (see the red point in the Figure 11 at $t = 103$ s), the skill started the calibration process (at $t = 113$ s in Figure 11). The attention manager activated the calibration detector and waited until the calibration process was successful (at $t = 114$ s in Figure 10). As with the play zone detection, when the robot could not perform the calibration, the Mini informed the user that it could not play at that moment and ended the game.

Once the calibration was completed, the robot was ready to play and provided instructions to the user on how to play the game (at $t = 115$ s in Figure 11). At that moment, the attention manager activated the face detector and a few seconds later (at $t = 132$ s in Figure 11), it activated the detector that recognises tangram pieces. Although two sustained attention detectors were active simultaneously (face and tangram pieces), there was no punctual attention in this case, so the robot always paid more attention to the user's face than to the board due to the preset weights ($\omega_{audio} = 2.0$ and $\omega_{vision} = 1.0$). When the Mini finished giving the instructions, the attention manager deactivated the face detector (at $t = 136$ s in Figure 11) and the robot started paying more attention to the pieces on the board.

Additionally, during the tangram game, the user could touch the robot at any time to ask for a hint. In this case study, that occurred at $t = 143$ s in Figure 10. At this point, the attention manager deactivated the tangram pieces detector and activated the face detector to look at the user while giving a clue (at $t = 144$ s in Figure 11). When the robot finished the clue, it switched its attention from the user to the game pieces (at $t = 157$ in Figure 11).

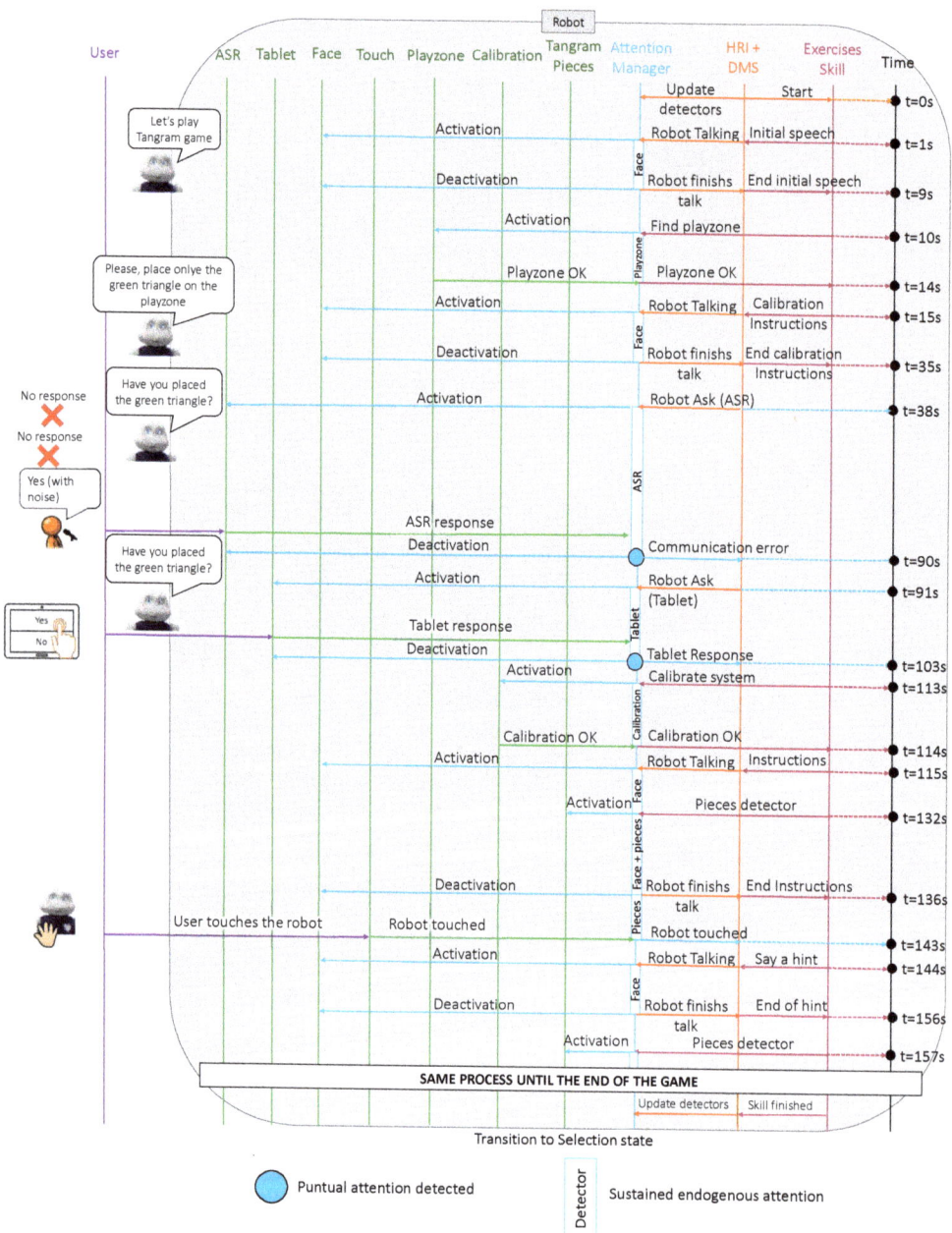

**Figure 10.** A sequence diagram of the connections between the different software blocks in the Mini during the execution of the tangram game.

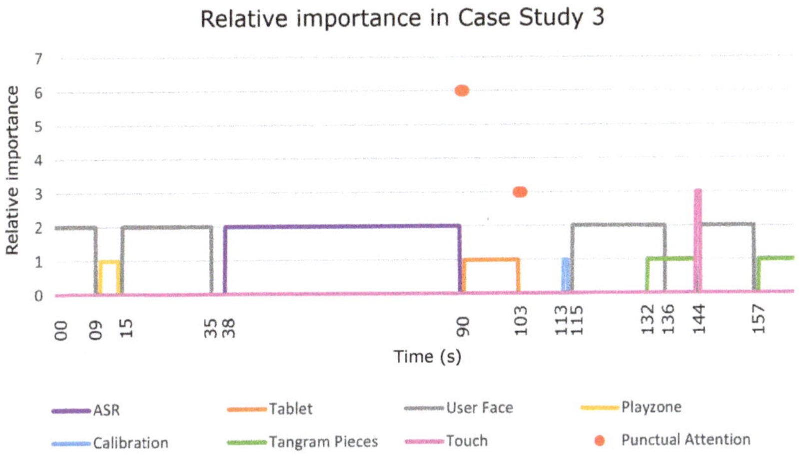

**Figure 11.** The relative importance of the detectors during sustained and punctual attention in case study 3. The lines show the sustained attention of the different detectors that were used during the case study. The red dots show the punctual attention that was detected. The first dot was produced by the user's response via ASR and the second dot was produced by the user's response via the tablet.

## 7. Discussion

In this study, we tested whether an endogenous attention system allowed our robot to identify the most salient task-related stimulus, regardless of the modality (sight, sound or touch), in real time. In addition, we verified whether the robot could prioritise interaction-related stimuli to achieve the most natural HRI possible using bio-inspired techniques, such as sustained and punctual attention. Finally, enabling the robot to manage the activation and deactivation of the detectors was a priority to reduce the computational load and produce the correct functioning of the robot. We tested these requirements in three different scenarios, all of which proved satisfactory results.

In the first two case studies, we tested the system's ability to identify the most salient stimulus to perform the given task during the selection stage. In this part of the case study, two possible stimuli stood out in a sustained way: the ASR and the tablet. In the first case study, although the ASR was more salient in a sustained way, the tablet was more relevant when the user answered the questions using the tablet and not using their voice. We tested whether the proposed attention system could identify the most relevant stimulus when there was more than one possible task-related stimulus in the environment and whether it could react correctly to sustained and punctual attention. Moreover, in these two case studies, we also checked that the robot could correctly control the activation and deactivation of the detectors, both in the selection phase and in the execution of the skill itself. This reduced the computational load of the system and ensured the correct function of our social robot.

On the other hand, in case study 2, we verified that the system could adapt to a change in communication channel due to an error in the voice recognition system. This adaptation showed the system's adaptability in real time as this change did not cause any delays in the attention system's response and there were no additional problems due to the error during the interaction.

Focusing on case study 3, it is instantly noticeable that this case study was more complex than the previous cases. Firstly, we tested how the system reacted to stimuli from three sensory modalities (sight, sound and touch). The results from this case study showed that the robot had an excellent adaptation to the different sensory modalities and that their detection and localisation did not cause any delays in the system. Furthermore, in this case, we also verified whether the system could prioritise the stimuli that were necessary for the interaction over the stimuli that were related to the task itself. This prioritisation allowed the robot to interact with the user in the most natural way possible as it prioritised human–robot interaction.

During the development of the tests, we researched other endogenous attention models in the literature to compare to our system. However, we did not find any multimodal implementations that could allow this comparison. Furthermore, many of the models that have been presented in the literature are not purely endogenous, but rather use exogenous information to guide endogenous attention. Another problem was the lack of multimodal attention systems, including visual, auditory and tactile stimuli. Therefore, we focused on performing different tests on our system to check that it was working correctly.

Despite the advantages of the proposed system, there were also some limitations. Firstly, the integration of detectors for the stimuli of different modalities allowed us to make the robot capable of performing more complex tasks that included visual, auditable and tactile stimuli. However, the system was unable to analyse whether stimuli from different sensory modalities belonged to the same stimulus (e.g., detecting a person and their associated voice at the same time). We hope to explore this in future work by including a fusion level that merges stimuli from different modalities and relates them to each other.

On the other hand, although the results that were obtained in the case studies were satisfactory, using preset weights for the relative importance of each sensory modality did not allow the robot to adapt to its task. Furthermore, these preset values meant that the robot always reacted in the same way when performing a task, which could lead to errors on certain occasions. Therefore, in future work, we hope to include a machine learning level that directly adjusts these weights using deep learning techniques. This automatic learning would take into account past experiences and current activity so we could able to customise the system to the user that it interacts with and the robot's tasks.

Finally, the case studies allowed us to check the correct functioning of the system during different skills. The research team conducted the tests in 2021. Unfortunately, due to the current COVID-19 situation, we could not test this system with elderly people. Moreover, although the implementation phase is complete, the system needs to be validated in a controlled environment before long-term testing with real users. Despite this, we hope to perform tests with the target population to check the correct functioning of the system. In addition, we plan to carry out long-term tests that would allow us to check that the system responds correctly in stressful situations.

## 8. Conclusions

This work presented a bio-inspired endogenous attention architecture for social robots. The architecture aims to detect the most relevant stimulus in a given environment and consider the necessary adaptions for the robot's behaviour in order to complete the current task. Moreover, the system's reaction time aims to be similar to that of a human for the same process. Therefore, the model aims to work in real time. The system also considers stimuli from various sensory modalities. The ego-spheric representation provides a 3D view of the stimuli within the robot's environment.

Our experimental results showed that our system met the requirements that were set out in Section 4. Firstly, we developed a communication process with the DMS that allowed the robot to know the current task at all times. Moreover, as the results show, the adaptation of the attention manager was fast and always less than one second, thus avoiding possible bottlenecks that could be produced by our architecture. In the same way, the attention manager could process information from different sensory modalities, including visual, auditory and tactile detectors. We added a detector block to the software that allows for the individual preprocessing of environment information and works in parallel to avoid delays in the system.

The results also indicated that the system could work with both sustained and punctual attention in the three case studies. The system correctly sorted the stimuli according to salience using the preset weights and modified them when there was a punctual stimulus.

In addition, with the previously fixed relative importance values, the attention manager selected the most important stimuli for a more natural human–robot interaction in a sustained way.

A machine learning level would dynamically adjust these weights in future work by considering past experiences and current activity. Moreover, a fusion level would merge the stimuli from different modalities to allow the robot to understand the relationships between the detections. Finally, we also hope to perform long-term tests with elderly people to check the functioning of the system and its adaptation to a stressful environment.

A critical architecture requirement was real-time responses. The case studies demonstrated that the stimulus localisation and saliency classification worked in less than one second in all cases.

As for the system's modularity, we verified that the architecture also satisfied this requirement using different algorithms. In the first two case studies, the system only used interaction detectors, while in the last case study, we added endogenous and mixed detectors and we achieved the correct output in all cases. These results also demonstrated that the system could activate and deactivate the detectors when necessary, thus efficiently managing its resources.

**Author Contributions:** Conceptualisation, S.M.-V., J.C.C. and M.A.S.; methodology, S.M.-V., J.C.C. and J.J.G.-M.; software, S.M.-V., J.J.G.-M. and J.S.-S.; validation, S.M.-V., J.J.G.-M. and J.S.-S.; formal analysis, S.M.-V., J.C.C. and J.S.-S.; investigation, S.M.-V., J.C.C. and J.J.G.-M.; resources, J.C.C. and M.A.S.; data curation, S.M.-V., J.J.G.-M. and J.S.-S.; writing—original draft preparation, S.M.-V., J.C.C., J.J.G.-M. and J.S.-S.; writing—review and editing, S.M.-V., J.C.C. and M.A.S.; visualisation, S.M.-V. and J.C.C.; supervision, J.C.C. and M.A.S.; project administration, J.C.C. and M.A.S.; funding acquisition, M.A.S. All authors have read and agreed to the published version of the manuscript.

**Funding:** The research received funding from the following projects: Robots Sociales para Estimulación Física, Cognitiva y Afectiva de Mayores (ROSES), grant number RTI2018-096338-B-I00, which was funded by the Ministerio de Ciencia, Innovación y Universidades; Robots Sociales para Mitigar la Soledad y el Aislamiento en Mayores (SOROLI), grant number PID2021-123941OA-I00, which was funded by the Agencia Estatal de Investigación (AEI) at the Ministerio de Ciencia e Innovación. This publication was part of the R&D&I project PLEC2021-007819, which was funded by MCIN/AEI/10.13039/501100011033 and by the European Union NextGenerationEU/PRTR.

**Institutional Review Board Statement:** Not applicable.

**Informed Consent Statement:** Not applicable.

**Data Availability Statement:** Not applicable.

**Conflicts of Interest:** The authors declare no conflicts of interest.

# References

1. Talsma, D.; Senkowski, D.; Soto-Faraco, S.; Woldorff, M.G. The multifaceted interplay between attention and multisensory integration. *Trends Cogn. Sci.* **2010**, *14*, 400–410. doi: 10.1016/j.tics.2010.06.008. [CrossRef]
2. Broadbent, D. *Perception and Communication*; Pergamon Press: Elmsford, NY, USA, 1958.
3. Stein, B.; Jiang, W.; Stanford, T. Multisensory integration in single neurons of the midbrain. *Handb. Multisens. Process.* **2004**, *15*, 243–264.
4. Frintrop, S.; Rome, E.; Christensen, H.I. Computational visual attention systems and their cognitive foundations: A survey. *ACM Trans. Appl. Percept. (TAP)* **2010**, *7*, 1–39. [CrossRef]
5. Corbetta, M.; Shulman, G.L. Control of goal-directed and stimulus-driven attention in the brain. *Nat. Rev. Neurosci.* **2002**, *3*, 201–215. [CrossRef]
6. Rosenbaum, D.A. Human movement initiation: Specification of arm, direction, and extent. *J. Exp. Psychol. Gen.* **1980**, *109*, 444. [CrossRef]
7. Carretié, L. Exogenous (automatic) attention to emotional stimuli: A review. *Cogn. Affect. Behav. Neurosci.* **2014**, *14*, 1228–1258. doi: 10.3758/s13415-014-0270-2. [CrossRef] [PubMed]
8. Corbetta, M.; Patel, G.; Shulman, G.L. The reorienting system of the human brain: From environment to theory of mind. *Neuron* **2008**, *58*, 306–324. [CrossRef] [PubMed]
9. Koch, C.; Ullman, S. Shifts in selective visual attention: Towards the underlying neural circuitry. In *Matters of Intelligence*; Springer: Berlin, Germany, 1987; pp. 115–141.

10. Herbranson, W.T. Selective and divided attention in comparative psychology. In *APA Handbook of Comparative Psychology: Perception, Learning, and Cognition*; APA Handbooks in Psychology®; American Psychological Association: Washington, DC, USA, 2017; Volume 2, pp. 183–201. doi: 10.1037/0000012-009. [CrossRef]
11. Treisman, A.M.; Gelade, G. A feature-integration theory of attention. *Cogn. Psychol.* **1980**, *12*, 97–136. [CrossRef]
12. Simon, J.R. Effect of ear stimulated on reaction time and movement time. *J. Exp. Psychol.* **1968**, *78*, 344. [CrossRef]
13. Cherry, E.C. Some experiments on the recognition of speech, with one and with two ears. *J. Acoust. Soc. Am.* **1953**, *25*, 975–979. [CrossRef]
14. Neisser, U.; Becklen, R. Selective looking: Attending to visually specified events. *Cogn. Psychol.* **1975**, *7*, 480–494. [CrossRef]
15. Sohlberg, M.M.; Mateer, C.A. Effectiveness of an attention-training program. *J. Clin. Exp. Neuropsychol.* **1987**, *9*, 117–130. [CrossRef] [PubMed]
16. McKay, K.E.; Halperin, J.M.; Schwartz, S.T.; Sharma, V. Developmental analysis of three aspects of information processing: Sustained attention, selective attention, and response organization. *Dev. Neuropsychol.* **1994**, *10*, 121–132. [CrossRef]
17. Fisher, A.V. Selective sustained attention: A developmental foundation for cognition. *Curr. Opin. Psychol.* **2019**, *29*, 248–253. [CrossRef]
18. Shulman, G.L.; d'Avossa, G.; Tansy, A.P.; Corbetta, M. Two attentional processes in the parietal lobe. *Cereb. Cortex* **2002**, *12*, 1124–1131. [CrossRef]
19. Rogers, R.D.; Monsell, S. Costs of a predictable switch between simple cognitive tasks. *J. Exp. Psychol. Gen.* **1995**, *124*, 207. [CrossRef]
20. Castellotti, S.; Montagnini, A.; Del Viva, M.M. Information-optimal local features automatically attract covert and overt attention. *Sci. Rep.* **2022**, *12*, 9994. [CrossRef]
21. Melício, C.; Figueiredo, R.; Almeida, A.F.; Bernardino, A.; Santos-Victor, J. Object detection and localization with Artificial Foveal Visual Attention. In Proceedings of the 2018 Joint IEEE 8th International Conference on Development and Learning and Epigenetic Robotics (ICDL-EpiRob), Tokyo, Japan, 17–20 September 2018; pp. 101–106. [CrossRef]
22. Tsotsos, J.K. *A Computational Perspective on Visual Attention*; MIT Press: Cambridge, MA, USA, 2011.
23. Itti, L.; Borji, A. Computational models: Bottom-up and top-down aspects. *arXiv* **2015**, arXiv:1510.07748.
24. Meibodi, N.; Abbasi, H.; Schubö, A.; Endres, D.M. A model of selection history in visual attention. In Proceedings of the Annual Meeting of the Cognitive Science Society, Vienna, Austria, 21–29 July 2021; Volume 43.
25. Zhu, Y.; Zhai, G.; Min, X.; Zhou, J. The prediction of saliency map for head and eye movements in 360 degree images. *IEEE Trans. Multimed.* **2019**, *22*, 2331–2344. [CrossRef]
26. Yang, Z.; Huang, L.; Chen, Y.; Wei, Z.; Ahn, S.; Zelinsky, G.; Samaras, D.; Hoai, M. Predicting goal-directed human attention using inverse reinforcement learning. In Proceedings of the Proceedings of the IEEE/CVF Conference on Computer Vision and Pattern Recognition, Seattle, WA, USA, 13–19 June 2020; pp. 193–202.
27. Fang, K.; Xiang, Y.; Li, X.; Savarese, S. Recurrent autoregressive networks for online multi-object tracking. In Proceedings of the 2018 IEEE Winter Conference on Applications of Computer Vision (WACV), Tahoe, NV, USA, 12–15 March 2018; pp. 466–475.
28. Adeli, H.; Zelinsky, G. Deep-BCN: Deep networks meet biased competition to create a brain-inspired model of attention control. In Proceedings of the IEEE Conference on Computer Vision and Pattern Recognition Workshops, Salt Lake City, UT, USA, 18–22 June 2018; pp. 1932–1942.
29. Ballard, D.H.; Hayhoe, M.M.; Pelz, J.B. Memory representations in natural tasks. *J. Cogn. Neurosci.* **1995**, *7*, 66–80. [CrossRef]
30. Wang, W.; Shen, J. Deep visual attention prediction. *IEEE Trans. Image Process.* **2017**, *27*, 2368–2378. [CrossRef]
31. Zhou, H.; Desimone, R. Feature-based attention in the frontal eye field and area V4 during visual search. *Neuron* **2011**, *70*, 1205–1217. [CrossRef] [PubMed]
32. Milanese, R.; Gil Milanese, S.; Pun, T. Attentive mechanisms for dynamic and static scene analysis. *Opt. Eng.* **1995**, *34*, 2428–2434. [CrossRef]
33. Navalpakkam, V.; Itti, L. Modeling the influence of task on attention. *Vis. Res.* **2005**, *45*, 205–231. [CrossRef]
34. Wolfe, J.M. Guided search 2.0 a revised model of visual search. *Psychon. Bull. Rev.* **1994**, *1*, 202–238. [CrossRef] [PubMed]
35. Cave, K.R. The FeatureGate model of visual selection. *Psychol. Res.* **1999**, *62*, 182–194. [CrossRef] [PubMed]
36. Borji, A.; Ahmadabadi, M.N.; Araabi, B.N. Cost-sensitive learning of top-down modulation for attentional control. *Mach. Vis. Appl.* **2011**, *22*, 61–76. [CrossRef]
37. Beuter, N.; Lohmann, O.; Schmidt, J.; Kummert, F. Directed attention-a cognitive vision system for a mobile robot. In Proceedings of the RO-MAN 2009-The 18th IEEE International Symposium on Robot and Human Interactive Communication, Toyama, Japan, 27 September–2 October 2009; pp. 854–860.
38. Yu, Y.; Gu, J.; Mann, G.K.; Gosine, R.G. Development and evaluation of object-based visual attention for automatic perception of robots. *IEEE Trans. Autom. Sci. Eng.* **2012**, *10*, 365–379. [CrossRef]
39. Borji, A.; Itti, L. Exploiting local and global patch rarities for saliency detection. In Proceedings of the 2012 IEEE Conference on Computer Vision and Pattern Recognition, Providence, RI, USA, 16–21 June 2012; pp. 478–485.
40. Salichs, M.A.; Castro-González, Á.; Salichs, E.; Fernández-Rodicio, E.; Maroto-Gómez, M.; Gamboa-Montero, J.J.; Marques-Villaroya, S.; Castillo, J.C.; Alonso-Martín, F.; Malfaz, M. Mini: A New Social Robot for the Elderly. *Int. J. Soc. Robot.* **2020**, *12*, 1231–1249. [CrossRef]
41. Velazquez Navarro, E.; Gonzalez-Diaz, S.; Alonso-Martin, F.; Castillo, J.; Castro-Gonzalez, A.; Malfaz, M.; Salichs, M. Social Robot Mini as a platform for developing multimodal interaction games. In Proceedings of the Actas de las Jornadas Nacionales de Robótica, Spanish National Robotics Conference, Alicante, Spain, 13–14 June 2019; pp. 214–220.

42. Menendez, C.; Marques-Villarroya, S.; Castillo, J.C.; Gamboa-Montero, J.J.; Salichs, M.A. A computer vision-based system for a tangram game in a social robot. In Proceedings of the International Symposium on Ambient Intelligence, L'Aquila, Italy, 7–9 October 2020; pp. 61–71.
43. González-Díaz, S.; Velázquez Navarro, E.; Alonso-Martín, F.; Castro-Gonzalez, A.; Castillo, J.; Malfaz, M.; Salichs, M. Social Robot Mini as Information and Entertainment Platform. In Proceedings of the Actas de las Jornadas Nacionales de Robótica, Spanish National Robotics Conference, Alicante, Spain, 13–14 June 2019; pp. 92–97.
44. Alonso-Martín, F.; Salichs, M.A. Integration of a voice recognition system in a social robot. *Cybern. Syst. Int. J.* **2011**, *42*, 215–245. [CrossRef]
45. Marqués Villaroya, S.; Castillo, J.C.; Alonso Martín, F.; Maroto, M.; Gamboa, J.J.; Salichs, M.Á. Interfaces táctiles para Interacción Humano-Robot. In *XXXVIII Jornadas de Automática*; Servicio de Publicaciones de la Universidad de Oviedo: Oviedo, Spain, 2017; pp. 787–792.
46. Fernández-Rodicio, E.; Castro-González, Á.; Alonso-Martín, F.; Maroto-Gómez, M.; Salichs, M.Á. Modelling Multimodal Dialogues for Social Robots Using Communicative Acts. *Sensors* **2020**, *20*, 3440. [CrossRef] [PubMed]
47. Grondin, F.; Michaud, F. Lightweight and optimized sound source localization and tracking methods for open and closed microphone array configurations. *Robot. Auton. Syst.* **2019**, *113*, 63–80. [CrossRef]
48. Bodiroza, S.; Schillaci, G.; Hafner, V.V. Robot ego-sphere: An approach for saliency detection and attention manipulation in humanoid robots for intuitive interaction. In Proceedings of the 2011 11th IEEE-RAS International Conference on Humanoid Robots, Bled, Slovenia, 26–28 October 2011; pp. 689–694. doi: 10.1109/Humanoids.2011.6100900. [CrossRef]
49. Quigley, M.; Conley, K.; Gerkey, B.; Faust, J.; Foote, T.; Leibs, J.; Wheeler, R.; Ng, A.Y. ROS: An open-source Robot Operating System. In Proceedings of the ICRA Workshop on Open Source Software, Kobe, Japan, 12–17 May 2009; Volume 3, p. 5.
50. Tejero, J.P.; Soto-Rey, J.; González, J.J.R. Estudio del tiempo de reacción ante estímulos sonoros y visuales. *Eur. J. Hum. Mov.* **2011**, *2011*, 149–162.
51. Salichs, E.; Fernández-Rodicio, E.; Castillo, J.C.; Castro-González, Á.; Malfaz, M.; Salichs, M.Á. A social robot assisting in cognitive stimulation therapy. In Proceedings of the International Conference on Practical Applications of Agents and Multi-Agent Systems, Salamanca, Spain, 6–8 October 2018; pp. 344–347.

*Article*

# Benchmarking Object Detection Deep Learning Models in Embedded Devices

David Cantero [1,*], Iker Esnaola-Gonzalez [1], Jose Miguel-Alonso [2] and Ekaitz Jauregi [3]

1. TEKNIKER, Basque Research and Technology Alliance (BRTA), 20600 Eibar, Spain; iker.esnaola@basf.com
2. Department of Computer Architecture and Technology, University of the Basque Country UPV/EHU, 20018 San Sebastian, Spain; j.miguel@ehu.eus
3. Department of Languages and Information Systems, University of the Basque Country UPV/EHU, 20018 San Sebastian, Spain; ekaitz.jauregi@ehu.eus
* Correspondence: david.cantero@tekniker.es

**Abstract:** Object detection is an essential capability for performing complex tasks in robotic applications. Today, deep learning (DL) approaches are the basis of state-of-the-art solutions in computer vision, where they provide very high accuracy albeit with high computational costs. Due to the physical limitations of robotic platforms, embedded devices are not as powerful as desktop computers, and adjustments have to be made to deep learning models before transferring them to robotic applications. This work benchmarks deep learning object detection models in embedded devices. Furthermore, some hardware selection guidelines are included, together with a description of the most relevant features of the two boards selected for this benchmark. Embedded electronic devices integrate a powerful AI co-processor to accelerate DL applications. To take advantage of these co-processors, models must be converted to a specific embedded runtime format. Five quantization levels applied to a collection of DL models are considered; two of them allow the execution of models in the embedded general-purpose CPU and are used as the baseline to assess the improvements obtained when running the same models with the three remaining quantization levels in the AI co-processors. The benchmark procedure is explained in detail, and a comprehensive analysis of the collected data is presented. Finally, the feasibility and challenges of the implementation of embedded object detection applications are discussed.

**Keywords:** object detection; embedded devices; deep learning; benchmarking

## 1. Introduction

Deep Learning (DL) is a sub-field of Machine Learning (ML) based on the computation of multi-layer Artificial Neural Networks (ANN), also known as Deep Neural Networks (DNN) in reference to the presence of multiple internal processing layers. One of the applications where DL is proving most successful is computer vision, where impressive levels of performance are being achieved. This work discusses object detection technology, which is defined as a computer vision technique that enumerates the objects presented in an image and classifies each of the detected objects, assigning a confidence or probability of existence while locating them and squaring their position in the image. In the traditional computer vision approach, object detection algorithms were based on handcrafted sets of features explicitly programmed by the authors. However, an object may present a diversity of morphological appearances and could be deformed, present a large variety of shapes and/or be immersed in scenes with very different illumination levels and backgrounds. Furthermore, objects may be partially occluded by other objects, making it almost impossible to extract robust features manually. DL, on the other hand, uses a huge amount of detection examples and trains a DNN to automatically infer the appropriate detection features. This strategy has proven to be highly successful.

Even if DL is a computationally intensive task, modern embedded hardware devices are powerful enough to execute some of the most successful models. In addition, hardware manufacturers have developed powerful AI (Artificial Intelligence) co-processors, specifically designed to execute DL models. These co-processors provide considerable computing power with high power efficiency. As a result, more and more AI-based applications are implemented in smart embedded devices [1]. Many techniques have been developed to improve the deployment of DL models on such devices, starting from simplified training processes using pre-trained networks and fine-tuning the parameters in a process called Transfer Learning [2], to many model simplifications and transformations, such as quantization, model pruning, etc., to squeeze the model onto embedded devices [3]. Note that even if the models are executed on the embedded devices, all the previous stages in the DL workflow cited above take place in powerful host computers, usually equipped with dedicated high performance graphics processing units (GPUs).

Embedded devices are of paramount importance to bring DL capabilities to robotic applications [4]. To name just a few examples, in [5] the authors present a system that can detect and track multiple objects from aerial images taken by a flying robot, while in [6] a 3D-printed robotic arm is brain-controlled via embedded DL from sEMG sensors. Real-time human detection is an important sub-field of computer vision, of interest in areas ranging from industrial environments to autonomous driving. For a review of this task using DL on embedded platforms, the reader is referred to [7].

The goal of this article is to provide a review of the major challenges in the development of embedded DL applications. The article is divided into two main parts. The first part presents a detailed analysis of the main elements to be taken into account in any DL embedded application: Section 2 explains the motivation for the use of embedded hardware and the most important features to be taken into account when selecting embedded devices. A description of the devices chosen for this work is also included. In Section 3, ML framework requirements are evaluated for both embedded hardware devices and host computers. The embedded hardware libraries are intended to provide a specific runtime environment for the execution of inference based on DL models in specialized hardware co-processors. ML host frameworks, on the other hand, are usually powerful software packages designed to support the whole DL application development workflow. Since the compatibility of both frameworks is mandatory, only a few options are feasible, so the selection is, as explained, quite straightforward. Section 4 describes some of the most successful and modern object detection models available and how they are handled by the selected ML framework.

The second part of the article carries out a benchmark of embedded hardware platforms based on the ML framework and previously identified models. Each model must be converted from its original format to an embedded-friendly format. Hardware co-processors support INT8 arithmetic operations, so model conversion also involves some kind of model quantization. Five quantization levels are considered for this work, as described in Section 5. After conversion, models are deployed in the embedded devices, and their inference performance is measured and tested. Section 6 describes the benchmark procedure and analyzes the obtained results. Finally, Section 7 states the conclusions of this work, and Section 8 enumerates some reflections about future lines of work.

## 2. AI at the Edge: Intelligent Embedded Systems

Edge computing is a distributed computing architecture where most data processing is executed by hardware devices close to the source of the data. As opposed to cloud computing, where large and powerful central facilities receive huge amounts of data from remotely connected sensors and compute complex and performance-demanding algorithms, edge computing brings the computation to devices with limited resources.

Related to cloud computing, the Internet of Things (IoT) paradigm, which consists of physical things equipped with electronic components and ubiquitous intelligence that allow them to connect, interact and exchange data [8], has contributed to the deployment of millions of connected devices in almost any imaginable scenario. Similarly, the Industry 4.0 paradigm has made available multi-sensory data of industrial processes that allow complex algorithms to control and optimize the performance of industrial plants [9].

The current trend is to move data processing from the cloud to the edge. In particular, ML algorithms are being increasingly deployed in embedded devices [10]. There are many reasons why computing at the edge is preferable to computing at the cloud [11]. On the one hand, the amount of data traffic increases together with the number of deployed devices. On the other hand, data transmission and processing in remote systems introduces a delay that in some cases is unacceptable. Additionally, there may be security issues if private or sensitive information needs to be transmitted from local facilities to an external data center [12].

In the literature, edge devices are vaguely defined. Even if the premise is always that the processing is located near the source of the data, this could refer to both a computing network infrastructure located in the same facilities as sensors or an embedded device with a tiny micro-controller. In the present work, edge devices are understood to be embedded devices that usually incorporate sensor data acquisition hardware and are able to autonomously execute data processing algorithms and make some "smart" decisions.

### 2.1. Selection of Embedded AI Hardware Devices

The first challenge to benchmarking the performance of a DL model in an embedded device is to select the appropriate hardware device itself. There are hundreds of hardware devices that claim to have a design oriented to the execution of ML algorithms. In fact, many modern micro-controllers are actually able to run a set of ML algorithms [13,14], but since one of the goals of this work is to deploy machine vision DL algorithms, a powerful enough device should be selected. On average, the number of operations required to compute a complete inference from an input image is around some tens of billions of operations or Giga-Operations (GOPS) [15]. Since a video sequence has around 30 to 60 frames per second, it is estimated that the minimum computational power an embedded device must have is around one Tera-Operations per second (TOPS). This requirement rules out most general-purpose micro-controllers, for example those based on the widely used ARM CortexM architecture, and also many application processors, including those based on the ARM CortexA architecture. Even some processors based on the x86 architecture are not powerful enough. To reach those figures, it is necessary to select a processor with a specific integrated mathematical co-processor. Due to the great success of DL, modern embedded hardware devices have begun to integrate powerful AI co-processors to perform DL computations. There are three main solutions to integrate a DL-oriented co-processor in embedded hardware: (i) use a general-purpose processor that already integrates a co-processor in the same semiconductor die; (ii) include a separate Application Specific Integrated Circuit (ASIC) designed for DL inference together with the general purpose processor in the embedded hardware design; or (iii) use a programmable logic device (CPLD or FPGA) to implement custom co-processor hardware [16]. The design of a math accelerator circuit for DL model inference is outside the scope of this work, and therefore the third solution is rejected in favor of the first two. Based on these criteria, the embedded hardware devices selected for this work are described in the next sub-sections.

### 2.2. NXP i-MX8M-PLUS Application Processor

The first hardware platform selected is the i-MX8M-PLUS processor. It is an NXP heterogeneous multi-core processor for high-performance applications focused on video processing and DL (https://www.nxp.com/products/processors-and-microcontrollers/arm-processors/i-mx-applications-processors/i-mx-8-processors/i-mx-8m-plus-arm-cortex-a53-machine-learning-vision-multimedia-and-industrial-iot:IMX8MPLUS, accessed on 11 July 2021). The embedded System on Chip (SoC) from Variscite shown in Figure 1 and the matching evaluation kit were used in this work.

**Figure 1.** iMX 8M Plus System on Module. Image from https://www.variscite.com/ (accessed on 2 September 2021).

From a DL application development perspective, the most interesting component of this board is the embedded Neural Processing Unit (NPU) with 2.3 TOPS of computing power. It is also quite remarkable that the NPU is integrated onto the same die as the general-purpose processors and shares the high-speed internal memory bus. This architecture helps speed up the DNN inference as the data interchanged between both computing units are optimized. The NPU is a Vivante VIP8000 specifically designed for being embedded in processors of the i-MX family. It works with 8-bit integer data types (INT8) rather than 32-bit floating-point data (FLOAT32). As will be seen in Section 5, this means that the DNN needs to be transformed (quantized) before being executed in the NPU. NXP provides the entire ecosystem of tools to manage the entire workflow pipeline, including the design, deployment and inference of neural networks. The processor also features a powerful image-processing pipeline, camera interfaces and a comprehensive set of communication peripherals.

### 2.3. Google Coral Dev Board with EdgeTPU Module

The other hardware platform considered in this work is the Coral Dev Board. This is an evaluation kit for the EdgeTPU AI accelerator module (see Figure 2), an ASIC with a PCI or high-speed USB communication interface that performs 4 TOPS while drawing 2 W of power. It also uses INT8 operands, and it is designed to add DNN inference ability to general-purpose processors.

(a)                (b)

**Figure 2.** (a) EdgeTPU AI accelerator module; (b) Coral Deep Learning embedded hardware with EdgeTPU AI accelerator module. Images from https://coral.ai/products/dev-board/ (accessed on 2 September 2021).

The Coral Dev board integrates an NXP i-MX8-MINI processor from the i-MX8 family designed for industrial applications. It is slightly less powerful than the i-MX8M-PLUS, with fewer image peripherals and interfaces and without the integrated AI co-processor—that role is played by the EdgeTPU. Note that the two devices selected for this work are partially compatible, as both use processors from the i-MX8 family. This was, as a matter of fact, one of the reasons they were chosen. However, Google provides its own tool set for both the EdgeTPU and the i-MX8-MINI SoC, based on a Mendel Linux distribution and TensorFlow Lite framework.

## 3. Deep Learning Frameworks

ML's success and popularity could not be understood without the existence of powerful and, at the same time, user-friendly application development frameworks. Some technology companies and universities have developed complete ML inference libraries for their own research purposes that they have ended up making public as open source software. Many ML algorithms are based on complex and quite cumbersome mathematical formulations that are not easy to implement. Frameworks simplify the development of such algorithms by exposing a high-level API to deal with complex calculations. In the case of DL networks, frameworks allow the implementation of a complete workflow, including defining the network architecture, training and optimization, model performance testing and model deployment into the final embedded devices.

There are many frameworks to choose from, and in general there are a lot of resources available on the web for almost all of them, but some frameworks have gained popularity among programmers and offer better support for application development. In [17], some of the most popular DL frameworks are classified by user access statistics to GitHub repositories. These frameworks demand considerable computing power, and they run on powerful computers usually complemented with GPUs [18]. Some of the processes involved in DL applications, such as model training and validation, require a large amount of memory and computational power. For that reason, they still run on high-end computing systems, and rarely on embedded devices.

Each framework uses its own model formats and APIs to build and implement DL applications. If the model is going to run in an embedded device, the framework must be supported by the embedded software distribution. This in fact determines the selection of the framework in the host (high-end) computer because the software of the host and the device must be compatible. To deal with this challenge, a standard interoperability library called Open Neural Network Exchange (ONNX) (https://onnx.ai/, accessed on 20 July 2021) was designed. Many embedded software distributions support this standard, allowing the selecting of the host framework without worrying about embedded device compatibility issues, as shown in Figure 3. Furthermore, this means that, at least theoretically, any model developed using any ML framework could be deployed into any embedded device by adequately converting the format of the model. In reality, embedded software distributions present strong restrictions, even more so if the embedded hardware integrates design-specific AI co-processors, so interoperability is far from total. A main issue is that ONNX is not widely supported by all embedded devices, and hardware manufactures provide specific libraries to deploy DNN in their co-processors that support a limited, if not unique, model format. For this reason, in the following sections the frameworks and libraries available in the selected embedded devices are revised.

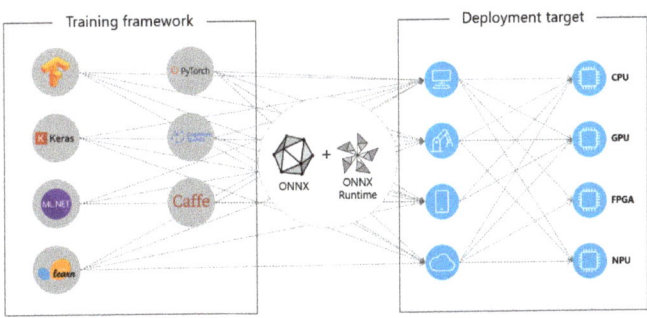

**Figure 3.** Interoperability of different frameworks by using ONNX.

## 3.1. Yocto Distribution and eIQ Machine Learning Framework for NXP i-MX8M Processors

The Yocto Project (https://www.yoctoproject.org/, accessed on 20 July 2021) is an open-source collaborative project that helps developers create custom Linux-based systems regardless of hardware architecture. NXP (the manufacturer of the i-MX8M-PLUS processor) provides a software release based on the Yocto Project framework. It can be used to build images for any i-MX8M board.

The compilation process downloads and installs many libraries and packages to create the binary image of a functional Linux distribution for the board. This binary image contains all the resources NXP provides to create an embedded ML application. In particular, the eIQ development environment supports these six run-time environments (inference engines): ArmNN, TensorFlow Lite, ONNX Runtime, PyTorch, OpenCV and DeepView$^{TM}$RT. To fully exploit the potential of the board, the framework selected must be supported by the internal NPU processor. Figure 4 shows the supported eIQ inference engines across the i-MX computing units.

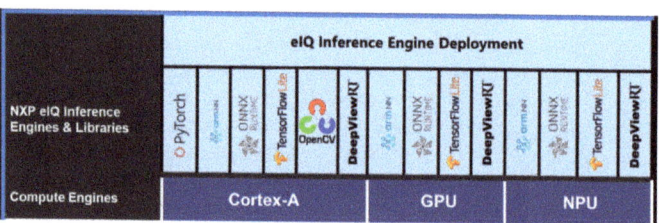

**Figure 4.** i-MX8 Deep Learning runtime environments supported by embedded computing units.

Pytorch and OpenCV are not supported by the embedded NPU and are directly discarded. A user guide (https://www.nxp.com/design/software/embedded-software/i-mx-software/embedded-linux-for-i-mx-applications-processors:IMXLINUX, accessed on 20 July 2021) explains the capabilities of all inference engines. For reasons that will become apparent in the next subsection, the most suitable runtime environment for this work is TensorFlow Lite (https://www.TensorFlow.org/lite/guide, accessed on 20 July 2021). As the name suggests, this is a lightweight version of the TensorFlow library for mobile, IoT and embedded devices. It is a runtime package that provides a way to run Deep Neural Networks on a specific hardware processor.

## 3.2. Mendel Linux and TensorFlow Lite in Coral Dev Board

The Coral Dev Board uses a Mendel Linux distribution maintained by Google. Unlike NXP Linux distributions, Coral Mendel Linux is specifically designed for this evaluation board kit, so there is no need to configure and compile the kernel or install any software packages or libraries. Everything is already available in a binary image that can be downloaded from https://coral.ai/docs/dev-board/get-started/ (accessed on 20 July 2021). The Coral Dev Board has a complete runtime ready to deploy DL models on its EdgeTPU AI co-processor unit. This co-processor was designed by Google to deploy TensorFlow models in embedded hardware, so the use of TensorFlow and its variant TensorFlow Lite is mandatory. TensorFlow Lite models must be off-line processed with a specific tool named "EdgeTPU Compiler" before being deployed in the EdgeTPU AI co-processor.

## 3.3. Host PC Setup

The host computer is an essential part of the whole development ecosystem. For this work, a host PC running Ubuntu 18.04 64-bit is used. The ML framework installed in the host is TensorFlow 2.5.0. The selection was straightforward, as both embedded devices support the TensorFlow Lite runtime. It comprises many functionalities, but the only one used in this work is the ability to convert object detection models into "lite" formats suitable for embedded systems. The TensorFlow programming interface is mainly written for Python, and it was decided to use this language to write all the model conversion scripts.

TensorFlow (and TensorFlow Lite) can be integrated with Python and C/C++ applications. It was decided to use Python to develop all the necessary scripts for the benchmarks described in this paper.

## 4. Object Detection Models

Object detection models are specialized ANN architectures designed to solve the computer vision task of object identification and localization in a digital image. From the model architecture perspective, object detection models inherit the feature extraction backbone from classification models. It is common to implement an object detection model by reusing a classification model such as VGG16, Mobilenet or Resnet, trained on a very large image dataset. The backbone used in embedded devices must be carefully selected, as the number of layers in the models varies greatly. Integration of the classification and localization heads in the model defines two separate solutions: two-stage models and one-stage models, in reference to the number of functional parts that the model contains. In the case of two-stage models, the first stage generates region proposals for object detection, and the second stage computes each proposed region and extracts both the classification result and the bounding boxes. Compared to one-stage models (which perform all functions together) two-stage models tend to have higher accuracy, although at a higher computational cost [19]. One of the first and most representative two-stage models is R-CNN [20], whose region proposal stage proposes around 2000 regions from the input image.

One-stage models use a feed-forward architecture in which everything is inferred in a single pass by applying a single neural network to the entire image. This approach results in significantly lower accuracy than two-stage detectors, but also higher detection speed. One of the first one-stage detectors was YOLO [21].

The TensorFlow library is accompanied by auxiliary libraries that complement its functionalities. Of particular interest for DL is the TensorFlow models repository (https://github.com/TensorFlow/models, accessed on 30 July 2021), also called the TensorFlow model zoo. This repository contains models for many DL applications, such as natural language processing, speech recognition and object detection. The model git repository version 2.5.0 was cloned (in accordance with the TensorFlow version). Inside the "models" directory, the "official" folder includes the code and models directly maintained by Google. The "research" folder contains some state-of-the-art technologies maintained by the developers themselves. The "object_detection" directory inside the "research" folder contains the libraries, code and models that have been used for hardware benchmarking. A brief explanation and an installation procedure can be found in https://github.com/TensorFlow/models/blob/master/research/object_detection/g3doc/tf2.md (accessed on 30 July 2021). The TensorFlow model zoo contains several types of object detection model architectures, which are described in the following paragraphs.

*4.1. CenterNet*

CenterNet (https://github.com/xingyizhou/CenterNet, accessed on 15 September 2021) is a one-stage object detection network that infers object position by assigning one point to every object rather than a square [22]. The size and even the pose of the object are calculated afterwards using a regression network. This strategy increases the accuracy of the network while maintaining fast inference time.

*4.2. Single Shot Multibox Detection (SSD)*

SSD networks [23] are widely used in embedded devices. They were the first one-stage networks, along with YOLO networks, that achieved accuracy similar to that of two-stage networks. Combined with the "mobilenet" backbone, it is the most supported network in TensorFlow Lite, mainly because it was developed by Google Research (among other researchers from academia) and it is a lightweight network suitable for deployment in embedded devices.

SSD networks usually come with a specialized component named a Feature Pyramid Network (FPN) [24] designed to improve the detection performance with objects at different scales. Usually object detection networks function quite poorly with very small or very big objects (in terms of the number of pixels that an object occupies in the image). FPNs solve this problem, increasing detection accuracy but also increasing processing time.

### 4.3. EfficientDet

The EfficientDet [25] DNN describes an improved one-stage network architecture that can be optimized and scaled to obtain a complete family of neural networks. Depending on the available computing resources and requirements, it is possible to select the most adequate member of the family. EfficientDet-D0 is the least resource demanding network of the family, and it should be adequate for embedded devices. The backbone used as feature extractor is called EfficientNet, hence its name.

### 4.4. Faster R-CNN

Faster R-CNN [26] is a two-stage object detection network. This architecture incorporates a new first-stage region proposal that improves network performance, achieving inference times comparable to those of single-stage networks while maintaining high accuracy. It is the latest of consecutively improved architectures, starting with R-CNN, then Fast-RCNN and finally Faster-RCNN. Some enhancements are also applied to the Faster R-CNN architecture to improve both inference speed and result accuracy [27,28].

### 4.5. Mask R-CNN

Mask R-CNN is an object segmentation model [29]. Object segmentation is a technique that, instead of detecting the object inside the image, categorizes each individual pixel of the image as belonging to a particular class. The goal is to obtain all the pixels belonging to a given class in the image, being able to draw the silhouette and the exact contour of an object, not only the surrounding square. In this sense, object segmentation can be seen as an improvement over object detection. Some architecture enhancements are available in the literature [30].

## 5. Model Conversion for Embedded Hardware Devices

The Design and Training stages of a DL model are almost always accomplished using a powerful host computer. The host computer includes an installation of a full ML framework with a set of packages and libraries to support and facilitate the whole DL application development workflow. The embedded devices, on the other hand, contain a runtime environment designed only and specifically to run a DL model inference.

In the TensorFlow environment, a model is described by a computational graph containing both the node connections and the weights or parameters of each node. The model is usually defined as a code file containing the API function calls necessary to build the model, for example using Keras API (https://keras.io/getting_started/, accessed on 15 September 2021). The model is built sequentially by adding a series of computational layers that fully describe the model architecture. However, at this point, the model is not functional because it does not yet contain the value of the weights, which are computed in the training process. Weights are stored in separated files named checkpoints. A checkpoint can be stored and reloaded at any time. This allows comparing the performance of different training stages, or retraining some of the model layers to accomplish an object detection task different from the one the model was previously trained for. Once the model is created, it is possible to save the computational graph and the weights all together in a single file format named "SavedModel" format using a specific TensorFlow API function call. A brief tutorial on TensorFlow model formats is available in https://www.TensorFlow.org/tutorials/keras/save_and_load (accessed on 11 July 2021).

For the TensorFlow Lite runtime environment, models created in TensorFlow must be converted using a specific library. This process modifies the model format appropriately to adapt it to run efficiently on the specific AI co-processors. Conversions mainly affect model weights, input tensors and output tensors. In general, TensorFlow models by default use floating-point parameters, which are appropriate for high-performance CPUs and GPUs, but embedded AI accelerators normally are restricted to work with integers only. Converting from float to integer types is called quantization.

In this work, five different quantization levels are considered based on the TensorFlow Lite optimization guide (https://www.TensorFlow.org/lite/performance/model_optimization, accessed on 11 July 2021). A brief description of the quantization levels is presented in Table 1, assigning to each level a numerical value. Note that the TensorFlow Lite conversion with no quantization has (properly) a quantization level 0. In the rest of this work, models with quantization levels 0 and 1 will be referred to as CPU models since they will run entirely on the main processor. In contrast, level 2, 3 and 4 models are intended to be executed in the specialized AI co-processor and will be referred to as co-processor models. An important part of this work is to measure the performance advantages of co-processor models over CPU models when an AI accelerator is available.

**Table 1.** Model quantization (optimization) levels used in this work.

| Level | Input | Weights | Output | Description |
|---|---|---|---|---|
| 0 | float | float | float | No quantization (all data is FLOAT32) |
| 1 | float | int8 | float | Quantization of model weights |
| 3 | float | int8 | float | Quantization of weights and internal variables using a representative dataset. Input and output layers remain in FLOAT32 |
| 3 | int8 | int8 | float | Quantization of input tensor uses the representative dataset |
| 4 | int8 | int8 | int8 | Full integer conversion. All computation is intended to be done in embedded AI co-processor |

*5.1. Model Conversion Issues*

The model conversion workflow is depicted as a block diagram in Figure 5. Models downloaded from the TensorFlow model zoo are already trained. The parameters in the trained checkpoint files are exported into a "SavedModel" file, and afterward model conversion is applied. Five conversion Python scripts were implemented to obtain the five corresponding TensorFlow Lite models, one per quantization level. These models are ready to be deployed in the i-MX8M-PLUS processor, but for the EdgeTPU module an extra compilation step must be done using a specific compiler developed by Google named "edgetpu_compiler". Therefore, after this compilation another five quantized models are obtained.

There are more than 80 models available In the TensorFlow model zoo (https://github.com/TensorFlow/models/blob/master/research/object_detection/g3doc/tf2_detection_zoo.md, accessed on 30 July 2021). Table 2 lists the nine models selected to be used in the present work. The name of each model describes the architecture, the input tensor size and the dataset used for training (all models are trained using COCO 2017 dataset). Some of the models integrate a Feature Pyramid Network (FPN) component, which improves the detection of objects at different scales in the image. Note that all the object detection architectures from the TensorFlow model zoo are represented except for Mask R-CNN. This model is in fact an object segmentation model with very different inference results and computation requirements, not comparable with the others, and for this reason it was not included in the benchmark. The justification of the selection of the rest of the models will become clear in the following subsections. For a given network, a total of ten optimized embedded ".tflite" models are generated (five for i-MX8M-PLUS and another five for

EdgeTPU). Considering the nine selected DL models, 90 embedded models were obtained. However, application of the conversion scripts was not always completed successfully. In Figure 6, all the issues found when trying to convert checkpoint files to TensorFlow Lite formats are listed. The next paragraphs explain each of them.

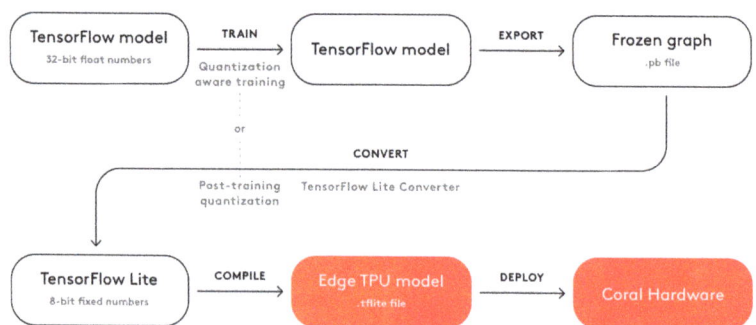

**Figure 5.** DL model conversion workflow using TensorFlow and TensorFlow Lite.

**Table 2.** Models used in the hardware benchmark.

| No. | Model Name |
|---|---|
| 1 | ssd_mobilenet_v2_320x320_coco17_tpu-8 |
| 2 | centernet_mobilenet_v2_fpn_512x512_coco17_od |
| 3 | ssd_mobilenet_v2_fpnlite_640x640_coco17_tpu-8 |
| 4 | efficientdet_d1_coco17_tpu-32 |
| 5 | ssd_mobilenet_v2_fpnlite_320x320_coco17_tpu-8 |
| 6 | ssd_mobilenet_v1_fpn_640x640_coco17_tpu-8 |
| 7 | ssd_resnet50_v1_fpn_640x640_coco17_tpu-8 |
| 8 | ssd_resnet101_v1_fpn_640x640_coco17_tpu-8 |
| 9 | faster_rcnn_resnet50_v1_640x640_coco17_tpu-8 |

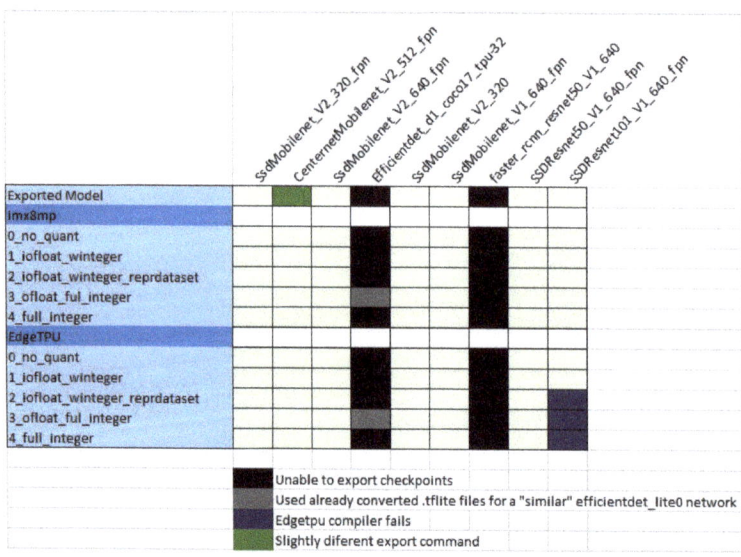

**Figure 6.** TensorFlow Lite model conversion issues.

#### 5.1.1. Unable to Export Checkpoint Files

The simplest way to convert a model is to use a "SavedModel" format from TensorFlow. It is possible to download any model from the TensorFlow model zoo in "SavedModel" format and also the training checkpoint files. Once uncompressed, it contains the saved_model directory with the ".pb" file, together with the checkpoint directory with checkpoint files. There is also a configuration file ".config" describing the model architecture. Unfortunately, this default "SavedModel" format is not suitable for conversion to TensorFlow Lite format because all object detection models have internal operations not supported by TensorFlow Lite. Instead, the object detection model code library provides a specific Python script to create a valid "SavedModel" from checkpoint files called "export_tflite_graph_tf2.py". However, this script supports neither EfficientDet nor Faster R-CNN network architectures. This means that approximately one half of the networks in the model zoo are in fact not suitable for use in embedded devices.

Some alternative model repositories were reviewed to try to overcome this problem, but the models must indeed fulfill so many constrains to be used with TensorFlow Lite that in the end only TensorFlow models were valid. A Lite version of "EfficientDet" model already converted to ".tflite" format was found at https://tfhub.dev/TensorFlow/efficientdet/lite0/detection/1 (accessed on 15 September 2021). It was also compiled for EdgeTPU and was included in the benchmark with the name "Efficientdet_lite0_320".

#### 5.1.2. Experimental CenterNet Model Export

CenterNet models checkpoint export fails when using TensorFlow library versions older than 2.4.0. Starting with this version, support for these networks was added. However, the export command requires small modifications compared with the command provided in the TensorFlow model optimization guide. The specific conversion command should be consulted in an example Jupyter Notebook at https://github.com/TensorFlow/models/blob/master/research/object_detection/colab_tutorials/centernet_on_device.ipynb (accessed on 15 September 2021). In the export command, the model size is also modified from its original value to 320 × 320. The model name is modified to reflect this change in the figures herein.

#### 5.1.3. EdgeTPU Compiler Fails

Co-processor models of "ssd_resnet101_v1_640_fpn" could not be compiled to EdgeTPU format. The compiler does not provide any information about the reasons for this failure. The network is by far the largest in the benchmark (more than 200 MB), so it is assumed that it in some way exceeds the capacity of the EdgeTPU module (or of the compiler itself).

### 5.2. Converted Model Size Analysis

Three files describe each model before the conversion: checkpoint file, original saved model file and exported saved model file, obtained from the checkpoint file after the execution of "export_tflite_graph_tf2.py", as explained above. Figure 7 displays the size of such files. The size range is from approximately 20 MB to more than 220 MB. The exported saved model file and the original saved model are similar, with the former slightly bigger than the last, and the checkpoint file is some MB smaller than the other two, except for CenterNet network. All other networks have a single-shot detection "SSD" architecture, and this could explain the difference.

The models are sorted by ascending size of the exported file (gray column in the figure). There are no TensorFlow files for "EfficientDet" network, so it was positioned in its corresponding position, attending to quantization level 3. This model order will be maintained in the rest of the document.

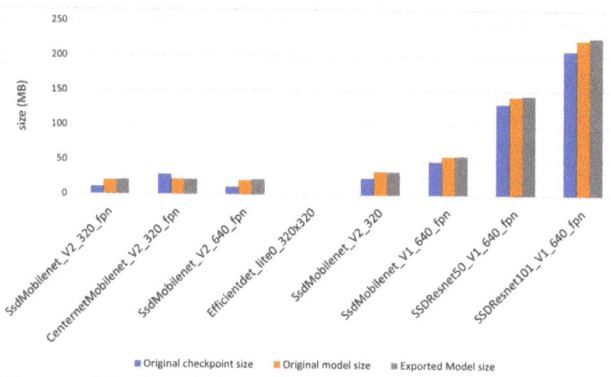

**Figure 7.** TensorFlow file sizes for object detection models.

The converted TensorFlow Lite model file sizes are shown in Figure 8 for i-MX8M-PLUS and in Figure 9 for EdgeTPU. The names of the quantized model files start with a number indicating the quantization level. The converted model without quantization (level 0) is smaller than the original model when the model itself is small, but exceeds the original model size considerably for the largest models. The other converted files present some kind of optimization. Starting from quantization level 1, model files present a type conversion of the network weights. Its size is, as expected, four times smaller than the model without quantization. Level 2, 3 and 4 models are slightly larger than those of level 1 (some hundreds of kilobytes) to include quantization of the inner intermediate layers and activation functions. There is no significant difference between the converted models for i-MX8M-PLUS and EdgeTPU devices.

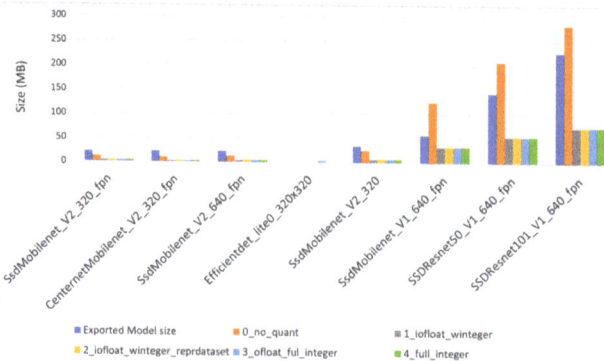

**Figure 8.** TensorFlow Lite converted file sizes for i-MX8M-PLUS.

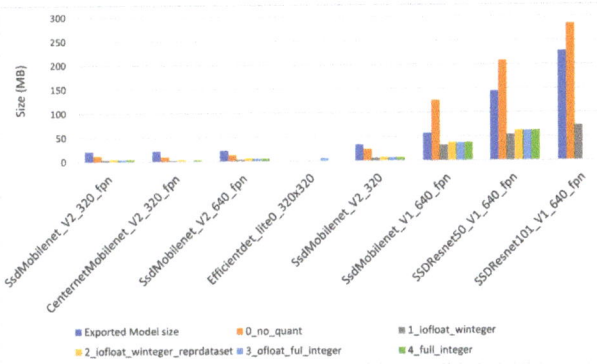

**Figure 9.** TensorFlow Lite converted file sizes for EdgeTPU.

## 6. Embedded Hardware Benchmarks

Once the embedded models are created, the next step is to execute them on the embedded hardware devices. To run the inference, a Python script is implemented using the TensorFlow Lite runtime environment API. Each of the quantized models is slightly different from the rest, so it is mandatory to write ten different Python scripts to execute all of them. The benchmark has two parts: (1) verification of the correctness of the model inference by examination of the obtained results, and (2) measurement of the models computation times. Three computation times are of interest when measuring the performance of the selected hardware devices:

1. **Warm up time.** This is the time the devices use to initialize their specific AI co-processor. Usually, the first inference is used for this initialization in addition to the inference itself. The device is not functional until the warm-up finishes.
2. **Auxiliary (image) processing time.** This is the time the CPU needs to access the image, resize and maybe re-scale it, load it into the input tensor and, after inference, get results and store a new image with bounding boxes around the detected objects.
3. **Model inference time.** This accounts for the time used to execute the mathematical model's operations, from the input tensor initialization to the access of the output results. Ideally, all the model operations should belong to the AI co-processor, but actually, due to limitations in model conversion and model deployment, some operations are delegated to the general-purpose CPU.

*6.1. Model Inference Issues*

Many issues were identified during the benchmark test. The following sections explain each of the errors or malfunctions detected, pointing out which models fall into each category. Figure 10 collects all of them.

6.1.1. Unable to Execute the Model

In this category two types of issues arise. In the first type, there is no model to be tested because it was not possible to create it. This is the case for EfficientDet and Faster-RCNN models, which are not supported by the export script "export_tflite_graph_tf2.py"; the same applies to the co-processor models for the "SSD_Resnet101" model for EdgeTPU. The second type of errors affects i-MX8M-PLUS with SSDResnet models that are not quantized (level 0). They have the biggest size of all models, and in addition, since they are not deeply optimized, they are executed almost completely in the CPU. This exceeds the memory or hardware capacity of the i-MX8M-PLUS processor and results in a fatal execution error.

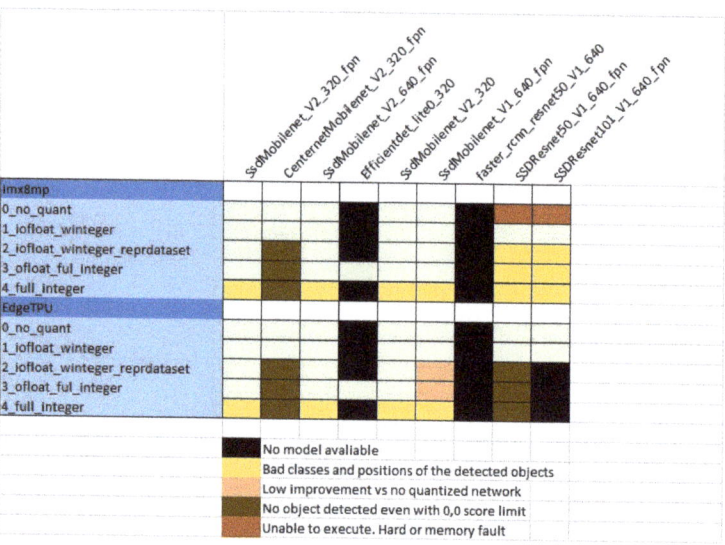

**Figure 10.** Inference issues.

6.1.2. Bad Inference Results

In the case of "SSD_Mobilenet" models, the objects detected by level 4 models have wrong class and position values. In this optimization level, the output tensor is converted to INT8 type. The level 3 model with FLOAT32 output tensors behaves correctly, so the error may be due to the quantization process or even to an internal quantization factor that is not being taken into account.

Similar behavior is observed in the case of "SSDResnet" co-processor models for i-MX8M-PLUS. However, in this case, the detection scores are also very low (<10%), indicating that the problem is even worse.

Finally, some models do not detect any objects. This is the case for all co-processor models for CenterNet in both devices and for "SSDResnet50" models only in the EdgeTPU module. Figure 11 shows a bad inference output results image.

**Figure 11.** Bad inference results. Neither object location squares nor object class labels are correct.

### 6.1.3. No Inference Time Improvement

The level 2 and 3 models for "SSDMobilenet_V1" in EdgeTPU detect correct object classes and position, but they present almost the same inference time as the CPU models. Currently, the EdgeTPU compiler cannot partition the model more than once, so as soon as an unsupported operation occurs, that operation and everything after it executes on the CPU, even if supported operations occur later. See https://coral.ai/docs/edgetpu/models-intro/ (accessed on 15 September 2021) for a more detailed explanation. This could explain this anomalous behavior.

### 6.1.4. Input Tensor Value Range

The input tensor value range is not the same for all network models. The "SSD_Mobilenet" networks present a FLOAT32 range of $[-1, 1]$ and a quantized UINT8 input of $[0, 255]$. All the other networks have the same $[0, 255]$ input range for both float and integer models. This supposes a small modification in the inference script for quantization levels 0, 1 and 2 for models with float input tensors.

### 6.1.5. Good Inference Results

It is not difficult to identify an incorrect behavior described in previous paragraphs because the errors are very evident. However, in general, inference results vary slightly among models and even among quantization levels. Usually, some models detect some object in an image that other models do not detect but fail to detect an object in another image. The detection scores vary from model to model and, because a limit score of 50% was imposed in the test, the objects near the limit may or may not be detected. Inference results are measured by visual inspection rather than by using a function that calculates the possible error. If these results are satisfactory, it is understood that the model is globally correct. Figure 12 shows good inference results for some test images.

**Figure 12.** Good inference results. The squares correctly locate object positions and object labels correctly identify object classes.

### 6.2. Analysis of the Computation Times

All the benchmark tests were conducted using the same image dataset of twenty images taken from the COCO set. The inference script loops for each image in the dataset and stores the computation times. The average values of all computation times are analyzed in the next sections.

### 6.2.1. Warm Up Time Analysis

Warm up times for the i-MX8M-PLUS are displayed in Figure 13. The figure shows clearly how the warm-up time increases with model size. It is also evident that the co-processor models present much larger times than the other CPU models. This could be easily explained by taking into account that the latter are executed completely in the CPU, so AI co-processor initialization is not necessary, while the former are deployed in the AI co-processor.

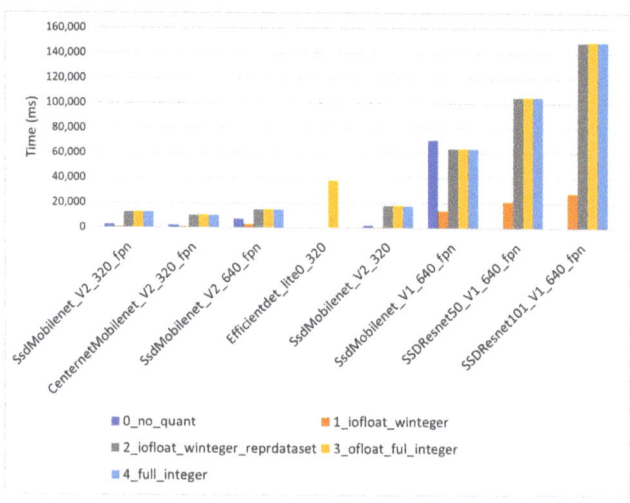

**Figure 13.** i-MX8M-PLUS warm up times.

The warm-up times vary for co-processor models from approximately 10 s to about 150 s. For small, non-quantized models it is smaller than 10 s, but when model size increases, the warm-up time is extremely long. In fact, the largest model raises an execution error. Quantization level 1 presents warm-up times from some seconds to around 25 s. All these figures represent a considerable amount of time, which must be considered in application design and development.

In the EdgeTPU module, the warm-up times behave differently than in the i-MX8M-PLUS (see Figure 14). The warm-up time for co-processor models is nearly the same as that of any other inference time, showing no significant overhead in EdgeTPU module initialization. For small models, the warm-up time is in the order of hundreds of milliseconds, making a specific initialization stage unnecessary. However, the EdgeTPU did not behave well when the model size increased, showing warm-up times of more than 10 s. Indeed, the largest co-processor models do not run in the EdgeTPU module.

### 6.2.2. Auxiliary Processing Time Analysis

Auxiliary processing times are fairly homogeneous in all network architectures. For i-MX8M-PLUS (Figure 15), the values vary between 20 and 40 ms with no correlation with model size. However, correlation with model quantization level is observed. The models with float input tensors (levels 0, 1 and 2) present notably larger times than those with quantized INT8 input tensors. This is more evident in "SSD_Mobilenet" networks. It is also observed that in the models with a large input size of 640 × 640, the difference is even bigger. The explanation is straightforward. The "SSD_Mobilenet" models need a preparatory scale operation (those models have a float $[-1, 1]$ input range) that involves floating-point operations in the input image. The cost of these operations increases with the size of the input tensor. The difference ranges form 4–5 ms for 320 × 320 input tensors up to 15 ms for sizes of 640 × 640. This time difference is not very high, but, especially in real time applications, should not be neglected.

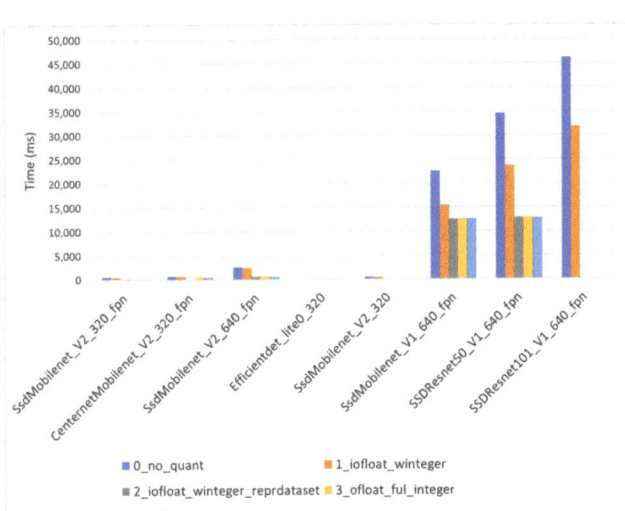

**Figure 14.** EdgeTPU warm up times for large models.

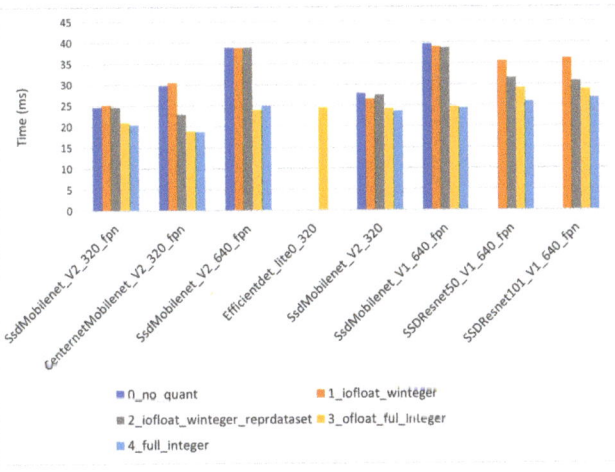

**Figure 15.** i-MX8M-PLUS auxiliary processing times.

Auxiliary processing times in the EdgeTPU are slightly larger (around 5 ms) than those in the i-MX8M-PLUS due to the slightly smaller computing power of the Coral Dev general purpose processor. However, the times behave exactly in the same way as explained above.

6.2.3. i-MX8M-PLUS Inference Time Analysis

The DL model inference time is the most relevant parameter to be analyzed in order to measure the performance of the embedded hardware and the feasibility of the deployment of DL object detection applications. Both devices' inference times are analyzed independently, starting here with the i-MX8M-PLUS processor, and the results are compared afterwards.

The inference times for the i-MX8M-PLUS strongly depend on quantization level. As expected, CPU models have considerably longer inference times than co-processor models. CPU models' inference times in Figure 16 range from 500 ms to around 25 s. The quantization level 0 inference time for "SSD_Mobilent_V1" presents an outlier value exceeding one minute. This points to even longer inference times for "SSD_Resnet" networks, but those models do not work on the i-MX8M-PLUS. The co-processor models' inference times in Figure 17 range from 20 ms to near 800 ms. Note that the timescale in the figure is 100 times lower than in the previous figure above. The yellow line in the figure represents the quantization level 3 models' inference time and is used later to compare results between hardware devices.

Attending to the inference times, it is clear that "ssd_mobilenet_v2_320" should be moved to first place, and "ssd_mobilenet_v2_640 × 640" should be move back one position behind "efficientdet_lite0_320". This means that the inference times cannot be directly inferred from model size; rather, network complexity should be taken into account. Sorted by ascending inference time, "SSD_Mobilenet_V2" is followed by networks with Feature Pyramid Network (FPN), which introduces computation complexity, and afterward the models with size 640 × 640 are positioned as expected at the end. It is important to note that there is no significant difference in the inference times between co-processor models with different quantization levels.

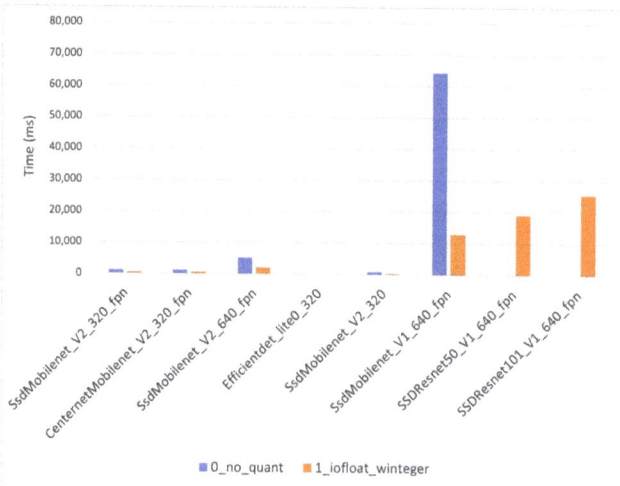

**Figure 16.** i-MX8M-PLUS inference time for CPU models.

Note also that even if they appear in the figure above, CenterNet and "SSD_Resnet" Network do not obtain good inference results. The inference time figures were included in the benchmark because the CPU models worked properly, and the obtained inference times are also coherent with model size and complexity.

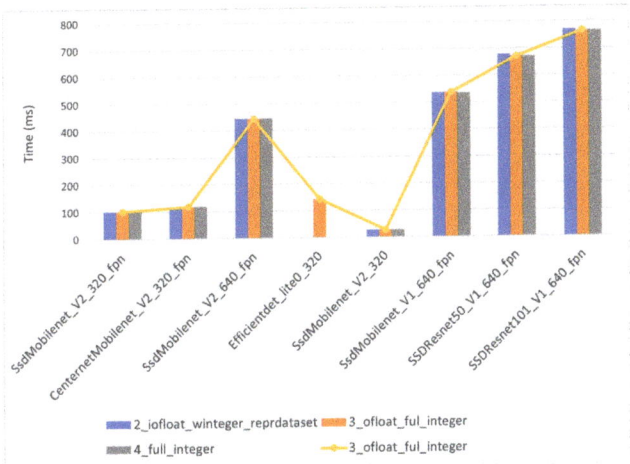

**Figure 17.** i-MX8M-PLUS inference for co-processor models.

### 6.2.4. EdgeTPU Inference Time Analysis

Inference times for the EdgeTPU module behave nearly in the same way as those of the i-MX8M-PLUS. The times for CPU models (Figure 18) are considerably longer than those for co-processor models (Figure 19). However, the CPU models did not present the anomalous behavior for large models, and all of them were correctly executed on the Coral Dev Board.

In the case of co-processor models, for large models, there is no time reduction compared with CPU models, and those models are omitted in the inference time analysis. The yellow line in the Figure 19 belongs to the quantization level 3 models, as was the case for the i-MX8M-PLUS. The fastest model is, as in the case for the i-MX8M-PLUS processor, the "ssd_mobilenet_v2_320" model, with inference time below 20 ms. The "eficiendet_lite0_320" model, with 145 ms inference time, overtakes the "centernet_Mobilenet_320", with more than 500 ms, and "ssd_mobilenet_V2_640", with 650 ms inference time.

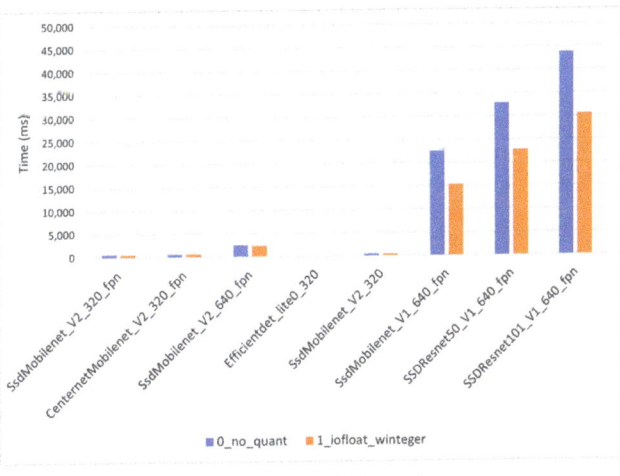

**Figure 18.** EdgeTPU inference time for CPU models.

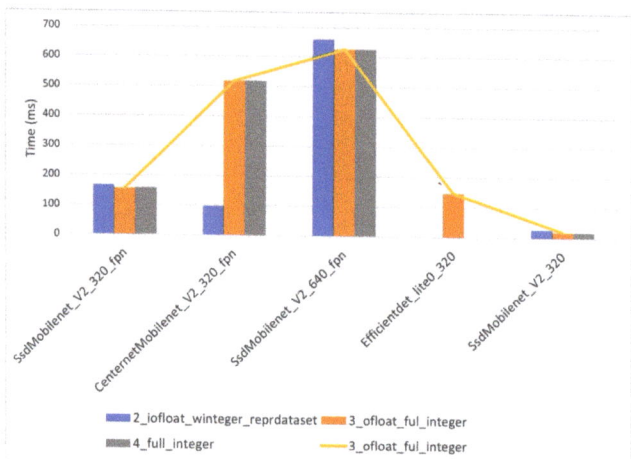

**Figure 19.** EdgeTPU inference time for co-processor models.

#### 6.2.5. i-MX8M-PLUS vs. EdgeTPU Inference Time Comparison

A performance improvement factor is calculated by dividing the inference times of the quantization level 1 model by the inference time of the corresponding model with quantization level 3. The improvement factor for the i-MX8M-PLUS processor increases monotonically with model size, as can be observed in Figure 20. Its value varies from 5 for smaller models up to more than 30 for the largest model, "ssd_resnet_101_V1".

For the EdgeTPU module, the performance improvement factor presents a value of around 4, except for the network "ssd_mobilenet_v2_320", which obtains a value of 23. The values are below those of the i-MX8M-PLUS processor, and these results are even worse taking into account that the inference times for quantized level 1 models in the Coral Dev board are longer (around 10%) than the corresponding values in the i-MX8M-PLUS processor due to the computing power differences in the general purpose ARM CPUs of both devices.

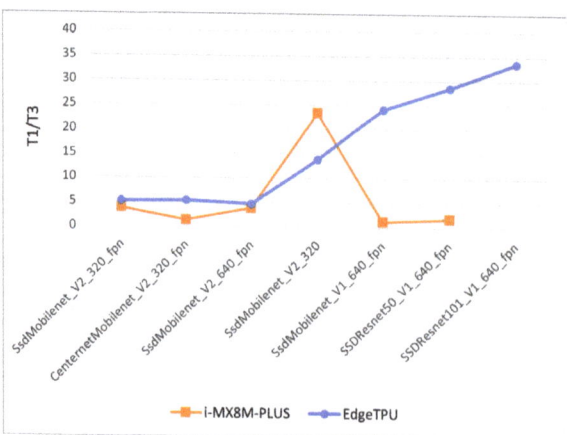

**Figure 20.** Inference time improvement factor calculated using quantization levels 1 and 3.

In Figure 21, the inference times for quantization level 3 models for both devices are displayed. In the case of the EdgeTPU, only the first, small models are depicted because the last three models do not have valid inference times. The i-MX8M-PLUS processor shows better performance than the EdgeTPU Coral Dev board for the first three models and almost the same performance for the next two. Taking into account that the EdgeTPU has 4 TOPS computing power and the i-MX8M-PLUS has 2.3 TOPS, these results suggest that the i-MX8M-PLUS processor is more efficient than the EdgeTPU module when deploying and running DL models.

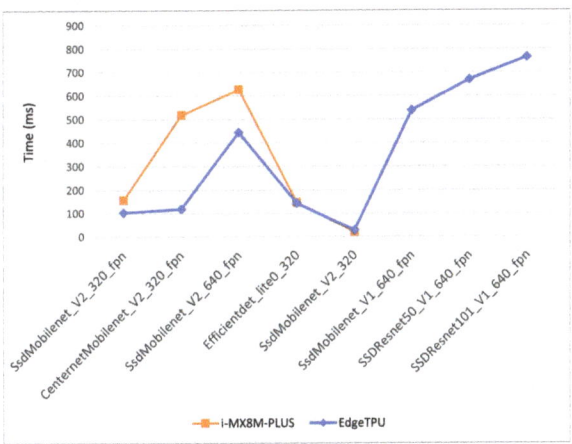

**Figure 21.** i-MX8M-PLUS vs. EdgeTPU inference times for quantized level 3 models.

This better performance is confirmed by looking at the behavior of the largest models. In the i-MX8M-PLUS processor, the inference time is kept under one second, with a improvement factor of up to 30, while the EdgeTPU module presents times over 10 s and improvement factors below 2.

## 7. Conclusions

The first effect related to AI at the edge paradigm is the emergence of many embedded devices with specialized AI co-processors to execute deep neural network inferences. In this work, after a detailed review of the available embedded hardware devices, two of them were selected to demonstrate and evaluate the feasibility of the deployment of DL object detection models in resource constrained devices: Variscite i-MX8M-PLUS Board and EdgeTPU Coral Dev Board. Requirements to select a device for this analysis included: (1) it must belong to an important and reliable manufacturer, and (2) it must offer a strong development community supporting the tools and applications. The devices selected were designed by NXP and Google. NXP is one of the most successful industrial processor manufacturers, and Google could be the most important player in the AI arena. A large portion of this work was devoted to setting up the hardware devices—understanding what libraries and packages needed to be installed and the appropriate tools to use. One of the main goals of the work was to learn and understand the workflow of AI application development, and it can be concluded that the success of this task depends considerably on the selection of the development framework.

The AI framework used to develop and deploy DL networks in embedded devices was TensorFlow, together with TensorFlow Lite. As a first workflow stage, TensorFlow models need to be converted into TensorFlow Lite format. Even if an easy-to-use tool is provided by TensorFlow Lite to convert the models, the conversion is not trivial because of a number of incompatibilities between both frameworks. Many mathematical operations deeply hidden in the layers of the neural networks are not supported by the Lite version runtime, and the conversion of many model architectures remains still unsolved.

All four main model architectures for object detection in the TensorFlow model repository were considered: "CenterNet", "SSD", "EfficientDet" and "Faster R-CNN". However, in the early stages we realized that TensorFlow Lite conversion of some of the models was impossible. As a matter of fact, only "SSD" and "CenterNet" architectures are compatible with the current TensorFlow Lite converter; thus, a set of seven models were finally selected: six "SSD" with different feature extractor backbones and one "CenterNet". Further, an "EfficientDet" model already converted to TensorFlow Lite format was added to test as many architectures as possible.

AI co-processors are very specialized hardware units that only accept eight-bit integers as operands, so the models must also be quantized. Five quantization levels were defined in accordance with the capabilities of the TensorFlow Lite library API. After executing model quantization scripts, 35 models for each device were compiled, plus the 2 already converted, giving a total of 72 models.

It is not easy to understand the quality of the converted model to guess how the model should be deployed in the AI co-processor. As a guideline, in the case of the i-MX8M-PLUS, the inference script returns a list of unsupported operations in the initial execution stage, while in the case of the EdgeTPU, a log file is created when the TensorFlow Lite model is compiling, with the number of operations mapped to both the EdgeTPU and the CPU.

The benchmark consisted of executing all the converted models, verifying correct behavior and measuring the model inference time. Many issues were detected during this process. Some converted models did not detect the validation image objects the same was as the original model; others simply did not run in the embedded devices. The number of models with correct behavior was considerably shortened. Only forty of the initial seventy-two models provided acceptable results. If only quantized models with representative datasets are considered, the number decreases to only 16 models, 2 of them belonging to an "EfficientDet_lite0" network not created by the "standard" workflow. Finally, only the four "SSD_Mobilenet" frameworks were proven to be valid for embedded devices. Again, the problems rely on the efficiency and quality of the converted models and the ability of the embedded runtime to fit the models into specialized hardware.

Both hardware devices, the i-MX8M-PLUS and the EdgeTPU, were able to execute the quickest object detection models in approximately 20 ms. The auxiliary CPU processing time spent another 25 ms. The whole inference time supposes nearly 50 ms, or 20 frames per second. The inference times increased up to 100 ms for more complex network models and even more to 500–800 ms when input image size increased. Even if the EdgeTPU claims to have almost double computing power, this benchmark demonstrates that the i-MX8M-PLUS device performed slightly better in general. The performance improvement of co-processor models compared with CPU models is about 10 times in the i-MX8M-PLUS and 5 or even worse in the EdgeTPU.

A few quick calculations were carried out to determine the quality of the AI co-processor inference time results. The i-MX8M-PLUS processor integrates four ARM Cortex-A53 cores at 1.8 GHz. Assuming (to obtain a very raw estimate of computing power) that the cores are able to execute one operation per clock, the maximum theoretical computing processing power should be around 10 Giga-operations per second (GOPS) for floating-point operations. Compared to the AI co-processor's 2.3 TOPS, the theoretical optimal improvement factor should be in the order of 100. The calculation is based on very imprecise and simplified assumptions, and the actual number should be lower than the theoretical number. Even though, the improvement factor of 5 to 15 obtained for most of the small "SSD_mobilent" networks is quite far from that figures. Once again, the converted model is not competent to be efficiently executed in the AI co-processor. The models are partitioned when unsupported operations are found, and many operations are delegated back to the general purpose CPU, slowing down the total inference performance.

In general, the feeling about the current state of object detection for embedded devices is that many aspects of performance depend on the efficiency of the software frameworks on both the host computer and the embedded device, and on their ability to extract maximum performance from the embedded hardware co-processors. Those libraries are now under construction and continuous modifications. Nearly every month, NXP releases a new version of the Yocto framework for the i-MX processor family (at least two new versions were released since the first benchmark test was accomplished). Coral also releases new compiler tools, API libraries and trained models periodically. In the case of TensorFlow and TensorFlow Lite, even if the libraries were updated many times along the development of the benchmark, new releases are now available to be downloaded. The repository of models is updated every day (there are continuous commits to the research repository), and an official version is released synchronized with every TensorFlow release.

## 8. Future Work

It should be clear after reading the previous sections that many issues remain open and unsolved. The present work does not make a quantitative assessment of the (numerical) performance of the converted models. Performance correctness is decided by visual inspection of the detected objects and correct object classification. Even if this approach easily detects catastrophic failures (such as those shown in Figure 11), subtle performance variations are undetected. A means to measure the error should be included as part of the inference script. There is a straightforward error computation standard defined by the COCO dataset, called mean average precision (mAP), specifically defined for object detection. This error metric is in fact available in TensorFlow, but needs to be implemented from scratch in embedded devices. It would be interesting to investigate whether different levels of quantization introduce noticeable errors, or whether certain network architectures are more sensitive to quantization processes. We plan to carry out a quantitative evaluation of these aspects in a future paper.

One of the main constraints imposed on the work was the requirement of using pre-built models from the TensorFlow model zoo. TensorFlow provides the possibility to implement the model using a flexible API at different levels of abstraction. It would be illustrative to build the standard object detection models used in this work, or even other similar ones, and to investigate how those models behave after quantization in the embedded devices considered here. The final objective should be to learn if there is a way to optimize model deployment by defining model internal operations and layer connections using supported operations of the TensorFlow Lite embedded runtime. Furthermore, additional model sources besides TensorFlow should be investigated. The ONNX model exchange should allow the import of models from other AI frameworks. The EdgeTPU is only supported by TensorFlow Lite runtime libraries, but the i-MX8M-PLUS has some other supported frameworks, such as DeepViewRT, armNN or the previously mentioned ONNX.

Finally, more hardware devices should be considered. The two embedded boards considered in this work shared many hardware specifications. Both have an NXP i-MX family processor, integrate an integer tensor processor and rely on TensorFlow Lite libraries as a runtime. In order to have a more global view of the hardware performance, different types of embedded devices should be tested. At the beginning of the present work, a third hardware platform called Jetson Nano was pre-selected to be included in the benchmark. The Jetson Nano Nvidia AI platform integrates a floating point arithmetic AI co-processor and uses other specialized libraries called TensorRT. The board was successfully launched, and some preliminary tests have been performed, but the software framework is quite different from the one used with the other two boards, and significant work is needed to implement the inference processes.

**Author Contributions:** Conceptualization, investigation and writing, D.C. as part of his PhD research; methodology, overall supervision and writing, including review and editing, I.E.-G., J.M.-A. and E.J. All authors have read and agreed to the published version of the manuscript.

**Funding:** This work has received support from the following programs: PID2019-104966GB-I00 (Spanish Ministry of Science and Innovation), IT-1244-19 (Basque Government), KK-2020/00049, KK-2021/00111 and KK-2021/00095 (Elkartek projects 3KIA, ERTZEAN and SIGZE, funded by the SPRI-Basque Government) and the AI-PROFICIENT project funded by European Union's Horizon 2020 research and innovation program under grant agreement no. 957391.

**Institutional Review Board Statement:** Not applicable.

**Informed Consent Statement:** Not applicable.

**Data Availability Statement:** Not applicable.

**Conflicts of Interest:** The authors declare no conflict of interest.

## References

1. Merenda, M.; Porcaro, C.; Iero, D. Edge machine learning for ai-enabled iot devices: A review. *Sensors* **2020**, *20*, 2533. [CrossRef] [PubMed]
2. Weiss, K.; Khoshgoftaar, T.M.; Wang, D. A survey of transfer learning. *J. Big Data* **2016**, *3*, 1–40. [CrossRef]
3. Murshed, M.S.; Murphy, C.; Hou, D.; Khan, N.; Ananthanarayanan, G.; Hussain, F. Machine learning at the network edge: A survey. *ACM Comput. Surv.* **2021**, *54*, 1–37. [CrossRef]
4. Pena, D.; Forembski, A.; Xu, X.; Moloney, D. Benchmarking of CNNs for low-cost, low-power robotics applications. In Proceedings of the RSS 2017 Workshop: New Frontier for Deep Learning in Robotics, Rhodes, Greece, 15–16 July 2017; pp. 1–5.
5. Hossain, S.; Lee, D. Deep learning-based real-time multiple-object detection and tracking from aerial imagery via a flying robot with GPU-based embedded devices. *Sensors* **2019**, *19*, 3371. [CrossRef] [PubMed]
6. Lonsdale, D.; Zhang, L.; Jiang, R. 3D printed brain-controlled robot-arm prosthetic via embedded deep learning from sEMG sensors. In Proceedings of the 2020 International Conference on Machine Learning and Cybernetics (ICMLC), Adelaide, Australia, 2 December 2020; pp. 247–253.
7. Rahmaniar, W.; Hernawan, A. Real-time human detection using deep learning on embedded platforms: A review. *J. Robot. Control* **2021**, *2*, 462–468.
8. Gubbi, J.; Buyya, R.; Marusic, S.; Palaniswami, M. Internet of Things (IoT): A vision, architectural elements, and future directions. *Future Gener. Comput. Syst.* **2013**, *29*, 1645–1660. [CrossRef]
9. Lasi, H.; Fettke, P.; Kemper, H.G.; Feld, T.; Hoffmann, M. Industry 4.0. *Bus. Inf. Syst. Eng.* **2014**, *6*, 239–242. [CrossRef]
10. Véstias, M.P.; Duarte, R.P.; de Sousa, J.T.; Neto, H.C. Moving deep learning to the edge. *Algorithms* **2020**, *13*, 125. [CrossRef]
11. Shi, W.; Cao, J.; Zhang, Q.; Li, Y.; Xu, L. Edge computing: Vision and challenges. *IEEE Internet Things J.* **2016**, *3*, 637–646. [CrossRef]
12. Cao, K.; Liu, Y.; Meng, G.; Sun, Q. An overview on edge computing research. *IEEE Access* **2020**, *8*, 85714–85728. [CrossRef]
13. Branco, S.; Ferreira, A.G.; Cabral, J. Machine learning in resource-scarce embedded systems, FPGAs, and end-devices: A survey. *Electronics* **2019**, *8*, 1289. [CrossRef]
14. Ajani, T.S.; Imoize, A.L.; Atayero, A.A. An overview of machine learning within embedded and mobile devices–optimizations and applications. *Sensors* **2021**, *21*, 4412. [CrossRef]
15. Bianco, S.; Cadene, R.; Celona, L.; Napoletano, P. Benchmark analysis of representative deep neural network architectures. *IEEE Access* **2018**, *6*, 64270–64277. [CrossRef]
16. Imran, H.A.; Mujahid, U.; Wazir, S.; Latif, U.; Mehmood, K. Embedded development boards for edge-AI: A comprehensive report. *arXiv* **2020**, arXiv:2009.00803.
17. Zacharias, J.; Barz, M.; Sonntag, D. A survey on deep learning toolkits and libraries for intelligent user interfaces. *arXiv* **2018**, arXiv:1803.04818.
18. Dai, W.; Berleant, D. Benchmarking contemporary deep learning hardware and frameworks: A survey of qualitative metrics. In Proceedings of the 2019 IEEE First International Conference on Cognitive Machine Intelligence (CogMI), Los Angeles, CA, USA, 12–14 December 2019; pp. 148–155.
19. Zhao, Z.Q.; Zheng, P.; Xu, S.; Wu, X. Object detection with deep learning: A review. *IEEE Trans. Neural Netw. Learn. Syst.* **2019**, *30*, 3212–3232. [CrossRef]
20. Girshick, R.; Donahue, J.; Darrell, T.; Malik, J. Region-based convolutional networks for accurate object detection and segmentation. *IEEE Trans. Pattern Anal. Mach. Intell.* **2015**, *38*, 142–158. [CrossRef]
21. Redmon, J.; Divvala, S.; Girshick, R.; Farhadi, A. You only look once: Unified, real-time object detection. In Proceedings of the IEEE Conference on Computer Vision and Pattern Recognition, Las Vegas, NV, USA, 27–30 June 2016; pp. 779–788.
22. Zhou, X.; Wang, D.; Krähenbühl, P. Objects as points. *arXiv* **2019**, arXiv:1904.07850.
23. Liu, W.; Anguelov, D.; Erhan, D.; Szegedy, C.; Reed, S.; Fu, C.Y.; Berg, A.C. Ssd: Single shot multibox detector. In Proceedings of the European Conference on Computer Vision, Amsterdam, The Netherlands, 11–14 October 2016; pp. 21–37.
24. Lin, T.; Dollár, P.; Girshick, R.B.; He, K.; Hariharan, B.; Belongie, S.J. Feature Pyramid Networks for Object Detection. In Proceedings of the IEEE Conference on Computer Vision and Pattern Recognition, Honolulu, HI, USA, 21–26 July 2016; pp. 2117–2125.
25. Tan, M.; Pang, R.; Le, Q.V. Efficientdet: Scalable and efficient object detection. In Proceedings of the IEEE/CVF Conference on Computer Vision and Pattern Recognition, Seattle, WA, USA, 13–19 June 2020; pp. 10781–10890.
26. Ren, S.; He, K.; Girshick, R.; Sun, J. Faster R-CNN: Towards real-time object detection with region proposal networks. *Adv. Neural Inf. Process. Syst.* **2015**, *28*, 1–9. [CrossRef]
27. Cao, C.; Wang, B.; Zhang, W.; Zeng, X.; Yan, X.; Feng, Z.; Liu, Y.; Wu, Z. An Improved Faster R-CNN for Small Object Detection. *IEEE Access* **2019**, *7*, 106838–106846. [CrossRef]
28. Chu, J.; Guo, Z.; Leng, L. Object Detection Based on Multi-Layer Convolution Feature Fusion and Online Hard Example Mining. *IEEE Access* **2018**, *6*, 19959–19967. [CrossRef]

29. He, K.; Gkioxari, G.; Dollár, P.; Girshick, R.B. Mask R-CNN. In Proceedings of the IEEE International Conference on Computer Vision (ICCV), Venice, Italy, 22–29 October 2017; pp. 2961–2969.
30. Zhang, Y.; Chu, J.; Leng, L.; Miao, J. Mask-Refined R-CNN: A Network for Refining Object Details in Instance Segmentation. *Sensors* **2020**, *20*, 1010. [CrossRef] [PubMed]

*Article*

# Robot System Assistant (RoSA): Towards Intuitive Multi-Modal and Multi-Device Human-Robot Interaction

Dominykas Strazdas *, Jan Hintz, Aly Khalifa, Ahmed A. Abdelrahman, Thorsten Hempel and Ayoub Al-Hamadi *

Neuro-Information Technology, Otto-von-Guericke-University Magdeburg, 39106 Magdeburg, Germany; jan.hintz@ovgu.de (J.H.); aly.khalifa@ovgu.de (A.K.); ahmed.abdelrahman@ovgu.de (A.A.A.); thorsten.hempel@ovgu.de (T.H.)
* Correspondence: dominykas.strazdas@ovgu.de (D.S.); ayoub.al-hamadi@ovgu.de (A.A.-H.)

**Abstract:** This paper presents an implementation of RoSA, a Robot System Assistant, for safe and intuitive human-machine interaction. The interaction modalities were chosen and previously reviewed using a Wizard of Oz study emphasizing a strong propensity for speech and pointing gestures. Based on these findings, we design and implement a new multi-modal system for contactless human-machine interaction based on speech, facial, and gesture recognition. We evaluate our proposed system in an extensive study with multiple subjects to examine the user experience and interaction efficiency. It reports that our method achieves similar usability scores compared to the entirely human remote-controlled robot interaction in our Wizard of Oz study. Furthermore, our framework's implementation is based on the Robot Operating System (ROS), allowing modularity and extendability for our multi-device and multi-user method.

**Keywords:** augmented reality; activity recognition; cooperative systems; facial recognition; gesture recognition; human-robot interaction; interactive systems; robot control; speech recognition

**Citation:** Strazdas, D.; Hintz, J.; Khalifa, A.; Abdelrahman, A.A.; Hempel, T.; Al-Hamadi, A. Robot System Assistant (RoSA): Towards Intuitive Multi-Modal and Multi-Device Human-Robot Interaction. *Sensors* **2022**, *22*, 923. https://doi.org/10.3390/s22030923

Academic Editor: José María Martínez-Otzeta

Received: 23 December 2021
Accepted: 19 January 2022
Published: 25 January 2022

**Publisher's Note:** MDPI stays neutral with regard to jurisdictional claims in published maps and institutional affiliations.

**Copyright:** © 2022 by the authors. Licensee MDPI, Basel, Switzerland. This article is an open access article distributed under the terms and conditions of the Creative Commons Attribution (CC BY) license (https:// creativecommons.org/licenses/by/ 4.0/).

## 1. Introduction

Recently, collaborative robotics (cobots) has experienced increasing popularity as it is targeted to be a more flexible and general task type of robot [1]. Compared to conventional industrial robots, cobots share their work area with humans and interact directly. This requires new standards of safety and interaction interfaces. Typically, robots are instructed by buttons, knobs, joysticks, specific speech, and gestures commands, teaching through touching and guiding or dedicated teaching panels. Either way, the handling of the robot requires prior knowledge and is not intuitive for untaught users. Therefore, the interaction interface must shift to a more human-centered and adaptive relation to enable the use of cobots in an unconstrained environment with varying tasks and interchanging human collaboration partners.

There have been multiple promising research approaches tackling flexible Human-Robot Interaction (HRI) scenarios [2–5], but most of their methods are driven by the introductions of new techniques instead of focusing on the needs and characteristics of interaction patterns from a human perspective.

With the aim of a better understanding of human behavior in robotic interaction scenarios, we carried out an extensive Wizard of Oz (WoZ) study [6] to examine common communication intuitions of untaught human interaction partners. In addition, we worked out human key actions to approach human-robot interactions. On this basis, we conceptualized and implemented a new multi-modal robotic system called "RoSA" (Robot System Assistant) that tackles the challenge of intuitive and user-centered human-robot interaction by facilitating multiple input streams such as speech, gesture, face, body, and object recognition. The use of a wide range of perception capabilities combined with speech processing ensures a robust scene and interaction state estimation and leads to an efficient and intuitive human-robot collaborative task performance.

In this paper, we tackle the challenge of turning the results from our prior WoZ study into a fully autonomous robotic systems that can handle interaction with untaught human partners without external control. First, we derive a concept that enables the robot visual and acoustic perception of potential human interaction partners and the scene understanding. We show which features are crucial for perceiving interpretable indications to derive meaningful instructions for the robot. We continue with an in-depth analysis of each module and its interplay with the core system and present further implementation details. Finally, the evaluation of our RoSA system is conducted via a separate study similar to our previous Wizard of Oz study, but this time with a fully autonomous robotic system.

## 2. Related Work

In our earlier Wizard of Oz (WoZ) study [6], we reviewed different interaction modalities required for an intuitive HRI. The participants were permitted to use different features like gestures, speech, mimics, and gaze without any limitations to communicate with a cobot and execute different tasks like cube stacking. We tricked the subjects into thinking they were interacting with a state-of-the-art artificial intelligence controlling a cobot, whereas in fact, we were remote controlling the cobot based on the participant's instructions and a strict set of rules. This principle is commonly called a Wizard of Oz experiment. Figure 1 shows multiple views of the subjects interacting and giving orders through speech or gestures.

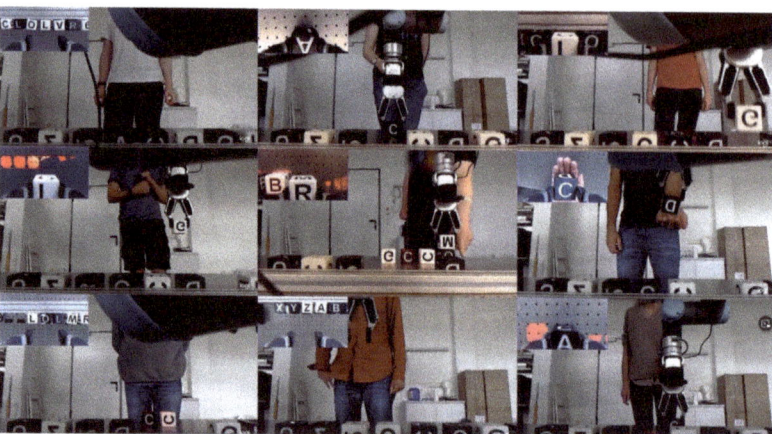

**Figure 1.** Previous field study for natural human-robot interaction using the Wizard of Oz method. A video summary can be found here: https://youtu.be/JL409R7YQa0 (accessed on 18 January 2022).

It was shown that 97% of the 36 subjects used speech and 75% used pointing gestures to solve the given tasks. Most of the subjects preferred the path planning to be done by the robot assistance system and did not want to guide the robot directly but give more complex commands.

In regard to safety, by complying with the standards, ISO 10218-1/2 [7] and ISO/TS 15066 [8] on HRI, the danger to the user is minimized. As Pasinetti et al. [9] have shown, time-of-flight (ToF) cameras can be used to detect the operator and, in combination with virtual barriers, slow or stop the robot when safety guidelines are infracted. In addition, one of our previous implementation of a gadget-less HRI concept Robo-HUD suggests that a non-contact approach also contributes to a safe HRI [10]. An attention module based on head posture estimation was introduced to monitor attention, allowing intended user actions.

It also introduced an *Attention Module* based on head pose estimation to monitor awareness, enabling the interaction only when intended. This module allows users to switch between workstations without logging in or out when used in a multi-device scenario.

Many implementations already exist in this area [11–19]. Magrini et al. [20] proposed a system that ensures human safety in a robotic cell. It is based on the method of Pasinetti et al., which uses ToF cameras to localize users in real time. In addition, their system enables gesture recognition for low-level robot control (e.g., start/stop). These gestures are two-handed gestures that must be clearly shown above the head. This type of gesture is already built into the Kinect V2 software and is known to be relatively robust. Additionally, facial recognition can be used to personalize and create a long-term user experience [21].

Based on these findings and due to the lack of a state-of-the-art implementation of an intuitive multi-modal overall concept, the Robot System Assistant (RoSA) concept was created [22].

## 3. Framework Overview

### 3.1. Concept

We worked out a concept to assemble all important information streams into synergetic interplay. The overview can be seen in Figure 2. It consists of seven modules (face, attention, speech, gesture, robot, scene, and cube) communicating through middleware to a core unit, the interaction module, which is responsible for the logic and actions of the system. The modular approach allows an independent development and evaluation of each necessary component. While the functionality for most of the modules is self-explanatory, the *Scene Module* and *Cube Module* require a more detailed introduction. The *Scene Module* is responsible for the data storage from the virtual objects and contains the constraints, positions, and calibration of the system. In this module, a digital twin of the scene gets depicted.

The idea of using cubes as an interaction object was carried over from the experimental design of preceding studies in order to stay consistent and to allow a direct comparison with the WoZ study. The cube module takes care of the cube logic and their detection.

For scalability purposes and to evaluate a realistic scenario, the modules designed to be able to run as multiple instances, allowing a multi-device setup. The individual devices, the middleware, and functions of the mentioned modules are described in detail in Sections 4 and 5.

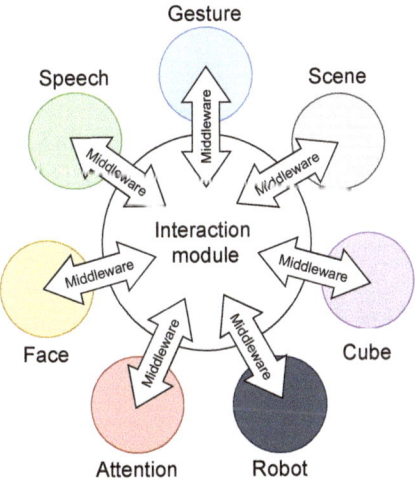

**Figure 2.** Concept: The interaction module connects through middleware to other modules.

## 3.2. Features

In the previous studies RoSA was a system that was capable of speech, gesture, face, body, and attention recognition based on the "wizard", the operator controlling the robot. To automate the system and combine the cobot with artificial intelligence, a set of necessary features was elaborated, which is shown in Table 1.

**Table 1.** Extracted feature stream.

| Stream | Feature | Description | Methods |
|---|---|---|---|
| Face | Face embedding | 512 features $\in [0, 1]$ | ArcFace [23] |
| | Facial expression | 7 features $\in [0, 1]$ | Residual Masking Network [24] |
| | Face box | 4 features for each $\in (x, y)$ (in pixels) | RetinaFace [25] |
| | Face center | 1 features for each $\in (x, y)$ (in pixels) | post processed |
| | Facial landmark | 5 features for each $\in (x, y)$ (in pixels) | RetinaFace [25] |
| | No. detected faces | 1 feature $\in \mathbb{Z}_{>0}$ | post processed |
| | Face Id | 1 feature $\in \mathbb{Z}_{>0}$ | post processed (cosine similarity) |
| Head | Head angles | 3 features [yaw, pitch, roll] (in degrees) | Im2pose [25] |
| Gaze | Gaze direction | 2 features [yaw, pitch] (in degrees) | Gaze360 [25] |
| | Attention visual | 1 feature $\in 0, 1$ | post processed |
| Speech | Wakeword | 1 feature $\in 0, 1$ | Piccovoice [26] |
| | Voice Activity Detection | 1 feature $\in 0, 1$ | Deepspeech [27] |
| | Speech-to-text | n features $\in$ "spoken text" | WebRCT [28] |
| | Natural Language Processing | 2 features $\in$ [intent, entity] | RASA [29] |
| Distance | 3D head position | 3 features [x, y, z] | post processing using kinect |
| | Face distance | 1 feature (in meter) | post processed |
| Gesture | Hand Pose | 4 Features (Open, Closed, Finger, None) | Kinect for Windows SDK 2.0 |
| Body | Body Joints | 26 Features [x, y, z] | Kinect for Windows SDK 2.0 |
| Object | Cube Location | 4 Features (Letter, Color, Bounding Box, Angle) | CubeDetector [30] |

The detected features are depicted in Figure 3 showing an exemplary situation of a user interacting with the system.

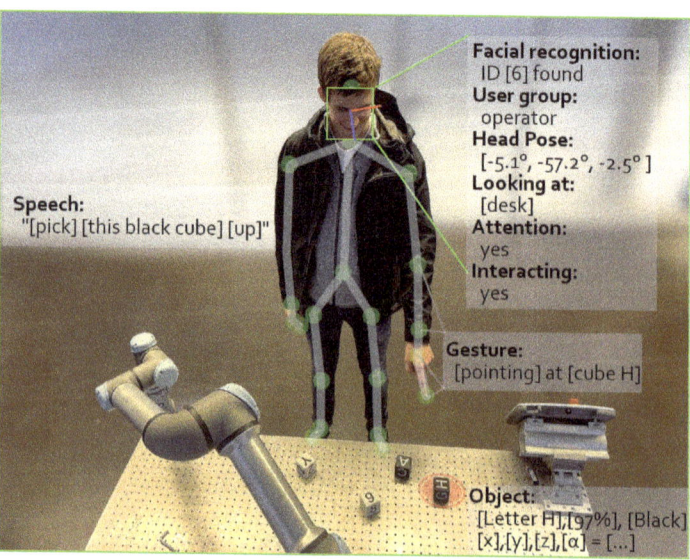

**Figure 3.** User pointing at a letter Cube. Multiple detected features are displayed for clarification.

## 4. System Setup

### 4.1. Hardware

We chose a modular structure in the form of so-called workstations (WS) to be capable of evaluating the system in a multi-device scenario and to simplify the development and the integration of additional hardware. A WS is mainly characterized by its hardware, which forms a closed system.

For consistency and comparison purposes, we did not change the hardware used in the previous studies. As cobot, the UR5e industrial robot equipped with an *RG6* gripper was used. The robot was bolted to a sturdy metal table. A TV, with a ToF (Kinect) camera on top, was placed behind. A projector, lighting the scene from above, was used to illuminate the objects and the metal table for visual feedback. The same cubes, as in the previous WoZ study, were used. We refer to this cube-related setup as workstation 1 (WS1).

The second workstation (WS2) can be used for registration and questionnaires, featuring a smart screen with a touch function and also a ToF (Kinect) camera. Every workstation utilizes microphones (from Kinect) and speakers (from TVs). The current setup can be seen in Figure 4.

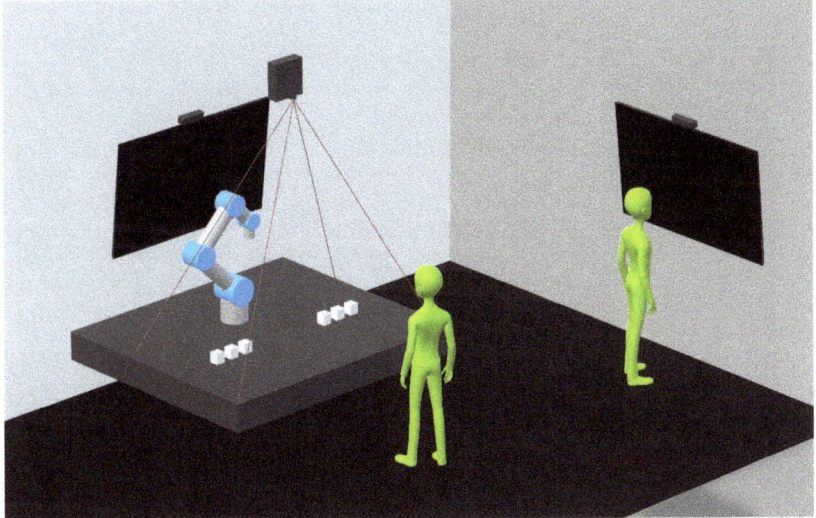

**Figure 4.** A schematic overview of the workstations.

### 4.2. Middleware

The development of a distributed system consisting of heterogeneous devices from different manufacturers is a non-trivial task that, among other things, must ensure communication between numerous devices. The Robot Operating System (ROS) is used for communication between the hardware components as well as the individual software modules. For communication within the *Speech Module*, the Message Queuing Telemetry Transport (MQTT) protocol was implemented. Using ROS as Middleware grants a direct machine-to-machine communication interface and allows an easy integration of additional workstations. Each module can be run as an independent node multiple times on multiple devices.

For communication between modules, a set of custom ROS messages were made: *Body*, *Joint*, *Face*, *RobotAction*, *CubeAction*, and *CubeMessage*. The code is open source [31].

## 5. Modules

### 5.1. Scene Module

This module creates a virtual scene that contains all virtual objects and their relations, enabling the interaction and management of the objects. The scene can be an exact or an abstract representation of the real environment, or it can create an entirely new one.

The *Scene Module* also includes the calibration of the system's input and output devices. Recognition algorithms can then be used to associate real objects in the virtual world. In the context of HRI, the *Scene Module* is also used for collision calculations and as a database for object positions.

The table on which the experimental setup of WS1 is located serves as the basis for the scene and calibration. This table is provided with threads at regular intervals of 2.5 cm, which form a grid. This grid allows for calibration between the robot and scene, as well as proper alignment of the projector. In this case, the calibration is an adjustment and scaling of the respective coordinate systems in relation to each other. The point where the projection of the pixel [0,0] meets the table surface serves as the origin of the coordinate system in the virtual world. This is scaled according to the grid.

All virtual objects are defined by a point that marks their position in the virtual world in Cartesian coordinates.

#### 5.1.1. Skeleton

In addition to the cubes, the data of the Kinect skeleton is also converted into virtual objects, consisting of 26 [x,y,z] points. This allows the user to interact with virtual objects such as security planes or augmented user interfaces.

#### 5.1.2. Spot

The pointing gesture creates a so-called *Spot* at the point where the line, through the elbow and wrist of the Kinect skeleton, intersects the surface of the grid. It contains the position of the intersection point [x,y,0]. To help the stacking of the cubes, the spot jumps to the nearest grid position. The *Spot* can be seen in Figure 5.

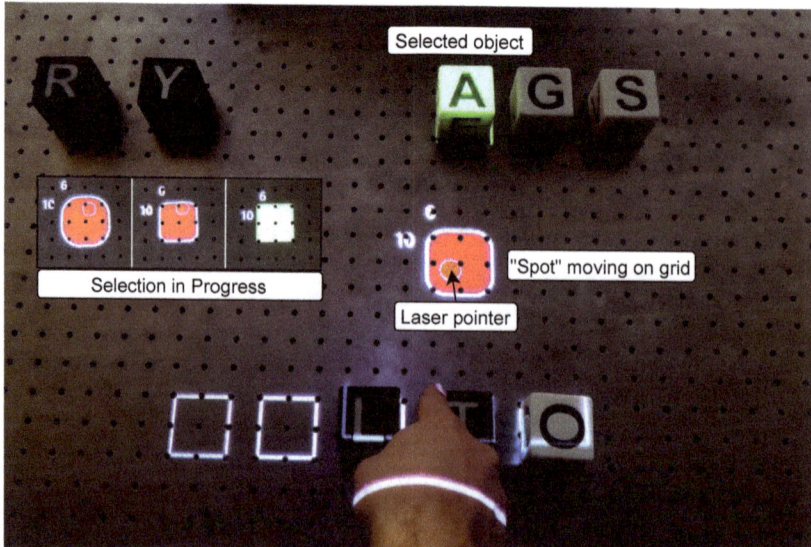

**Figure 5.** Selection in progress: The borders of the grid rectangle around the laser pointer are narrowing. In the next step, the selection would be moved from *Cube A* to the new coordinates.

#### 5.1.3. Visual Feedback

Through monitors or projection, virtual objects can appear in reality. A selected virtual cube is illuminated with a green rectangle, and a space on the grid is illuminated with a white rectangular frame. Transformations are applied to the projection so that the projected objects visually match the real objects. The exact position of the *Spot* is projected onto the table like a laser pointer. The process of selecting an object and the projected laser pointer can also be seen in Figure 5.

#### 5.1.4. User

The user database includes the ID, names, facial features, and session status of the subjects. WS1 and WS2 can both access this database to address the user by name, or to retrieve the last session status for the activated subject. This personalized experience contributes to the intuitiveness of the system.

### 5.2. Robot Module

The *Robot Module* is responsible for path planning and managing robot actions. Path planning is an important feature of a robot assistant. For this task, the definition of start and end points, from now called *Source* and *Destination*, is crucial.

The *Robot Module* sends commands, so called *RobotActions* to the robot. The robot can move to the discrete position in the virtual grid and grasp real objects. For this purpose, there are four basic operations within the robot program (see Figure 6). These can be defined as individual modes for programming the cobot. These sub-routines can be called directly via Real-Time Data Exchange (RTDE)—a protocol developed by Universal Robots for fast communication with the robot. A thread observes a particular register and jumps to the corresponding mode called by the *Robot Module*. When performing the *pick* or *place* operation, the appropriate mode must be specified, as well as the desired coordinates of the object. If a cube is to be *taken* or *given*, only the mode and position of the hand are necessary. The four basic operations, require only either a *Source* or *Destination* (i.e., object or hand position), since the respective counterpart, logically results itself as an actuation of the gripper. These operations execute only a single step. The more complex commands like *picking and placing* are further explained in Section 5.3.3. In addition to the four basic modes of interaction with the cubes, there are five more modes that complete the robot program.

- Abort: Motion is aborted;
- Home: Robot goes to initial position;
- Sleep: Robot goes to idling position;
- Toggle gripper: Opens or closes the gripper;
- Greet: Robot performs nodding motion.

To ensure that all commands are registered correctly, the robot changes its state to *busy* during an action. When a robot action is finished, the robot changes its mode to *Ready* and continues to execute the *RobotActions* that have queued up.

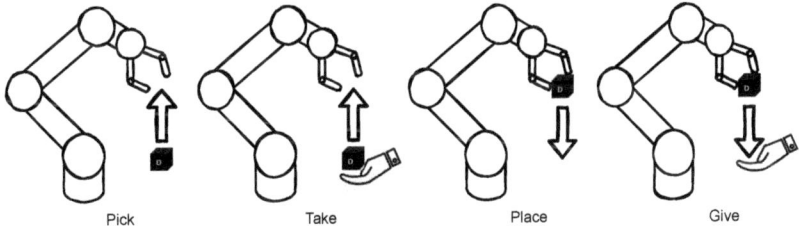

**Figure 6.** Basic robot operations: Pick up, take from user, place on table, and give to user.

## 5.3. Cubes Module

### 5.3.1. Physical Cube

The interaction at WS1 is primarily with $5 \times 5 \times 5$ [cm] cubes, that can be positioned exactly on the $2.5 \times 2.5$ [cm] grid of the table. The cubes have a letter on each side and are 3D-printed from lightweight and robust Polylactic Acid (PLA), allowing for safe interaction. The cube model can be downloaded from Thingiverse [32].

Each real cube is represented by a virtual object, which is given the attributes: Letter, color, and position. The cubes are unique and thus can be identified by one or the combination of the attributes. In summary, the cube data is stored as an ROS message of type:

```
CubeMessage { letter [A-Z], color [black/white], position [x,y,z] }.
```

When a cube is handed over, it is assigned the position [0,0,99] until it is reassigned a coordinate on the table by an interaction with the robot. If a cube is in the gripper, it is assigned the coordinates [0,0,-1]. All other possibilities [x,y,z] correspond to a position on the grid, where x, y, and z take real values.

### 5.3.2. Cube Detection

For an unconstrained interaction, the robot must be continuously aware of the position and order of the cubes placed on the desk. A straightforward approach would be the use a visual tracking method to follow the cube operations and update its current position accordingly. However, occlusions caused by the robot and cubes that are leaving the field of vision can heavily harm the tracking state and lead to interaction discontinuities. We therefore follow the approach of continuous detection and recognition of each cube and adapt a deep neural network for this purpose.

Figure 7 shows an example image of our cube detector in action. It demonstrates that even the almost 180° degree rotation of the cube "T" can be detected precisely with a high confidence (shown as a number next to the letter).

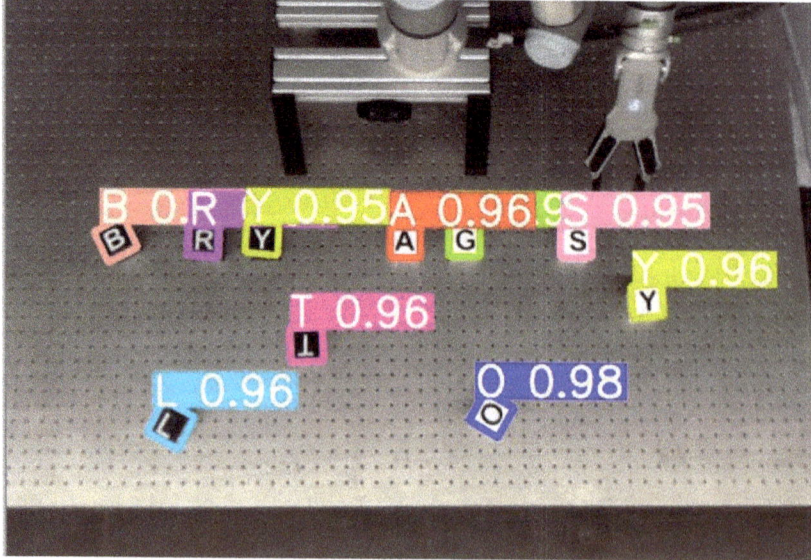

**Figure 7.** The cube detection in action. It detects also the rotated orientation of the cubes.

As there is no public dataset available that would fit our needs for the cube detection task, so we generated our own synthetic dataset. We cropped representative cube images for each letter and color and randomly placed them on arbitrary image backgrounds. Additionally, random rotations and scaling as well as different levels of blurring are applied to further augment the variance of the data. As a result, we received 100,000 randomly generated training images with corresponding annotation. Beside the ground truth letter and localization, we also annotated the rotation of the cubes as this information is crucial for accurate grasping by the robot.

For implementation, we use the Yolov5 networks as a backbone and change the number of output neurons to the number of letters on the cubes. Instead of regressing the cubes rotation directly, we found that classifying the angle leads to more stable results. Therefore, 90 classes are added for each degree of rotation. We used rotated bounding boxes for calculating the Intersection over Union to further help the network to learn the rotation. The cube detection network is open source [30].

5.3.3. Cube Logic

The Cube Module is an indispensable additional module that is used to check the interactions with the cubes, convert them into an *RobotAction*, and change the virtual cubes in the scene. Figure 8 shows the build of *CubeAction*, a ROS Message type consisting of two *CubeMessages* describing a source and destination. The task of the *Cube Module* is to find the transition from source to destination and convert it to a *RobotAction*, if the transition is valid and reasonable according to the *Scene Module*.

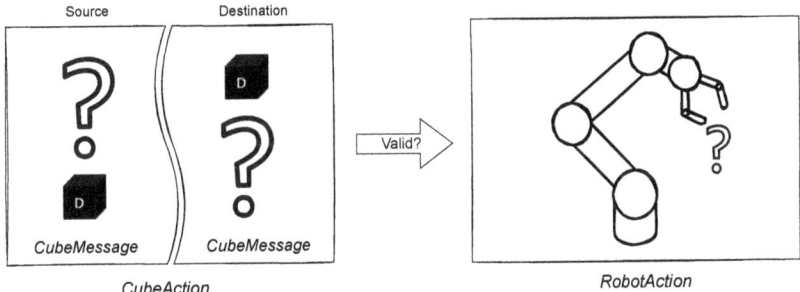

**Figure 8.** Conversion of *CubeAction* to *RobotAction*.

Every *CubeAction* has a source and destination. If either of the *CubeMessages* equals the robot gripper, then this *CubeAction* corresponds to a basic *RobotAction* discussed previously in Section 5.2. It is a single step of the robot moving the cube to or from the gripper. In the previous WoZ study, the users requested the robot to move the cubes between positions, disregarding the intermediate steps. Thus, the necessity of complex or combined instructions became eminent.

Two basic *CubeActions* can be chained together, if they have the same intermediate position, i.e., the gripper. Figure 9 depicts this concept.

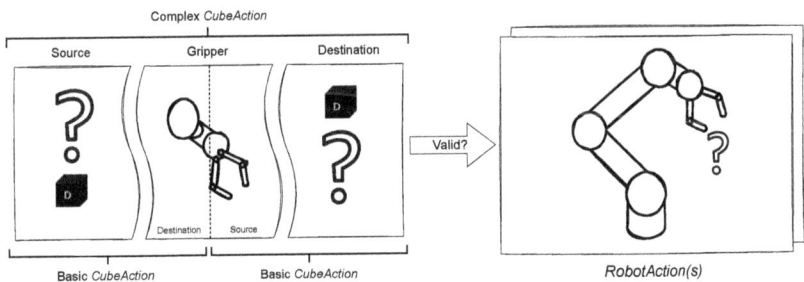

**Figure 9.** *CubeAction* can be simple or complex, corresponding to multiple *RobotActions*.

For the *Interaction Module* and *Speech Module* to function correctly and as expected, it is necessary to consider, identify, and group all possible cube manipulations and their corresponding *RobotActions*. Thanks to the complex *CubeAction* concept, it is possible to list all possible combinations. Figure 10 shows the four basic robot operations (see Figure 6) *Pick*, *Take*, *Place*, and *Give* depicted as either source or destination *CubeMessage*.

**Figure 10.** Basic operations Source: {*Pick, Take*} and Destination: {*Place, Give*}.

These operations can be combined to form four complex operations, as depicted in Figure 11. *Pick-Place* is used to move a cube from one position to another. *Take-Place* is used to receive the cube from the user and place it on the table. *Pick-Give* is used to give a cube after picking it up. *Take-Give* action is logically exclusive, since the object does not change its position from the system's perspective and thus can be neglected.

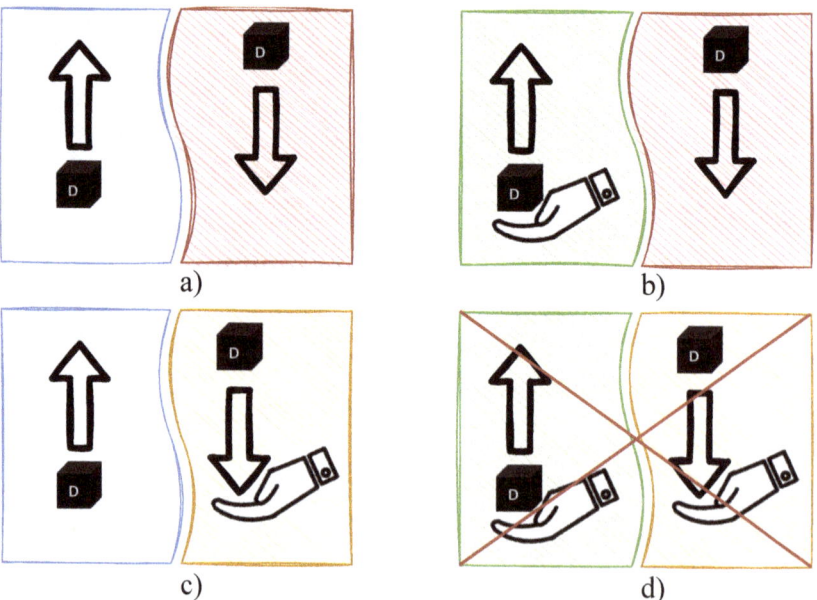

**Figure 11.** Complex operations: (**a**) Pick-Place, (**b**) Take-Place, (**c**) Pick-Give, and (**d**) Take-Give.

The source and destination cube may be specified by any combination of the attributes position, color, and letter. The *Cube Module* then checks the start and target position by matching them with the cubes stored in the scene. If no position but color and/or letter was specified as a start cube, then the corresponding position is retrieved from the scene and the information is added to the *CubeMessage*. Contradictory *CubeMessages* are filtered out. When a *CubeAction* is verified, it is converted into one or, in the case of complex operations, two *RobotActions* and sent to the *Robot Module*.

The four basic operations always include the gripper (`position [0,0,-1]`) as a source or destination. An example of a *CubeAction* for *Place*, that puts the cube that is currently in the gripper to `position [12,12,1]`, could look like this:

```
Src. { letter [], color [], position [0,0,-1] }
Dst. { letter [], color [], position [12,12,1] }.
```

This message would be then checked by the *Scene Module* and then converted to a *Robot Action* and enqueued by the *Robot Module*. After the successful movement of the robot, the *Robot Module* would report the new position, and the corresponding cube in the *Scene Module* would be updated:

```
CubeMessage { letter "M", color "white", position [12,12,1] }.
```

All other cube manipulations are executed in the same manner. The remaining challenge for the *Interaction Module* is to combine the correct information from the *Speech Module* and *Gesture Module*. A complex *Cube Action Pick-Place*, for picking a black cube "A" and placing it on top of cube "B" could look like this:

```
Src. { letter "A", color "black", position [4,8,1] }
Dst. { letter "B", color [], position [6,10,2] }.
```

In addition to checking whether a field is already occupied, the fields in the immediate vicinity are also checked. If these are also occupied, it is possible to place a cube on top between two cubes, thus stacking a pyramid. If no neighboring cubes are present, a cube can only be placed directly on another one. Vice versa, whether there is another cube above the selected one is also checked.

The module is also able to move the cubes back to their initial position. For this purpose, corresponding *Cube Actions* are created based on the scene. The order of the stacked blocks is also taken into account.

*5.4. Face Module*

The human identification serves both security and personalization of the data presented. Moreover, face verification is a vital identity authentication technology used in more and more mobile and embedded applications. Our system benefits from face verification to achieve high fidelity and confidence for user authentication and authorization to control the robot for crucial tasks.

The *Face Module* is subdivided into two main parts: Face detection and face recognition. The face detection can detect the location of the face in any input stream (image or video frames). The output is the bounding box coordinates and facial landmarks of the detected faces. On the other hand, face recognition is a process that compares multiple faces to identify which face belongs to the same person. This identification process can be done by comparing the feature vector of the detected face with the stored face feature vectors.

5.4.1. Detection Part

We used the RetinaFace [25] light-weight model, based on MobileNet-0.25 [33], as the pre-trained model for face detection, which employs a multi-task learning strategy to simultaneously predict face score, face box, and five facial landmarks. The network was pre-trained on a WIDER FACE dataset [34].

5.4.2. Recognition Part

After the detection bounding boxes are obtained, the filtered boxes are fed into the recognition part. Before the next steps, the Practical Facial Landmark Detector (PFLD) [35] is used to align the detected faces. The aligned faces are used for both face recognition and facial expression recognition.

For face recognition, the deep CNN used was MobileFaceNet [36] which uses less than 1 million parameters and is specifically tailored for high-accuracy real-time face verification on mobile and embedded devices. This is less accurate than its counterparts, but it is real-time capable. The network was trained on the refined MS-Celeb-1M dataset using the loss function Arcface [23]. This loss function produces much better discriminative features compared to others. The extracted facial features are compared against each other using cosine similarity. If the faces are similar enough, they are assigned the same ID (see Figure 12). The IDs in the image section are passed to the *Interaction Module* along with their position. Persons who cannot be assigned an ID from the database are assigned the ID −1.

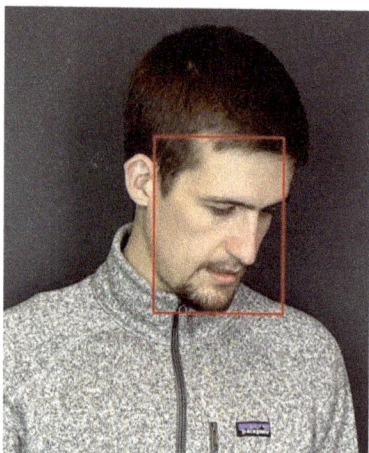 

**Figure 12.** Face detection, face recognition, and facial expression recognition.

*5.5. Attention Module*

Various features can be used to estimate the user's intention to interact with the robot (attention) and the intention breakdown during the interaction (attention breakdown). The most common features that can be used to estimate engagement and disengagement in HRI include gaze, head pose, face, posture, speech, and distance [37–41]. Using more features to estimate user attention could increase the accuracy of the attention algorithm. However, it will increase the computational cost, which will have a bad impact on the overall system.

The direction of a person's gaze has a regulative function for the interaction taking place and allows conclusions to be drawn about the willingness to interact [42]. The direction of gaze can be recognized by the position of the head but also by the pupils. The latter gives more precise information about the current visual focus. If no fast or small gaze changes are required (e.g., when reading) and the object viewed is in the middle range of the visual angle or the person is further away, the orientation of the head offers a possibility for approximation [43].

Considering attention in the context of a technical system, a POI can be defined. If a person turns away too far from the POI, the person likely has no longer potential engagement. If the camera is in the POI and the user's head is in the center of the frame, a deviation in the yaw-pitch-roll angle results from the subject turning away, thus breaking the engagement. However, when the relative position in the image plane is changed, a transformation must take place to determine whether that person is looking at the POI.

In our algorithm, we fused the head posse features with the gaze features through a rule-based classifier to estimate the person's attention while interacting with the robot. For the head pose features, we used the img2pose method proposed by Albiero et al. [44]. This method outperforms many current state-of-the-art models in terms of accuracy and real-time capability. The method does not use the elaborate detected bounding boxes and landmarks, but uses a Faster-R-CNN-based model that computes the 6-Degrees of Freedom pose for all faces in the photo (see Figure 13). The model used, was trained using the WIDER FACE dataset. For the gaze features, we used the gaze360 method proposed by Kellnhofer et al. [45]. They uniquely take a multi-frame input (to help resolve single frame ambiguities) and employs a pinball regression loss for error quantile regression to provide an estimate of gaze uncertainty. This method is trained on 3D gaze in-the-wild dataset, which make it robust to diverse physically unconstrained scenes.

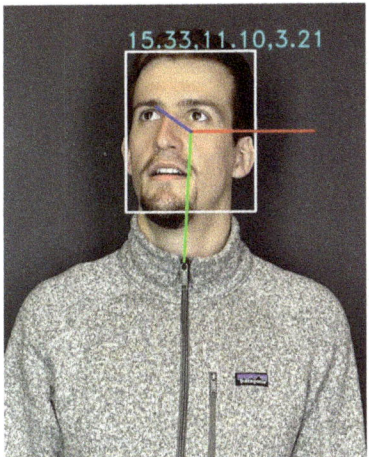 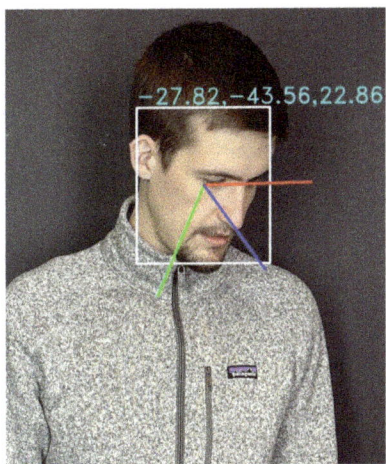

**Figure 13.** Head pose detection.

An algorithm converts the resulting head pose and gaze features into a person-based visual attention score for each person in the scene. We fused these scores together through our algorithm for outputting a final score for each person in the scene. If the predefined threshold of visual attention focus is exceeded, the person is recognized as attentive and can interact with the robot. If the person turns away, the visual attention focus decreases over time. If a threshold value is undershot, this person is no longer detected as attentive.

### 5.6. Gesture Module

The pointing gesture uses the skeletal data of the forearm provided by the Kinect in combination with the "index finger pointing" gesture (Kinect Lasso gesture). According to the concept of "Laws of Linear HRI" [12], a line is formed from the two joints, elbow, and wrist, of the recognized human and the intersection with the plane of the table is determined, the aforementioned *spot*.

The lack of direct user feedback can be solved by adding a laser pointer feature that directly corresponds to the pointing position. This way, the user does not have to wait for the robot to fulfill the task, as in the case of the original implementation by Williams et al. [46]. Using the *spot*, the user is currently pointing at, as real-time feedback, further helps to increase the overall accuracy of the pointing gesture.

The laser spot responds to the objects stored in the *Scene Module* and wraps around the object being pointed at. This feedback is intended to simplify the handling of the gesture. This concept is shown in Figure 14. If the pointing position is not changed and the gesture is not canceled, the object will be selected after a given time and highlighted in green by the projector. If the gesture is held further, the selection is sent to the *Interaction Module*.

**Figure 14.** Pointing gesture: User points to cube [S], which is enveloped by the laser pointer.

The grid given by the virtual positions, from the *Scene Module*, facilitates selection and positioning. The circle adapts its shape to the grid when the user dwells on a position. The virtual object corresponding to the highlighted coordinate can now be used as either the source or destination. The selection done with the *spot* is saved in the scene.

### 5.7. Speech Module

The *Speech Module*, along with the *Gesture Module*, is an important part of RoSA, that directly interacts with the user. As shown by Haeb-Umbach et al. [47], most established speech assistants consist of the modules shown in Figure 15. The individual components are explained in more detail below.

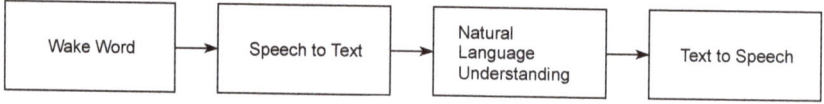

**Figure 15.** Common structure of speech assistants.

#### 5.7.1. Wake-Word Detection

Piccovoice was used to implement wake word detection. This application provides an online service for training personalized wake words. Furthermore, Piccovoice has a lower error rate compared to others [26]. The disadvantage of this implementation is that the software is only partially open-source and the use of the personalized wake words is limited to a 30-day license at a time.

#### 5.7.2. Voice Activity Detection (VAD)

Voice Activity Detection (VAD) is intended to prevent loud noises or the like from being interpreted as speech, e.g., after the system has been woken up by using the wake word [48]. In addition to activating STT, detection can be used as an abort criterion for the process. If a pause in speech exceeds a certain period of time, the sentence is terminated. We utilize a VAD developed by Google as part of the WebRTC project [28], which was intended to provide new standards for real-time communications with video, voice, and generic data support. It uses multiple frequency band features with a pre-trained Gaussian mixture model classifier.

5.7.3. Speech-to-Text (STT)

For privacy reasons, an offline service based on Mozilla's Deepspeech [49] is used. This open-source STT engine uses methods from Baidu's research [27]. We used Deepspeech German, a pretrained network by Agarwal et al. [50]. The training data is based on Common Voice, a project started by Mozilla to collect speech data. It is also an open-source project that people donate their voice to, reading out sentences or validating audio transcripts. Since incorrect recognition of the STT can lead to difficulties in further processing, the vocabulary is adapted to that of the Natural Language Understanding (NLU) module.

To improve user experience and system accuracy, we introduced a visual feedback that displays the recognized spoken words to the user in real time. As suggested by Schurick et al. [51] this approach can greatly reduce the necessary time for speech input by a factor of three.

5.7.4. Natural Language Understanding (NLU)

The system uses the open-source RASA solution to extract intent from text provided by the STT. RASA [29] consists of loosely coupled modules that combine a set of natural language processing and machine learning libraries into a unified API. It strives to balance adaptability and usability [52]. Braun et al. [53] show that the performance of Rasa NLU is compelling compared to several closed-source solutions. The NLU pipeline used consists of the SpacyNLP with the German language corpus, Tokenizer Featurizer, and EntityExtractor. First, the text is segmented by the Tokenizer, the Featurizer generates features for entity extraction and intention classification, and the EntityExtractor extracts information objects. The DIETClassifier is used to classify the intention. The classified intention and extracted information objects are sent to the *Interaction Module* as *Cube Action*.

Interaction with the robot works using *Cube Actions* (see Section 5.3.3). These can be categorized into the already mentioned, four basic and three complex operations. For each of these commands, there is a voice command. For example, the command "give me the white block with the letter A" combines the operations *pick* and *give*. The *Cube Action* passed through ROS looks like this:

```
Src.  { letter  "A",   color  "white"  position  []       }
Dst.  { letter  [],    color  [],      position  [0,0,99] }.
```

In addition to voice-only commands, there is also the option of combining voice and gesture. The user uses the pointing gesture, selects a position and simultaneously specifies the desired block. "Place the black cube A here". The *CubeAction* would look like this:

```
Src.  { letter  "A",     color  "black",  position  []  }
Dst.  { letter  "spot",  color  [],       position  []  }.
```

Since the *Speech Module* has no information about user's gesture input, the *Interaction Module* has to fill in the gaps and update the information using the *Gesture Module* and *Scene Module*.

By combining speech and gestures, ambiguous cases such as this can arise: "Give me the block", which can mean handing over a block that has already been grabbed, but also *pick up* and *hand over* the block that is currently pointed to. The *CubeAction* passed via ROS nonetheless looks like this:

```
Src.  { letter  "spot",  color  [],  position  []       }
Dst.  { letter  [],      color  [],  position  [0,0,99] }.
```

At this point, the *Interaction Module* must decide according to the *Scene Module* and *Cube Module* which action to perform.

From the four basic and three more complex operations, explained in Section 5.3.3, there is a set of 14 possible operations for manipulating the cubes using speech or speech and gesture combined. Since as many variations of the speech commands as possible are to be covered (e.g., instead of "cube", "block", or "square block"), the file generated from example sentences for training the NLU comprises 60,000 lines.

#### 5.7.5. Text-to-Speech (TTS)

The Windows Speech Application Programming Interface [54] is used for text to speech synthesis. This interface allows the user to make speaker variations such as:

- Audio Pitch: Determines the pitch (relative height or depth) of the speech synthesis.
- Include sentence boundary metadata: Determines whether sentence boundary metadata is added to a SpeechSynthesisStream object.
- Punctuation silence: Length of silence added after punctuation in SpeechSynthesis before another utterance begins.
- Speech rate: Sets the tempo, including pauses of the speech synthesis.

An advantage of this TTS system is that it is already integrated into Microsoft operating systems and is freely accessible since Windows Vista.

### 5.8. Interaction Module

The *Interaction Module* is adapted to the framework conditions of the individual workstations, since they require different user inputs and are to act independently of each other. At the beginning, the readiness for interaction is evaluated with features from the *Attention Module* and Kinect-stream. If a person is attentive, the corresponding workstation is activated and visual feedback is shown. Users identified by face recognition are signed in. The user-interface changes accordingly to the currently active user, settings, and last state of the workstation are restored. Users are signed in as long as they are attentive and recognized by the system. Identities assigned by the *Face Module* are applied to the Kinect skeleton with the largest intersection over union using the bounding box provided. Now, the user has a workstation bound *Active Session* until the user logs out (see Figure 16). This process is active on all workstations.

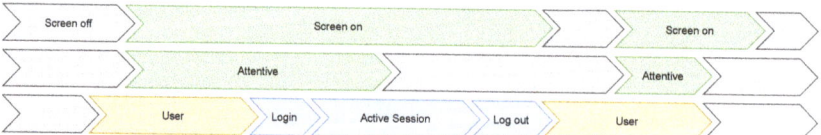

**Figure 16.** *Active Session* flow diagram.

#### 5.8.1. WS1: Cubes and Cobot

If the WS is activated by the *Attention Module*, the user is informed via the screen that the authentication process is running, as long as no ID is available. If the user is still an unknown person, they will be prompted to register at WS2. Registered users are greeted with their stored name and a nod (quick up and down motions of the gripper) of the robot. At the first log-in, RoSA introduces itself to the user and runs through a basic tutorial. Once an *Active Session* is started at the workstation, the system can be interacted with. To issue a voice command, the wake word must be used. This activates the STT and the screen displays an icon of an earpiece to show the user that the system is now listening. The transcribed text is shown on the screen. Another icon indicates whether the use of a pointing gesture is possible. The gesture input is paused when the robot is already in motion so as to avoid problems with occlusion by the robot. Input via the pointing gesture results from holding the gesture in one position for a short time. The currently selected cube is also displayed on the screen.

### 5.8.2. WS2: Registration

The interface of WS2 is based on ROS-QT (RQT), a development environment for visualization. The user is provided with an interface for registration. The data collected includes a name, preferred hand, and face recording. To record the face, the subject is asked to look first frontally and then once to the right and once to the left. The recorded data is stored in the database. Once this process is complete, an *Active Session* is started and the first part of the survey can begin. Upon successful completion of the first survey, a brief tutorial on RoSA and the tasks are presented. The user is prompted to perform the tasks on WS1. Finally, the second part of the questionnaire can be completed. When using WS2, the current progress of each user's survey and tasks is saved. When logging in again, the display jumps to the last session. The user input comes from touch screen or a keyboard.

## 6. Experimental Studies

In order to prove the concept and to gather insight about necessary improvements for the system, a pilot study was conducted in the same manner as the previous WoZ study. During the study, data were collected from 11 subjects (2 ♀ | 9 ♂) aged between 20 and 34 years. Five of these subjects had already participated in the previous RoSA study. The procedure of the study is as follows:

- Informed consent;
- WS2: Registration and collection of sociodemographic data;
- WS1: Collaborative tasks with robot;
    1. Have RoSA give you a block.
    2. Spell a specific word with alternating color of blocks.
    3. Build a 3-2-1-Pyramid with black-white-black layers.
- WS2: Questionnaires;
- WS1: Benchmark: Data collection for module assessment.

## 7. Results

Within the scope of this work, a functional robot system assistant is created. The system includes eight modules that communicate with each other. The system is activated by the *Attention Module* when a user shows enough willingness to interact. The *Face Module* allows a personalized user experience. The user is addressed personally and is shown personalized content. Both the voice and *Gesture Modules* can be used for intuitive operation of the robot. With the *Cube Module* and *Scene Module*, the system can interact with its environment. In addition, RoSA provides the user with auditory and visual feedback. The individual functions are explained to the user in a short tutorial, in the form of a self-introduction by RoSA, to start a natural dialogue.

### 7.1. Time

Table 2 summarizes the time needed for the completion of the collaborative tasks. Time was started as soon as the task was known and the user gave the first command, and stopped as soon as the task was declared complete by the experimenter.

Table 2. Time needed to accomplish the tasks.

| Variables | Fastest | Slowest | Mean |
|---|---|---|---|
| Task 1 | 00:00:16 | 00:04:30 | 00:01:21 |
| Task 2 | 00:02:51 | 00:26:36 | 00:12:56 |
| Task 3 | 00:02:36 | 00:36:39 | 00:11:06 |
| Total | 00:06:07 | 01:07:24 | 00:25:20 |

All subjects were able to solve the tasks in collaboration with the robot in under two hours.

## 7.2. Questionnaires

The questionnaires (SUS [55], UMUX [56], PSSUQ [57], and ASQ [58]) were completed after the experiment. A module-specific questionnaire was then taken to additionally evaluate each module.

The summary of the user satisfaction questionnaires is presented in Table 3. The questionnaires were evaluated using the methods described in the literature for each individual. This mostly consisted of an alternating weighting of the questions, from which the mean value was calculated. To make the questionnaires comparable with each other, the scores were normalized by bringing them to the same scoring range of [0–100].

**Table 3.** Results of usability questionnaires.

| Variablen | SUS [55] | UMUX [56] | PSSUQ [57] | ASQ [58] |
|---|---|---|---|---|
| Answer Range | 1–5 | 1–7 | 1–7, NA | 1–7, NA |
| Score Range | 0–100 | 0–100 | 1–7 | 1–7 |
| No. of Questions | 10 | 4 | 16 | 3 |
| Normalized Score | 72.27 | 57.57 | 62.90 | 64.06 |
| | Total Avg. Score: 64.2 | | | |

## 7.3. Modules

Users were also asked to rate each module according to their personal satisfaction on a scale of one, very dissatisfied, to seven, very satisfied (see Figure 17).

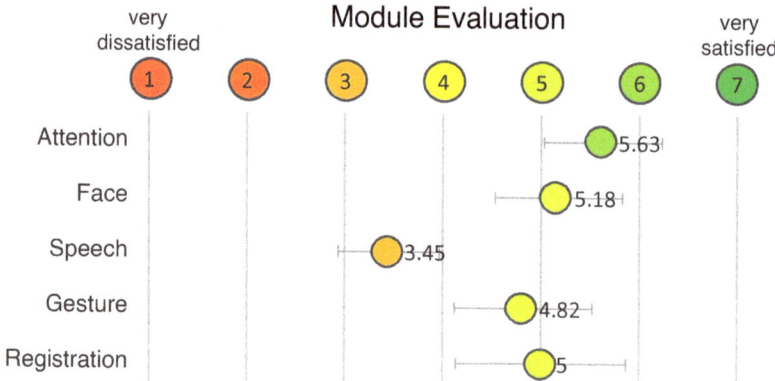

**Figure 17.** User rating of individual modules.

The test subjects were less satisfied with the *Gesture Module* and especially with the *Speech Module*. This is why we decided to evaluate these modules were tested independently of each other and outside the experiment to avoid external sources of error. The evaluation is based on the data that was collected during the benchmark phase at the end of the initial experiment with the cubes. The benchmark was run after the user had completed the questionnaires and already rated the system in so as to not bias the test subject.

## 7.4. Speech Module Evaluation

To evaluate the *Speech Module* the test subjects were asked to read out sentences displayed. For example: "Give me the white block with an A".

The performance of a speech recognition system is measured by the Word Error Rate (WER):

$$WER = \frac{S + I + D}{N} \qquad (1)$$

where S is the number of words incorrectly replaced, I represents the number of additional words inserted, D is the number of words deleted, and N is the number of words correctly transcribed [59]. During the interaction, it was often the case that RoSA did not understand or misunderstood the subject. The error rate of the wake word detection was 33%.

The STT worked well overall. However, in some cases, words could not be detected because they were not part of the previously defined vocabulary. Short words like "yes" or single letters were often not recognized. Within the benchmark, the STT had a WER of 28.6%. The NLU module had an intent error rate of 27.3% and an entity error rate of 47.7%

### 7.5. Gesture Module Evaluation

For the evaluation of the pointing gestures, users were asked to point at targets highlighted on the screen in front of them, without any additional pointing feedback. The participants were asked to hold the gesture for two seconds. A 13-dot calibration pattern, as commonly used for eye-trackers, was used.

To estimate the pointing position, an intersection of a line, formed by two joints and a plane 2.5 m in front of the participant, was used. For each target, the timespan of one second, or equivalently 30 frames were evaluated and the resulting intersection points calculated. The spread, or the overall deviation of the positions from the calculated mean for each target, can be used as an estimation of the pointing quality.

As implied by the Laws of Linear HRI [12], any two joints can be used. The resulting intersection POIs that were calculated using joint pairs "Elbow Right—Hand Tip Right" as implemented in the experiment and "Shoulder Right—Hand Tip Right", an alternative pointing method using the same skeletal data, can be seen in Figure 18.

The overall mean average of the position deviations is 3.88 cm for Elbow–Hand and 0.66 cm for Shoulder–Hand showing a possible way of decreasing the spread between the consecutively calculated pointing positions.

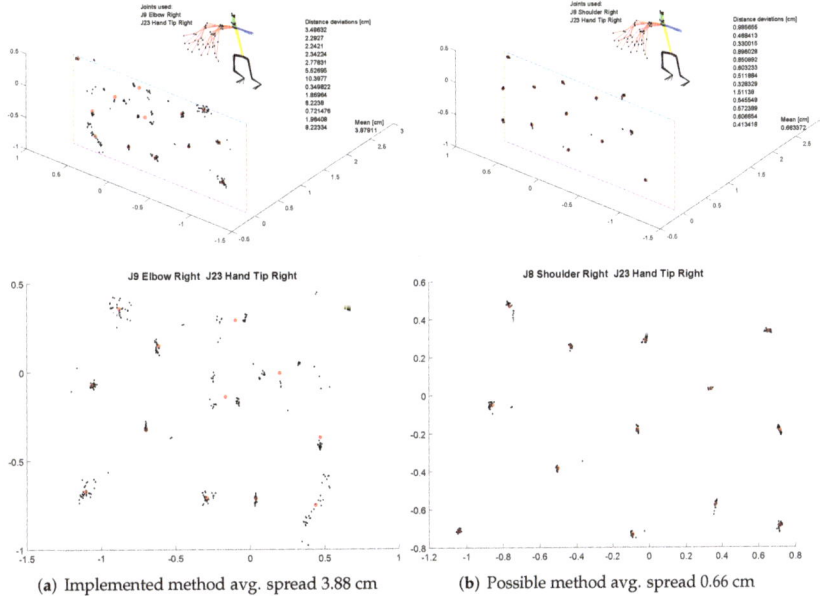

(a) Implemented method avg. spread 3.88 cm       (b) Possible method avg. spread 0.66 cm

**Figure 18.** Evaluation of the spread of the resulting Points of Interest (POIs) at a 2.5-m distance. Calculated mean positions are depicted as red circle crosses.

## 8. Discussion

In order to put the user data in perspective, they are compared with those of the WoZ study (see RoSA study [6]). It is important to note that the WoZ experiment was conducted under idealized conditions, demonstrating a close to flawless system adapting directly to the user's preferred method of interaction.

Almost half of the participants in this study had already taken part in the WoZ study. Although the operating concepts could be individualized and freely chosen by the users and thus differed from the current scenario, an influence of the previous study cannot be ruled out.

Nonetheless, the participants were invited to aid in the discussion and evaluation of the system. Unfortunately, the sample size is still too small and the group of test subjects too homogeneous to be able to draw generally valid conclusions. However, qualitative statements can already be made about the system.

### 8.1. Efficiency

The system's efficiency can be assessed by the time needed for the tasks. Table 4 shows the comparison to the WoZ study. It should be noted that during the WoZ study, the pyramid had to be built twice in the third task. However, across all tasks, the subjects in the WoZ study were faster.

Table 4. Comparison efficiency.

| Variables | Task 1 | Task 2 | Task 3 | Total |
|---|---|---|---|---|
| Results | 00:01:21 | 00:12:56 | 00:11:06 | 00:25:20 |
| WoZ study [6] | 00:01:46 | 00:07:57 | 00:09:52 | 00:19:35 |
| Deviation | 00:00:25 | −00:04:59 | −00:01:08 | −00:05:45 |

The results in Table 2 show a large variance. The fastest user only needed one-sixth of the time for each task compared to the average user. Different times are required depending on the modality used. Since the pointing gesture can only express basic operations, hybrid or speech-based solution approaches are faster in theory. This is also true for the first task. The fastest person completed the task, using a voice command, within 16 s.

However, the more complex tasks showed that the error-proneness of the voice assistant caused severe delays. The person who completed the entire experiment the fastest used pointing gestures exclusively.

As explained earlier, errors occurred more frequently during the execution of the experiment. These negatively influenced the time needed for a task.

### 8.2. Usability

These aspects are also reflected in the user experience. The overall user satisfaction of the system turns out to be "satisfactory". RoSA, in the current state, has visible deficiencies in the area of language that should be addressed.

Putting this study in context with the WoZ experiment, we find that the assessment is consistently unfavorable (see Table 5). This can be attributed to the lack of learning ability, higher susceptibility to errors, lower range of functions, and lack of personalization.

Table 5. Comparison usability.

| Variables | SUS [55] | UMUX [56] | PSSUQ [57] | ASQ [58] |
|---|---|---|---|---|
| Results | 72.27 | 57.57 | 62.90 | 64.06 |
| WoZ study [6] | 79.24 | 71.53 | 73.70 | 71.60 |
| Deviation | 6.97 | 13.96 | 10.8 | 7.54 |
| Avg. deviation: 9.82 | | | | |

## 8.3. Modules

The connection of the individual components via middleware has weaknesses. Especially with large amounts of data such as video or audio data, high latency times can occur in the system. These lead to errors in modules that operate in a time-critical manner, such as attention monitoring.

The head pose recognition is insufficient to determine gaze direction and derive the user's attention in certain scenarios. In addition to gaze direction, aspects such as proxemics and linguistic interaction should play a more important role. Drawing on more modalities increases the system's robustness against failures such as loss of face detection.

Users were generally satisfied to very satisfied with attention detection. This module, in isolation, was one of the most robust. However, it is dependent on face recognition.

The face recognition faced problems with lighting conditions, covering of the face, and significant deviation of the angles. This would sometimes lead to loss of tracking and active session. The majority of subjects were satisfied with the registration process, while two subjects were rather dissatisfied. Both subjects had problems with the registration process because they were logged in as another user even before registration. The registration process was thus skipped and had to be initiated manually. The described classification error can be reduced by an improved initial recording of facial features; thus, different faces can be better differentiated.

As the first module in the speech processing pipeline, robust wake word detection is important. Experiments have shown that one third of the activations were not detected. Therefore, Wake-word activation needs to be improved by training on audio data. These problems might also be caused by the fact that for a system to accept the Wake-word, the user has to be logged in and attentive. In retrospect, this feature hinders the system's intuitiveness, as the use of the Wake-word implies readiness for interaction. The inaccuracy of the STT is due to the reduction of the vocabulary by the known words. Likewise, the VAD sometimes causes very short words such as "yes" or single letters to be truncated. The benchmark results show that the NLU was partially able to compensate for STT errors by classifying the correct intent. However, due to the fact that the extraction of the entities had a high error rate and the created *Action* thus also contained errors, the commands could not be executed. The speech module is functional, but needs to be revised in its structure and handling to achieve good to very good results.

All users used the pointing gesture. This was necessary because stacking the pyramid was not possible using voice commands. However, this may also be due to the low error-proneness of the module. The support of the pointing gesture by the "laser pointer" was consistently seen as a relief. Users asked for the specification of two endpoints (start and end position) as a feature of the pointing gesture. Additionally, it was noted that when cubes were stacked on top of each other, the selection had to be made using the lowest position. One user described the laser pointer feedback as "shaky". This refers to the pointer jumping back and forth between two grid positions. This user mainly used the wrist and index finger line for pointing. However, the implemented pointing gesture uses the elbow and wrist. Thus, the resulting line is inaccurate and leads to a more difficult selection. This is also shown by the data plots of the pointing gesture, showing that the pointing estimation using the shoulder leads to far better results. It could be possible to use an alternative algorithm using multiple joins (head, shoulder, elbow, hand, fingertip) instead of only two, to further increase the pointing estimation.

For more intuitive interaction, the dialog guidance should be further deepened. Currently, the system cannot fill in missing information on its own. For example, the dialog terminates if a command is not understood or only partially understood. Feedback is given to the user: "I didn't understand that". One solution would be for the system to output the specific error message. More intuitively, the missing information could be rephrased into a question.

## 9. Conclusions and Future Work

The presented implementation of the RoSA system is the first step from a simulated concept towards a real and functioning system.

The developed system meets the requirements for intuitiveness, which is confirmed by the study conducted. However, due to the limited number of participants, this study is only suitable as a pilot study to find errors in the system and optimize it. Furthermore, it shows how and what kind of data and streams could be gathered in future studies to further improve the system. The data from the benchmark was used to evaluate the system in its current state, but could be used to develop and test new methods. For example, the pointing gesture data could be used for a complex algorithm using multiple joints or for machine learning. Furthermore, the attention module could be enhanced by adding new features like body posture and distance.

The next upcoming studies and the updated RoSA system will include mobile robotics as additional non-stationary workstations to allow a higher adaptability of the system to real life scenarios and to improve the overall natural communication. For this, the system will be extended by the mobile robots Tiago (WS3) and Ari (WS4). These were developed by PAL-Robotics and can be seamlessly integrated through the ROS middleware.

For the future development of natural and collaborative human-robot interaction, a system is needed that can be further developed in a modular fashion and iteratively improved through ongoing studies and regular evaluations.

With RoSA, such a system has been created.

**Author Contributions:** Conceptualization, D.S. and A.A.-H.; methodology, D.S., J.H., A.K., A.A.A. and T.H.; software, D.S., J.H., A.K., A.A.A. and T.H.; validation, D.S., J.H., A.K. and A.A.A.; investigation, D.S., J.H., A.K. and A.A.A.; resources, A.A.-H.; writing—original draft preparation, D.S., J.H., A.K., A.A.A. and T.H.; writing—review and editing, D.S., J.H., A.K., A.A.A., T.H. and A.A.-H.; visualization, D.S., J.H., A.K., A.A.A. and T.H.; supervision, A.A.-H.; project administration, D.S. and A.A.-H.; funding acquisition, A.A.-H. All authors have read and agreed to the published version of the manuscript.

**Funding:** This research was funded by the Federal Ministry of Education and Research of Germany (BMBF) RoboAssist no. 03ZZ0448L, Robo-Lab no. 03ZZ04X02B within the Zwanzig20 Alliance 3Dsensation, and DFG-Project Vitalparameter (in review).

**Institutional Review Board Statement:** The study was conducted according to the guidelines of the Declaration of Helsinki. Ethical approval was done by Ethik Kommision der Otto-von-Guericke Universtiät (IRB00006099, Office for Human Research) 157/20 on 23 October 2020.

**Informed Consent Statement:** Informed consent was obtained from all subjects involved in the study. Also, written informed consent has been obtained from the participants to publish this paper.

**Conflicts of Interest:** The authors declare no conflict of interest. The funders had no role in the design of the study; in the collection, analyses, or interpretation of data; in the writing of the manuscript, or in the decision to publish the results.

## References

1. Rusch, T.; Ender, H.; Kerber, F. Kollaborative Robotikanwendungen an Montagearbeitsplätzen. *HMD Prax. Wirtsch.* **2020**, *57*, 1227–1238. [CrossRef]
2. Hasnain, S.K.; Mostafaoui, G.; Salesse, R.; Marin, L.; Gaussier, P. Intuitive human robot interaction based on unintentional synchrony: A psycho-experimental study. In Proceedings of the 2013 IEEE Third Joint International Conference on Development and Learning and Epigenetic Robotics (ICDL), Osaka, Japan, 18–22 August 2013; pp. 1–7. [CrossRef]
3. Reardon, C.M.; Haring, K.S.; Gregory, J.M.; Rogers, J.G. Evaluating Human Understanding of a Mixed Reality Interface for Autonomous Robot-Based Change Detection. In Proceedings of the IEEE International Symposium on Safety, Security, and Rescue Robotics, SSRR 2021, New York, NY, USA, 25–27 October 2021; pp. 132–137. [CrossRef]
4. Szafir, D. Mediating Human-Robot Interactions with Virtual, Augmented, and Mixed Reality. In *Virtual, Augmented and Mixed Reality. Applications and Case Studies*; HCII 2019. Lecture Notes in Computer Science; Springer: Cham, Switzerland, 2019; Volume 11575.

5. Al, G.A.; Estrela, P.; Martinez-Hernandez, U. Towards an intuitive human-robot interaction based on hand gesture recognition and proximity sensors. In Proceedings of the 2020 IEEE International Conference on Multisensor Fusion and Integration for Intelligent Systems (MFI), Karlsruhe, Germany, 14–16 September 2020; pp. 330–335. [CrossRef]
6. Strazdas, D.; Hintz, J.; Felßberg, A.M.; Al-Hamadi, A. Robots and Wizards: An Investigation Into Natural Human–Robot Interaction. *IEEE Access* **2020**, *8*, 207635–207642. [CrossRef]
7. International Organization for Standardization. *ISO 10218-1:2011 Robots and Robotic Devices—Safety Requirements for Industrial Robots*; ISO: Geneva, Switzerland, 2011.
8. International Organization for Standardization. *ISO/TS 15066:2016 Robots and Robotic Devices—Collaborative Robots*; ISO: Geneva, Switzerland, 2016.
9. Pasinetti, S.; Nuzzi, C.; Lancini, M.; Sansoni, G.; Docchio, F.; Fornaser, A. Development and Characterization of a Safety System for Robotic Cells Based on Multiple Time of Flight (TOF) Cameras and Point Cloud Analysis. In Proceedings of the 2018 Workshop on Metrology for Industry 4.0 and IoT, Brescia, Italy, 16–18 April 2018; pp. 1–6. [CrossRef]
10. Strazdas, D.; Hintz, J.; Al-Hamadi, A. Robo-HUD: Interaction Concept for Contactless Operation of Industrial Cobotic Systems. *Appl. Sci.* **2021**, *11*, 5366. [CrossRef]
11. Rogalla, O.; Ehrenmann, M.; Zollner, R.; Becher, R.; Dillmann, R. Using gesture and speech control for commanding a robot assistant. In Proceedings of the 11th IEEE International Workshop on Robot and Human Interactive Communication, Berlin, Germany, 27 September 2002; pp. 454–459. [CrossRef]
12. Tölgyessy, M.; Dekan, M.; Duchoň, F.; Rodina, J.; Hubinský, P.; Chovanec, L. Foundations of Visual Linear Human–Robot Interaction via Pointing Gesture Navigation. *Int. J. Soc. Robot.* **2017**, *9*, 509–523. [CrossRef]
13. Alvarez-Santos, V.; Iglesias, R.; Pardo, X.M.; Regueiro, C.V.; Canedo-Rodriguez, A. Gesture-based interaction with voice feedback for a tour-guide robot. *J. Vis. Commun. Image Represent.* **2014**, *25*, 499–509. [CrossRef]
14. Fang, H.C.; Ong, S.K.; Nee, A.Y. A novel augmented reality-based interface for robot path planning. *Int. J. Interact. Des. Manuf.* **2014**, *8*, 33–42. [CrossRef]
15. Ong, S.K.; Yew, A.W.; Thanigaivel, N.K.; Nee, A.Y. Augmented reality-assisted robot programming system for industrial applications. *Robot. -Comput.-Integr. Manuf.* **2020**, *61*, 101820. [CrossRef]
16. Gadre, S.Y.; Rosen, E.; Chien, G.; Phillips, E.; Tellex, S.; Konidaris, G. End-user robot programming using mixed reality. In Proceedings of the 2019 International Conference on Robotics and Automation (ICRA), Montreal, QC, Canada, 20–24 May 2019; [CrossRef]
17. Kousi, N.; Stoubos, C.; Gkournelos, C.; Michalos, G.; Makris, S. Enabling human robot interaction in flexible robotic assembly lines: An augmented reality based software suite. *Procedia CIRP* **2019**, *81*, 1429–1434. [CrossRef]
18. Stetco, C.; Muhlbacher-Karrer, S.; Lucchi, M.; Weyrer, M.; Faller, L.M.; Zangl, H. Gesture-based contactless control of mobile manipulators using capacitive sensing. In Proceedings of the 2020 IEEE International Instrumentation and Measurement Technology Conference (I2MTC), Dubrovnik, Croatia, 25–28 May 2020; [CrossRef]
19. Mühlbacher-Karrer, S.; Brandstötter, M.; Schett, D.; Zangl, H. Contactless Control of a Kinematically Redundant Serial Manipulator Using Tomographic Sensors. *IEEE Robot. Autom. Lett.* **2017**, *2*, 562–569. [CrossRef]
20. Magrini, E.; Ferraguti, F.; Ronga, A.J.; Pini, F.; De Luca, A.; Leali, F. Human-robot coexistence and interaction in open industrial cells. *Robot. -Comput.-Integr. Manuf.* **2020**, *61*, 1–55. [CrossRef]
21. Irfan, B.; Lyubova, N.; Garcia Ortiz, M.; Belpaeme, T. Multi-modal open-set person identification in hri. In Proceedings of the 2018 ACM/IEEE International Conference on Human-Robot Interaction Social, Chicago, IL, USA, 5–8 March 2018.
22. Strazdas, D.; Hintz, J.; Khalifa, A.; Al-Hamadi, A. Robot System Assistant (RoSA): Concept for an intuitive multi-modal and multi-device interaction system. In Proceedings of the 2021 IEEE 2nd International Conference on Human-Machine Systems (ICHMS), Magdeburg, Germany, 8–10 September 2021; pp. 1–4. [CrossRef]
23. Deng, J.; Guo, J.; Xue, N.; Zafeiriou, S. Arcface: Additive angular margin loss for deep face recognition. In Proceedings of the IEEE/CVF Conference on Computer Vision and Pattern Recognition, Long Beach, CA, USA, 15–20 June 2019; pp. 4690–4699.
24. Luan, P.; Huynh, V.; Tuan Anh, T. Facial Expression Recognition using Residual Masking Network. In Proceedings of the 2020 25th International Conference on Pattern Recognition (ICPR), Milan, Italy, 10–15 January 2021; pp. 4513–4519.
25. Deng, J.; Guo, J.; Ververas, E.; Kotsia, I.; Zafeiriou, S. Retinaface: Single-shot multi-level face localisation in the wild. In Proceedings of the IEEE/CVF Conference on Computer Vision and Pattern Recognition, Seattle, WA, USA, 13–19 June 2020; pp. 5203–5212.
26. Picovoice. Porcupine. 2020. Available online: https://github.com/Picovoice/porcupine (accessed on 22 December 2021).
27. Hannun, A.; Case, C.; Casper, J.; Catanzaro, B.; Diamos, G.; Elsen, E.; Prenger, R.; Satheesh, S.; Sengupta, S.; Coates, A.; et al. Deepspeech: Scaling up end-to-end speech recognition. *arXiv* **2014**, arXiv:1412.5567.
28. Google. WebRTC. 2020. Available online: https://webrtc.org (accessed on 22 December 2021).
29. RASA. RASA. 2020. Available online: https://rasa.com/open-source/ (accessed on 22 December 2021).
30. Hempel, T. RoSA: Cube Detector, 2021. Available online: https://github.com/thohemp/cube_detector/tree/v1.0.0 (accessed on 17 December 2021). [CrossRef]
31. Strazdas, D.; Khalifa, A.; Hempel, T. DoStraTech/rosa_msgs: Initial Release, 2021. Available online: https://github.com/DoStraTech/rosa_msgs/tree/v0.1-alpha (accessed on 15 December 2021). [CrossRef]
32. Bramel, J. Alphabet Play Blocks. Available online: https://www.thingiverse.com/thing:2368270 (accessed on 22 December 2021).

33. Howard, A.G.; Zhu, M.; Chen, B.; Kalenichenko, D.; Wang, W.; Weyand, T.; Andreetto, M.; Adam, H. Mobilenets: Efficient convolutional neural networks for mobile vision applications. *arXiv* **2017**, arXiv:1704.04861.
34. Yang, S.; Luo, P.; Loy, C.C.; Tang, X. Wider face: A face detection benchmark. In Proceedings of the IEEE Conference on Computer Vision and Pattern Recognition, Las Vegas, NV, USA, 27–30 June 2016; pp. 5525–5533.
35. Guo, X.; Li, S.; Yu, J.; Zhang, J.; Ma, J.; Ma, L.; Liu, W.; Ling, H. PFLD: A practical facial landmark detector. *arXiv* **2019**, arXiv:1902.10859.
36. Chen, S.; Liu, Y.; Gao, X.; Han, Z. Mobilefacenets: Efficient cnns for accurate real-time face verification on mobile devices. In *Chinese Conference on Biometric Recognition*; Springer: Berlin/Heidelberg, Germany, 2018; pp. 428–438.
37. Foster, M.E.; Gaschler, A.; Giuliani, M. Automatically Classifying User Engagement for Dynamic Multi-party Human–Robot Interaction. *Int. J. Soc. Robot.* **2017**, *9*, 659–674. [CrossRef]
38. Vaufreydaz, D.; Johal, W.; Combe, C. Starting engagement detection towards a companion robot using multimodal features. *Robot. Auton. Syst.* **2016**, *75*, 4–16. [CrossRef]
39. Anzalone, S.M.; Boucenna, S.; Ivaldi, S.; Chetouani, M. Evaluating the engagement with social robots. *Int. J. Soc. Robot.* **2015**, *7*, 465–478. [CrossRef]
40. Li, L.; Xu, Q.; Tan, Y.K. Attention-based addressee selection for service and social robots to interact with multiple persons. In *Proceedings of the Workshop at SIGGRAPH Asia*; Association for Computing Machinery: New York, NY, USA, 2012; pp. 131–136.
41. Richter, V.; Carlmeyer, B.; Lier, F.; Meyer zu Borgsen, S.; Schlangen, D.; Kummert, F.; Wachsmuth, S.; Wrede, B. Are you talking to me? Improving the robustness of dialogue systems in a multi party HRI scenario by incorporating gaze direction and lip movement of attendees. In Proceedings of the Fourth International Conference on Human Agent Interaction, Biopolis, Singapore, 4–7 October 2016; pp. 43–50.
42. Ellgring, J.H. *Nonverbale Kommunikation*; Universität Würzburg: Würzburg, Germany, 1986.
43. Murphy-Chutorian, E.; Trivedi, M.M. Head pose estimation in computer vision: A survey. *IEEE Trans. Pattern Anal. Mach. Intell.* **2008**, *31*, 607–626. [CrossRef]
44. Albiero, V.; Chen, X.; Yin, X.; Pang, G.; Hassner, T. img2pose: Face alignment and detection via 6dof, face pose estimation. In Proceedings of the IEEE/CVF Conference on Computer Vision and Pattern Recognition, Nashville, TN, USA, 20–25 June 2021; pp. 7617–7627.
45. Kellnhofer, P.; Recasens, A.; Stent, S.; Matusik, W.; Torralba, A. Gaze360: Physically unconstrained gaze estimation in the wild. In Proceedings of the IEEE/CVF International Conference on Computer Vision, Seoul, Korea, 27–28 October 2019; pp. 6912–6921.
46. Williams, T.; Hirshfield, L.; Tran, N.; Grant, T.; Woodward, N. Using augmented reality to better study human-robot interaction. In *International Conference on Human-Computer Interaction*; Springer: Cham, Switzerland, 2020; pp. 643–654.
47. Haeb-Umbach, R.; Watanabe, S.; Nakatani, T.; Bacchiani, M.; Hoffmeister, B.; Seltzer, M.L.; Zen, H.; Souden, M. Speech Processing for Digital Home Assistants: Combining Signal Processing with Deep-Learning Techniques. *IEEE Signal Process. Mag.* **2019**, *36*, 111–124. [CrossRef]
48. Ramirez, J.; Górriz, J.M.; Segura, J.C. Voice activity detection. fundamentals and speech recognition system robustness. *Robust Speech Recognit. Underst.* **2007**, *6*, 1–22.
49. Mozzilla. Deepspeech. 2020. Available online: https://github.com/mozilla/DeepSpeech (accessed on 22 December 2021).
50. Agarwal, A.; Zesch, T. German End-to-end Speech Recognition based on DeepSpeech. In *Preliminary Proceedings of the 15th Conference on Natural Language Processing (KONVENS 2019): Long Papers*; German Society for Computational Linguistics & Language Technology: Erlangen, Germany, 2019; pp. 111–119.
51. Schurick, J.M.; Williges, B.H.; Maynard, J.F. User feedback requirements with automatic speech recognition. *Ergonomics* **1985**, *28*, 1543–1555. [CrossRef]
52. Bocklisch, T.; Faulkner, J.; Pawlowski, N.; Nichol, A. Rasa: Open source language understanding and dialogue management. *arXiv* **2017**, arXiv:1712.05181.
53. Braun, D.; Hernandez Mendez, A.; Matthes, F.; Langen, M. Evaluating Natural Language Understanding Services for Conversational Question Answering Systems. In *Proceedings of the 18th Annual SIGdial Meeting on Discourse and Dialogue*; Association for Computational Linguistics: Saarbrücken, Germany, 2017; pp. 174–185. [CrossRef]
54. Microsoft. Speechsynthesis. 2020. Available online: https://docs.microsoft.com/de-de/uwp/api/windows.media.speechsynthesis (accessed on 22 December 2021).
55. Brooke, J. SUS: A quick and dirty usability scale. *Usability Eval. Ind.* **1995**, *189*, 4–7.
56. Finstad, K. The usability metric for user experience. *Interact. Comput.* **2010**, *22*, 323–327. [CrossRef]
57. Lewis, J.R. Psychometric Evaluation of the PSSUQ Using Data from Five Years of Usability Studies. *Int. J. -Hum.-Comput. Interact.* **2002**, *14*, 463–488. [CrossRef]
58. Lewis, J. Psychometric evaluation of an after-scenario questionnaire for computer usability studies: The ASQ. *SIGCHI Bull.* **1991**, *23*, 78–81. [CrossRef]
59. Malik, M.; Muhammad, K.; Mehmood, K.; Makhdoom, I. Automatic speech recognition: A survey. *Multimed. Tools Appl.* **2021**, *80*, 9411–9457. [CrossRef]

*Review*

# RANSAC for Robotic Applications: A Survey

José María Martínez-Otzeta [1,*], Itsaso Rodríguez-Moreno [1], Iñigo Mendialdua [2] and Basilio Sierra [1]

[1] Department of Computer Science and Artificial Intelligence, University of the Basque Country, 20018 Donostia-San Sebastián, Spain
[2] Department of Languages and Information Systems, University of the Basque Country, 20018 Donostia-San Sebastián, Spain
\* Correspondence: josemaria.martinezo@ehu.eus

**Abstract:** Random Sample Consensus, most commonly abbreviated as RANSAC, is a robust estimation method for the parameters of a model contaminated by a sizable percentage of outliers. In its simplest form, the process starts with a sampling of the minimum data needed to perform an estimation, followed by an evaluation of its adequacy, and further repetitions of this process until some stopping criterion is met. Multiple variants have been proposed in which this workflow is modified, typically tweaking one or several of these steps for improvements in computing time or the quality of the estimation of the parameters. RANSAC is widely applied in the field of robotics, for example, for finding geometric shapes (planes, cylinders, spheres, etc.) in cloud points or for estimating the best transformation between different camera views. In this paper, we present a review of the current state of the art of RANSAC family methods with a special interest in applications in robotics.

**Keywords:** RANSAC; feature matching; transformation matrix; shape detection; object recognition; robotic systems; real time

## 1. Introduction

The Random Sample Consensus algorithm, commonly known by its acronym RANSAC, was developed by Fischler and Bolles more than forty years ago as a novel approach to the robust estimation of the parameters of a model in regression analysis [1]. It addresses situations where there is a high percentage of outliers in the data, which hinders the parameter estimation task. While other approaches, such as least squares linear regression, use all the available data to produce a model which could be later refined, RANSAC creates several models in sequence, each time choosing, from the available data, a random sample of the minimum size needed to create a model. After each step in the sequence, the support of the model is calculated, typically splitting the data into inliers or outliers, the former are data points whose measure of fitness with respect to the model fall below a certain threshold, and the latter are those which do not comply with that requirement. For example, if the task is to estimate the parameters of a plane in a point cloud, RANSAC would sample three points, the minimum needed to define a plane (provided they are not all collinear with each other), then would compute the euclidean distance of all the points in the point cloud to such a plane, and then the percentage of inliers according to some threshold. After some number of iterations, RANSAC would return the plane with the highest support, thus the plane with more points close to it.

Several questions and practical issues arise from this simple definition. It is not obvious at all how to choose a reasonable threshold to discriminate between inliers and outliers, or how to determine the number of iterations through the process of creating and evaluating a model. The computational cost could also skyrocket if the whole dataset has to be checked against the hypothesis model at each iteration. For these reasons, researchers have devised a number of variations over this vanilla RANSAC trying to address these issues.

The range of problems for which RANSAC is well-suited is very wide, although in robotic applications they usually pertain to two big classes: shape detection and feature matching. Geometric models of simple shapes such as planes, spheres, cylinders, etc., are well understood and relatively straightforward to implement, while of relevant practical interest, due to the fact that human-made objects and some natural ones are close to those shapes. Therefore, the ability to detect these geometric structures in 3D data is of great importance for environment understanding in indoor or outdoor robotic navigation and/or mapping. A high-performant feature-matching procedure is also very desirable for finding the right transformation between different views of a scene. Features are extracted from several scenes taken from different points of view, and the task is defined as finding the matrix transformation between images that minimizes the distance between matching features.

Other machine learning approaches, such as deep neural networks [2,3], while extremely successful for object detection, can be difficult to fine-tune or interpret. Their main advantage is that they could be readily applied to problems in which there is no restriction on the object shapes, but a RANSAC implementation can be more efficient when the shape is simple enough to be expressed as a model with a few parameters. As a result, the fields of application of these two kinds of methods are usually different, although it is possible to apply RANSAC as a preprocessing step before applying deep learning techniques. For example, a deep learning application for point cloud segmentation could benefit from RANSAC filtering the floor or the walls of the scene.

Several studies have been carried out on the performance of RANSAC methods. In [4], the authors classify the RANSAC variants into three types, depending on the intended improvement over the vanilla version: accuracy of the model, computational speed, and robustness with respect to the choice of the number of iterations and the threshold value. They analyze the performance of ten variants on a line fitting synthetic data and on a 2D homography estimation on real data, comparing also with Least Median of Squares [5] and projection-based M-estimator [6]. Their results show the existence of a trade-off of accuracy and robustness over computing time. In [7], the authors present a comparison of eleven variants on the problem of finding planes in 3D point clouds. Their findings suggest a trade-off similar to those noticed in [4].

Our goal is to update the list of RANSAC variants with more recent development and, at the same time, present practical applications that could be of interest to the robotics practitioner. The reader is also informed of existing open-source software that could be of interest. Due to the plethora of RANSAC variants developed by the community, it is impossible to list them all. We tried our best to present the most used, cited, or influential ones in further developments. Some algorithms that might deserve a mention, such as RAMOSAC [8] or KALMANSAC [9], were left out, as they were specifically designed for one application (target tracking in those mentioned above) and lack the generality of applications of those presented in Section 3.

In this paper, we present a survey of RANSAC-like methods with a focus on shape detection and image matching for robotic applications. First, we describe the vanilla RANSAC algorithm in detail, along with several variations that have been devised to try to address some of its limitations. Then, we review some recent applications, also pointing out some open-source software that could be of interest to the researchers interested in this field. Finally, we discuss the current state of the art and present our conclusions.

## 2. RANSAC

Fischler and Bolles's Random Sample Consensus (RANSAC) algorithm is a general parameter estimation method designed to handle data where a high percentage of outliers is present [1]. RANSAC was created by members of the computer vision community [10], in contrast to many other popular robust estimating methods also embraced by that community, as in the case of M-estimators [11] and Least Median of Squares [12].

It is a resampling method that produces potential solutions by using the fewest possible observations to estimate the model's underlying parameters. RANSAC employs the smallest possible set of observations and then expands this set with consistent data points, in contrast to typical sampling strategies that use all the available data to generate an initial solution proposal and then refine the model deleting outliers.

The vanilla RANSAC pseudocode is described in Algorithm 1. The condition that must meet a point to be an inlier with respect to the model being evaluated is that the point "fits well". As the most common applications of RANSAC are related to finding shapes or matching features between different views, it is usual to refer to the euclidean distance from a point to the model. However, any other function that measures the concordance between the model and the point might be used. As it can be observed in Algorithm 1, the time complexity of the vanilla RANSAC algorithm is linear in the product of the number of iterations by the number of data points, and therefore it could be computationally expensive.

---

**Algorithm 1:** RANSAC algorithm

$nData \leftarrow$ number of data points
$nP \leftarrow$ smallest number of points required by the model
$nI \leftarrow$ number of iterations
$t \leftarrow$ maximum distance (threshold) of a point to the model to be an inlier
$bestInliers \leftarrow 0$;
$bestModel \leftarrow NULL$;
**while** *Not all iterations done* **do**
    Draw $nP$ points randomly;
    Fit a model $M$ to those points;
    $nInliers \leftarrow 0$;
    **for** *each point in the data* **do**
        $d \leftarrow$ distance from the point to the model;
        **if** $d < t$ **then**
            $nInliers \leftarrow nInliers + 1$
    **if** $nInliers > bestInliers$ **then**
        $bestInliers \leftarrow nInliers$;
        $bestModel \leftarrow M$
**return** $bestModel$

---

In order to tackle this problem, a very common modification of this algorithm consists of defining a measure of goodness of a model such that, if in any of the iterations a model good enough is found, the procedure terminates and returns that model as the solution. For example, it would be possible to compute the percentage of inliers over the total number of data points and decide that if a model exceeds some predefined threshold, the model is suitable for our goal and the search ends.

When it is not clear what could be a criterion for considering a model good enough, or when all the models are of low quality, but we are interested in the best among them, the natural question that arises is how to choose the number of iterations. Let us denote the inlier ratio over the total data points as *iRatio*. The probability that in a set of $nP$ randomly chosen points all are inliers is of $iRatio^{nP}$, and therefore the probability of this situation not happening in an iteration is of $1 - iRatio^{nP}$, and in $nI$ iterations of $(1 - iRatio^{nP})^{nI}$. As we want this probability to be small, $nI$ has to be chosen such that $(1 - iRatio^{nP})^{nI} < \alpha$, where $\alpha$ might be 0.05 for a probability of 95% of randomly drawing all inlier points in some iteration. After some algebraic manipulation, we find the following equation:

$$nI = \frac{log(\alpha)}{log(1 - iRatio^{nP})}$$

Unfortunately, the ratio of inlier points is often not known in advance and the usefulness of the previous result is limited. In Section 3.2, several approaches focused on improving the speed of the algorithm are described, empirically showing that it is often possible to run RANSAC well below its theoretical time complexity. Some authors even claim constant time model evaluation [13].

## 2.1. Matching Images

One common problem in computer vision is finding the right correspondences between features of two images of the same scene taken by different cameras. Epipolar geometry is the branch of geometry that helps to formalize the relationship between cameras, points in the 3D space, and their projections in the camera views.

In Figure 1, a typical situation is shown where two cameras (or the same camera at different timestamps) observe a point P located at some distance in the space. The camera centers are $c_1$ and $c_2$, where the projections of the point P in the image planes are $p_1$ and $p_2$, respectively. The line connecting the two camera centers is referred to as the baseline, while the plane defined by $c_1$, $c_2$, and $P$ is the epipolar plane. The intersection points between the baseline and the two image planes are denoted as the epipoles $e_1$ and $e_2$. The epipolar lines are defined by the intersection of the epipolar plane with the image planes, and they intersect the baseline at the epipoles. When the two image planes are parallel, then the epipoles $e_1$ and $e_2$ are located at infinity.

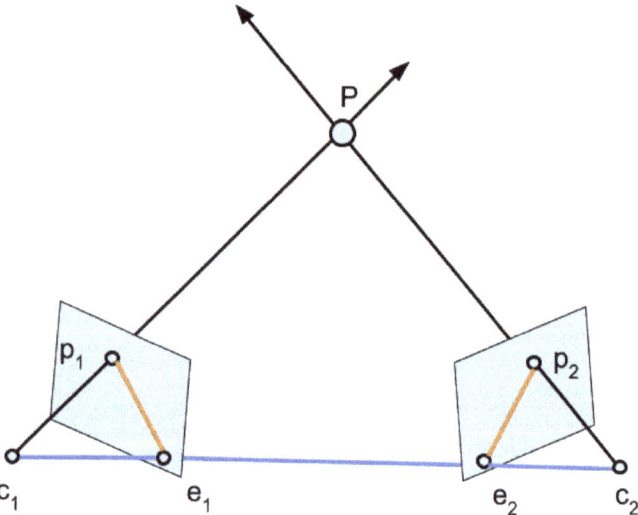

**Figure 1.** The general setup of epipolar geometry. The planar region defined by the points $P$, $c_1$, and $c_2$ is the epipolar plane. The blue line is the baseline, while the two orange lines are the epipolar lines.

The transformation from one view to the other is given by a homography matrix that is necessary to estimate. To do so, typically, some algorithm for finding features, such as SIFT [14], is employed, and then a model fitting procedure is conducted.

## 2.2. Finding 2D/3D Shapes

Basic geometric shapes are the constituent parts of multiple objects in the environments in which robots usually perform their tasks. Walls, floors, and ceilings are examples of planes located in indoor settings. Doors, windows, tables, and stairs are also composed of planes, while pipes could be an example of cylinders. In urban outdoors, roads, buildings, or traffic lights also resemble geometric shapes. Even in agricultural areas, the crops use to be organized in lines or trees could be modeled as conical structures. In Figure 2, several objects in a room can be seen, including the walls and columns. There is also noise in the point cloud, a typical circumstance that could be addressed using RANSAC to detect shapes.

**Figure 2.** Visualization of a point cloud of a room.

## 3. RANSAC Variants

It is difficult to classify the plethora of RANSAC variants for several reasons. First of all, sometimes, a previously known variant is employed as part of a pipeline with a small tweak dependent on the specific problem or the rest of the pipeline. Methods that clearly fall into this category are described in the section corresponding to applications. Secondly, some methods try to improve the vanilla RANSAC in more than one dimension, for example, achieving better accuracy while at the same time yielding a lower computational load and being faster. This paper does not intend to provide an exhaustive catalog of all the existing variants but to present a state-of-the-art survey of the most commonly used, along with some applications in the robotic field.

Following the classifications previously made in [4,7], we divide the methods into four types, depending on which is the main area they intend to improve: accuracy, speed, robustness, and optimality. As mentioned above, sometimes the distinction is not clear-cut, but when possible, we followed the previous surveys. When the authors have developed a method that is specifically designed for a concrete application, and not as a general estimation method, we present it in the Applications section. A summary of the variants grouped by areas is presented in Table 1.

**Table 1.** RANSAC variants grouped by the metric they aim to improve.

| Focus on | RANSAC Variant |
|---|---|
| Accuracy | MSAC (M-estimator SAC) [15]<br>MLESAC (Maximum Likelihood SAC) [16]<br>MAPSAC (Maximum A Posterior Estimation SAC) [17]<br>LO-RANSAC (Locally Optimized RANSAC) [18]<br>QDEGSAC (RANSAC for Quasi-degenerate Data) [19]<br>Graph-Cut RANSAC [20] |

Table 1. Cont.

| Focus on | RANSAC Variant |
|---|---|
| Speed | NAPSAC (N Adjacent Points SAmple Consensus) [21]<br>Randomized RANSAC with $T_{d,d}$ test [22]<br>Guided-MLESAC [23]<br>RANSAC with bail-out test [24]<br>Randomized RANSAC with Sequential Probability Ratio Test [25]<br>PROSAC (Progressive Sample Consensus) [26]<br>GASAC (Genetic Algorithm SAC) [27]<br>1-point RANSAC [28]<br>GCSAC (Geometrical Constraint SAmple Consensus) [29]<br>Latent RANSAC [13] |
| Robustness | AMLESAC [30]<br>u-MLESAC [31]<br>Recursive RANSAC [32]<br>SC-RANSAC (Spatial Consistency RANSAC) [33]<br>NG-RANSAC (Neural-Guided RANSAC) [34]<br>LP-RANSAC (Locality-preserving RANSAC) [35] |
| Optimality | Optimal Randomized RANSAC [36]<br>Optimal RANSAC [37] |

*3.1. Accuracy-Focused Variants*

3.1.1. MSAC (M-Estimator SAC)

MSAC [15] is the first iteration of a family of RANSAC variants that uses maximum likelihood estimates for the parameter models. The authors tackle the problem of the estimation of the trifocal tensor which, given correspondences between points in two images, determines the position of such points in a third image. Their results show an increase in accuracy with the downside of degradation of computational performance. They claim their method could be used for any other problem in computer vision, mentioning the fundamental matrix estimation task as an example.

3.1.2. MLESAC (Maximum Likelihood SAC)

The RANSAC algorithm looks for the model that maximizes the number of inliers. MLESAC [16] computes the log-likelihood of the model, taking into account the distribution of outliers, and uses random sampling to maximize it. The authors show the usefulness of their approach by deriving the log-likelihood for the problem of estimating the fundamental matrix in a two-view problem and implementing the algorithm, obtaining good results.

3.1.3. MAPSAC (Maximum A Posterior Estimation SAC)

A Bayesian approach is presented in [17], from the same authors of MSAC and MLESAC, with the aim of improving over their previous maximum likelihood formulations. They develop MAPSAC to obtain a robust Maximum A Posterior (MAP) estimate of the problem of the least square fitting of an arbitrary manifold, and in particular, of lines or planes. A new method for approximating the posterior probability of a model, called GRIC, is derived, and it is theoretically and empirically demonstrated that it is more accurate than AIC [38], BIC [39], and MDL [40] in tasks where a large number of latent variables is present.

3.1.4. LO-RANSAC (Locally Optimized RANSAC)

In [18], the authors introduce Locally Optimized RANSAC (LO-RANSAC) to address the empirical observation that the number of samples needed to find an optimal solution with a given probability is significantly higher than the amount predicted by the theory [41].

They realize that a commonly held assumption is incorrect: that a model computed from a sample composed only of inliers has to be consistent with the whole set of inliers.

In spite of the computed model not being optimal, in practice, it is sufficiently close to the optimal model for a local optimization method to approach or even find it. After the optimization step, the model covers a greater amount of inliers, and therefore reduces the number of steps of the RANSAC method, making it approach to its theoretical value. The optimization strategy guarantees to keep the number of used samples very low, hence, accordingly, the extra time spent in each step is almost negligible.

Their proposed optimization algorithm consists of the following steps:

1. Define a threshold $\theta$ and a number of optimization iterations $I$.
2. In each step of the RANSAC method, the samples are selected only from the data points that are consistent with the model created in the previous step.
3. Take all data points with error smaller than $I \times \theta$ and compute new model parameters according to a linear algorithm. Reduce $I$ by one and iterate until the threshold is $\theta$.

This local optimization step is only performed when the number of inliers in the current RANSAC step is greater than the previous maximum. The number of points from a randomly drawn sample that are consistent with a given model is a random variable with usually unknown probability density function. As this density function is the same for all samples with the same discrete cardinality, the probability that the $k$th sample will be the best among the already drawn samples is $\frac{1}{k}$. Therefore, the average number of samples which are the best so far is a sequence of $n$ samples:

$$\sum_{1}^{n} \frac{1}{n} \leq \int_{1}^{n} \frac{1}{n} dx + 1 = \log n + 1$$

The authors perform experiments on epipolar geometry and homography estimation and find that the empirical results are close to this theoretical average, while speeding up the RANSAC procedure by two to three times. The number of inliers of the solution found, which can be thought as a proxy for its overall quality, is increased in the range of 10–20%.

3.1.5. QDEGSAC (RANSAC for Quasi-Degenerate Data)

Sometimes, when confronted with the problem of computing the fundamental matrix for image matching coming from different views, the data do not provide enough constraints to find a unique solution, but only up to a set of solutions that all of them could explain the data. To tackle this problem, in [19], QDEGSAC is presented to cope with the problem of degenerate data. The authors develop a hierarchical RANSAC over the number of present constraints that do not require problem-dependant tests. They claim results similar to other approaches that leverage knowledge about the degeneracy source.

3.1.6. Graph-Cut RANSAC

A modification of LO-RANSAC is presented in [20], where the graph-cut algorithm [42] is applied in the local optimization step to the best model obtained till that moment. The motivation is to separate inliers and outliers. The authors test its adequacy to several computer vision problems with synthetic and real data and claim it to be more geometrically accurate and at the same time easy to implement. Additional improvements have been presented in [43,44], where USAC [45] and MAGSAC++ [46] robust estimators are included in the algorithm.

3.2. Speed-Focused Variants

3.2.1. NAPSAC (N Adjacent Points SAmple Consensus)

The premise behind NAPSAC [21] is that inliers tend to be closer among them than with respect to the outliers. Therefore, instead of picking completely random samples to generate the models, a better strategy could take into account this fact. The authors suggest taking an initial point randomly and then finding the number of points lying within a hypersphere of radius $r$ centered on that point. If the number of points in such a hypersphere is fewer than the minimal set needed to estimate the parameters of the model, then fail and choose another initial point; otherwise, select the initial point and other points uniformly from the set of points inside the hypersphere until the minimal number needed to estimate the model have been selected. The authors derive optimal values for the radius of the hypersphere if the inliers are perturbed by Gaussian noise and the outliers are distributed uniformly in the hypersphere.

### 3.2.2. Randomized RANSAC with $T_{d,d}$ Test

The hypothesis test step in the RANSAC algorithm is often very costly. If a plane has to be fitted against a point cloud containing millions of points, the distance to every point has to be computed to assess the goodness of the hypothesized plane model. To deal with this issue, in [22], the authors propose an algorithm that only evaluates a fraction of the data points. They define a $T_{d,d}$ test that is passed if all the $d$ randomly selected data points are consistent with the hypothesis currently being tested. The optimal value of $d$ is computed with the following expression

$$d \approx \frac{\ln(\frac{\ln \epsilon (t_M+1)}{N(\ln \delta - \ln \epsilon)})}{\ln \delta},$$

where $t_M$ is the time necessary to compute the parameters of the model from a sample, $\delta$ is the probability that a data point is consistent with a random model, and $\epsilon$ is the fraction of outliers in the data. As $d$ has to be an integer, $d_{opt}$ is chosen as the number in $\{\lfloor d \rfloor, \lceil d \rceil\}$, which minimizes the previous expression, provided it is greater than zero. One drawback of this approach is that an estimation of the fraction of outliers in the original data is needed.

### 3.2.3. Guided-MLESAC

A limitation of the maximum-likelihood estimation in MLSAC is that it does not take into account possible knowledge about the prior probabilities of the parameters of the model to be estimated. In [23], the authors propose guided-MLESAC, where a good estimation of the prior probabilities is shown to give an order of magnitude speed improvement in the problem of finding correspondences between features in images taken from different views. After a theoretical analysis and experiments to compute the priors, their conclusions are that, with little extra computation, it is possible to leverage quality measures provided by image matcher software to derive confidence in the validity of a match and incorporate them as priors. This knowledge can be useful to select more probable hypotheses and also to compute more accurately the cost of fitting.

### 3.2.4. RANSAC with Bail-Out Test

In [24], the author, inspired by the randomized RANSAC with $T_{d,d}$ test, describes a modification of the RANSAC procedure that allows the scoring process to be terminated earlier, and therefore could save computational time. He first defines a trivial early bail test, in which a hypothesis is not further checked against the remaining data if its current score cannot improve the score of the best hypothesis tested so far. Then, he proceeds to propose a test in which a randomly selected subset of size $n$ is evaluated, and its fraction of inliers ($\epsilon_n$) is computed. If $\epsilon_n$ is clearly smaller than the best $\epsilon_{best}$ found so far, it is very unlikely that evaluating the rest of the points would produce a better result than $\epsilon_{best}$. In the paper, estimates for the probability of $\epsilon$ improving over $\epsilon_{best}$ when evaluating the remaining points are derived under the supposition that the number of inliers contained in a subset of size $n$ follows a hypergeometric distribution. Experiments show a significant reduction in the number of evaluations with respect to the $T_{d,d}$ test approach.

### 3.2.5. Randomized RANSAC with Sequential Probability Ratio Test

For randomized models to work properly, usually an estimate of the fraction of inliers in the data is needed. In [25], another approach that does not require such knowledge is presented. The authors base their work on Wald's theory of sequential decision making [47], deriving a process to generate a solution with confidence $1 - \eta$, where $\eta$ is a probability decided by the user. Wald's Sequential Probability Test Ratio is based on the likelihood ratio

$$\lambda_i = \prod_{n=1}^{i} \frac{p(x_n|H_b)}{p(x_n|H_g)} = \lambda_{i-1} \frac{p(x_i|H_b)}{p(x_i|H_g)},$$

where $H_g$ is the hypothesis that the model is good, i.e., computed from a sample composed only of inliers, and $H_b$ corresponds to the alternative hypothesis that the model is considered bad. The variable $x_n$ is equal to 1 if the n-th data point is consistent with the evaluated model, and 0 otherwise. The probability $p(1|H_g)$ that a random point is consistent with a good model is approximately the percentage of inliers in the original data, represented as $\epsilon$, and the probability of being coherent with a bad model is modeled as a Bernouilli distribution with parameter $\delta = p(1|H_b)$. Given that the majority of all models tested by RANSAC are bad in the former sense, $\delta$ can be estimated as the average percentage of consistent data points in rejected models. On the other hand, a lower bound on $\epsilon$ is given by the size of the largest support for the considered models so far. Experiments show that this method is from 2.8 to 10 times faster than RANSAC and up to 4 times faster than the Randomized RANSAC with $T_{d,d}$ test.

### 3.2.6. PROSAC (Progressive Sample Consensus)

PROSAC [26] establishes a rank of promising data points (and therefore, of promising hypotheses or models) according to some measure of the quality of the data. As the hypothesis testing procedure advances, the confidence in the adequacy of the quality scores decreases, and the sampling strategy is shifting toward the original RANSAC. The more promising samples are drawn at the earlier stages, but in further steps, data points with lower quality scores are gradually incorporated until all the original samples have a nonzero probability of being drawn. The authors employ PROSAC for the task of estimating the correspondences between features in two images from different camera views. They claim to achieve significant time savings over RANSAC, in the order of hundreds of times, due to the fact that good hypotheses are generated early on in the sampling process.

### 3.2.7. GASAC (Genetic Algorithm SAC)

In [27], the authors propose an approach based on genetic algorithms, where a population of sets of parameters evolves to yield a solution. They tackle the problem of finding the fundamental matrix associated with different views of a scene, but the method can be adapted to any problem usually solved by RANSAC. The individuals of the genetic pool are characterized by a chromosome in which the model parameters are encoded and are subject to the usual crossover and mutation operators. A number of evaluations of an order of magnitude less than with the usual RANSAC are reported, when tested on several image-matching problems.

### 3.2.8. One-Point RANSAC

While multiple works have analyzed the way to reduce the number of data points against which to test the hypothesized models, some researchers have worked on how to reduce the number of data points needed to generate a hypothesis. In [28], the authors propose a method to generate a hypothesis from just a data point by leveraging a priori information from an extended Kalman filter [48]. Experiments are performed in two scenarios: the first one is a six-degree-of-freedom motion estimation from a monocular sequence, and the second one is a robot trajectory estimation combining wheel odometry and monocular vision. The authors claim results comparable to other visual odometry methods.

### 3.2.9. GCSAC (Geometrical Constraint SAmple Consensus)

Geometric constraints could be used to select good samples to generate models that would be further tested for consistency against the data. In [29], the authors present a method that searches for such samples following two criteria: the selected samples must be consistent with the estimated model according to an inlier ratio evaluation and, at the same time, they must satisfy geometrical constraints of the object we are looking for. Experiments were performed for cylinder fitting in several datasets, one of them synthetic, the second one consisting of data obtained in their laboratory, and the third one from public datasets. Their results demonstrate better accuracy than MLSAC and real-time performance.

### 3.2.10. Latent RANSAC

An attempt to evaluate a RANSAC-generated model in constant time, independently of the size of the data set, is presented in [13]. The authors' insight is that the correct hypotheses form clusters in the latent parameter domain. From this observation, an approach similar to the randomized version of the generalized Hough transform [49] can be applied to find those clusters, claiming that only two votes are necessary to succeed in the search. The fast localization of the pairs of similar hypotheses is possible thanks to an adaptation of the random grids search technique [50]. Therefore, the computationally demanding hypothesis verification stage only takes place after the discovery of a similar pair of them, and it is shown that this event is very rare when the hypotheses are incorrect. The authors perform experiments on three different types of problems on both synthetic and real data: camera localization, 3D rigid alignment, and 2D-homography estimation. They claim an improvement in speed without degradation in accuracy.

## 3.3. Robustness-Focused Variants

### 3.3.1. AMLESAC

A noise-adaptive variant of MLESAC [16] is presented in [30]. It applies the sampling strategy of MLESAC, and also searches for the model that maximizes the likelihood. The improvement over it is the simultaneous estimation of the percentage of inliers ($\gamma$) and the standard deviation of the noise affecting the inliers ($\sigma$). This is achieved by defining as a function of $\gamma$ and $\sigma$ the log-likelihood of all points under a hypothesis $\theta_k$ and selecting the values that maximize it. Then, the likelihood of $\theta_k$ using all the data and the previously estimated values for $\gamma$ and $\sigma$ is computed. All the process is repeated $M$ times, where $M$ was obtained based on a prior estimate of $\gamma$, and the hypothesis $\theta$ with the highest likelihood among the $\theta_k$ obtained in each iteration is returned. Then, the model is refined by applying nonlinear minimization using point-based parameterization [51]. Experiments on synthetic and real data show that AMLESAC outperforms previous methods for the pose estimation task without relying on the knowledge of noise parameters.

### 3.3.2. u-MLESAC

Another method based on MLESAC is u-MLESAC [31]. As in MLESAC, it can be decomposed into four steps: sampling of the data, estimation of the parameters, estimation of the variables of the error model, and evaluation of the parameters according to the maximum likelihood criterion. The novelties of u-MLESAC are the estimation both of the variance of the error model and of the number of iterations. The variance $\sigma$ of the error model is estimated by the expectation–maximization (EM) algorithm [52], and the number of iterations is computed from the condition that all the sampled data are inliers and within a desired error tolerance $\beta$. The number $t$ of iterations is computed as

$$t = \frac{\log \alpha}{\log \left(1 - k^m \gamma^m\right)},$$

where $m$ is the number of sampled data points, $\gamma$ is the inlier ratio, and $k = \text{erf}(\frac{\beta}{\sqrt{(2)}\sigma})$, with $erf$ as the Gauss error function. Experiments with line fitting tasks showed high accuracy and robustness in different data distributions.

### 3.3.3. Recursive RANSAC

The standard RANSAC algorithm assumes that all the data are available at the start of the estimation process. To tackle the problem of data appearing sequentially, in [32], the Recursive RANSAC algorithm is presented. The authors point out that, as the recursive least-squares algorithm (RLS) [53] is the extension of the least-squares method to sequential data, Recursive RANSAC is the recursive version of RANSAC, with the added capability of being able to track multiple signals simultaneously. This approach makes use of RANSAC to estimate models that fit the current observations with previous observations. When an observation is an inlier to a model, the model is updated by means of recursive least squares. Experiments with simulated data show that Recursive RANSAC is more accurate than RLS, Hough transform [54], and batch RANSAC [55] when the task is to estimate the parameters of a single random line. Another simulation of multiple signals tracking in the task of geolocating stationary ground objects using aerial sensors shows promising results.

### 3.3.4. SC-RANSAC (Spatial Consistency RANSAC)

The authors of SC-RANSAC [33] present a robust and efficient method to detect points that are clearly outliers, with the aim of removing them and therefore increasing the inlier ratio in the data given to RANSAC. To detect those outliers, the method takes advantage of spatial relations between corresponding points in two images. This approach can also be seen as a preprocessing step for other RANSAC variants. Experiments performed over standard datasets of real images show improvements in computational time and also in accuracy over other methods such as RANSAC and PROSAC. These advantages are especially noticeable when the percentage of outliers is high.

### 3.3.5. NG-RANSAC (Neural-Guided RANSAC)

The field of neural networks permeates every area of machine learning and parameter estimation nowadays, and RANSAC research is not an exception. NG-RANSAC [34] is a RANSAC variant that uses prior information to guide the search of model hypotheses, with the aim of increasing the probability of finding sets with no outliers or very few of them. In other approaches, the prior information is obtained by heuristic methods that make use of hand-designed descriptors, built from the domain knowledge of the researcher. In contrast, NG-RANSAC uses neural networks to navigate through the set of hypotheses. Self-supervision of the process is achieved using the inlier percentage as part of the training data, and the addition of a differentiable version of RANSAC allows for further improvements. Experiments on fundamental matrix estimation, camera relocalization, and horizon line estimation achieve state-of-the-art results.

### 3.3.6. LP-RANSAC (Locality-Preserving RANSAC)

In [35], the authors integrate a locality-preserving constraint into the RANSAC workflow, with the goal of pruning unreliable hypotheses before the scoring loop and also guiding nonuniform sampling to generate and score more promising models earlier. Experiments on public datasets yield more accurate and stable solutions than other state-of-the-art methods, this advantage being more evident when there is a low inlier percentage. The locality-preserving constraint is derived from the work in [56], which observed that in two images of the same scene taken under different points of view, the absolute distance between two feature points may change greatly, but their relative location is much better preserved due to physical constraints. The guided sampling strategy makes use of the locality-preserving scores to assign more weight to more promising areas in the search space.

## 3.4. Optimality-Focused Variants

### 3.4.1. Optimal Randomized RANSAC

In [36], the authors present a randomized version of RANSAC and prove that it is optimal regarding a probability estimated by the user. The time spent to arrive at a solution is close to the minimum possible and better than any deterministic strategy. In fact, the algorithm is the fastest possible, in the average case, among all randomized algorithms when the proportion of inliers is known in advance. The algorithm is a version of Randomized RANSAC with SPRT test [25], the improvement being that the optimal decision threshold for deciding if a model is good or bad is also derived and not left to the user as a parameter to set up.

### 3.4.2. Optimal RANSAC

Another algorithm that finds the optimal model in nearly every run in some kinds of problems is presented in [37]. The authors present an approach with some similarities to LO-RANSAC [18], as both of them conduct repeated resampling on the set of tentative inliers performing iterative estimation of the model. The differences are: the optimization is performed only when the tentative set has more than five inliers, in order to avoid little promissory sets when there is a low inlier ratio; when a larger set is found when resampling, the resampling starts again with that set, so the set will grow until the largest set is found thanks to iterative re-estimation and rescoring; the iteration process continues until the set no longer changes, which yields a high probability that the found set is optimal; a pruning step is finally performed with a low tolerance, in order to preserve only the best inliers; the model is recomputed from the remaining inliers in each iteration. Experiments with line finding in aerial images show optimal solutions in more than 99.95% of the cases.

## 4. Applications

RANSAC or any of its variants have been used in many different kinds of applications as a solution to implement shape detectors or to find the correspondence between features extracted from images taken from different points of view. In this section, we present some of the research described in the literature. We organized the applications into three groups: image matching, shape detection, and hardware acceleration. The first two groups correspond to the two main areas in which RANSAC has been applied in robotic applications, while in the third one, we present all the implementations that make use of parallelism or special hardware. As far as the hardware acceleration does not rely on a specific kind of application, we deemed it appropriate to group them in a dedicated section.

### 4.1. Image Matching

Finding the correspondence between feature points in two images is a problem of great interest in the field of robotics, with applications in indoor as well as outdoor environments. Simultaneous Localization and Mapping (SLAM) is the family of techniques that allow a mobile robot to build a map of the environment and localize inside it at the same time [57]. To deal with SLAM in dynamic environments, in [58], the authors present a variant of RANSAC, called multilevel-RANSAC (ML-RANSAC), to classify objects into static or dynamic. The main advantage of their approach is that it can address both static and dynamic objects in SLAM and detect and track moving objects without the need of splitting the problem. The ML-RANSAC method takes as input in time step $n$ the estimated state and covariance at step $n-1$ according to an Extended Kalman Filter, the sensors measurements at time step $n$, a threshold value to decide if a detected object is a track of a previous one, and the maximum number of desired iterations. It is also needed to provide the kinematic models of the robot, stationary and dynamic objects, as well as the observation model. ML-RANSAC outputs the estimated state and the number of static and moving objects, as well as the covariance of all these entities at the time $n$. Experiments performed in simulation and with a Pioneer P3-DX robot in an indoor dynamic setting show that this approach can reliably estimate the robot's pose while building the map and keeping track of moving objects. In [59], the authors present further developments over their previous work, adding a layer of object detection and classification using machine learning, more precisely convolutional neural networks (CNN). A CNN is trained to detect doors and people, and the pipeline is tested in a real environment, yielding promising results.

The problem of removing erroneous or redundant matches in SLAM has also been tackled with RANSAC. In [60]; the authors present GMS-RANSAC, an algorithm to remove the mismatches based on oriented fast and rotated brief (ORB) in SLAM [61]. The key idea behind the grid-based motion statistics (GMS) algorithm [62] is the realization that adjacent pixels in images taken from different points of view share a similar motion, and that those relationships can be defined as smoothing constraints and be combined into a statistical framework to reject erroneous matching. Therefore, good correspondences are associated with a high number of similar neighbors in a 3D region. The main problem with this approach is that when there are few points in each 3D grid, the confidence is low and the number of errors could increase. The addition of RANSAC to the method allows for more robust results when the dataset is challenging to GMS. Experiments on public datasets show an average correction rate of 28.81% over the GMS algorithm.

Another problem of interest for robotic applications such as camera calibration, scene tracking, or robot navigation is the detection of vanishing points in images [63]. Due to perspective, lines that are parallel in 3D space appear to converge to a point called a vanishing point when projected in a 2D space. In [64], the authors propose a new RANSAC variant, called under-parameterized RANSAC (UPRANSAC) which, combined with the Hough transform, is able to detect vanishing points in uncalibrated monocular images in real time. The degrees of freedom of a vanishing point are found first by applying UPRANSAC to choose a hypothetical inlier and compute a portion of the degrees of freedom and then executing the voting scheme associated with a 1D Hough transform to find the remaining degrees of freedom along the extension line of the previously hypothesized inlier. Vanilla RANSAC selects two edges as a hypothetical pair of inliers and needs both of them to be right to fit a correct model of vanishing points, while UPRANSAC has a higher likelihood of finding one inlier and, therefore, is more reliable in this task. Experiments on public datasets show high accuracy and real-time performance.

Dense alignment between two images is the goal of the work described in [65]. The authors start from the observation that parametric and nonparametric alignment methods have different strengths that are complementary to each other. Then, they propose a two-stage method, where a feature-based parametric coarse alignment is followed by a nonparametric fine alignment. The coarse alignment is performed by RANSAC estimating the transformation matrix from deep features, and the fine alignment is learned at the pixel level in an unsupervised way by a deep neural network that tries to optimize a standard structural similarity metric between the two given images. The deep features for the coarse alignment stage are the conv4 layer of a ResNet-50 network, while in the fine alignment stage, the goal is to find a flow that warps the image source into an image similar to the target, with that similarity being measured by structural similarity [66]. The authors claim good results on a range of tasks, including unsupervised optical flow on KITTI [67], dense correspondence on HPatches [68], and two-view geometry estimation on YFCC100M [69], among others.

A parallel robot, also called parallel manipulator, or generalized Stewart platform, is a mechatronics device that supports a single platform using several serial chains controlled by a computer system [70]. The parallel aspect of the robot is not related to its geometric appearance but to the fact that several actuators could work in parallel, affecting the platform at the same time. In [71], the authors present a system that employs Harris-SIFT [72] and RANSAC to detect the pose of a parallel robot with three degrees of freedom which was developed by them. Harris-SIFT combines the Harris corner detection algorithm with SIFT, but their results could contain mismatches that are tackled by the RANSAC step. The RANSAC algorithm is customized for this problem by substituting the pure random sampling by sampling in separate grids and, also, by performing an efficient model validation strategy that can detect invalid models without checking all the input data. Experiments report that when compared with unmodified RANSAC, the average matching time decreases by 63.45%, the average matching accuracy increases by 15.66%, and the average deviation in pose detection decreases in all the coordinate axes.

The robotics subfield of unmanned aerial vehicles (UAVs) can also greatly benefit from advances in image matching. In [73], the authors propose a method that combines RANSAC with SURF [74] for the problem of matching images and test it on aerial images taken from UAVs. As in previously mentioned works, RANSAC is employed to refine the matches found by another method. In this case, SURF is the method used to detect features, and the authors find that it compares favorably to using SURF, SIFT, or ORB alone, although their experimental setup only includes a pair of aerial images. In [75], the authors present Prior Sampling and Sample Check RANSAC (PSSC-RANSAC), which incorporates prior knowledge of the sampling goodness coming from three different sources: texture magnitude, spatial consistency, and feature similarity. This prior sampling should possibly generate more correct samples. Furthermore, prior information on the collection of sample subsets is used to check them and rule out incompatible arrangements of subsets, yielding further improvements in speed. Their experiments on a dataset composed of images taken from online sources and collected by themselves show improvements over standard RANSAC and SVH-RANSAC [76]. Target tracking and following from a multirotor UAV is the subject of the research carried out in [77]. The paper presents an end-to-end architecture that combines: image acquisition to obtain the data to compute the transformation matrix, Recursive RANSAC to perform target tracking, a track selection process, and a controller for the target-following task. The physical setup is composed of a monocular camera, an inertial measurement unit, an altitude sensor, and an embedded computer, all of them into a multirotor UAV with a flight control unit. The system works under the assumption that the target is moving on a surface close to planar and with a velocity approximately constant. Their results in simulation suggest that the proposed pipeline is effective and robust to target modeling errors. Another use of the data collected by UAVs is the creation of digital surface models (DSM). In [78], the authors propose a RANSAC modification to improve image matching with a special interest in the quality of the photogrammetry needed to create digital surface models. They enhance RANSAC using an iterative least-squares-based loop, a similarity termination criterion, and a post-processing step. In the locally iterative least-squares-based loop, all inliers found in the previous iteration are used to recompute the model parameters. Then, a least-square solution to improve the model is applied, and the number of inliers is counted in each step until that number does not change. The loop stops when a predetermined maximum number of iterations is achieved, or if the inliers ratio is higher than a good enough threshold. This iterative least-squares-based loop improves the stability, convergence rate, and number of inliers of the found solution. Another termination criterion is defined for the RANSAC loop: if the similarity of the sets of inlier points between two consecutive RANSAC iterations is greater than 95%, the loop stops, saving running time. Finally, a post-processing step is performed to remove outliers in the final model. The authors find favorable comparison with RANSAC over a set of four aerial images taken by themselves.

The next research does not fall into the category of image matching, but it is also of interest to any application in need of data for training a model. Image augmentation is the process of generating images similar to those present in a dataset, through several transformations, often with the aim of providing more training data to machine learning algorithms [79]. In [80], the authors propose a hybrid RANSAC algorithm to create a mosaic from several single images. They take images from similar areas and perform feature matching using RANSAC, using the location of those features to blend the pictures to create new ones. The authors claim their method is well-suited for aerial photos and report an increase in image augmentation data compared with other techniques.

*4.2. Shape Detection*

One of the key capabilities needed for a robotic system able to interact in an intelligent way with its environment is the potential to detect and identify objects. This is in itself a vast field of research, fueled in the last years by the great interest of the big technological actors and the advent of deep learning. YOLO [81] represented a big leap for object detection in 2D images, and 3D versions have been proposed [82–84]. Other approaches based on 3D descriptors [85–87] or other deep learning architectures [88–90] have also been the subject of research. These approaches do not rely on a priori knowledge about the objects to be recognized, but in human-made objects, it is usual that familiar geometric shapes are prevalent. Even in nature, flat terrain or water reservoirs can be roughly characterized as planes.

An example of simple shape detection in nature can be found in [91], where the authors develop a method for monitoring the water level in a river for a flood warning system. A UAV records aerial images, and a dense point cloud is obtained from them using photogrammetric software. They fit the river surface plane using RANSAC, but to estimate the water level, a time-invariant reference in the scene with a known altitude is needed. They take a point in the road over a bridge with precise altitude information, but if this were not available, a recognizable feature in the scene should be used as a reference, at least to estimate water level change between point clouds taken at different times. Experiments are performed with data collected on ten separate dates over the course of a month, with different water levels. Testing different image resolutions, the authors find that low-resolution images provide a more detailed point cloud due to the fact that the alignment software detects more matching points. They speculate that this could be because detailed river flow and tree branch movements with the wind introduce undesirable noise in the images. A linear regression of the calculated water level against the reference water level shows $R^2 = 0.98$ for a slope of 0.95 and a standard deviation of 0.37 m. Another application in nature is the detection and delineation of trees presented in [92]. The LiDAR data are captured in an area of 1796 hectares during the flight of a Cessna. The models that they use to fit a tree are of a paraboloid, a cone, or another one that they call a shape-shifter, which is an interpolation between a cone and a paraboloid, performing filtering of local maxima before executing RANSAC. To compute the height of each tree, they use Hardy's multiquadric method [93] to reconstruct the ground surface beneath the tree canopy. The authors report that their method, when applied to terrain with a mix of different tree species and is densely populated, yields tree counts similar to the inventory performed directly on the field. The difference is attributed mainly to small trees not detected by the LiDAR but that contribute less to the total counts.

Aerial imagery also has applications in urban areas in order to locate spaces of interest. One of the sources of renewable energy of great interest nowadays is solar energy. To find suitable places to place solar panels, in [94], the authors are interested in the analysis of the inclination and plane parameters of the roofs in an urban area. They apply RANSAC to data obtained from aerial photogrammetry and LIDAR data of three buildings taken by a UAV at a height of 80 m. The experiment shows that, while LIDAR data are less accurate than aerial photogrammetry, sometimes trees occlude parts of the roofs, circumstances in which LIDAR performs better than photogrammetry. The authors point out the importance of an accurate data source and find that irregular roof shapes are not detected correctly.

Another use case for urban areas is autonomous driving. If a vehicle is going to successfully navigate through a city, it is of paramount importance to correctly detect the traffic lanes. In [95], a real-time method for detecting all the lane markers in an image is presented. After filtering the image using selective oriented Gaussian filters, a RANSAC line fitting step provides initial guesses to another proposed fast RANSAC algorithm for fitting Bezier splines. A post-processing step to better localize and extend the spline is applied, with excellent results at a real-time rate of 50 Hz. In [96], the lane detection process starts with the application of inverse perspective mapping to change the camera perspective to a bird's-eye view. This transforms the problem of detecting lanes into finding parallel lines separated by a fixed and given distance. Candidate lanes are found applying the Hough transform, and the results are further refined with RANSAC. A Kalman filter helps to remove minor perturbations. Experiments on the streets and highways around Atlanta in various traffic conditions show that the approach achieves good performance. In [97], the authors introduce ridgeness, a low-level image descriptor, which assigns high values along the center lines of the lane markings and low values close to the boundaries in the longitudinal direction. Then, RANSAC is applied using ridgeness and orientation as input to find the hyperbolas which correspond to the projection of the actual lane markings. They claim good results under different driving circumstances and straight and curved lanes. In [98], the model is refined to detect left and right lanes simultaneously, and extra information is returned: lane width, lane curvature, vehicle yaw angle, and lateral offset with respect to the lane medial axis. In [99], the images obtained by the car camera are split into two areas: a far-field area and a near-field area. In the near-field area, the lanes are detected by the Hough transform for lines, while in the far-field area, the lanes are observed as curves and are therefore detected by RANSAC using a hyperbolic model. Experiments under different driving conditions yield good results. A survey of advances in vision-based lane detection, covering works in which RANSAC was employed, is presented in [100].

Cable inspection is one of the tasks that autonomous underwater vehicles must perform. In [101], Crossline Correction Nonlinear RANSAC (CCNL-RANSAC) is presented to tackle the problem of detecting objects with a shape similar to a curved line. As underwater imagery often suffers from blurring, low contrast, nonuniform illumination, and noise, their approach performs a preprocessing step in order to improve the quality of the acquired images. Afterward, an adaptive edge detector based on Canny [102] and Otsu's method [103] is run, and then CCNL-RANSAC is applied. CCNL-RANSAC integrates a preliminary inlier estimation module with a nonlinear fitting model and a final crossline correction procedure to remove false positives that could arise. Experiments with images collected in a boat tank at a university facility show that the algorithm can detect underwater curved-line objects with a success rate of 95% to a distance of 21 m.

Plane detection is one of the most usual applications of RANSAC, which is also of interest for autonomous driving or any navigation in urbanized terrain. When fitting several planes from point cloud data, it is possible that sometimes a spurious plane that shares inliers from other legitimate planes is erroneously detected. This is a usual fact when detecting curbs or ramps in urban scenery. CC-RANSAC [104] addresses this issue by changing the way that the fitness of a candidate plane is computed: instead of counting the total number of inliers, it only considers those that lie in the largest connected components. CC-RANSAC fails if two areas of the scene corresponding to planar surfaces are too close to each other, because the connected components of the two areas could join together. NCC-RANSAC [105] overcomes some of the limitations of CC-RANSAC, performing a check of the normal vectors in the area to find if they are coherent with the fitted plane. After obtaining a collection of candidate planes, a recursive clustering process is performed to grow each one of the candidates. The authors validate the robustness of their approach with a probabilistic model and obtain a very high rate of success. In indoor or outdoor mobile robotics, ground detection is a common task. In [106], the authors take advantage of the fact that all the other objects in the scene are always above the ground to define an asymmetric kernel as the score function for RANSAC. The ground parameter is estimated by maximum likelihood estimation, where the log-likelihood is modeled as an asymmetric Gaussian kernel. Experiments show that the proposed model is fast as well as robust.

A general method aimed to improve the plane segmentation process in point clouds is described in [107]. After downsampling the point cloud using the voxel grid method, the authors estimate the normal at each point and refine such estimate employing the Mean Shift algorithm [108]. After that, RANSAC, with the constraints given by the normals, is applied to find the plane. Experiments with data acquired by the authors show good results and practical value in industrial settings. In [109], the authors present a Python library for segmenting assets in an industrial indoor scene. They use RANSAC for plane segmentation and then employ parallelism and perpendicularity between the detected planes, along with the sensor orientation, to find the ground, ceiling, and walls of a room. Other elements in the scene can also be detected by analyzing further relationships. In [110], pairwise orthogonal planes are defined as a primitive shape and then detected directly by RANSAC. The parameters of the shape are a point and two unit orthogonal normals, and they formulate the problem of refining each candidate model as a nonlinear least-square optimization task, which is solved by employing the Levenberg–Marquardt algorithm [111]. The candidate models are generated by RANSAC. Experiments on Stanford 3D large-scale dataset [112] show that the method is efficient, even for extracting small planes. Moreover, the authors claim that their approach can also be adapted to deal with other geometry structures.

Building Information Modeling (BIM) is a process that involves the generation and management of digital information about the physical and functional characteristics of buildings. As that information is not directly available from structures in which BIM was not present in the design and building process, it is important to generate BIMs from existing places in an efficient manner [113]. In [114], the authors present a method to apply RANSAC iteratively, where each iteration takes as input the inlier set of the previous one, to automatically extract the height and the layout of a room. They report promising results in a cloud extracted from the ISPRS dataset [115], although they point out that a limitation of their model is that it is limited to rooms with polygonal layouts and flat surfaces.

### 4.3. Hardware Acceleration

Making software run faster is one of the main goals driving the new technology industry. From the point of view of pure software engineering research into algorithmic theory, computational gains could be achieved that are to some extent independent of the underlying hardware. However, advances in hardware also make it possible to leverage the new capabilities of modern processors and dedicated processing units to achieve running speeds orders of magnitude above a single CPU. It is possible to write parallel software somehow independent of the hardware, but this approach has been gradually superseded by the advent of GPUs and TPUs: dedicated graphical processors and tensor processors, respectively [116]. GPUs were designed for the acceleration of graphics rendering computations but now have a prominent place in artificial intelligence research and applications. TPUs have been specifically designed for tensor operations in deep learning. FPGAs [117] are programmable hardware that could very efficiently perform a specific task and are suitable for embedded devices.

Several steps of the RANSAC procedure are very suitable for parallelization. For example, the generation of hypotheses or the scoring of those hypotheses against the input data. We suspect that some straightforward implementations do not have enough relevance for publication as a research article, and that is the reason for the lack of description of such obvious variations in the literature. However, some articles deserve a mention in this section.

In [118], the authors directly implement RANSAC in hardware using Verilog, a hardware description language (HDL) that is used to model electronic systems. Their design implements random sampling by using the multiple-input signature register (MISR) and the index register. At the same time, the matrix triangularization operation needed by the forward elimination is implemented by a systolic array [119], which is a piece of hardware specifically built for fast and efficient implementations of regular algorithms that perform the same task with different data at different timestamps. The authors report speeds, in simulated hardware, of 30 frames (1024 × 1024 pixels) per second for computing the homography between pairs of images.

RANSAC has been implemented in FPGAs. In [120], an implementation for real-time affine geometry estimation is introduced. The main task chosen to be accelerated was the fitness scoring function, where the authors claim that the speed-up factor increases with input data size. Another layer of acceleration was implemented over the iterations of the full RANSAC workflow, therefore increasing the probability of obtaining good estimation results in the same running time. Experiments with video frames extracted from the Unmanned Aerial Vehicle Database [121] show increases in the speed of about 11.4 times for 100 data points, the system being able to handle a video stream of 30 fps. In [122], another FPGA implementation for the same problem is described, with their architecture able to reject false correspondences between similar images. Three modules are defined: transformation matrix calculator, inliers count calculator, and RANSAC controller. The transformation matrix calculator computes the affine transformation parameters from three samples from the set of feature matches, the inliers count calculator computes the number of inliers for the current transformation matrix, and the RANSAC controller reads samples from the array of initial matches and stores them in the array of random samples. The execution of RANSAC takes a number of clock cycles equal to the number of selected random samples. The authors report a running time of less than 23 ms for the processing of 128 initial matches, with a supported video streaming rate of at least 43 fps. Their architecture has been tested in simulation and on hardware (Altera Cyclone IV). These same authors later present another FPGA implementation for real-time SIFT [123] matching and RANSAC to improve over the previous solutions to the problem of identifying the correct correspondences between feature points between consecutive video frames [124]. The feature descriptors from each frame are stored, and when a new feature is extracted from the next frame, its descriptor is compared with those corresponding to the previous frame. If the matching criterion is fulfilled, then the coordinates of the match are stored in on-chip RAM. A moving window of size 16 is defined in the shift register structure to store and shift the feature descriptors. This facilitates the fit in the processing pipeline of the matching procedure, by supporting a standard number of parallel comparisons between features. Several sets of moving windows are defined concurrently, allowing for an effective size of 128. Using Altera Cyclone IV again, they achieve a processing rate of 40 fps for VGA resolution (640 × 480).

It is also possible to parallelize RANSAC using APIs that could access to the parallel capabilities of modern processors or GPUs. In [125], the task of fitting a plane in a 3D point cloud is implemented in three different paradigms: OpenMP, POSIX threads, and CUDA. Their goal is to analyze the relative performance of these three approaches over a collection of point clouds collected by the same authors in an indoor environment. The point clouds cover different spaces: living room, kitchen, hallway, saloon, room, and furniture. In addition to three usual metrics in evaluating search strategies (Precision, Recall, and F-Score), they also report other two standard metrics in parallelism (Runtime and Speedup). In their study, they find that CUDA over NVidia GPUs is the best option, with good results in all the metrics. POSIX threads are a better option than OpenMP if the researcher is willing to program to a low level to profit from the fine control that OpenMP does not allow for, it being too high-level.

## 5. Software

Researchers working on RANSAC variants have, sometimes, made their code available to the community independently, giving rise to a fragmented ecosystem with implementations in several languages that often lack maintenance. At the same time, popular computer vision libraries have implemented RANSAC variants for shape detection or image matching. In this section, we present available software that could be of interest to the robotics practitioner willing to test RANSAC capabilities.

### 5.1. OpenCV

OpenCV is a library of functions aimed to tackle common tasks in computer vision, especially focused on real-time performance. The project was started by Intel, and it is now released under the open-source Apache 2 license. It is cross-platform and offers C++ and Python APIs. Some operations support GPU accelerations. Robot Operating System (ROS) (http://wiki.ros.org/ accessed on 23 November 2022) provides an easy way to integrate

OpenCV calls in ROS developments (http://wiki.ros.org/vision_opencv accessed on 23 November 2022).

The documentation is not very exhaustive, but the *calib3d* module, which is in charge of finding the transformation matrix between different points of view, provides several RANSAC-related implementations (https://docs.opencv.org/4.x/d1/df1/md__build_master-contrib_docs-lin64_opencv_doc_tutorials_calib3d_usac.html accessed on 23 November 2022).

Choices for the sampling method in the general RANSAC procedure include the standard sampling of RANSAC, or the alternatives of PROSAC, NAPSAC, or progressive-NAPSAC [126]. The score method could also be set to the standard of RANSAC, or those of MSAC, MAGSAC or the least median of squared error distances. A local optimization step using LO-RANSAC, Graph-Cut RANSAC, or the sigma consensus of MAGSAC++ is available to the user. Finally, sequential probability ratio test (SPRT) verification evaluates a model on randomly drawn points using statistical properties obtained from the probability of a point being inlier, the average number of output models, etc. This could speed up the process in a significant manner because a bad model could be rejected without computing the error for every point.

*5.2. Point Cloud Library*

The Point Cloud Library (PCL) is an open project for 2D and 3D image and point cloud processing [127]. It is released under the three-clause BSD license, which permits research and commercial use free of any fees. The authors claim to have implemented state-of-the-art algorithms in the areas of registration, feature estimation, filtering, surface reconstruction, segmentation, and model fitting. Some examples of their capabilities that could be of interest to the robotic community are outlier filtering from noisy point clouds, scene segmentation, and geometric descriptors computation. It is written in C++ and has been compiled on Linux, macOS, Windows, and Android. As with OpenCV, ROS permits integration of PCL in ROS-based software (http://wiki.ros.org/pcl/ accessed on 23 November 2022).

PCL implements several sample consensus methods applied to different models (https://pointclouds.org/documentation/group__sample__consensus.html accessed on 23 November 2022). It is possible to call the different methods by their corresponding individual implementation or to create a segmentation of objects and pass the method and model types as parameters. In Table 2, we show all the available methods along with the name they receive according to the PCL API. The geometric models, along with the number of coefficients needed to describe them, are shown in Table 3.

**Table 2.** List of sample consensus methods available in PCL (as of 1.12.1 version).

| API Name | Method |
|---|---|
| SAC_RANSAC | RANdom SAmple Consensus |
| SAC_LMEDS | Least Median of Squares |
| SAC_MSAC | M-Estimator SAmple Consensus |
| SAC_RRANSAC | Randomized RANSAC |
| SAC_MLESAC | Maximum LikeLihood Estimation SAmple Consensus |
| SAC_PROSAC | PROgressive SAmple Consensus |

The four coefficients of the plane model must be provided in Hessian normal form: $n_x$, $n_y$, $n_z$, and $d$, where $n_x$, $n_y$, and $n_z$ are the components of the unit normal vector, and $d$ is the signed distance from the plane to the origin. The sign of $d$ determines the side of the plane on which the origin is located. If $p > 0$, the origin is in the half-space determined by the direction of the normal, and if $p < 0$, it is in the other half-space. The six coefficients of the line model are given by a point on the line ($p$) and the direction of the line ($d$) as $[p_x, p_y, p_z, d_x, d_y, d_z]$. The 2D circle's three coefficients are given by its center ($c$) and radius ($r$) as: $[c_x, c_y, r]$, while the seven coefficients of the 3D circle are given by its center ($c$), radius

($r$), and normal ($n$) as $[c_x, c_y, c_z, r, n_x, n_y, n_z]$. The four coefficients of the sphere are given by its 3D center ($c$) and radius ($r$) as: $[c_x, c_y, c_z, r]$. In the case of the cylinder model, the seven coefficients must be given by a point on its axis ($p$), the axis direction ($d$), and a radius ($r$), as $[p_x, p_y, p_z, d_x, d_y, d_z, r]$, while for a cone model, its seven coefficients correspond to a point of its apex ($a$), the axis direction ($d$), and the opening angle ($o$), as: $[a_x, a_y, a_z, d_x, d_y, d_z, o]$. For parallel lines and parallel and perpendicular planes, the extra constraints, in addition to the model coefficients, are the axis with respect to being parallel or perpendicular and the maximum angular deviation tolerated. For a normal plane, the surface normals at each tentative inlier point are computed and have to be parallel to the normal of the tentative plane, within a maximum specified angular deviation. The normal sphere model adds additional surface normal constraints. The normal parallel plane restricts the normal plane, with the constraint that such a normal plane has to be parallel to a given axis. There are plans to implement a torus model as well as parallel lines.

The PCL maintainers do not plan (as of version 1.12.1) to provide widespread GPU support due to the difficult integration of NVidia libraries with their CI/CD practices.

**Table 3.** List of models available in PCL (as of 1.12.1 version).

| API Name | Model | Coefficients | Constraints |
|---|---|---|---|
| SACMODEL_PLANE | Plane | 4 | No |
| SACMODEL_LINE | Line | 6 | No |
| SACMODEL_CIRCLE2D | Circle | 3 | No |
| SACMODEL_CIRCLE3D | Circle | 7 | No |
| SACMODEL_SPHERE | Sphere | 4 | No |
| SACMODEL_CYLINDER | Cylinder | 7 | No |
| SACMODEL_CONE | Cone | 7 | No |
| SACMODEL_PARALLEL_LINE | Line | 6 | Yes |
| SACMODEL_PERPENDICULAR_PLANE | Plane | 4 | Yes |
| SACMODEL_NORMAL_PLANE | Plane | 4 | Yes |
| SACMODEL_NORMAL_SPHERE | Sphere | 4 | Yes |
| SACMODEL_PARALLEL_PLANE | Plane | 4 | Yes |
| SACMODEL_NORMAL_PARALLEL_PLANE | Plane | 4 | Yes |
| SACMODEL_STICK | Line | 6 | Yes |

*5.3. Other Software*

Some of the authors of the variants or applications mentioned so far have provided a software implementation of their algorithms. In Table 4, there is a list of available implementations. It is worth noticing that GraphCut-RANSAC has also been implemented in OpenCV.

Open3D [128] aims to be an alternative to PCL, focusing on making its use easy and also providing the capability of rapid prototyping. It implements a plane segmentation function that relies on RANSAC, although it is not very sophisticated in the current version (0.16), not permitting other geometric models apart from the plane. Open3D is the library behind indoor3D, the library for processing 3D data from indoor scenes [109]. Another Python library not mentioned so far is pyRANSAC-3D [129], employed, for example, in [91]. The characteristics of these libraries, along with those previously presented, are summarized in Table 5.

Table 4. Software implementations of RANSAC variants.

| RANSAC Variant | Language | Code |
| --- | --- | --- |
| GraphCut-RANSAC [44] | C++ | https://github.com/danini/graph-cut-ransac accessed on 23 November 2022 |
| GCSAC [29] | C++ | http://mica.edu.vn/perso/Le-Van-Hung/GCSAC/index.html accessed on 23 November 2022 |
| Latent RANSAC [13] | C++ | https://github.com/rlit/LatentRANSAC accessed on 23 November 2022 |
| Optimal RANSAC [37] | Matlab | https://www.cb.uu.se/~aht/code.html accessed on 23 November 2022 |
| RANSAC-Flow [65] | Python (PyTorch) | https://github.com/XiSHEN0220/RANSAC-Flow accessed on 23 November 2022 |

Table 5. Libraries with RANSAC APIs.

| Library | Language | URL |
| --- | --- | --- |
| OpenCV [130] | C++, Python | https://opencv.org/ accessed on 23 November 2022 |
| PCL [127] | C++ | https://pointclouds.org/ accessed on 23 November 2022 |
| Open3D [128] | C++, Python | http://www.open3d.org/ accessed on 23 November 2022 |
| pyRANSAC-3D [129] | Python | https://github.com/leomariga/pyRANSAC-3D/ accessed on 23 November 2022 |
| indoor3D [109] | Python | https://github.com/rsait/indoor3d accessed on 23 November 2022 |

## 6. Discussion and Conclusions

Methods from the RANSAC family have been widely studied and applied to problems arising in robotic applications. The interested researcher could easily implement a simple model with open-source software, but even programming a RANSAC variant from scratch should not be such a daunting task as with other models, as they are comparatively simpler than, for example, deep learning approaches.

In the literature in general, and in the open source tools in particular, we miss more general support of parallel implementations. In the deep learning era, GPU acceleration is widespread, and RANSAC could benefit greatly from these techniques, as the method iterations or the model score function could be parallelized rather easily.

Plane fitting is the most common shape detection task, likely due to the simplicity of the geometric model and its ubiquity in human-made structures. While detectors of other simple geometric shapes, such as spheres, cones, or cylinders have also been implemented in open-source software, the field could also benefit from research into modeling more complex shapes.

Other machine learning methods such as deep neural networks are nowadays used in all kinds of areas due to their unquestionable performance, but RANSAC is conceptually simpler and easier to implement when the model to recognize is known in advance and suitable for parameterization. For example, if a robotics application needs to recognize objects with no shape restriction, deep neural networks would be the default choice, while if those shapes are known in advance and simple enough to parameterize, RANSAC could be a good option. RANSAC could also be used as a preprocessing step for filtering the floor, ceiling, or walls of a point cloud taken indoors, before deep learning takes charge of segmenting the remaining data.

In contrast with the trial-and-error approach of hyperparameter tuning of deep neural networks, theoretical results have been achieved of RANSAC parameters depending on the a priori knowledge of the data distribution. For example, the number of iterations needed for a given probability of randomly drawing all inlier points in some iteration can be computed if the ratio of inlier points is known in advance.

In short, the main conclusions could be summarized as:

1. RANSAC is a good alternative to deep learning approaches when the model whose parameters we want to estimate is known in advance, which is the case, e.g., of shape matching of simple objects in many robotic applications.
2. Theoretical analysis of the probability of estimating the model parameters is possible, and this can lead to optimal use of resources in embedded devices or real-time applications.

3. Open-source implementations of RANSAC variants are available for the robotics community.
4. Research in parallelization and hybrid approaches with deep learning methods could be promising.

**Author Contributions:** Conceptualization, J.M.M.-O. and I.M.; methodology, I.R.-M. and B.S.; investigation, J.M.M.-O., I.R.-M., I.M. and B.S.; writing—original draft preparation, J.M.M.-O.; writing—review and editing, J.M.M.-O., I.R.-M., I.M. and B.S.; supervision, I.M. and B.S.; project administration, I.M.; funding acquisition, I.M. and B.S. All authors have read and agreed to the published version of the manuscript.

**Funding:** This work has been partially funded by the Basque Government, Spain, under Research Teams Grant number IT1427-22 and under ELKARTEK LANVERSO Grant number KK-2022/00065; the Spanish Ministry of Science (MCIU), the State Research Agency (AEI), the European Regional Development Fund (FEDER), under Grant number PID2021-122402OB-C21 (MCIU/AEI/FEDER, UE); and the Spanish Ministry of Science, Innovation and Universities, under Grant FPU18/04737.

**Institutional Review Board Statement:** Not applicable.

**Informed Consent Statement:** Not applicable.

**Data Availability Statement:** Not applicable.

**Conflicts of Interest:** The authors declare no conflict of interest.

# References

1. Fischler, M.A.; Bolles, R.C. Random sample consensus: A paradigm for model fitting with applications to image analysis and automated cartography. *Commun. ACM* **1981**, *24*, 381–395. [CrossRef]
2. Voulodimos, A.; Doulamis, N.; Doulamis, A.; Protopapadakis, E. Deep learning for computer vision: A brief review. *Comput. Intell. Neurosci.* **2018**, *2018*, 7068349. [CrossRef] [PubMed]
3. Guo, Y.; Liu, Y.; Oerlemans, A.; Lao, S.; Wu, S.; Lew, M.S. Deep learning for visual understanding: A review. *Neurocomputing* **2016**, *187*, 27–48. [CrossRef]
4. Choi, S.; Kim, T.; Yu, W. Performance evaluation of RANSAC family. In Proceedings of the British Machine Vision Conference, London, UK, 7–10 September 2009; pp. 81.1–81.12.
5. Rousseeuw, P.J. Least median of squares regression. *J. Am. Stat. Assoc.* **1984**, *79*, 871–880. [CrossRef]
6. Subbarao, R.; Meer, P. Subspace estimation using projection based M-estimators over Grassmann manifolds. In Proceedings of the European Conference on Computer Vision, Graz, Austria, 7–13 May 2006; Springer: Berlin/Heidelberg, Germany, 2006; pp. 301–312.
7. Zeineldin, R.A.; El-Fishawy, N.A. A survey of RANSAC enhancements for plane detection in 3D point clouds. *Menoufia J. Electron. Eng. Res.* **2017**, *26*, 519–537. [CrossRef]
8. Strandmark, P.; Gu, I.Y. Joint random sample consensus and multiple motion models for robust video tracking. In Proceedings of the Scandinavian Conference on Image Analysis, Oslo, Norway, 15–18 June 2009; Springer: Berlin/Heidelberg, Germany, 2009; pp. 450–459.
9. Vedaldi, A.; Jin, H.; Favaro, P.; Soatto, S. KALMANSAC: Robust filtering by consensus. In Proceedings of the Tenth IEEE International Conference on Computer Vision (ICCV'05), Beijing, China, 17–21 October 2005; Volume 1, pp. 633–640.
10. Derpanis, K.G. Overview of the RANSAC Algorithm. *Image Rochester NY* **2010**, *4*, 2–3.
11. Hoseinnezhad, R.; Bab-Hadiashar, A. An M-estimator for high breakdown robust estimation in computer vision. *Comput. Vis. Image Underst.* **2011**, *115*, 1145–1156. [CrossRef]
12. Shapira, G.; Hassner, T. Fast and accurate line detection with GPU-based least median of squares. *J. Real-Time Image Process.* **2020**, *17*, 839–851. [CrossRef]
13. Korman, S.; Litman, R. Latent RANSAC. In Proceedings of the IEEE Conference on Computer Vision and Pattern Recognition, Salt Lake City, UT, USA, 18–23 June 2018; pp. 6693–6702.
14. Lowe, D.G. Distinctive image features from scale-invariant keypoints. *Int. J. Comput. Vis.* **2004**, *60*, 91–110. [CrossRef]
15. Torr, P.H.; Zisserman, A. Robust parameterization and computation of the trifocal tensor. *Image Vis. Comput.* **1997**, *15*, 591–605. [CrossRef]
16. Torr, P.H.; Zisserman, A. MLESAC: A new robust estimator with application to estimating image geometry. *Comput. Vis. Image Underst.* **2000**, *78*, 138–156. [CrossRef]
17. Torr, P.H.S. Bayesian model estimation and selection for epipolar geometry and generic manifold fitting. *Int. J. Comput. Vis.* **2002**, *50*, 35–61. [CrossRef]
18. Chum, O.; Matas, J.; Kittler, J. Locally optimized RANSAC. In Proceedings of the Joint Pattern Recognition Symposium, Magdeburg, Germany, 10–12 September 2003; Springer: Berlin/Heidelberg, Germany, 2003; pp. 236–243.

19. Frahm, J.M.; Pollefeys, M. RANSAC for (quasi-) degenerate data (QDEGSAC). In Proceedings of the 2006 IEEE Computer Society Conference on Computer Vision and Pattern Recognition (CVPR'06), New York, NY, USA, 17–22 June 2006; Volume 1, pp. 453–460.
20. Barath, D.; Matas, J. Graph-cut RANSAC. In Proceedings of the IEEE Conference on Computer Vision and Pattern Recognition, Salt Lake City, UT, USA, 18–23 June 2018; pp. 6733–6741.
21. Myatt, D.R.; Torr, P.H.; Nasuto, S.J.; Bishop, J.M. NAPSAC: High noise, high dimensional robust estimation-it's in the bag. In Proceedings of the British Machine Vision Conference (BMVC), Cardiff, UK, 2–5 September 2002; Volume 2, p. 3.
22. Matas, J.; Chum, O. Randomized RANSAC with Td, d test. *Image Vis. Comput.* **2004**, *22*, 837–842. [CrossRef]
23. Tordoff, B.J.; Murray, D.W. Guided-MLESAC: Faster image transform estimation by using matching priors. *IEEE Trans. Pattern Anal. Mach. Intell.* **2005**, *27*, 1523–1535. [CrossRef] [PubMed]
24. Capel, D.P. An Effective Bail-out Test for RANSAC Consensus Scoring. In Proceedings of the British Machine Vision Conference (BMVC), Oxford, UK, 5–8 September 2005; Volume 1, p. 2.
25. Matas, J.; Chum, O. Randomized RANSAC with sequential probability ratio test. In Proceedings of the Tenth IEEE International Conference on Computer Vision (ICCV'05), Beijing, China, 17–21 October 2005; Volume 2, pp. 1727–1732.
26. Chum, O.; Matas, J. Matching with PROSAC-progressive sample consensus. In Proceedings of the 2005 IEEE Computer Society Conference on Computer Vision and Pattern Recognition (CVPR'05), San Diego, CA, USA, 20–25 June 2005; Volume 1, pp. 220–226.
27. Rodehorst, V.; Hellwich, O. Genetic algorithm sample consensus (GASAC)-a parallel strategy for robust parameter estimation. In Proceedings of the 2006 IEEE Conference on Computer Vision and Pattern Recognition Workshop (CVPRW'06), New York, NY, USA, 17–22 June 2006; p. 103.
28. Civera, J.; Grasa, O.G.; Davison, A.J.; Montiel, J.M. 1-Point RANSAC for extended Kalman filtering: Application to real-time structure from motion and visual odometry. *J. Field Robot.* **2010**, *27*, 609–631. [CrossRef]
29. Le, V.H.; Vu, H.; Nguyen, T.T.; Le, T.L.; Tran, T.H. Acquiring qualified samples for RANSAC using geometrical constraints. *Pattern Recognit. Lett.* **2018**, *102*, 58–66. [CrossRef]
30. Konouchine, A.; Gaganov, V.; Veznevets, V. AMLESAC: A new maximum likelihood robust estimator. In Proceedings of the GraphiCon, Novosibirsk, Russia, 20–24 June 2005; Volume 5, pp. 93–100.
31. Choi, S.; Kim, J.H. Robust regression to varying data distribution and its application to landmark-based localization. In Proceedings of the 2008 IEEE International Conference on Systems, Man and Cybernetics, Singapore, 12–15 October 2008; pp. 3465–3470.
32. Niedfeldt, P.C.; Beard, R.W. Recursive RANSAC: Multiple signal estimation with outliers. *IFAC Proc. Vol.* **2013**, *46*, 430–435. [CrossRef]
33. Fotouhi, M.; Hekmatian, H.; Kashani-Nezhad, M.A.; Kasaei, S. SC-RANSAC: Spatial consistency on RANSAC. *Multimed. Tools Appl.* **2019**, *78*, 9429–9461. [CrossRef]
34. Brachmann, E.; Rother, C. Neural-guided RANSAC: Learning where to sample model hypotheses. In Proceedings of the IEEE/CVF International Conference on Computer Vision, Seoul, Republic of Korea, 27 October–2 November 2019; pp. 4322–4331.
35. Wang, G.; Sun, X.; Shang, Y.; Wang, Z.; Shi, Z.; Yu, Q. Two-view geometry estimation using RANSAC with locality preserving constraint. *IEEE Access* **2020**, *8*, 7267–7279. [CrossRef]
36. Chum, O.; Matas, J. Optimal randomized RANSAC. *IEEE Trans. Pattern Anal. Mach. Intell.* **2008**, *30*, 1472–1482. [CrossRef]
37. Hast, A.; Nysjö, J.; Marchetti, A. Optimal RANSAC-towards a repeatable algorithm for finding the optimal set. *J. WSCG* **2013**, *21*, 21–30.
38. Akaike, H. A new look at the statistical model identification. *IEEE Trans. Autom. Control* **1974**, *19*, 716–723. [CrossRef]
39. Schwarz, G. Estimating the dimension of a model. *Ann. Stat.* **1978**, *6*, 461–464. [CrossRef]
40. Rissanen, J. Modeling by shortest data description. *Automatica* **1978**, *14*, 465–471. [CrossRef]
41. Tordoff, B.; Murray, D.W. Guided sampling and consensus for motion estimation. In Proceedings of the European Conference on Computer Vision, Copenhagen, Denmark, 28–31 May 2002; Springer: Berlin/Heidelberg, Germany, 2002; pp. 82–96.
42. Boykov, Y.; Veksler, O. Graph cuts in vision and graphics: Theories and applications. In *Handbook of Mathematical Models in Computer Vision*; Springer: Boston, MA, USA, 2006; pp. 79–96.
43. Barath, D.; Valasek, G. Space-Partitioning RANSAC. *arXiv* **2021**, arXiv:2111.12385.
44. Barath, D.; Matas, J. Graph-cut RANSAC: Local optimization on spatially coherent structures. *IEEE Trans. Pattern Anal. Mach. Intell.* **2021**, *44*, 4961–4974. [CrossRef]
45. Raguram, R.; Chum, O.; Pollefeys, M.; Matas, J.; Frahm, J.M. USAC: A universal framework for random sample consensus. *IEEE Trans. Pattern Anal. Mach. Intell.* **2012**, *35*, 2022–2038. [CrossRef]
46. Barath, D.; Noskova, J.; Ivashechkin, M.; Matas, J. MAGSAC++, a fast, reliable and accurate robust estimator. In Proceedings of the IEEE/CVF Conference on Computer Vision and Pattern Recognition, Seattle, WA, USA, 13–19 June 2020; pp. 1304–1312.
47. Wald, A. *Sequential Analysis*; Courier Corporation: New York, NY, USA, 1947.
48. Ribeiro, M.I. Kalman and extended Kalman filters: Concept, derivation and properties. *Inst. Syst. Robot.* **2004**, *43*, 46.
49. Xu, L.; Oja, E.; Kultanen, P. A new curve detection method: Randomized Hough transform (RHT). *Pattern Recognit. Lett.* **1990**, *11*, 331–338. [CrossRef]

50. Aiger, D.; Kokiopoulou, E.; Rivlin, E. Random grids: Fast approximate nearest neighbors and range searching for image search. In Proceedings of the IEEE International Conference on Computer Vision, Sydney, NSW, Australia, 1–8 December 2013; pp. 3471–3478.
51. Torr, P.; Zisserman, A. Robust computation and parametrization of multiple view relations. In Proceedings of the Sixth International Conference on Computer Vision (IEEE Cat. No. 98CH36271), Bombay, India, 7 January 1998; pp. 727–732.
52. Moon, T.K. The expectation-maximization algorithm. *IEEE Signal Process. Mag.* **1996**, *13*, 47–60. [CrossRef]
53. Engel, Y.; Mannor, S.; Meir, R. The kernel recursive least-squares algorithm. *IEEE Trans. Signal Process.* **2004**, *52*, 2275–2285. [CrossRef]
54. Illingworth, J.; Kittler, J. A survey of the Hough transform. *Comput. Vis. Graph. Image Process.* **1988**, *44*, 87–116. [CrossRef]
55. Shan, Y.; Matei, B.; Sawhney, H.S.; Kumar, R.; Huber, D.; Hebert, M. Linear model hashing and batch RANSAC for rapid and accurate object recognition. In Proceedings of the 2004 IEEE Computer Society Conference on Computer Vision and Pattern Recognition, 2004. CVPR 2004, Washington, DC, USA, 27 June–2 July 2004; Volume 2, p. II.
56. Ma, J.; Zhao, J.; Jiang, J.; Zhou, H.; Guo, X. Locality preserving matching. *Int. J. Comput. Vis.* **2019**, *127*, 512–531. [CrossRef]
57. Macario Barros, A.; Michel, M.; Moline, Y.; Corre, G.; Carrel, F. A comprehensive survey of visual SLAM algorithms. *Robotics* **2022**, *11*, 24. [CrossRef]
58. Bahraini, M.S.; Bozorg, M.; Rad, A.B. SLAM in dynamic environments via ML-RANSAC. *Mechatronics* **2018**, *49*, 105–118. [CrossRef]
59. Bahraini, M.S.; Rad, A.B.; Bozorg, M. SLAM in dynamic environments: A deep learning approach for moving object tracking using ML-RANSAC algorithm. *Sensors* **2019**, *19*, 3699. [CrossRef]
60. Zhang, D.; Zhu, J.; Wang, F.; Hu, X.; Ye, X. GMS-RANSAC: A Fast Algorithm for Removing Mismatches Based on ORB-SLAM2. *Symmetry* **2022**, *14*, 849. [CrossRef]
61. Campos, C.; Elvira, R.; Rodríguez, J.J.G.; Montiel, J.M.; Tardós, J.D. ORB-SLAM3: An accurate open-source library for visual, visual–inertial, and multimap SLAM. *IEEE Trans. Robot.* **2021**, *37*, 1874–1890. [CrossRef]
62. Bian, J.; Lin, W.Y.; Matsushita, Y.; Yeung, S.K.; Nguyen, T.D.; Cheng, M.M. GMS: Grid-based motion statistics for fast, ultra-robust feature correspondence. In Proceedings of the IEEE Conference on Computer Vision and Pattern Recognition, Honolulu, HI, USA, 21–26 July 2017; pp. 4181–4190.
63. Kroeger, T.; Dai, D.; Van Gool, L. Joint vanishing point extraction and tracking. In Proceedings of the IEEE Conference on Computer Vision and Pattern Recognition, Boston, MA, USA, 7–12 June 2015; pp. 2449–2457.
64. Wu, J.; Zhang, L.; Liu, Y.; Chen, K. Real-time vanishing point detector integrating under-parameterized RANSAC and Hough transform. In Proceedings of the IEEE/CVF International Conference on Computer Vision, Montreal, QC, Canada, 10–17 October 2021; pp. 3732–3741.
65. Shen, X.; Darmon, F.; Efros, A.A.; Aubry, M. RANSAC-flow: Generic two-stage image alignment. In Proceedings of the European Conference on Computer Vision, Glasgow, UK, 23–28 August 2020; Springer: Cham, Switzerland, 2020; pp. 618–637.
66. Wang, Z.; Bovik, A.C.; Sheikh, H.R.; Simoncelli, E.P. Image quality assessment: From error visibility to structural similarity. *IEEE Trans. Image Process.* **2004**, *13*, 600–612. [CrossRef]
67. Geiger, A.; Lenz, P.; Stiller, C.; Urtasun, R. Vision meets robotics: The KITTI dataset. *Int. J. Robot. Res.* **2013**, *32*, 1231–1237. [CrossRef]
68. Balntas, V.; Lenc, K.; Vedaldi, A.; Mikolajczyk, K. HPatches: A benchmark and evaluation of handcrafted and learned local descriptors. In Proceedings of the IEEE Conference on Computer Vision and Pattern Recognition, Honolulu, HI, USA, 22–25 July 2017; pp. 5173–5182.
69. Thomee, B.; Shamma, D.A.; Friedland, G.; Elizalde, B.; Ni, K.; Poland, D.; Borth, D.; Li, L.J. YFCC100M: The new data in multimedia research. *Commun. ACM* **2016**, *59*, 64–73. [CrossRef]
70. Merlet, J.P. *Parallel Robots*; Springer Science & Business Media: Dordrecht, The Netherlands, 2005; Volume 128.
71. Gao, G.Q.; Zhang, Q.; Zhang, S. Pose detection of parallel robot based on improved RANSAC algorithm. *Meas. Control* **2019**, *52*, 855–868. [CrossRef]
72. Zhao, Q.; Zhao, D.; Wei, H. Harris-SIFT algorithm and its application in binocular stereo vision. *J. Univ. Electron. Sci. Technol. China Pap.* **2010**, *4*, 2–16.
73. Li, X.; Ren, C.; Zhang, T.; Zhu, Z.; Zhang, Z. Unmanned aerial vehicle image matching based on improved RANSAC algorithm and SURF algorithm. *Int. Arch. Photogramm. Remote Sens. Spat. Inf. Sci.* **2020**, *42*, 67–70. [CrossRef]
74. Bay, H.; Ess, A.; Tuytelaars, T.; Van Gool, L. Speeded-up robust features (SURF). *Comput. Vis. Image Underst.* **2008**, *110*, 346–359. [CrossRef]
75. Zheng, J.; Peng, W.; Wang, Y.; Zhai, B. Accelerated RANSAC for accurate image registration in aerial video surveillance. *IEEE Access* **2021**, *9*, 36775–36790. [CrossRef]
76. Wang, Y.; Zheng, J.; Xu, Q.Z.; Li, B.; Hu, H.M. An improved RANSAC based on the scale variation homogeneity. *J. Vis. Commun. Image Represent.* **2016**, *40*, 751–764. [CrossRef]
77. Petersen, M.; Samuelson, C.; Beard, R.W. Target Tracking and Following from a Multirotor UAV. *Curr. Robot. Rep.* **2021**, *2*, 285–295. [CrossRef]
78. Salehi, B.; Jarahizadeh, S. Improving the UAV-derived DSM by introducing a modified RANSAC algorithm. *Int. Arch. Photogramm. Remote Sens. Spat. Inf. Sci.* **2022**, *43*, 147–152. [CrossRef]

79. Shorten, C.; Khoshgoftaar, T.M. A survey on image data augmentation for deep learning. *J. Big Data* **2019**, *6*, 60. [CrossRef]
80. Cherian, A.K.; Poovammal, E. Image Augmentation Using Hybrid RANSAC Algorithm. *Webology* **2021**, *18*, 237–254. [CrossRef]
81. Redmon, J.; Divvala, S.; Girshick, R.; Farhadi, A. You only look once: Unified, real-time object detection. In Proceedings of the IEEE Conference on Computer Vision and Pattern Recognition, Las Vegas, NV, USA, 27–30 June 2016; pp. 779–788.
82. Ali, W.; Abdelkarim, S.; Zidan, M.; Zahran, M.; El Sallab, A. YOLO3D: End-to-end real-time 3d oriented object bounding box detection from lidar point cloud. In Proceedings of the European Conference on Computer Vision (ECCV) Workshops (Part III), Munich, Germany, 8–14 September 2018; pp. 718–728.
83. Takahashi, M.; Ji, Y.; Umeda, K.; Moro, A. Expandable YOLO: 3D object detection from RGB-D images. In Proceedings of the 2020 21st IEEE International Conference on Research and Education in Mechatronics (REM), Cracow, Poland, 9–11 December 2020; pp. 1–5.
84. Simony, M.; Milzy, S.; Amendey, K.; Gross, H.M. Complex-YOLO: An Euler-region-proposal for real-time 3D object detection on point clouds. In Proceedings of the European Conference on Computer Vision (ECCV) Workshops (Part I), Munich, Germany, 8–14 September 2018; pp. 197–209.
85. Hana, X.F.; Jin, J.S.; Xie, J.; Wang, M.J.; Jiang, W. A comprehensive review of 3D point cloud descriptors. *arXiv* **2018**, arXiv:1802.02297.
86. Chen, J.; Fang, Y.; Cho, Y.K. Performance evaluation of 3D descriptors for object recognition in construction applications. *Autom. Constr.* **2018**, *86*, 44–52. [CrossRef]
87. Kasaei, S.H.; Ghorbani, M.; Schilperoort, J.; van der Rest, W. Investigating the importance of shape features, color constancy, color spaces, and similarity measures in open-ended 3D object recognition. *Intell. Serv. Robot.* **2021**, *14*, 329–344. [CrossRef]
88. Liang, M.; Yang, B.; Wang, S.; Urtasun, R. Deep continuous fusion for multi-sensor 3D object detection. In Proceedings of the European Conference on Computer Vision (ECCV), Munich, Germany, 8–14 September 2018; pp. 641–656.
89. Qi, S.; Ning, X.; Yang, G.; Zhang, L.; Long, P.; Cai, W.; Li, W. Review of multi-view 3D object recognition methods based on deep learning. *Displays* **2021**, *69*, 102053. [CrossRef]
90. Li, Y.; Yu, A.W.; Meng, T.; Caine, B.; Ngiam, J.; Peng, D.; Shen, J.; Lu, Y.; Zhou, D.; Le, Q.V.; et al. Deepfusion: Lidar-camera deep fusion for multi-modal 3d object detection. In Proceedings of the IEEE/CVF Conference on Computer Vision and Pattern Recognition, New Orleans, LA, USA, 21–24 June 2022; pp. 17182–17191.
91. Giulietti, N.; Allevi, G.; Castellini, P.; Garinei, A.; Martarelli, M. Rivers' Water Level Assessment Using UAV Photogrammetry and RANSAC Method and the Analysis of Sensitivity to Uncertainty Sources. *Sensors* **2022**, *22*, 5319. [CrossRef]
92. Tittmann, P.; Shafii, S.; Hartsough, B.; Hamann, B. Tree detection and delineation from LiDAR point clouds using RANSAC. In Proceedings of the SilviLaser, Hobart, TAS, Australia, 16–20 October 2011; pp. 1–23.
93. Hardy, R.L. Multiquadric equations of topography and other irregular surfaces. *J. Geophys. Res.* **1971**, *76*, 1905–1915. [CrossRef]
94. Gönültaş, F.; AtiK, M.E.; Duran, Z. Extraction of roof planes from different point clouds using RANSAC algorithm. *Int. J. Environ. Geoinform.* **2020**, *7*, 165–171. [CrossRef]
95. Aly, M. Real time detection of lane markers in urban streets. In Proceedings of the 2008 IEEE Intelligent Vehicles Symposium, Eindhoven, The Netherlands, 4–6 June 2008; pp. 7–12.
96. Borkar, A.; Hayes, M.; Smith, M.T. Robust lane detection and tracking with RANSAC and Kalman filter. In Proceedings of the 2009 16th IEEE International Conference on Image Processing (ICIP), Cairo, Egypt, 7–10 November 2009; pp. 3261–3264.
97. Lopez, A.; Canero, C.; Serrat, J.; Saludes, J.; Lumbreras, F.; Graf, T. Detection of lane markings based on ridgeness and RANSAC. In Proceedings of the 2005 IEEE Intelligent Transportation Systems, Vienna, Austria, 16 September 2005; pp. 254–259.
98. López, A.; Serrat, J.; Canero, C.; Lumbreras, F.; Graf, T. Robust lane markings detection and road geometry computation. *Int. J. Automot. Technol.* **2010**, *11*, 395–407. [CrossRef]
99. Tan, H.; Zhou, Y.; Zhu, Y.; Yao, D.; Wang, J. Improved river flow and random sample consensus for curve lane detection. *Adv. Mech. Eng.* **2015**, *7*, 1687814015593866. [CrossRef]
100. Xing, Y.; Lv, C.; Chen, L.; Wang, H.; Wang, H.; Cao, D.; Velenis, E.; Wang, F.Y. Advances in vision-based lane detection: Algorithms, integration, assessment, and perspectives on ACP based parallel vision. *IEEE/CAA J. Autom. Sin.* **2018**, *5*, 645–661. [CrossRef]
101. Yang, K.; Yu, L.; Xia, M.; Xu, T.; Li, W. Nonlinear RANSAC with crossline correction: An algorithm for vision-based curved cable detection system. *Opt. Lasers Eng.* **2021**, *141*, 106417. [CrossRef]
102. Ding, L.; Goshtasby, A. On the Canny edge detector. *Pattern Recognit.* **2001**, *34*, 721–725. [CrossRef]
103. Otsu, N. A threshold selection method from gray-level histograms. *IEEE Trans. Syst. Man Cybern.* **1979**, *9*, 62–66. [CrossRef]
104. Gallo, O.; Manduchi, R.; Rafii, A. CC-RANSAC: Fitting planes in the presence of multiple surfaces in range data. *Pattern Recognit. Lett.* **2011**, *32*, 403–410. [CrossRef]
105. Qian, X.; Ye, C. NCC-RANSAC: A fast plane extraction method for 3-D range data segmentation. *IEEE Trans. Cybern.* **2014**, *44*, 2771–2783. [CrossRef]
106. Choi, S.; Park, J.; Byun, J.; Yu, W. Robust ground plane detection from 3D point clouds. In Proceedings of the 2014 14th IEEE International Conference on Control, Automation and Systems (ICCAS 2014), Gyeonggi-do, Republic of Korea, 22–25 October 2014; pp. 1076–1081.
107. Yue, W.; Lu, J.; Zhou, W.; Miao, Y. A new plane segmentation method of point cloud based on Mean Shift and RANSAC. In Proceedings of the 2018 IEEE Chinese Control And Decision Conference (CCDC), Shenyang, China, 9–11 June 2018; pp. 1658–1663.

108. Comaniciu, D.; Meer, P. Mean shift: A robust approach toward feature space analysis. *IEEE Trans. Pattern Anal. Mach. Intell.* **2002**, *24*, 603–619. [CrossRef]
109. Martínez-Otzeta, J.M.; Mendialdua, I.; Rodríguez-Moreno, I.; Rodriguez, I.R.; Sierra, B. An Open-source Library for Processing of 3D Data from Indoor Scenes. In Proceedings of the 11th International Conference on Pattern Recognition Applications and Methods (ICPRAM 2022), Online, 3–5 February 2022; pp. 610–615.
110. Wu, Y.; Li, G.; Xian, C.; Ding, X.; Xiong, Y. Extracting POP: Pairwise orthogonal planes from point cloud using RANSAC. *Comput. Graph.* **2021**, *94*, 43–51. [CrossRef]
111. Moré, J.J. The Levenberg-Marquardt algorithm: Implementation and theory. In *Numerical Analysis*; Springer: Berlin/Heidelberg, Germany, 1978; pp. 105–116.
112. Armeni, I.; Sax, S.; Zamir, A.R.; Savarese, S. Joint 2D-3D-semantic data for indoor scene understanding. *arXiv* **2017**, arXiv:1702.01105.
113. Xiong, X.; Adan, A.; Akinci, B.; Huber, D. Automatic creation of semantically rich 3D building models from laser scanner data. *Autom. Constr.* **2013**, *31*, 325–337. [CrossRef]
114. Capocchiano, F.; Ravanelli, R. An original algorithm for BIM generation from indoor survey point clouds. In Proceedings of the International Archives of the Photogrammetry, Remote Sensing & Spatial Information Sciences, Enschede, The Netherlands, 10–14 June 2019; pp. 769–776.
115. Khoshelham, K.; Tran, H.; Díaz-Vilariño, L.; Peter, M.; Kang, Z.; Acharya, D. An evaluation framework for benchmarking indoor modelling methods. *Int. Arch. Photogramm. Remote Sens. Spat. Inf. Sci.* **2018**, *42*, 297–302. [CrossRef]
116. Wang, Y.E.; Wei, G.Y.; Brooks, D. Benchmarking TPU, GPU, and CPU platforms for deep learning. *arXiv* **2019**, arXiv:1907.10701.
117. Kuon, I.; Tessier, R.; Rose, J. FPGA architecture: Survey and challenges. *Found. Trends Electron. Des. Autom.* **2008**, *2*, 135–253. [CrossRef]
118. Dung, L.R.; Huang, C.M.; Wu, Y.Y. Implementation of RANSAC algorithm for feature-based image registration. *J. Comput. Commun.* **2013**, *1*, 46–50. [CrossRef]
119. Gentleman, W.M.; Kung, H. Matrix triangularization by systolic arrays. In Proceedings of the Real-Time Signal Processing IV, Arlington, VA, USA, 4–7 May 1982; SPIE: Philadelphia, PA, USA, 1982; Volume 298, pp. 19–26.
120. Tang, J.W.; Shaikh-Husin, N.; Sheikh, U.U. FPGA implementation of RANSAC algorithm for real-time image geometry estimation. In Proceedings of the 2013 IEEE Student Conference on Research and Developement, Putrajaya, Malaysia, 16–17 December 2013; pp. 290–294.
121. Dantsker, O.D.; Caccamo, M.; Vahora, M.; Mancuso, R. Flight & ground testing data set for an unmanned aircraft: Great planes avistar elite. In Proceedings of the AIAA Scitech 2020 Forum, Orlando, FL, USA, 6–10 January 2020; p. 780.
122. Vourvoulakis, J.; Lygouras, J.; Kalomiros, J. Acceleration of RANSAC algorithm for images with affine transformation. In Proceedings of the 2016 IEEE International Conference on Imaging Systems and Techniques (IST), Chania, Greece, 4–6 October 2016; pp. 60–65.
123. Lowe, D.G. Object recognition from local scale-invariant features. In Proceedings of the Seventh IEEE International Conference on Computer Vision, Kerkyra, Greece, 20–27 September 1999; Volume 2, pp. 1150–1157.
124. Vourvoulakis, J.; Kalomiros, J.; Lygouras, J. FPGA-based architecture of a real-time SIFT matcher and RANSAC algorithm for robotic vision applications. *Multimed. Tools Appl.* **2018**, *77*, 9393–9415. [CrossRef]
125. Hidalgo-Paniagua, A.; Vega-Rodríguez, M.A.; Pavón, N.; Ferruz, J. A comparative study of parallel RANSAC implementations in 3D space. *Int. J. Parallel Program.* **2015**, *43*, 703–720. [CrossRef]
126. Barath, D.; Ivashechkin, M.; Matas, J. Progressive NAPSAC: Sampling from gradually growing neighborhoods. *arXiv* **2019**, arXiv:1906.02295.
127. Rusu, R.B.; Cousins, S. 3D is here: Point Cloud Library (PCL). In Proceedings of the IEEE International Conference on Robotics and Automation (ICRA), Shanghai, China, 9–13 May 2011; pp. 1–4.
128. Zhou, Q.Y.; Park, J.; Koltun, V. Open3D: A modern library for 3D data processing. *arXiv* **2018**, arXiv:1801.09847.
129. Mariga, L. pyRANSAC-3D. 2022. Available online: https://github.com/leomariga/pyRANSAC-3D (accessed on 23 November 2022).
130. Bradski, G. The OpenCV library. *Dr. Dobb's J. Softw. Tools Prof. Program.* **2000**, *25*, 120–123.

**Disclaimer/Publisher's Note:** The statements, opinions and data contained in all publications are solely those of the individual author(s) and contributor(s) and not of MDPI and/or the editor(s). MDPI and/or the editor(s) disclaim responsibility for any injury to people or property resulting from any ideas, methods, instructions or products referred to in the content.

MDPI
St. Alban-Anlage 66
4052 Basel
Switzerland
www.mdpi.com

*Sensors* Editorial Office
E-mail: sensors@mdpi.com
www.mdpi.com/journal/sensors

Disclaimer/Publisher's Note: The statements, opinions and data contained in all publications are solely those of the individual author(s) and contributor(s) and not of MDPI and/or the editor(s). MDPI and/or the editor(s) disclaim responsibility for any injury to people or property resulting from any ideas, methods, instructions or products referred to in the content.